Lecture Notes in Computer Science 937

Edited by G. Goos, J. Hartmanis and J. van Leeuwen

Advisory Board: W. Brauer   D. Gries   J. Stoer

Springer

Berlin
Heidelberg
New York
Barcelona
Budapest
Hong Kong
London
Milan
Paris
Tokyo

Zvi Galil Esko Ukkonen (Eds.)

Combinatorial Pattern Matching

6th Annual Symposium, CPM 95
Espoo, Finland, July 5-7, 1995
Proceedings

 Springer

Series Editors

Gerhard Goos
Universität Karlsruhe
Vincenz-Priessnitz-Straße 3, D-76128 Karlsruhe, Germany

Juris Hartmanis
Department of Computer Science, Cornell University
4130 Upson Hall, Ithaca, NY 14853, USA

Jan van Leeuwen
Department of Computer Science, Utrecht University
Padualaan 14, 3584 CH Utrecht, The Netherlands

Volume Editors

Zvi Galil
Computer Science Department, Columbia University
New York, NY 10027, USA

Esko Ukkonen
Department of Computer Science, University of Helsinki
P.O. Box 26, FIN-00014, Helsinki, Finland

CR Subject Classification (1991): F.2.2, I.5.4, I.5.0, I.7.3, H.3.3, E.4, G.2.1, J.3

ISBN 3-540-60044-2 Springer-Verlag Berlin Heidelberg New York

CIP data applied for

© Springer-Verlag Berlin Heidelberg 1995
Printed in Germany

Typesetting: Camera-ready by author
SPIN: 10486313   06/3142-543210 - Printed on acid-free paper

Foreword

The papers contained in this volume were presented at the Sixth Annual Symposium on Combinatorial Pattern Matching (CPM 95), held July 5–7, 1995, at Hanasaari, a congress center located in Espoo, Finland, a few kilometers from Helsinki. They were selected from 44 papers submitted in response to the call of papers. In addition, invited lectures were given by Uzi Vishkin (On a Technique for Parsing a String) and H. W. Mewes (Genome Analysis: Pattern Search in Biological Sequences).

Combinatorial Pattern Matching addresses issues of searching and matching strings and more complicated patterns such as trees, regular expressions, extended expressions, etc. The goal is to derive non-trivial combinatorial properties for such structures and then to exploit these properties in order to achieve improved performance for the corresponding computational problems.

In recent years, a steady flow of high-quality research on this subject has changed a sparse set of isolated results into a full-fledged area of algorithmics with important applications. This area is expected to grow even further due to the increasing demand for speed and efficiency that comes especially from molecular biology, but also from areas such as information retrieval, pattern recognition, compiling, data compression, and program analysis. The objective of annual CPM gatherings is to provide an international forum for the research in combinatorial pattern matching.

The general organization and orientation of CPM Conferences is coordinated by a Steering Committee composed of A. Apostolico, M. Crochemore, Z. Galil, and U. Manber.

The first five meetings were held in Paris (1990), London (1991), Tucson (1992), Padova (1993), and Pacific Grove (1994). After the first meeting a selection of the papers appeared as a special issue of *Theoretical Computer Science.* Since the third meeting the proceedings have appeared as volumes 644, 684, and 807 of the present series.

CPM 95 was organized by the Department of Computer Science at the University of Helsinki. The Local Organizing Committee of CPM 95 consisted of P. Kilpeläinen, J. Kärkkäinen, E. Sutinen, and J. Tarhio. The conference was supported in part by the Academy of Finland, Ministry of Education (Finland), and the University of Helsinki. The efforts of all are gratefully acknowledged.

Helsinki and New York, April 1995

Zvi Galil
Esko Ukkonen

Program Committee

A. Ehrenfeucht	P. Pevzner
Z. Galil, *co-chair*	I. Simon
D. Gusfield	J. Storer
U. Manber	E. Ukkonen, *co-chair*
M. Paterson	M. Wegman

Additional Referees

R. Agrawalla
D. Breslauer
M. Farach
P. Feofiloff
J. Field
R. Giancarlo
L. Guerra
M. Jain
P. Kilpeläinen
G. Kock
Y. Kohayakawa
A. Pereira do Lago

J. Meidanis
W. Miller
K. Murphy
K. Park
T. Przytycka
R. Ravi
J. Setubal
J. Soares
S. W. Song
P. Stelling
J. Tarhio
Y. Wakabayashi

Table of Contents

Computing similarity between RNA strings[*]

Vineet Bafna [†]   *S. Muthukrishnan* [‡]   *R. Ravi* [§]
       DIMACS   DIMACS   Princeton University

Abstract

Ribonucleic acid (RNA) strings are strings over the four-letter alphabet $\{A, C, G, U\}$ with a secondary structure of base-pairing between $A-U$ and $C-G$ pairs in the string[1]. Edges are drawn between two bases that are paired in the secondary structure and these edges have traditionally been assumed to be noncrossing. The noncrossing base-pairing naturally leads to a tree-like representation of the secondary structure of RNA strings.

In this paper, we address several notions of similarity between two RNA strings that take into account both the primary sequence and secondary base-pairing structure of the strings. We present efficient algorithms for exact matching and approximate matching between two RNA strings. We define a notion of alignment between two RNA strings and devise algorithms based on dynamic programming. We then present a method for optimally aligning a given RNA string with unknown secondary structure to one with known sequence and structure, thus attacking the structure prediction problem in the case when the structure of a closely related sequence is known. The techniques employed to prove our results include reductions to well-known string matching problems allowing wild cards and ranges, and speeding up dynamic programming by using the tree structures implicit in the secondary structure of RNA strings.

Keywords: RNA structure, edit distances, approximate matching, string algorithms, trees.

---

[*]Research supported by DIMACS (Center for Discrete Mathematics and Theoretical Computer Science), a National Science Foundation Science and Technology Center under NSF contract STC-8809648.

[†]Address: DIMACS Center, P. O. Box 1179, Piscataway, NJ 08855. Email: bafna@dimacs.rutgers.edu.

[‡]Address: DIMACS Center, P. O. Box 1179, Piscataway, NJ 08855. Email: muthu@dimacs.rutgers.edu.

[§]Address: DIMACS, Department of Computer Science, Princeton University, NJ 08544. Email: ravi@cs.princeton.edu.

[1]We reserve the term RNA string when the sequence as well as the structure is given and use RNA sequence to denote only the primary sequence.

1 Introduction

A variety of string matching problems are motivated by the analysis of DNA or protein sequences. An example is the problem of computing the similarity between two sequences such as the edit distance to transform one into another using insertions, deletions and substitutions of characters [13, 20, 21]. When comparing RNA sequences, usually much more is known about the secondary structure of base pairing between nucleotides in the sequence. A bonded pair of bases is usually represented as an edge between the two complementary bases involved in the bond; traditional models of RNA secondary structure[27] assume that every base participates in at most one such pair, and that the edges representing the paired bases are noncrossing or noninterleaving along the length of the string. The secondary structure can be represented by a nesting tree whose nodes correspond to edges of the pairing and parenthood in the tree represents the immediate enclosure relation between the two edges (see fig. 1).

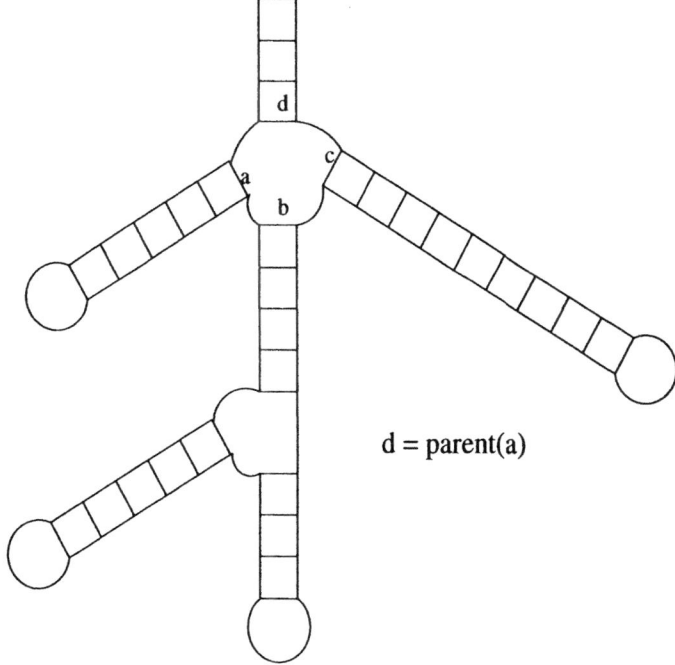

d = parent(a)

Figure 1: Secondary structure of an RNA sequence

Consequently, much of the work on comparing the secondary structures of two RNA strings have been modeled as problems of comparing two trees [7, 22, 25, 26]. In this paper, we study several problems in computing the similarity between two RNA strings that take into consideration *both* the primary sequence and secondary base-pairing information provided with the strings. We also investigate the problem of inferring the secondary structure of an RNA

	Problem	Running time
String	Symmetric Exact Matching (SEM)	$O(n+m)$
Matching	Containment Exact Matching (CEM)	$O(n \text{ polylog } m)$
	Symmetric K-mismatches problem (SKM)	$O(nm^{2/3} \text{ polylog } m)$
	Containment K-mismatches problem (CKM)	$O(n\sqrt{m} \text{ polylog } m)$
Alignment	Global similarity or RNA edit distance	$O(n^2 m^2)$
	Longest Common RNA Substring (RLCS)	$O(n^2 m^2)$
	Shortest Common RNA Superstring (RSCS)	$O(n^2 m^2(n^2 + m^2))$
Structure	Inferring structure through alignment	$O(n^2 m^2 + nm^3)$

Table 1: Summary of results.

sequence when a closely related RNA string is given.

Standard notions of similarity between two sequences have been formulated as problems of exact and approximate string matching, finding a longest common subsequence (LCS) or a shortest common supersequence (SCS) of the sequences, and computing optimal alignments under general scoring functions. We formulate the corresponding versions of these problems between two RNA strings and present efficient algorithms for computing them. In the context of RNA strings, we would like to match symbols as well as edges in the two strings. An edge in one of the strings is said to be matched if the two symbols in the other string that are aligned to its endpoints are connected by an edge in the second string. Two variants of exact and approximate matching problems arise, depending on whether edges in both the pattern and text strings or only the edges in the pattern are required to be matched: we call these variants the *symmetric* and *containment* variants respectively. We observe that the action-at-a-distance effect of aligning edges significantly increases the complexity of RNA string similarity algorithms. Furthermore, problems like LCS and SCS that are computationally identical for sequences turn out to be quite different in the context of RNA strings.

Finally, we solve the problem of inferring the secondary structure of an RNA sequence when a closely related RNA string is given by aligning the sequence with the given RNA string. The alignment not only maximizes the common characters as in traditional sequence alignment but also favors the alignment of the endpoints of an edge in the related string with complementary base pairs in the input string.

Results

Table 1 describes the list of problems addressed in this paper and the time complexity of the algorithms we propose for solving them. In the table, n and m represent the sizes of the two input strings.

String matching. The first four entries of the table are analogous to exact and approximate matching problems in standard strings. In such problems, a pattern RNA string and a typically longer text RNA string are given, and the task is to compute the positions in the text where the pattern occurs with at most $k(\geq 0)$ mismatches. We obtain our bounds by reducing these RNA string matching problems to well-known string matching problems allowing wild cards and ranges.

Alignment. The global similarity between two RNA strings is defined as a weighted sum of sequence similarity and structural similarity. Our algorithm for computing global similarity can be used to compute an edit distance between two RNA strings that is a weighted sum of sequence edit cost and edge-mismatch costs.

The next two entries in Table 1 are analogous to the LCS and SCS problem for two RNA strings. A longest common RNA subsequence (RLCS) of two RNA strings is one whose sequence is a subsequence of the two given strings and the sum of the common characters in the sequences as well as the number of common edges in the two strings that are matched to each other is maximum. This definition, as well as our algorithm, can be extended to allow an arbitrary weighted sum of the number of common characters and the number of matched edges in the two strings. The shortest common RNA supersequence (RSCS) is an RNA string of which both the given strings are RNA subsequences such that the total number of characters plus edges in the supersequence is minimum. Again, the extension to weighting edges and characters differently is straightforward. Somewhat surprisingly, RLCS and RSCS seem to be computationally very different problems. We can reduce RLCS to computing an optimal alignment by defining the scoring function appropriately. On the other hand our algorithm for computing the RSCS has running time that is two orders of magnitude more.

Structure. The problem of inferring structure through alignment, takes as input an RNA sequence and an RNA string and computes an alignment of the sequence with the string to maximize a weighted sum of the sequence similarity and the number of edges in the RNA string whose endpoints are aligned with complementary base-pairs in the RNA sequence. This alignment can be used as a model of commonality between the RNA string and the unknown RNA sequence. As more and more RNA secondary sequences become available, our formulation of the prediction problem appears increasingly relevant.

Related work

Comparison methods. First we review work on comparison methods developed to estimate distances between RNA secondary structures. Since secondary structures can be represented as trees, there are several papers [7, 22, 25, 26] addressing comparisons of trees. Tree edits are discussed and efficient algorithms are derived in [22, 25, 26] while a new notion of tree alignment is proposed and algorithms developed in [7]. Even though these comparison methods compute distances only between secondary structures, the worst-case running time of estimating this distance (tree edits or alignments) is quadric; the running time of our alignment algorithm that takes in account both sequence and structure compares favorably with this estimate.

Tree based comparison methods do not appear to generalize when comparing both the secondary structure as well as the underlying sequence of two RNA strings. The difficulty seems to be in the local definition of the operations of tree edit and tree alignment: A single edit operation, for instance, can either create a new node or delete an existing node and reconnect its children to its parent. Suppose we include characters from the sequence in a tree representation of the secondary structure to represent an RNA string. The deletion of a character that is an allowable tree edit operation does not take into account that there may be an edge in the secondary structure with this character as one endpoint. Therefore, such edit operations have no semantic meaning in the process of converting one RNA string to another, and illustrate the weakness of purely tree-based methods of edit. Our algorithms use a combination of sequence edit computation and a tree-based computation to align the edges in the secondary structure to overcome this difficulty.

Prediction methods. Prediction of RNA secondary structure of a single RNA molecule from its sequence has been widely studied in the past [3, 4, 10, 14, 23, 24, 28, 29]. Most of this work use adaptations of dynamic programming to solve variants of the problem that take into account different energy assignments for the different secondary structure primitives such as stacked pairs, hairpins, multiloops, interior loops and bulges. Sankoff [17] considers prediction of secondary structure common to two RNA sequences also taking into account alignment of the sequences. The key difference from our approach is that he does not assume that the secondary structure of either of the sequences is given, but instead computes an alignment and a most likely folding that is common to both the sequences. Sankoff's algorithm more carefully models the energy functions for different kinds of loops in the structure (such as stacked pairs, multiloops etc.); the running time of his algorithm for two sequences is two orders of magnitude higher. In contrast to the above work, our version of the prediction problem assumes that more information is available in the form of the sequence and the secondary structure of an RNA string that is closely related to the one whose structure is to be predicted.

Another related line of work is [6, 15, 16] that uses stochastic context free grammars to model a family of related RNA strings. Other related work appears in [8, 11, 18, 19].

In the next section, we describe our results on RNA string matching; We then turn to alignment of RNA strings and its variants in Section 3. In Section 4, we describe our algorithm for structure prediction given a related RNA string. We conclude with some open issues.

2   RNA string matching

In this section, we explore string matching problems in the context of RNA strings. We formally define the variants of string matching problems that arise when we deal with RNA strings. The first problem we consider is the basic problem of exact matching.

Symmetric Exact Matching.(SEM) Given a text RNA string t and a pattern RNA string p of length n and m respectively, determine all those positions where p occurs in t, that is, all locations i in t where,

1. *Strings match*, that is, $t[i] \cdots t[i + m - 1] = p[1] \cdots p[m]$, and

2. *Secondary structures are identical*, that is, for any bonding pair $(i + j - 1, i + k - 1)$, $0 \leq j < k \leq m - 1$, (j, k) is a bonding pair as well and vice versa. Note that the existence of some bonding pair $(i + j, i + k)$, where $0 \leq j < k$ and $k > m - 1$ does not affect p occurring in t at i. Similarly for bonding pairs $(i + j, i + k)$, where $0 \leq k \leq m - 1$ and $j < 0$.

Containment Exact Matching.(CEM) It is defined just as SEM except that the secondary structure of p and that of the substring of t of length m beginning at i need not be identical. It is only required that the secondary structure of p contained in the latter. Formally, p occurs in t at i if strings match as before and

2. For any bonding pair (j, k), $1 \leq j < k \leq m$, $(i + j - 1, i + k - 1)$ is a bonding pair as well, but not necessarily vice versa.

The other basic problem we consider is the k-mismatches problem for the RNA strings defined as below.

Symmetric K-mismatches problem.(SKM) Given a text RNA string t and a pattern RNA string p of length n and m respectively, and a parameter k, determine all those positions where p occurs in t with at most K mismatches. We say that p occurs in t with at most K mismatches at i in t if there exist integers K_1 and K_2 such that $K_1 + K_2 \leq K$ and in addition,

1. *There are K_1 string mismatches.* That is, $t[i] \cdots t[i+m-1]$ and $p[1] \cdots p[m]$ differ in K_1 positions.

2. *There are K_2 secondary structures mismatches.* That is, there are K_2 pairs of text and pattern positions $i + j - 1, i + k - 1$, $0 \leq j < k \leq m - 1$ and j, k where precisely one of $(i + j - 1, i + k - 1)$ or (j, k) is a bonding pair. Note again that the existence of some bonding pair $(i + j, i + k)$, where $0 \leq j < k$ and $k > m-1$ does not affect the number of mismatches between the secondary structure of p and that of $t[i] \cdots t[i + m - 1]$. Similarly for bonding pairs $(i + j, i + k)$, where $0 \leq k \leq m - 1$ and $j < 0$.

Containment K-mismatches problem.(CKM) The CKM problem is defined the same way as SKM except for the mismatches in the secondary structure. Here we only consider the number of bonding pairs in p which mismatch, that is, those that do not fall on bonding pairs in t; the bonding pairs in t that do not fall on a bonding pair in p are not counted as mismatches.

2.1 Exact Matching with RNA strings

In this section, we sketch our results for SEM and CEM problems.

Theorem 2.1 The CEM problem can be solved in $O(n \operatorname{polylog} m)$ time.

Proof.(Sketch) Our algorithm works as follows. First we perform standard string matching (using, say [9]) to find all locations where p does not occur in t since the strings mismatch. This takes $O(n + m)$ time.

Now we look for mismatches in the secondary structure. We construct an instance of string matching with wild cards to solve this problem. Assume that the RNA strings are strings drawn from $\{A, C, G, U\}$; the strings we generate for string matching with wild cards will contain symbols from $\{A, C, G, U\}$ as well as integers. We generate a text string t' from t by replacing each position i by $k - i$ if (i, k) or (k, i) is a bonding pair, and by the character $t[i]$ if there is no bonding pair involving i. Generate p' from p by replacing each position i by $k - i$ if (i, k) or (k, i) is a bonding pair, and by the wild card ϕ if there is no bonding pair involving i.

We claim that matching p' in t' gives all mismatches between the secondary structure of p and that of the substrings of t. The proof, a case analysis, is omitted here. The algorithm therefore takes time $O(n \operatorname{polylog} m)$ using the bound in [5]. □

Theorem 2.2 The SEM problem can be solved in $O(n + m)$ time.

Proof. We prove that a simple modification of the Knuth-Morris-Pratt algorithm [9] suffices to solve the SEM problem. Let the match operation between two strings as defined for the SEM problem be denoted by $\cong$.

Lemma 2.3 The $\cong$ relation is transitive, that is, if $t[i] \cdots t[i + k - 1]$ matches $p[1] \cdots p[k]$, and some prefix $p[1] \cdots p[j]$, $j < k$, matches the suffix of $t[i] \cdots t[i + k - 1]$, then the string $p[1] \cdots p[j]$, matches the suffix of $p[1] \cdots p[k]$.

Proof. Follows from the definition of the $\cong$ relation. □

Now we modify the comparisons in the KMP algorithm to perform SEM. Note that in standard KMP, comparing two locations is merely testing for equality. The following subroutines for Compare($p[i]$,$t[j]$) for $j \geq i$, and Compare($p[i]$,$p[j]$) for $i \leq j$ implement our $\cong$ relation. Compare($p[i]$,$t[j]$)

> if $p[i] \neq t[j]$ return *unequal*
> if $p[i] = t[j]$, $(i - k, i)$ is a bonding pair in t, $0 \leq k \leq i$ but $(j - k, j)$ is not a bonding pair in p then return *unequal*
> if $p[i] = t[j]$, $(j - k, j)$ is a bonding pair in p, $0 \leq k \leq i$ but $(i - k, i)$ is not a bonding pair in t then return *unequal*
> > else return *equal*
> end

The Compare($p[i]$,$p[j]$) for $i \leq j$ operation is implemented the same way. That completes the description of the algorithm. We omit the proof of the correctness except to invoke Lemma 2.3 and to remark that the following is preserved in the preprocessing using the $\cong$ operation above: The largest prefix of $p[1] \cdots p[i+1]$ that matches its suffix under $\cong$ is the largest prefix $p[1] \cdots p[k]$ of $p[1] \cdots p[i]$ that matches $p[i = k+1] \cdots p[i]$ and also answers Compare($p[i+1]$,$p[k+1]$) equal. It is easy to see that the entire algorithm works in $O(n+m)$ time. □

2.2 K–mismatches with RNA strings

We will sketch algorithms for the SKM and the CKM problem.

Theorem 2.4 The CKM problem can be solved in $O(n\sqrt{m}\text{polylog}m)$ time.

Proof. As usual we solve the version of the problem in which for each text location we return the number of mismatches (in the string and the secondary structure) between p and the substring of t of length m. The case when the string considered by itself is easy; this is simply the standard problem of counting mismatches with strings and it takes $O(n\sqrt{m}\text{polylog}m)$ time [1]. In order to consider the secondary structure by itself, recall the reduction in Theorem 2.1. We generate t' and p' (with wild cards) as described there. It follows from an argument based on (nontrivial) case analysis (which we omit here) that our problem is reduced to solving the problem of counting the mismatches between p' and the substrings of t' (the number of mismatches of p' at a location i in t' is *exactly twice* the number of mismatches of p at i in t). This problem in turn can be solved in $O(n\sqrt{m}\text{polylog}m)$ time [1]. Finally, we can combine the number of mismatches in the two parts above in linear time and detect all locations in t where the total number of mismatches is at most K. □

Theorem 2.5 The SKM problem can be solved in $O(nm^{2/3}\text{polylog}m)$ time.

Proof. Suppose as usual without loss of generality that $n \leq 2m$. We will again determine for each position the number of mismatches between p and t. Doing this considering t and p as a string can be easily done in $O(nm^{1/2}\text{polylog}m)$ time; henceforth we only consider counting mismatches due to the secondary structure. We give a reduction from this problem to that of counting the mismatches between a text t' and pattern p' where t' contains symbols from an ordered alphabet set Σ and each position of p' matches sets of ranges of the symbols from Σ. We call this the *sets of ranges* problem.

An example of the sets of ranges problem. Let $p = a\langle[a-c]+[f-g]+[z]\rangle c$ and $t = ababcda$. The second position in p from the left matches symbols a, b, c, f, g, z. Therefore, the number of mismatches between p and t at the leftmost position is 1 and that at the 5th position from the left is 3. □

Now we describe the reduction. We generate t' from t by replacing each position i by $k - i$ if (i, k) or (k, i) is a bonding pair, and by the character X_t

if there is no bonding pair involving i. We generate p' from p by replacing each position i by $k - i$ if (i, k) or (k, i) is a bonding pair, and by

$$\langle [(m - i) \cdots (2m)] + [(-2m) \cdots (-i)] + X_t \rangle$$

if there is no bonding pair involving i.

We omit the lengthy case analysis proving that p' occurs at i in t' with $2k$ mismatches if and only if p occurs in t with k edge mismatches. This problem can be solved in $O(nm^{2/3}\text{polylog}m)$ time [12]. $\square$

3 Computing Alignment for RNA strings

In this section, we look at sequence-alignment problems in the context of RNA strings. Specifically, we will consider variants of the edit-distance, longest-common-subsequence and shortest-common-supersequence problems.

Following Zuker[27, 28, 29], assume a model in which there are no knots in the secondary structure. A secondary structure is denoted by the set S of all base-pairs which have formed bonds. For $(i, j) \in S$, h is accessible from (i, j) if $i < h < j$, and there is no pair $(k, l) \in S$, s.t. $i < k < h < l < j$. Define $(i, j) \in S$ as the *parent* of $(k, l) \in S$ if k, l are accessible from (i, j). Observe that each $(i, j) \in S$ has at most one parent, implying a forest on the elements of S. The definitions of sibling, child, leaf follow naturally.

Let $s[1 \ldots n]$ and $t[1 \ldots m]$ be two RNA strings over the alphabet $\sum = \{A, C, G, U\}$ with structure S_1 and S_2 respectively. For technical reasons, let $s[0] = t[0] =' -'$. An *alignment* of s and t is defined by a $2 \times m'$ matrix A, in which each row contains a string interspersed with spaces, and for all columns j, either $A[1, j] \neq' -'$ or $A[2, j] \neq' -'$. For $i \in \{1, 2\}$, define

$$gap[i, j] = \begin{cases} j & \text{if } A[i, j] =' -' \\ |\{l < j \text{ s.t. } A[i, l] =' -'\}| & \text{otherwise} \end{cases}$$

Intuitively, if $A[i, j] \neq' -'$, then $gap[i, j]$ is the number of gaps that were inserted in the *i*th string till position j in alignment A. Following standard terminology, position i in A has a *match* if $A[1, i] = A[2, i] \neq' -'$, an *insertion* if $A[1, i] =' -'$, a *deletion* if $A[2, i] =' -'$ and a *mismatch* otherwise. Additionally, for RNA strings a bonding pair occurs at positions i, j if $(i - gap[1, i], j - gap[1, j]) \in S_1$ and $(i - gap[2, i], j - gap[2, j]) \in S_2$. Intuitively, we would like to compute an alignment which maximizes both symbol and base-pair matches.

Formally, for elements $u, v \in \sum \cup \{'-'\}$, define $\gamma(u, v)$ to be the score associated with aligning u against v. For $1 \leq i < j \leq m$ and $1 \leq k < l \leq n$, let $\delta(i, j, k, l)$ be the score associated with aligning base-pairs (i, j) with (k, l).

Definition 3.1 The Global Alignment problem for RNA strings is defined as follows: Given RNA strings s and t, compute an alignment A (and the associated function gap) that maximizes

$$\sum_{1 \leq i \leq m+n} \gamma(s[i - gap[1, i]], t[i - gap[2, i]]) +$$
$$\sum_{1 \leq i < j \leq m+n} \delta(i - gap[1, i], j - gap[1, j], i - gap[2, i], j - gap[2, j])$$

Theorem 3.2 Algorithm AlignRNA (fig. 2) computes an optimum global alignment for Two RNA strings in $O(n^2m^2)$ time.

Note that edit-distance is the inverse problem of computing an alignment with a minimum number of insertions, deletions, mismatches and bonding pair mismatches. It follows that edit-distance can be computed in $O(n^2m^2)$ time.

Procedure $AlignRNA()$
begin
 for intervals (i_1, j_1), $1 \le i_1 < j_1 \le m$
  and (i_2, j_2), $1 \le i_2 < j_2 \le n$
(* Assume that the intervals are examined in lexicographically
 increasing order of widths *)

$$Align[i_1, j_1, i_2, j_2] = \max \left\{ \begin{array}{l} Align[i_1, j_1 - 1, i_2, j_2] + \gamma(s[j_1], '-') \\ Align[i_1, j_1, i_2, j_2 - 1] + \gamma('-', t[j_2]) \\ Align[i_1, j_1 - 1, i_2, j_2 - 1] + \gamma(s[j_1], t[j_2]) \end{array} \right.$$

 if there exist $i_1 \le k_1 < j_1, i_2 \le k_2 < j_2$
  s.t. $(k_1, j_1) \in S_1, (k_2, j_2) \in S_2$

$$Align[i_1, j_1, i_2, j_2] = \max \left\{ \begin{array}{l} Align[i_1, j_1, i_2, j_2], \\ Align[i_1, k_1 - 1, i_2, k_2 - 1] + \\ Align[k_1 + 1, j_1 - 1, k_2 + 1, j_2 - 1] \\ + \delta(k_1, j_1, k_2, j_2) + \gamma(s[k_1], t[k_2]) + \gamma(s[j_1], t[j_2]) \end{array} \right.$$

end

Figure 2: Computing optimal alignment for RNA strings

3.1   LCS and SCS of RNA strings

For sequences, LCS and SCS can easily be deduced from an alignment in which no mismatches occur. For i varying from 1 to $m + n$, the LCS is simply the sequence formed by concatenating non-space symbols that appear in both $A[1, i]$ and $A[2, i]$, while the SCS is the concatenation of non-space symbols that appear in $A[1, i]$ or $A[2, i]$. It is also easy to see that for strings of length m and n, if l is the length of the longest common subsequence and s is the length of the shortest common supersequence, then $l = m + n - s$. Therefore for sequences, the two problems are virtually identical.

 The notion of a subsequence and supersequence can be extended naturally to RNA strings as follows:

Definition 3.3 Let s and t be two RNA strings with structure S_1 and S_2 respectively. s is an RNA-supersequence of t if there exists an alignment A of s and t, such that for all i, $A[1, i] = s[i]$, and for all i, j $(i - gap[2, i], j - gap[2, j]) \in S_2$ implies $(i, j) \in S_1$. s is an RNA-subsequence of t if t is an RNA-supersequence of s.

Definition 3.4 Let s be an RNA string with structure S. Define the 'length' of s, denoted by $len(s)$, to be $|s| + |S|$.

From this, the definition of a shortest common RNA-supersequence (RSCS) and a longest common RNA-subsequence (RLCS) of two or more RNA sequences follows.

Theorem 3.5 Let s, t be two RNA strings over $\sum$ with structure S_1 and S_2 respectively. RLCS of s, t can be computed in $O(n^2 m^2)$ time.

Proof. (sketch): For elements $u, v \in \sum \cup \{'-'\}$, define

$$\gamma(u, v) = \begin{cases} 1 & u = v \\ 0 & u =' -' \text{ or } v =' -' \\ -\infty & \text{otherwise} \end{cases}$$

and

$$\delta(i, j, k, l) = \begin{cases} 1 & s[i] = t[k] \wedge s[j] = t[l] \\ & \wedge (i, j) \in S_1 \wedge (k, l) \in S_2 \\ -\infty & \text{otherwise} \end{cases}$$

We claim without proof that for these definitions of γ and δ, algorithm AlignRNA computes the length of RLCS of two strings. The claim on running time follows from theorem 3.2. $\square$

Note that the alignment obtained in the LCS computation does not always yield a common RNA supersequence. In fact, it is possible that for the alignment obtained, there exist indices $1 \leq i < j < k < l \leq m+n$ such that $(i - gap[1, i], k - gap[1, k]) \in S_1$ and $(j - gap[2, j], l - gap[2, l]) \in S_2$. Clearly a common RNA supersequence must have both base-pairs but they cannot be interleaved because of our assumption of an unknotted structure.

Our algorithm for computing RSCS must therefore enforce this condition. For an RNA-string s with structure S, define a spanning-interval $[i, j]$ as a substring $s[i \ldots j]$, such that no pair $(k, l) \in S$ satisfies $i \leq k \leq j < l$ or $k < i \leq l \leq j$. We compute the RSCS for each pair of spanning interval $[i_1, j_1]$ and $[i_2, j_2]$ in the two strings. If both j_1 and j_2 do not form bonds with another base-pair, then

$$RSCS[i_1, j_1, i_2, j_2] = \min \begin{cases} RSCS[i_1, j_1 - 1, i_2, j_2] + 1 \\ RSCS[i_1, j_1, i_2, j_2 - 1] + 1 \\ RSCS[i_1, j_1 - 1, i_2, j_2 - 1] + \gamma(s[j_1], t[j_2]) \end{cases}$$

where γ is defined as in RLCS. Otherwise, let $(k_1, j_1) \in S_1$ and $(k_2, j_2) \in S_2$. Then we have different cases depending whether (k_1, j_1) and (k_2, j_2) match or not in the RSCS of the two strings. (k_1, j_1) and (k_2, j_2) are matched only if $s[k_1] = t[k_2]$, and $s[j_1] = t[j_2]$. In that case,

$$\begin{aligned} RSCS[i_1, j_1, i_2, j_2] &= RSCS[i_1, k_1 - 1, i_2, k_2 - 1] + \\ &\quad RSCS[k_2 + 1, j_1 - 1, k_2 + 1, j_2 - 1] + 3 \end{aligned}$$

If (k_1, j_1) and (k_2, j_2) are not matched then (k_1, j_1) could be aligned against an arbitrary interval in $[i_2, j_2]$. Then,

$$RSCS[i_1, j_1, i_2, j_2] \ = \ \min_{x,y} \ \{RSCS[i_1, k_1 - 1, i_2, x - 1] +$$
$$RSCS[k_1, j_1, x, y] + len(t[y + 1 \ldots j_2])\}$$

where the minimization is over all $i_2 \le x < y \le j_2$ such that $[x, y]$ is a spanning interval of t. Likewise,

$$RSCS[i_1, j_1, i_2, j_2] \ = \ \min_{x,y} \ \{RSCS[i_1, x - 1, i_2, k_2 - 1] +$$
$$RSCS[x, y, k_2, j_2] + len(s[y + 1 \ldots j_1])\}$$

where $i_1 \le x < y \le j_1$ and $[x, y]$ is a spanning interval of s. Clearly, for each pair of spanning intervals, the complexity of computing RSCS is $O(n^2 + m^2)$, which implies theorem 3.6.

Theorem 3.6 Let s, t be two RNA strings over $\sum$ with structure S_1 and S_2 respectively. An RSCS of s, t can be computed in $O(n^2 m^2 (n^2 + m^2))$ time.

4 Inferring RNA structure via Alignment

Given two sequences $s[1 \ldots n]$ and $t[1 \ldots m]$, where s has a known structure S_1, we infer the structure of t by aligning the two sequences. This approach is useful if we know that the two sequences are functionally related and have similar structure. For all $1 \le i < j \le m + n$, base-pairs $i - gap[2, i]$ and $j - gap[2, j]$ form a bond in t only if they are complementary and $(i - gap[1, i], j - gap[1, j]) \in S$. We would like to find an alignment that maximizes the sequence alignment score as well as the number of bonds formed in t. The algorithm for computing RNA alignment can be easily modified to accomplish this.

Consider the algorithm *AlignRNA* (fig. 2). Note that when we are comparing intervals $[i_1, j_1]$ and $[i_2, j_2]$, and there exists $(k_1, j_1) \in S$, there may be $\Omega(m)$ complementary pairs (k_2, j_2) that (k_1, j_1) can align against, and we need to pick the pair that gives the best alignment. Therefore, the naive algorithm has complexity $O(n^2 m^3)$. In the following, we take advantage of the tree structure S of s to obtain an $O(n^2 m^2 + nm^3)$ algorithm.

Consider the forest defined by the elements of S and the function *parent*. First, we *binarize* the forest by introducing additional base-pairs in S to get S', so that each node in S' has at most 2 children. Procedure Binarize (fig. 3) accomplishes this, for a tree rooted at $(i, j) \in S$.

For the algorithm *InferStructure*(fig. 4), we assume that we already have the sets S and S' for s. Functions γ and δ are the costs for aligning bases and base-pairs respectively, as defined for RNA Alignment.

Theorem 4.1 The algorithm InferStructure() computes the optimal alignment for inferring structure in $O(m^2 n^2 + nm^3)$ time.

Procedure *Binarize(i,j)*
(* Assume that $(i,j) \in S$ has k children $\{(i_1,j_1),\ldots,(i_k,j_k)\}$ *)
begin
    for $1 \leq u < k$ **do**
        Binarize(i_u,j_u)
        $S' = S' \cup \{(i_1,j_u)\}$
        if $(u > 1)$
            parent$((i_1,j_{u-1})) = (i_1,j_u)$
            parent$((i_u,j_u)) = (i_1,j_u)$
    parent$(i_1,j_k) = (i,j)$
end

Figure 3: Binarizing an RNA structure tree

Procedure *InferStructure()*
begin
  for intervals (i_1,j_1), $1 \leq i_1 < j_1 \leq n$
    and intervals (i_2,j_2), $1 \leq i_2 < j_2 \leq m$
(* Assume that the intervals are examined in lexicographically
  increasing order of widths*)

$$Align[i_1,j_1,i_2,j_2] = \max \begin{cases} Align[i_1+1,j_1,i_2,j_2] + \gamma(s[i_1],{}'-') \\ Align[i_1,j_1,i_2+1,j_2] + \gamma('-',t[i_2]) \\ Align[i_1+1,j_1,i_2+1,j_2] + \gamma(s[i_1],t[i_2]) \\ Align[i_1,j_1-1,i_2,j_2] + \gamma(s[j_1],{}'-') \\ Align[i_1,j_1,i_2,j_2-1] + \gamma('-',t[j_2]) \\ Align[i_1,j_1-1,i_2,j_2-1] + \gamma(s[j_1],t[j_2]) \end{cases}$$

if $(i_1,j_1) \in S$ and
    $t[i_2]$ and $t[j_2]$ are complementary base-pairs

$$Align[i_1,j_1,i_2,j_2] = \max \begin{cases} Align[i_1,j_1,i_2,j_2], \\ \delta(i_1,j_1,i_2,j_2) + \gamma(s[i_1],t[i_2]) + \gamma(s[j_1],t[j_2]) \\ +Align[i_1+1,j_1-1,i_2+1,j_2-1] \end{cases}$$

else if $(i_1,j_1) \in S' - S$ and
    $(k,j_1)) = rightchild(i_1,j_1)$

$$Align[i_1,j_1,i_2,j_2] = \max \begin{cases} Align[i_1,j_1,i_2,j_2], \\ \max_{i_2<l<j_2}\{Align[i_1,k-1,i_2,l-1]+ \\ Align[k,j_1,l,j_2]\} \end{cases}$$

end

Figure 4: Inferring structure of an RNA string

Proof. (sketch): Observe that $|S'| = O(|S|)$, which implies that $|S'| = O(n)$. The first two conditional statements each take constant time and are executed for all possible $O(n^2m^2)$ intervals. The last conditional statement takes time $O(m)$, but is executed $O(|S'|m^2) = O(nm^2)$ times. $\qquad\square$

5 Conclusions

We have formulated and provided efficient algorithms for a number of problems on computing similarity between RNA strings. Among possible extensions, we mention two: careful modeling of energies of secondary structure primitives such as multiloops, bulges etc., and defining models that capture more general structures such as proteins. It is worthwhile to note that our algorithms for RNA string matching extend to structures that allow crossing edges as long as every base forms at most one bond.

References

[1] K. Abrahamson. Generalized string matching. *SIAM J. Comp.*, 1987, 1039-1051.

[2] A. Amir and M. Farach. Efficient 2-dimensional Approximate Matching of Non-rectangular Figures. *Proc of 2nd Ann ACM Symp on Discrete Algorithms*, 1991, 212-222.

[3] D. Eppstein, Z. Galil, R. Giancarlo, and G.F. Italiano, "Sparse dynamic programming I: Linear cost functions," *JACM*, Vol. 39, No. 3, 519-545 (1992).

[4] D. Eppstein, Z. Galil, R. Giancarlo, and G.F. Italiano, "Sparse dynamic programming II: Convex and concave cost functions," *JACM*, Vol. 39, No. 3, 546-567 (1992).

[5] M. Fischer and M. Paterson. String Matching and other Products. *SIAM-AMS Proceedings*, Vol. 7, 113-125, 1974.

[6] L. Grate, M. Hebster. R. Hughey, D, Haussler, I. S. Mian and H. Noller, "RNA modeling using Gibbs sampling and stochastic context free grammars," *Second Intl. Conf. on Intelligent Systems for Molecular Biology* (1994).

[7] T. Jiang, L. Wang and K. Zhang, "Alignment of trees - an alternative to tree edit," *Proc. Combinatorial Pattern Matching Conf. 94*, LNCS 807, 75-86 (1994).

[8] P. Kilpeläinen and H. Mannila, "Query primitives for tree-structured data," *Proc. Combinatorial Pattern Matching Conf. 94*, LNCS 807, 213-225 (1994).

[9] D. E. Knuth, J. H. Morris, and V. R. Pratt. Fast pattern matching in strings. *SIAM J. Computing*, 6:323–350, 1977.

[10] L. L. Larmore and B. Schieber, "On-line dynamic programming with applications to the prediction of RNA secondary structure," *Prof. First ACM-SIAM Symp. on Discrete Algorithms*, 503-512 (1990).

[11] S-Y Le, J. Owens, R. Nussinov, J-H. Chen, B. Shapiro and J. V. Maizel, "RNA secondary structures: comparison and determination of frequently recurring substructures by consensus," *CABIOS* Vol. 5, No. 3, 205-210 (1989).

[12] S. Muthukrishnan. New results and open problems related to nonstandard stringology. *Manuscript*, 1995.

[13] S. E. Needleman and C. D. Wunsch, "A general method applicable to the search for similarities in the amino-acid sequences of two proteins," *J. Mol. Bio.*, 48, 443-453 (1970).

[14] R. Nussinov, G. Pieczenik, J. R. Griggs and D. J. Kleitman, "Algorithms for loop matchings," *SIAM J. Appl. Math.*, 35, 68-82 (1978).

[15] Y. Sakakibara, M. Brown, I. S. Mian, R. Underwood, and D. Haussler, "Stochastic context free grammars for modeling RNA," *Proc. the Hawaii Intl. Conf. on System Sciences*, IEEE Computer Society Press, Los Alamitos, CA, (1994).

[16] Y. Sakakibara, M. Brown, R. Hughey, I. S. Mian, K. Sjölander, R. C. Underwood and D. Haussler, "Recent methods for RNA modeling using stochastic context-free grammars," *Proc. Combinatorial Pattern Matching Conf.*, LNCS 807, 289-306 (1994).

[17] D. Sankoff, "Simultaneous solution of the RNA folding, alignment and protosequence problems," *SIAM J. Appl. Math.* Vol. 45, No. 5, 810-825 (1985).

[18] B. A. Shapiro, "An algorithm for comparing multiple RNA secondary structures," *CABIOS*, Vol. 4, No. 3, 387-393 (1988).

[19] B. A. Shapiro and K. Zhang, "Comparing multiple RNA secondary structures using tree comparisons," *CABIOS* Vol. 6, No. 4, 309-318 (1990).

[20] T. F. Smith and M. S. Waterman, "The identification of common molecular subsequences," *J. Mol. Biol.* 147, 195-197 (1981).

[21] T. F. Smith and M. S. Waterman, "Comparison of biosequences," *Adv. in App. Math.* 2, 482-489 (1981).

[22] K-C Tai, "The tree to tree correction problem," *JACM*, Vol. 26, No. 3, 422-433 (1979).

[23] M. S. Waterman, "Secondary structure of single-stranded nucleic acids," *Studies in Foundations and Combinatorics, Advances in Mathematics supplementary studies* VOl. 1, Academic press, New York, 167-212 (1978).

[24] M. S. Waterman and T. F. Smith, "RNA secondary structure: a complete mathematical analysis," *Math. Biosci.* 42, 257-266 (1978).

[25] K. Zhang and D. Shasha, "Simple fast algorithms for the editing distance between trees and related problems, *SIAM J. Comput.* 18, 1245-1262 (1989).

[26] K. Zhang, R. Statman, and D. Shasha, "On the editing distance between unordered labeled trees," *Inform. Proc. Lett.* 42, 133-139 (1992).

[27] M. Zuker, "On finding all suboptimal foldings of an RNA molecule," *Science*, 244 7, 48-52 (1989).

[28] M. Zuker and D. Sankoff, "RNA secondary structures and their prediction," *Bull. Math. Biol.* 46, 591-621 (1984).

[29] M. Zuker and P. Stiegler, "Optimal computer folding of large RNA sequences using thermodynamics and auxiliary information," *Nucleic Acid Res.* 9, 133-148 (1981).

Of Chicken Teeth and Mouse Eyes, or Generalized Character Compatibility

Craig Benham[1] and Sampath Kannan[2] and Tandy Warnow[2]

[1] Department of Biomathematics, Mt. Sinai School of Medicine,
[2] Department of Computer and Information Sciences, University of Pennsylvania

Abstract. We propose a new model of computation for deriving phylogenetic trees based upon a generalization of qualitative characters. The model we propose is based upon recent experimental research in molecular biology. We show that the general case of determining perfect compatibility of generalized ordered characters is an *NP*-Complete problem, but can be solved in polynomial time for a special case.

1 Introduction

Given a species set S, a *qualitative character* is a function $c : S \rightarrow Q$, where Q represents the possible *states* of the character c. Examples of qualitative characters are *the number of legs for the members of the species*, which takes on non-negative values, *vertebrate*, which is boolean, and the nucleotide present at a particular position within a multiple alignment of DNA sequences. Constructing evolutionary history from qualitative characters, whether of morphological or biomolecular origin, is the standard practice of computational evolutionary biologists.[5] A classical problem in this area is the *Perfect Phylogeny Problem*, which was shown to be *NP-Complete* in [11, 4] and solvable in polynomial time when any of the associated parameters is fixed[10, 2, 7, 8]. In this paper we show that the model of qualitative characters currently in use is in fact too restrictive, and that a more general model needs to be used in order to take advantage of new technologies. In fact, we propose that it is possible that the interpretation of biological data sets, especially when using morphological characters, may need to be reconsidered.

The human genome, the complete DNA sequence containing the entire genetic endowment of an individual, consists of approximately 3×10^9 base pairs. Only about 5% of this DNA is genes, the regions that encode sequences of protein molecules[3]. The non-coding majority was originally called "junk" DNA to reflect the fact that it serves no discernable genetic purpose. Species vary widely in the amount of this non-functional DNA their genomes contain, some having much more of it than do humans. It can persist through evolution only because there is little or no selection pressure favoring streamlined genomes in multicelled organisms.

Recent work suggests that some of this non-coding majority of the genome consists of once active sequences inherited from a species' evolutionary antecedents which are not functional in their present form. These can be regulatory regions, which control the expression of genes, or pseudogenes, sequences which have all the attributes of genes but, perhaps because of mutations, are not presently expressed. Thus a genome is also a repository of information about a species' phylogenetic ancestry.

It has been shown experimentally that genetically encoded functions which were expressed in an evolutionary ancestor but are not expressed in its living descendents can, under appropriate circumstances, be restored to activity. The most striking example of this is the demonstration that embryonic chicken tissue can be induced to differentiate into a tooth structure [9]. (In this extraordinary experiment, a morphologically reptilian tooth was grown from embryonic chicken tissue implanted within the eye of a mouse, hence the title of this paper!) Although modern birds are well known not to have teeth, *Archaeopteryx*, their ancestral form, was toothed. Remarkably, genes coding for enamel synthesis that have been inactive in that role for almost 100 million years can still be induced to function. Thus, at least some of the ancestral DNA present in a genome has not been irreversibly degraded into inactivity, but rather is quiescent, potentially capable of being reactivated at some future time.

This fact has wide-ranging and important implications for evolutionary biology. In particular, the morphological characters on which phylogenetic tree constructions are based now must be regarded as having another possible state. In addition to the two standard conditions, either present or absent in a given species, a character now can also be dormant. That is, the genes encoding this attribute are present but quiescent. Although it is difficult to prove from the fossil record that a character must have arisen by reactivation from the dormant state, this could be the mechanism underlying iterative evolution, in which a morphological trait appears, becomes extinct, then reappears in a descendent species. One important example of this phenomenon is the sabre-toothed character in carnivores[6].

The algorithms used to construct phylogenetic trees cannot be applied when the characters involved can occur in three states - present, absent or dormant. More generally, we propose that proper modelling of qualitative characters may need to be much more general than even this simple model, so that ambiguity in the character state determination can be combined with definite information about the evolution between character states.

A *generalized* character (by contrast with the classical definition of character) will associate each species with a *subset* of character states. Thus, $\alpha(s)$ indicates that the *"true"* state of s with respect to the character α is known to be in the set $\alpha(s)$, but more precise information may not be available. In this way we can represent the uncertainty about the true character state arising, for example, from our inability to determine the genotype when all we know is the phenotype. A generalized character will also incorporate knowledge about transitions between character states. All the tree construction criteria that are used in the case of standard characters can be formulated for generalized characters. Therefore the problems that arise in this setting are generalizations of the corresponding problems for standard characters.

The particular question which we formulate in this paper is the question of *compatibility* of generalized characters. Classical characters are said to be *compatible* if a tree exists in which every state of every character occupies a subtree. This is a well-studied problem for classical characters, and is known to be *NP-Complete* in general[4, 11] but solvable in polynomial time when any of the associated parameters is fixed[10, 1, 8].

The rest of the paper is organized as follows: In Section 2 we define the precise mathematical formulation of generalized characters and the compatibility of these characters. In Section 3 we present polynomial time algorithms for special cases of the generalized character compatibility problem. In Section 4 we show that the general problem is *NP-Complete*. Open problems are presented in Section 5.

2 Generalized Characters

We generalize the notion of *qualitative characters* used in phylogenetic tree construction methods, to reflect the kind of information we can obtain from experiments of the type described above.

Let S be a species set. A *generalized character* is a pair $\hat{\alpha} = (\alpha, T_\alpha)$, such that the following two conditions hold:

1. α is a function $\alpha : S \to 2^{Q_\alpha}$, where $Q_\alpha \subset Z$ denotes the set of states of α.
2. $T_\alpha = (V(T_\alpha), E)$ is a rooted tree with the nodes bijectively labelled by the elements of the set Q_α.

That is, the character states for each species may not be entirely determined, but rather only limited to certain subsets of states for each character, while the permitted transitions may be restricted to a subset of the possible transitions. In this model we will presume that the permitted transitions form a rooted directed tree, T_α.

The problem of constructing phylogenies for species sets defined by generalized characters is different from the usual problem. Here, as with classical characters, we can define the *optimal* situation, which is a perfect phylogeny. With classical characters, a phylogeny T is a *perfect phylogeny* if for every state of every character, the nodes in the tree labelled with that state form a connected component. While such a phylogeny may not always exist, when it does it is optimal for many diverse criteria used to evaluate phylogenies, such as the *parsimony* and *compatibility* criteria. Determining whether a perfect phylogeny exists is known to be *NP-Complete* in general[11, 4] but can be solved in polynomial time when any of the associated parameters is fixed[8, 10, 1]. Given generalized characters, we can also define a notion of a perfect phylogeny as follows.

A *perfect phylogeny* for (S, C) where S is a species set and C is a set of generalized characters is a pair (T, c) where $T = (V(T), E(T))$ is a rooted tree and c is a function $c : V(T) \times C \to Z$ such that the following holds:

1. For each species $s \in S$ there is a leaf l_s in T such that for each $\hat{\alpha} \in C, c(l_s, \hat{\alpha}) \in \alpha(s)$.

2. For every $\hat{\alpha} \in C$ and $i \in Z$, $\{v \in T : c(v, \hat{\alpha}) = i\}$ is a connected component of T.
3. The trees $T_\alpha|c(V(T))$ and $T(\alpha)$ are isomorphic as node-labelled trees, where
   - $T_\alpha|c(V(T))$ is the minimum homeomorphic subtree of T_α induced by the states of α which label nodes in T, and
   - $T(\alpha)$ is the tree obtained from T by labelling the nodes of T only with their α-states, and then contracting edges having the same α-state at their endpoints.

Essentially, the function c indicates how to choose, for each α and $v \in V(T)$, a *specific* state for v. The choice for $c(s, \hat{\alpha})$ must lie within the set $\alpha(s)$, and each $s \in S$ is associated with a leaf l_s in T; thus the first condition. Just as in perfect phylogenies for "normal" characters, we require that the nodes labelled by the same state for the same character must form a connected component; hence, condition two. The constraints implied by the rooted trees T_α then give rise to condition three.

Thus, a perfect phylogeny for generalized characters differs from the usual perfect phylogeny by (1) permitting α to associate to each species in S a *subset* of the states of Q_α and by (2) allowing the instance of the problem to restrict the permitted transitions between character states to a subset of the possible transitions, such that the subset defines a rooted tree. Classical characters, by contrast, are characterized by the requirement that $|\alpha(s)| = 1$ for every $s \in S$, and yet the digraph of permitted transitions is complete.

We can define the **Generalized Character Compatibility Problem** as follows:

Input: A set S of species defined by a set of generalized characters C, along with a set of rooted trees $\{T_\alpha : \hat{\alpha} \in C\}$.

Question: Does a perfect phylogeny exist relative to the constraints listed above?

In this paper we will prove the following results:

1. When $\alpha(s)$ is a directed path in T_α for each $\hat{\alpha} = (\alpha, T_\alpha) \in C$, then we can determine whether a perfect phylogeny exists in polynomial time, and construct it if it does.
2. The general problem is *NP*-Complete, even when each character has at most five states.

3   Algorithms

We will present an algorithm for the case where the trees T_α may be arbitrary but each $\alpha(s)$ is a directed path in T_α starting at $start_\alpha(s)$ and ending at $end_\alpha(s)$. We will define a partial order on the nodes of T_α by saying $v \leq w$ if the path from the root of T_α to w passes through v. Thus the nodes of T_α (i.e. states of α) increase as you get further from the root. We define the following function:

Definition 1. $H(\alpha) = lca\{end_\alpha(s) : s \in S\}$, where lca denotes the *least common ancestor*.

Note the following:

Observation 1 *Let (T, c) be a perfect phylogeny for (S, C), and let $\hat\alpha \in C$. Then for all $\{s_1, s_2\} \subseteq S$, $c(lca_T(l_{s_1}, l_{s_2}), \hat\alpha) \le lca_{T_\alpha}(end_\alpha(s_1), end_\alpha(s_2))$.*

We then have the following lemma.

Lemma 1 *Suppose (T, c) is a perfect phylogeny rooted at r for the species set S with generalized character set C, and $\alpha(s)$ a directed path in T_α for every $s \in S, \hat\alpha \in C$. Then*

1. *$c(r, \hat\alpha) \le H(\alpha)$ for all α, and*
2. *There exists a perfect phylogeny (T', c') for (S, C) in which $c(r, \hat\alpha) = H(\alpha)$ for all $\hat\alpha \in C$.*

Proof: Consider an arbitrary $\hat\alpha \in C$. Recall that $H(\alpha) = lca_{T_\alpha}\{end_\alpha(s) : s \in S\}$. We will show that $c(r, \hat\alpha) \le H(\alpha)$.

By the definition of $H(\alpha)$, $\exists\{x, y\} \subseteq S$ s.t. $H(\alpha) = lca_{T_\alpha}\{end_\alpha(x), end_\alpha(y)\}$. Thus $c(r, \hat\alpha) \le c(lca_T(l_x, l_y), \hat\alpha) \le lca_{T_\alpha}(end_\alpha(x), end_\alpha(y)) = H(\alpha)$, where the second inequality follows from Observation 1.

We will now show that we can relabel every node $v \in V(T)$ resulting in a new assignment c' such that (T, c') is a perfect phylogeny satisfying $c'(r, \hat\alpha) = H(\alpha)$, proving our second assertion.

For any node $v \in V(T)$ such that $c(v, \hat\alpha) \le H(\alpha)$ set $c'(v, \hat\alpha) = H(\alpha)$. For all other nodes, c' and c are identical.

Suppose this pair (T, c') fails to be a perfect phylogeny. Then either $\exists s \in S$ s.t. $c'(l_s) \not\subseteq \alpha(s)$, some state of α now occupies more than one connected component, or the trees $T(\alpha)$ and $T_\alpha|c'(V(T))$ are not isomorphic. It is easy to see that the last two cases will not arise, so we will only consider the first case.

Consider a leaf l_x such that $c'(l_x, \hat\alpha) \not\subseteq \alpha(x)$. Hence l_x must be one of the nodes whose α-states was changed, since (T, c) was a perfect phylogeny. Hence $c(l_x, \hat\alpha) < H(\alpha)$. Note that $H(\alpha) = c'(l_x)$ so that there is only a problem if $H(\alpha) \not\subseteq \alpha(x)$. Since $H(\alpha) = lca\{end_\alpha(s) : s \in S\}$, clearly $H(\alpha) \le end_\alpha(x)$. Since T was a perfect phylogeny, $start_\alpha(x) \le c(l_x, \hat\alpha)$. Thus we have shown that $start_\alpha(x) \le c(l_x, \hat\alpha) \le c'(l_x, \hat\alpha) = H(\alpha) \le end_\alpha(x)$, and we are done.

Thus, (T, c') is also a perfect phylogeny and satisfies the conditions above for the character $\hat\alpha \in C$. Since $\hat\alpha$ was arbitrary, the lemma follows. ∎

We now need to define an equivalence relation on S.

Definition 2. Let $m \in Z^k$ be a vector, and let Q denote the relation where $(s_1, s_2) \in Q$ iff $\exists\hat\alpha \in C$ such that $\alpha(m) < lca(start_\alpha(s_1), start_\alpha(s_2))$. We then close Q under transitivity, and let the equivalence classes in Q be the sets of S/m.

The proof of the following lemma is straightforward and is omitted.

Lemma 2 *If (T, c) is a perfect phylogeny for (S, C), and m is a label of a node $v \in V(T)$, then species s, s' that are in the same equivalence class of S/m must be in the same subtree of T_v, the subtree of T rooted at v.*

Theorem 1 *Let $m = meet(S)$ be defined by $\alpha(m) = H(\alpha)$. If $m \notin S$, then S has a perfect phylogeny if and only if the following is true:*

- *S/m has at least two equivalence classes, and*
- *S_i has a perfect phylogeny for every equivalence class $S_i \in S/m$.*

Proof: We will first show that if S has a perfect phylogeny, then these conditions follow. By Lemma 1, if T is a perfect phylogeny for S, then for all $\hat{\alpha} \in C$, the root r of T has an α-state which lies between the root of T_α and $H(\alpha) = \alpha(m)$. If there is a perfect phylogeny for S, let T be one whose root r has at least two children. By Lemma 2, the subtrees off the root r define a partition of S which is refined by the partition S/m. Thus, S/m has at least two equivalence classes. The second condition follows since having a perfect phylogeny is a hereditary property.

For the converse, suppose that conditions (1) and (2) are true, and let $S_1, S_2, \ldots, S_k$ be the distinct equivalence classes. Let T_i be a perfect phylogeny for S_i. Inductively, we can set the root r_i for T_i to be defined recursively by $r_i = meet(S_i)$. We then define the tree T to have root $r = meet(S)$ with children r_i, where r_i is the root of tree T_i. ∎

This theorem indicates the following algorithm:

```
CONSTRUCTTREE(S, C)
begin algorithm
If |S| = 1 Return(S), else
   Compute m ← meet(S)
   Compute S/m
   If |S/m| = 1 Then
      If m ∉ S Then Return (Null)
         Else
            Let T' ← CONSTRUCTTREE(S − {m})
            Let T be the tree defined by
            root(T) = m and child(m) = T'
      {endif}
   Else
         let S/m have equivalence classes S¹, S², ..., Sᵏ
         Let Tⁱ = CONSTRUCTTREE(Sⁱ, C)
         If all Tⁱ ≠ null − tree then
            let T be tree defined by
            root(T)=m and children(m)={root(Tⁱ), i = 1, 2, ..., k}
         {endif}
   {endif}
```

Return(T)
end of algorithm

Analysis of running time To reduce to solving subproblems costs us only the amount of computing m, the meet of S, and then computing the equivalence relation S/m. Computing m costs us $O(nk)$ at worst and computing S/m costs us $O(nk)$ as well. Thus, reducing to subproblems costs only $O(nk)$, where $|S| = n$ and $|C| = k$. There are at most n recursive calls, for a total cost of $O(n^2k)$ time.

4   A Proof of NP-completeness

We now show that the problem is NP-complete. The reduction is from the **Quartet Consistency Problem (QCP)** which was shown to be NP-complete by Steel[11].

QCP is the following problem:

Problem: Quartet Consistency Problem.
Input: A set of quartets $\{((a_i, b_i), (c_i, d_i)) : i = 1, \ldots, k\}$.
Question: Is there a tree T containing nodes $\{a_i, b_i, c_i, d_i : i = 1, \ldots, k\}$ (and possibly other nodes) such that for each i, there is an edge e_i of the T whose removal separates $\{a_i, b_i\}$ from $\{c_i, d_i\}$.

In order to reduce QCP to our problem we introduce dummy nodes x, y, z, and a root node r in addition to the nodes mentioned in the instance of QCP. The character set we use is as follows:

We introduce a 3-state character $\hat{\alpha_0} = (T_{\alpha_0}, \alpha_0)$ such that

- $\alpha_0(r) = \{0\}$,
- $\alpha_0(t) = \{1\}$ for $t \in \{x, y, z\}$,
- $\alpha_0(t) = \{2\}$ for all $t \in S$, and
- $T_{\alpha_0} = 0 \to 1 \to 2$.

Lemma 3 α_0 *ensures that the paths* $P_{xy}, P_{yz},$ *and* P_{zx} *all separate* r *from every other node.*

Next for every quartet $((a_i, b_i), (c_i, d_i))$ we introduce a 5 state character $\hat{\alpha_i} = (\alpha_i, T_{\alpha_i})$ such that

- $\alpha_i(r) = \{0\}$,
- $\alpha_i(q) = \{1, 2\}, \forall q \in \{x, y, z\}$,
- $\alpha_i(q) = \{1, 4\}, \forall q \in \{a_i, b_i\}$,
- $\alpha_i(q) = \{2, 3\}, \forall q \in \{c_i, d_i\}$,
- $\alpha_i(q) = \{1, 2, 3, 4\}, \forall q \in S - \{a_i, b_i, c_i, d_i\}$.
- $T_{\alpha_i} = 0 \to 1 \to 3 \cup 0 \to 2 \to 4$.

The important point about the character $\hat{\alpha_i}$ is established in the following lemmas.

Lemma 4 *The root state of character α_i does not occur on the path from a_i to b_i nor does it occur on the path from c_i to d_i.*

Proof: Since each of x, y, and z have to be assigned a state out of the set $\{1, 2\}$ by the pigeonhole principle two of them have to be assigned the same state. Wlog assume that x and y are assigned the same state 1. Then since the path P_{xy} separates all other nodes from the root, the root state 0 does not occur on the paths mentioned in the lemma. ∎

Lemma 5 *Nodes a_i and b_i have to be assigned the same state of character α_i. Similarly, nodes c_i and d_i have to be assigned the same state of α_i.*

Proof: Suppose to the contrary that a_i gets state 1 and b_i gets state 4. Then since state 0 does not occur along the path from a_i to b_i there must be some edge along this path which violates the state derivation tree of character α_i and hence this cannot happen. A similar argument shows that c_i and d_i must have the same state. ∎

The above lemma holds the key to the overall proof.

Theorem 2 *The Generalized Character Compatibility Problem is NP-complete.*

Proof: We have shown how each instance of QCP yields an instance of the GCP problem. Now, suppose that there is a perfect phylogeny for the above set of characters. By lemma 5, this phylogeny satisfies the condition that for each i, the path from a_i to b_i does not intersect the path from c_i to d_i and hence this phylogeny is a solution to the quartet consistency problem.

Conversely suppose that we started with a YES-instance of QCP. Let T be the tree consistent with all the quartets. We will create a rooted tree T' from T, by adding nodes r, x, y, z, u, as follows. Let r be the root, with child u. The children of u are x, y, z, and v', where v' is any internal node in T. This tree T' contains the species set S as well as some additional species; if T' can be labelled so that it is perfect for this enlarged set of species, then it is also perfect for S as well. We will specify how to set the states of each node for each character α_i, so that our assertion is proved.

Since the tree T' is rooted, the changes of character state are oriented from the ancestral towards the derived. Thus, we need to ensure that for each of the characters α_i, there is at most one transition of each of the permitted types, and no transition which is not permitted (i.e. for α_0, there must be at most one $0 \to 1$ transition, and at most one $1 \to 2$ transition, while for α_i, $1 \le i \le q$, there can be at most one of each of the following transitions: $0 \to 1$, $0 \to 2$, $1 \to 3$ and $2 \to 4$).

We set the state of r to be 0 for every character α_i, $i = 0, 1, 2, \ldots, q$, where q is the number of quartets in the QCP instance. We now specify the settings for each of the remaining nodes in the tree.

For character α_0 we set $\alpha_0(p) = 2$ for all $p \in V(T)$, and $\alpha_0(u) = \alpha_0(x) = \alpha_0(y) = \alpha_0(z) = 1$. This setting has one $0 \to 1$ transition ($r \to u$) and one $1 \to 2$ transition ($u \to v'$), as required.

Now let us look at a character α_i, $1 \leq i \leq q$. Recall that the character α_i requires that $\alpha_i(r) = 0$, $\{\alpha_i(x), \alpha_i(y), \alpha_i(z)\} \subseteq \{1, 2\}$, $\{\alpha_i(a_i), \alpha_i(b_i)\} \subseteq \{1, 3\}$, $\{\alpha_i(c_i), \alpha_i(d_i)\} \subseteq \{2, 4\}$, and $\alpha_i(s) \neq 0, \forall s \in S - \{a_i, b_i, c_i, d_i\}$. The setting of $\alpha_i(p)$, for $p \in V(T) \cup \{u, x, y, z\}$ will be determined by how the paths P_1 from a_i to b_i and P_2 from c_i to d_i behave. These possibilities are as follows:

P_1 **separates** r **from** P_2: Let A be the set of nodes in T' on or below the path P_2, and let $B = V(T') - A - \{r\}$. Set $\alpha_i(a) = 3$ for all $a \in A$, and $\alpha_i(b) = 1$ for all $b \in B$. Note that this setting provides only one $0 \to 1$ transition ($r \to u$), and one $1 \to 3$ transition (from a node off the path P_2 to a node on the path P_2), and no forbidden transition occurs.

P_2 **separates** r **from** P_1: Let A be the set of nodes in T' on or below the path P_1, and let $B = V(T') - A - \{r\}$. Set $\alpha_i(a) = 4 \ \forall a \in A$, and $\alpha_i(b) = 2 \ \forall b \in B$. This setting has no forbidden transitions, one $0 \to 2$ transition (on the edge $r \to u$), and only one $2 \to 4$ transition, which occurs on an edge from a node off P_2 to a node on P_2.

Otherwise: In this case, neither of the paths P_1 and P_2 separates the other from the root. Let A be the nodes on or below the path P_2, and let $B = V(T') - A - \{r\}$. We will set $\alpha_i(a) = 3 \ \forall a \in A$ and $\alpha_i(b) = 1 \ \forall b \in B$. This setting has no forbidden transitions, one $0 \to 1$ transition (on the edge $r \to u$), and one $1 \to 3$ transition, which occurs on an edge from a node off P_2 to a node on P_2.

Thus, these assignments respect the derivation tree for the states and hence we have a perfect phylogeny for S, and therefore a YES-instance of our problem. ∎

5 Open Problems

The case that would be very interesting to solve and which is still open is the case where $Q_\alpha = 0, 1, 2$, T_α is the directed path $0 \to 1 \to, 2$, and $\alpha(s) \in \{\{0\}, \{1\}, \{2\}, \{0, 2\}\}$ for all $\hat{\alpha} \in C$. This is the case which arises from having a characteristic not be present at the root, arise, and then disappear, such as the chicken teeth situation.

The *NP*-Completeness proof we gave above does not imply that this particular variation of the Generalized Character Compatibility problem is also *NP*-Complete. This is an open problem and one of importance to the biological problem motivating this paper.

6 Acknowledgment

The work of the first author was supported in part by grant RO1-GM47012 from the National Institutes of Health. The second author's work was partially supported by funding from the Discrete Mathematics and Theoretical Computer Science (DIMACS) Center at Rutgers University. The third author's work was supported by a National Young Investigator Award from the National Science Foundation CCR-9457800.

References

1. R. Agarwala, and D. Fernández-Baca, (1994), *Fast and simple algorithms for perfect phylogeny and triangulating colored graphs*, DIMACS Tech Report # 94-51.
2. Agarwala, R. and D. Fernandez-Baca (1994), *A Polynomial-time Algorithm for the Perfect Phylogeny Problem when the Number of Character States is Fixed.* SIAM J. on Computing, Vol. 23 No. 6, pp. 1216-1224.
3. Alberts, B., Bray, D., Lewis, J., Raff, M., Roberts, K. and Watson, J.D. (1989) *The Molecular Biology of the Cell, 2nd Ed.*, Garland Publishing Inc., New York, pp. 485-488.
4. H. BODLAENDER, M. FELLOWS, AND T. WARNOW (1992), *Two strikes against perfect phylogeny*, Proceedings of the 19th International Colloquium on Automata, Languages, and Programming, Springer Verlag, Lecture Notes in Computer Science, pp. 273–283.
5. J. Felsenstein (1982), *Numerical methods for inferring evolutionary trees*, The Quarterly Review of Biology, Vol. 57, No. 4.
6. Janis, C. (1994) The Sabertooth's Repeat Performances *Nat'l. Hist.*, **103** pp. 78-82.
7. S. Kannan and T. Warnow (1994), *Inferring evolutionary history from DNA sequences*, SIAM J. on Computing, Vol. 23, Nov. 4, pp. 713-737.
8. S. Kannan and T. Warnow, *A fast algorithm for the computation and enumeration of perfect phylogenies*, Proceedings, ACM-SIAM Symposium on Discrete Algorithms, San Francisco, 1995, pp. 595-603.
9. Kollar, E.J. and Fisher, C. (1980) Tooth Induction in Chick Epithelium: Expression of Quiescent Genes for Enamel Synthesis *Science* **207**, pp. 993-995.
10. F.R. McMorris, T. Warnow, and T. Wimer (1994), *Triangulating vertex colored graphs*, SIAM J. Discrete Mathematics, Vol. 7 No. 2, pp. 296-306.
11. Michael Steel (1992), *The Complexity of Reconstructing Trees from Qualitative Characters and Subtrees*, Journal of Classification, Vol. 9, pp. 91-116.

Efficient String Matching on Coded Texts

Dany Breslauer[1] * Leszek Gąsieniec[23] **

[1] BRICS – Basic Research in Computer Science – Centre of the Danish National
Research Foundation, Department of Computer Science, University of Aarhus,
DK-8000 Aarhus C, Denmark.
[2] Instytut Informatyki, Uniwersytet Warszawski, 02–097 Warszawa, Poland.
[3] Département d'Informatique, Université du Québec à Hull, Hull, Québec J8X 3X7,
Canada.

Abstract. The so called "four Russians technique" is often used to
speed up algorithms by encoding several data items in a single memory cell. Given a sequence of n symbols over a constant size alphabet,
one can encode the sequence into $O(n/\lambda)$ memory cells in $O(\log \lambda)$ time
using $n/\log \lambda$ processors.
This paper presents an efficient CRCW-PRAM string-matching algorithm for coded texts that takes $O(\log\log(m/\lambda))$ time[4] making only
$O(n/\lambda)$ operations, an improvement by a factor of $\lambda = O(\log n)$ on the
number of operations used in previous algorithms. Using this string-
matching algorithm one can test if a string is square-free and find all
palindromes in a string in $O(\log\log n)$ time using $n/\log\log n$ processors.

1 Introduction

In the string-matching problem one is searching for occurrences of a pattern
string $\mathcal{P}[1..m]$ in a text string $\mathcal{T}[1..n]$. There exist several $O(n + m)$ time sequential string-matching algorithms that are used in a large variety of applications. Galil [23] published the first efficient parallel string-matching algorithm.
His algorithm takes $O(\log m)$ time and uses n processors in the concurrent-read
concurrent-write parallel random-access-machine model. If the symbols of the
input strings are taken from a constant size alphabet, then the number of processors is reduced to $n/\log m$, achieving an *optimal speedup*, or in other words
achieving a time-processor product that is equal to the running time of the
fastest sequential algorithm for the problem. (Notice that there is a trivial constant time parallel string-matching algorithm that uses nm processors. Our goal

---

* Partially supported by ESPRIT Basic Research Action Program of the EC under
contract #7141 (ALCOM II). Part of the research reported in the paper was carried
out while this author was visiting at the Istituto di Elaborazione dell'Informazione,
Consiglio Nazionale delle Ricerche, Pisa, Italy, with the support of the European Research Consortium for Informatics and Mathematics postdoctoral fellowship. Email:
dany@daimi.aau.dk.

** Supported by KBN grant 2-11-90-91-01 and EC Cooperative Action IC-1000 (project
ALTEC). Email: lechu@mimuw.edu.pl.

[4] Throughout the paper $\log n$ usually means $\max(1, \log_2 n)$.

is to design fast parallel algorithms that use few processors.) The saving is obtained by using the so called "four Russians technique", named after the work of Arlazarov et al. [8], where each block of $O(\log m)$ symbols is packed into a single memory cell to facilitate comparisons of many symbols in a single operation.

Vishkin [38] generalized Galil's algorithm and obtained an $O(\log m)$ time algorithm that uses only $n/\log m$ processors, regardless of the alphabet size. Breslauer and Galil [10] gave an $O(\log\log m)$ time string-matching algorithm that uses $n/\log\log m$ processors. Breslauer and Galil [11] proved that if $n = O(m)$, then this is the best time bound achievable by an optimal-speedup string-matching algorithm that has access to the input strings only by pairwise symbol comparisons.

Vishkin [39] presented an optimal-speedup string-matching algorithm that takes $O(\log^2 m)$ time for the pattern preprocessing and then only $O(\log^* m)$ time to find all occurrences of the pattern in the text. Galil [24] improved the text processing step to constant time. Goldberg and Zwick [26] presented an algorithm with a tradeoff between the time spent in the pattern preprocessing and the text processing steps. Recently, Crochemore et al. [16] discovered an algorithm that takes $O(\log\log m)$ time to preprocess the pattern and then constant time to find all occurrences of the pattern in the text. Crochemore et al. also gave a randomized version of their pattern preprocessing algorithm that takes only constant expected time. These algorithms access the input strings by pairwise symbol comparisons and do not require any special assumption on the alphabet size.

This paper gives a variant of Breslauer and Galil's [10] string-matching algorithm that takes $O(\log\log(m/\lambda))$ time making only $O(n/\lambda)$ operations, after the input strings are coded in $O(n/\lambda)$ memory cells. The parameter $\lambda = O(\log n)$. The input symbols, which are assumed to be taken from a constant size alphabet, are encoded in $O(\log\lambda)$ time using $n/\log\lambda$ processors. Notice that the encoding step dominates the number of processors used. Thus the new algorithm is inferior to the previously known parallel string-matching algorithms since it has the additional restriction on the alphabet size. However, the advantages of the algorithm become clear if the input strings are given in their coded form.

Apostolico, Breslauer and Galil gave efficient parallel algorithms for testing if a string is square-free and for finding all palindromes in a string [4, 5, 6]. Their algorithms share a similar structure, take $O(\log\log n)$ time utilizing $n\log n/\log\log n$ processors, and rely on a procedure that is used to solve several string-matching problems. Observing that it suffices to encode the input string only once and use the coded string as input to many string-matching problem instances, we improve the processor bounds of these algorithms and obtain optimal-speedup $O(\log\log n)$ time $n/\log\log n$-processor algorithms for the two problems. We assume that the reader is familiar with these algorithms and with the Breslauer-Galil string-matching algorithm.

The paper is organized as follows. Section 2 introduces the computation model. Section 3 describes how the input strings are encoded and how the coded strings are manipulated. The string-matching algorithm is given is Section 4 and its applications for testing if a string is square-free and for finding all palindromes in a string are given in Section 5. Concluding remarks and open problems are given in Section 6.

2 The computation model

The computation model we use in this paper is the *common concurrent-read concurrent-write parallel random-access-machine.* In this model, processors are allowed to read and write simultaneously at the same memory location. If many processors write to the same memory cell at the same time they are guaranteed to write the same value. The arithmetic operations $+$, $-$, $\times$, and integer division $/$ can be performed by each processor in constant time on any memory words. Notice that the memory words must be able to hold numbers which are as large as the lengths of the input strings.

The following lemma is often used in parallel algorithms. The claimed bounds hold also in the weaker exclusive-read exclusive-write parallel random-access-machine model.

Lemma 1. *(Ladner and Fischer [29]) Given a sequence $x_1, \ldots, x_h$, and an associative binary operation $\oplus$, one can compute the prefix sums $x_1 \oplus x_2 \oplus \cdots \oplus x_g$, for all $g = 1, \ldots, h$, in $O(\log h)$ time using $h/\log h$ processors.*

In the CRCW-PRAM model, certain computations can be carried out much faster.

Lemma 2. *(Fich, Ragde and Wigderson [20]) Given a collection of h integers from the range $1, \ldots, h$, it is possible to find their minima value in constant time using an h-processor CRCW-PRAM.*

The last lemma will be used mainly to find the leftmost non-zero entry in an array. We shall also use the following general theorem without going into the details of the assignment of processors to their tasks.

Theorem 3. *(Brent [9]) Any parallel algorithm of time t that consists of a total of x elementary operations can be implemented on p processors in $O(\lceil x/p \rceil + t)$ time.*

3 Encoding strings

Throughout the paper we assume that the input alphabet is $\Sigma = \{0, 1, \ldots, c-1\}$, for some fixed positive constant c. Since the memory words in our model are able to store numbers as large as n, where n is the length of the string $S[1..n]$ being encoded, we could represent at least $\lfloor \log_c n \rfloor$ symbols in each memory word as a number in base c that has the symbols as its digits.

The new string-matching algorithm takes advantage of the coded representation of strings in two ways: fast comparison of blocks of several symbols and table lookup of precomputed information. While the first use would benefit from packing as many symbols as possible in each memory word, the second might require a substantial use of computational resources (time, processors, space) to compute and store the tables. The balance is achieved by packing only $\lambda = \max(1, \lfloor \frac{1}{8} \log_c n \rfloor)$ symbols in each word. The parameters c and λ will be used throughout the paper.

Given a string $\mathcal{S}[1..n]$, we break the string into consecutive blocks of λ symbols and encode each block into a memory word. Thus, a string of length n is encoded into a sequence of $\lceil n/\lambda \rceil$ memory words. We shall continue to refer to the symbols, the indices and the length of the original string, using the encoded representation only when we wish to compare substrings fast or when we wish to look up some information that we have precomputed for the coded strings.

To manipulate the coded strings efficiently we extend the repertoire of operations supported by our model to include the powers c^h, for $h = 0, \ldots, \lambda$, and to support the modulo operation. The modulo operation can be implemented as $a \bmod b = a - b * \lfloor a/b \rfloor$, and the powers c^h are implemented by a table lookup.

Lemma 4. *Given a string $\mathcal{S}[1..n]$ over a constant size alphabet, one can encode the string into $O(n/\lambda)$ memory words in $O(\log \lambda) = O(\log \log n)$ time using $n/\log \lambda = O(n/\log \log n)$ processors.*

Using the encoded representation, we can save a factor of λ in the number of operations needed to compare two strings.

Lemma 5. *It is possible to compare two coded strings of original length l and to find the position of the first mismatch between them if they are not equal, in constant time and $O(\lceil l/\lambda \rceil)$ operations.*

4 String matching with coded strings

In this section we describe an algorithm that finds all occurrences of a pattern $\mathcal{P}[1..m]$ in a text $\mathcal{T}[1..n]$. The input strings are assumed to be given in their coded form with the coding parameter λ. The algorithm takes $O(\log \log(m/\lambda))$ time and makes $O(\lceil n/\lambda \rceil)$ operations. If the strings are not already coded, one can encode them as the single string $\mathcal{S}[1..n + m] = \mathcal{P}[1..m]\mathcal{T}[1..n]$.

Observe that for any text position t, $1 \leq t \leq n - m + 1$, where there is no occurrence of the pattern, there must be at least one text position $\mathcal{W}_t^T$, such that $\mathcal{T}[\mathcal{W}_t^T] \neq \mathcal{P}[\mathcal{W}_t^T - t + 1]$. The position $\mathcal{W}_t^T$ is called a *witness* for the non-occurrence of the pattern at text position t.

The output of the string-matching problem consists of a length n boolean vector whose entries indicate if there are any occurrences of the pattern starting at each of the corresponding text positions. This boolean vector will be encoded the same way as the input strings, with the same parameter λ, and the alphabet symbols 1 and 0. In addition to the boolean vector the algorithm provides witnesses for the non-occurrences of the pattern. Notice that since our algorithm makes only $O(\lceil n/\lambda \rceil)$ operations it is not possible to list all witnesses as in other string-matching algorithms.

The main idea in the new string-matching algorithm is that the witnesses are given implicitly where any specific witnesses can be computed from the output of the algorithm by a single processor in constant time whenever needed. The algorithm is otherwise similar to the parallel string-matching algorithm of Breslauer and Galil [10] with certain modifications that allow it to take advantage of coded strings in order to match short patterns by table lookup.

Theorem 6. *The string-matching problem with coded strings is solved in time* $O(\log\log(m/\lambda))$ *making* $O(\lceil n/\lambda \rceil)$ *operations and using* $O(\lceil n/\lambda \rceil)$ *space.*

We outline the structure of the algorithm next. Initially, there are $n - m + 1$ text positions at which an occurrence of the pattern might start. These positions are called *potential occurrences*. Using Lemma 5, one can verify in constant time making $O(\lceil m/\lambda \rceil)$ operations if any given potential occurrence is a real occurrence. However, verifying all $O(n)$ potential occurrences this way is too costly if the pattern is long. The strategy followed by most efficient parallel string-matching algorithms first eliminates many potential occurrences and then verifies which of the remaining potential occurrences are real occurrences.

Definition 7. A string $S[1..k]$ has a period of length p if $S[i] = S[i + p]$, for $i = 1, \cdots, k - p$.

The shortest non-zero period length of a string $S[1..k]$ is called *the period length* of $S[1..k]$. Denote by π the period length of the pattern $P[1..m]$. If p is not a period length of the pattern $P[1..m]$, then there must exist some pattern position W_p^P, such that $P[W_p^P] \neq P[W_p^P - p]$. The positions W_p^P are called *witnesses* for non-periods of the pattern. Notice that the witnesses W_p^P are defined for all $p = 1, \ldots, \pi - 1$.

Vishkin [38] suggested the *duel* method to eliminate potential occurrences efficiently. His method, which is described next, has been used in all efficient parallel string-matching algorithms afterward as well as in sequential and parallel two-dimensional matching algorithms [1, 13, 17, 25]. The idea in duels is that if there are two potential occurrence of the pattern at positions p and q of the text, such that $0 < q - p < \pi$, then since $P[W_{q-p}^P] \neq P[W_{q-p}^P - (q-p)]$, the text symbol $T[p + W_{q-p}^P - 1]$ can not be equal both to $P[W_{q-p}^P]$ and to $P[W_{q-p}^P - (q - p)]$. Therefore, text position $p + W_{q-p}^P - 1$ must be a witness for the non-occurrence of the pattern at text position p or at text position q (possibly at both positions) and the algorithm can eliminate one of the potential occurrences at p or at q by making a single pairwise symbol comparison.

Observe that if the pattern occurs at positions p and q of the text, such that $0 < q - p < m$, then it has a period of length $q - p$ and therefore $\pi \leq q - p$. Thus, there can be no more than n/π occurrences of the pattern in the text. Using duels, it is possible to eliminate efficiently potential occurrences that are close to each other, leaving at most n/π potential occurrences. Still, there might be too many occurrences to verify separately if the period length π is much smaller than the pattern length. In this case the algorithm must follow a different strategy. The algorithm proceeds in few steps:

1. If the pattern length $m \leq 2\lambda$, then the string-matching problem is solved by table lookup as described in Lemma 8.
2. If the pattern length $m > 2\lambda$, then the pattern preprocessing step described in Section 4.2 is invoked. It finds the period length of the pattern, π, and the witnesses $\mathcal{W}_p^{\mathcal{P}}$.
   (a) If the pattern is found to be non-periodic, namely, if $m \leq 2\pi$, then the algorithms finds the occurrences of the pattern directly, as described in Lemma 11.
   (b) If the pattern is periodic, namely, if $m > 2\pi$, then the algorithm only searches for occurrences of the non-periodic pattern prefix $\mathcal{P}[1..2\pi]$. This is done as described in Lemma 8 if this pattern prefix is short or as described in Lemma 11 if it is long.
   The algorithm then reconstructs from the occurrences of this pattern prefix and by matching some short pattern suffix, the occurrences of the complete pattern as described in Lemma 12.

In the description below we show how the algorithm computes the witnesses $\mathcal{W}_p^{\mathcal{P}}$ for non-periods of the pattern. We do not specify exactly how the witnesses $\mathcal{W}_t^{\mathcal{T}}$ for non-occurrences of the pattern can be computed since their computation is similar to the pattern witnesses and they can be easily reconstructed by tracing the steps of the algorithm.

4.1 Text processing

The saving in the number of processors used by the algorithm is achieved mainly by matching short patterns by table lookup.

Lemma 8. *One can find all occurrences of the pattern $\mathcal{P}[1..m]$, such that $m \leq d\lambda$, for some fixed constant $d \geq 1$, in the text $\mathcal{T}[1..n]$, in constant time making $O(\lceil n/\lambda \rceil)$ operations and using $O(\lceil n/\lambda \rceil)$ space.*

Non-periodic patterns In this section we describe how the string-matching algorithm deals with long non-periodic patterns. Namely $2\lambda < m \leq 2\pi$ and therefore $\pi > \lambda$.

Lemma 9. *(Lyndon and Schutzenberger [30]) If a string of length k has two periods of lengths p and q and $p + q \leq k$, then it also has a period of length $\gcd(p, q)$.*

Lemma 10. *If the pattern $\mathcal{P}[1..m]$ has period length $\pi \geq \lambda$, then it contains a substring $\mathcal{P}[z..z + 2\lambda - 1]$, called a synchronizing block, with period length that is at least λ.*

Proof. Recall that $m > 2\lambda$. Let $\hat{\pi}$ be the period length of the pattern prefix $\mathcal{P}[1..2\lambda]$. If $\hat{\pi} \geq \lambda$, then this prefix is the required substring. Otherwise, let $\mathcal{P}[1..l]$ be the longest prefix of the pattern whose period length is $\hat{\pi}$. By Lemma 9, the period length of $\mathcal{P}[l-2\lambda+2..l]$ is also $\hat{\pi}$ and the period length of $\mathcal{P}[l-2\lambda+2..l+1]$ is at least λ.

The pattern preprocessing described in the next section computes the period length of the pattern, the witnesses $\mathcal{W}_p^{\mathcal{P}}$ and a synchronizing block which are used in the next lemma.

Lemma 11. *The string matching problem with the coded pattern $\mathcal{P}[1..m]$ and text $\mathcal{T}[1..n]$, such that $2\lambda < m \leq 2\pi$, is solved in $O(\log\log(m/\lambda))$ time making $O(n/\lambda)$ operations and using $O(n/\lambda)$ space.*

Proof. The algorithm starts eliminating potential occurrences by finding all occurrences of the synchronizing block $\mathcal{P}[z..z+2\lambda-1]$ in the text using the table lookup in Lemma 8. The positions of the remaining potential occurrences are written into an array of size $O(n/\lambda)$. Notice that the witnesses for the non-occurrences of the potential occurrences eliminated in this step are given implicitly by matching the synchronizing block. The other witnesses that are computed later will be stored explicitly in an array.

The elimination of the remaining potential occurrences continues as in the algorithm of Breslauer and Galil [10]. Details are omitted.

Periodic patterns In this section we describe how the string-matching algorithm deals with long periodic patterns. Namely $m > 2\lambda$ and $m > 2\pi$. As mentioned above, in this case the general strategy of eliminating potential occurrences and verifying the remaining ones is too costly since there might be too many real occurrences. The algorithm searches only for occurrences of the pattern prefix $\mathcal{P}[1..2\pi]$, which is non-periodic by the following lemma, and then finds the occurrences of the whole pattern by "counting" consecutive occurrences of this prefix. Recall that the occurrences of $\mathcal{P}[1..2\pi]$ are found by Lemma 8, if $\pi \leq \lambda$, and by Lemma 11 otherwise.

Lemma 12. *Given the occurrences of the pattern prefix $\mathcal{P}[1..2\pi]$ in the text $\mathcal{T}[1..n]$, it is possible to find the occurrences of the entire pattern in constant time making $O(n/\lambda)$ operations and using $O(n/\lambda)$ space.*

4.2 Pattern preprocessing

The pattern preprocessing is invoked only if $m > 2\lambda$. It has to find the period length π of the pattern and the witnesses $\mathcal{W}_p^{\mathcal{P}}$. For technical reasons, the pattern preprocessing step computes only the witnesses $\mathcal{W}_p^{\mathcal{P}}$, for $p = 1, \ldots, \min(\lceil m/2 \rceil, \pi - 1)$. In addition, if $\pi \geq \lambda$, then the pattern preprocessing step finds also a synchronizing block.

Notice, that if the period length of the pattern $\pi > \lceil m/2 \rceil$, then it is not computed precisely. In this case the pattern is non-periodic and the period length π is not used by the algorithm.

Lemma 13. *The pattern preprocessing step with the coded pattern $\mathcal{P}[1..m]$, such that $m > 2\lambda$, takes $O(\log \log(m/\lambda))$ time making $O(m/\lambda)$ operations and using $O(m/\lambda)$ space.*

Proof. The pattern preprocessing step first finds a synchronizing block and then uses this block and witnesses that it has already computed to compute more witnesses in iterations that resemble the text processing step. The indices p for which the witnesses $\mathcal{W}_p^{\mathcal{P}}$ are not yet computed are called *potential period lengths*. The witnesses $\mathcal{W}_p^{\mathcal{P}}$, $p = 1, \ldots, \min(\lceil m/2 \rceil, \pi - 1)$, will be given implicitly, where any specific witness can be produced from the information computed in constant time by a single processor.

The pattern preprocessing uses a precomputed lookup table, similarly to the $\mathcal{SM}$ table from Lemma 8, that gives the boolean vector representing the *period lengths* and the witnesses for the non-periods of a short string. If the pattern length $m \leq 4\lambda$, then the pattern preprocessing step will be solved directly by this table lookup. Thus, from here on we assume that the pattern length $m > 4\lambda$.

Our first goal is to find a synchronizing block and to reduce the number of potential period lengths to $O(m/\lambda)$. Recall the constructive nature of the proof of Lemma 10. Using the precomputed table of period lengths of short strings, the algorithm finds the period length $\hat{\pi}$ of the pattern prefix $\mathcal{P}[1..2\lambda]$. If $\hat{\pi} \geq \lambda$, then the algorithm has found the synchronizing block $\mathcal{P}[1..2\lambda]$. Otherwise, if $\hat{\pi} < \lambda$, the algorithm checks if the whole pattern has period length $\hat{\pi}$, by Lemma 5. If $\hat{\pi}$ turns out to be the period length of the whole pattern, then the only information required from the pattern preprocessing step is this period length $\pi = \hat{\pi}$, and the pattern preprocessing is completed. Otherwise, the synchronizing block $\mathcal{P}[z..z + 2\lambda - 1]$ has been found.

If $z + \lambda - 1 > \lceil m/2 \rceil$, then by the construction of the synchronizing block in Lemma 10, the pattern prefix $\mathcal{P}[1..z + 2\lambda - 2]$ has period length $\hat{\pi}$ and $\mathcal{P}[1..z + 2\lambda - 1]$ does not have this period length. Thus, by Lemma 9, full occurrences of the pattern prefix $\mathcal{P}[1..2\lambda]$ that start in the first $\lceil m/2 \rceil$ positions of the pattern, start at positions $k\hat{\pi} + 1$. Matching the pattern prefix $\mathcal{P}[1..2\lambda]$ by Lemma 8, one obtains the witnesses $\mathcal{W}_p^{\mathcal{P}}$, except for the multiples $p = k\hat{\pi}$. The position $z + 2\lambda - 1$ where the period of length $\hat{\pi}$ terminates provides the witness $\mathcal{W}_{\hat{\pi}}^{\mathcal{P}}$, and since the pattern prefix $\mathcal{P}[1..z + 2\lambda - 2]$ has period length $\hat{\pi}$, $\mathcal{W}_p^{\mathcal{P}} = z + 2\lambda - 1$, for all the multiples $p = k\hat{\pi}$, such that $0 < p \leq \lceil m/2 \rceil$. Thus, the witnesses $\mathcal{W}_p^{\mathcal{P}}$ can be reproduced either by matching the pattern prefix $\mathcal{P}[1..2\lambda]$, by Lemma 8, if p is not a multiple of $\hat{\pi}$, or $\mathcal{W}_p^{\mathcal{P}} = z + 2\lambda - 1$ otherwise.

If $z + \lambda - 1 \leq \lceil m/2 \rceil$, then the algorithm finds all occurrences of the synchronizing block $\mathcal{P}[z..z + 2\lambda - 1]$ in the pattern, by Lemma 8. Observe that the witnesses to the non-occurrences correspond to witnesses $\mathcal{W}_p^{\mathcal{P}}$. The occurrences, which must be spaced at least λ positions apart, leave at most $O(m/\lambda)$ potential period lengths in the first half of the pattern. The positions of the remaining potential period lengths are written into an array and will be computed and stored explicitly as we show next. Observe that when a specific witness is called for, it can be either reproduced by matching the synchronizing block again or it will be stored explicitly in a table.

The computation of the remaining witnesses proceedings in the same fashion as the string-matching algorithm of Breslauer and Galil [10]. We sketch here only a non-optimal version of the algorithm making $O(m \log \log(m/\lambda)/\lambda)$ operations. The algorithm can be made optimal similarly to the algorithm of Breslauer and Galil.

The algorithm proceeds in iterations and maintains the invariant that at the beginning of iteration number i, there is at most one potential period length (yet-to-be-computed witness) in each block of length k_i, where,

$$k_i = m^{1-\frac{1}{2^i}} \cdot \lambda^{\frac{1}{2^i}} \quad \text{for } i = 0, \ldots, \log \log(m/\lambda).$$

Clearly, the invariant holds at the beginning of iteration number 0, since the potential period lengths remaining after the first part of the computation are space at least $k_0 = \lambda$ positions apart.

In iteration number i, there are at most k_{i+1}/k_i potential period length in each block of length k_{i+1}. The algorithm checks using Lemma 5, which of the potential period lengths in the first k_{i+1} block is a period length of the pattern prefix $\mathcal{P}[1..2k_{i+1}]$. Those potential period length which are eliminate have produced a witness, while the remaining potential period lengths, if any, are multiples of the shortest remaining period length, by Lemma 9. This computation takes constant time and $O(k_{i+1}/\lambda)$ operations for each potential period, or $O(k_{i+1}^2/k_i\lambda) = O(m/\lambda)$ operations in total.

If there are any potential period lengths remaining in the first k_{i+1} block, then the algorithm verifies whether the shortest one is the period length of the whole pattern by Lemma 5. If it is found to be the period length then the computation is complete.

Otherwise, the position at which this periodicity is terminated is a witness for all multiples of the shortest period in the first k_{i+1} block. Now, it remains only to eliminate all but at most one potential period length in each k_{i+1} block, before proceeding to the next iteration.

It is possible to eliminate all but at most one potential period length in each k_{i+1} block using duels, since at this point we have the witnesses $\mathcal{W}_p^{\mathcal{P}}$, for all $p = 1, \ldots, k_{i+1}$. The duels, however, are slightly different from those used in the text processing step, since occurrences might be overhanging: a duel that has to produce one of the witnesses $\mathcal{W}_i^{\mathcal{P}}$ or $\mathcal{W}_j^{\mathcal{P}}$, for $i < j < \lceil m/2 \rceil$, will normally produce the witness $i + \mathcal{W}_{j-i}^{\mathcal{P}} + 1$, if it is within the pattern; otherwise the duel produces the witnesses $\mathcal{W}_i^{\mathcal{P}} = \mathcal{W}_{j-i}^{\mathcal{P}} - j + i$ or $\mathcal{W}_j^{\mathcal{P}} = \mathcal{W}_{j-i}^{\mathcal{P}}$.

The duels are carried out in the same fashion as in the text processing step. However, we allow the algorithm to use $m/\lambda \log \log(m/\lambda)$ processors. The duels will take at most $O(\log \log(m/\lambda))$ time in the first two iterations of the pattern preprocessing, after which they take constant time since the number of remaining potential period lengths will be small enough relatively to the number of available processors.

The whole pattern preprocessing step described above takes $O(\log\log(m/\lambda))$ time. The overall number of operations used is $O(m/\lambda)$ except at the step that verifies if the shortest remaining potential period length in each iteration is the period length of the whole pattern. This step uses $O(m/\lambda)$ operation in each iteration and thus $O(m\log\log(m/\lambda)/\lambda)$ operations over all iteration. However, this step can be implemented more economically, making only $O(m/\lambda)$ operations [10].

5 Applications

In this section we present two application of the string-matching algorithm described above in reducing the number of processors used in known parallel algorithms for testing if a string is square-free and for finding all palindromes in a string. The reduction in the number of processors is achieved since the input string $S[1..n]$ has to be encoded only once while its encoded substrings are presented several times as input to the string-matching algorithm. Recall that the input string $S[1..n]$ is encoded with the parameter $\lambda = O(\log n)$.

5.1 Testing if a string is square-free

A non-empty string of the form xx is called a repetition. A *square* is defined as a repetition xx, where x is primitive, or in other words $x \neq v^h$ for all strings v and integers $h > 1$. Strings that do not contain any substring that is a repetition are called *repetition-free* or *square-free*. For example 'aa', 'abab' and 'baba' are the repetitions which are contained in the string 'baababa'. It is not difficult to verify that any string with at least four symbols over alphabets with two symbols contains a square. However, there exist infinite length strings on three letter alphabets that are square-free as shown by Thue [36, 37].

In the sequential setting, algorithms for testing if a string is square-free and for finding all repetitions in a string were designed by Apostolico and Preparata [7], Crochemore [14, 15], Kosaraju [28], Main and Lorentz [31, 32] and Rabin [34]. Main and Lorentz [31] proved that it is possible to find all repetition in a string in $O(n\log n)$ time using pairwise comparison of input symbols that test for equality. They have also shown that $\Omega(n\log n)$ equality tests are necessary even to decide if a string is square-free. Main and Lorentz [32] have shown using the "four Russians technique" that if the input alphabet has constant size, then it is possible to test if a string is square-free in $O(n)$ time. The same bound was obtained by Crochemore [15] using a different method. Notice that it is not possible to list all squares in $O(n)$ time since there might be too many squares [2, 14].

In the parallel setting, Crochemore and Rytter [18, 19] test if a string is square-free in $O(\log n)$ time using n processors and $O(n^{1+\epsilon})$ space. Apostolico [3] designed an algorithm that tests if a string is square-free and also detects all squares within the same time and processor bounds using only linear auxiliary space. If the input alphabet has constant size, then Apostolico's algorithm can use the "four Russians technique" to tests if a string is square-free in $O(\log n)$ time utilizing only $n/\log n$ processors.

Apostolico and Breslauer [4] gave a parallel implementation of the sequential algorithm of Main and Lorentz [32] to test if a string is square-free and find all squares in a string using equality tests in $O(\log \log n)$ time using $n \log n/\log \log n$ processors. If the input alphabet has constant size, then the number of processors used by their algorithm to test if a string is square-free can be reduced to $n/\log \log n$ by using the new string-matching algorithm. These bounds compare favorably also with the $O(\log n)$ time algorithm given by Apostolico [3] for testing if a string over a constant size alphabet is square-free. Notice that all the parallel algorithms mentioned above achieve an optimal speedup since their time-processor product is the same as the time complexity of the fastest known sequential algorithm under the same assumptions on the input alphabet.

Applying string matching algorithm introduced in section 4 to solution presented by Apostolico and Breslauer [4] we get the following result.

Theorem 14. *There exists an algorithm to test if a string $S[1..n]$ over a constant size alphabet is square-free in $O(\log \log n)$ time using $n/\log \log n$ processors and $O(n)$ space.*

5.2 Finding all palindromes in a string

Palindromes are symmetric strings that read the same forward and backward. Formally, a non-empty string w is a palindrome if $w = w^R$, where w^R denotes the string w reversed. It is convenient to distinguish between even length palindromes that are strings of the form $w = vv^R$ and odd length palindromes that are strings of the form $w = vav^R$, where v is an arbitrary string and 'a' is a single alphabet symbol.

Given a string $S[1..n]$, we say that there is an even palindrome of radius $\mathcal{R}$ *centered at* position k of $S[1..n]$, if $S[k-i] = S[k+i-1]$, for $i = 1, \ldots, \mathcal{R}$. We say that there is an odd palindrome of radius $\hat{\mathcal{R}}$ *centered on* position k of $S[1..n]$, if $S[k-i] = S[k+i]$, for $i = 1, \ldots, \hat{\mathcal{R}}$. The radius $\mathcal{R}$ (or $\hat{\mathcal{R}}$) is maximal if there is no palindrome of radius $\mathcal{R} + 1$ centered at (on) the same position. In this section we will be interested in computing the maximal radii $\mathcal{R}[k]$ and $\hat{\mathcal{R}}[k]$ of the even and the odd palindromes which are centered at (on) all positions k of $S[1..n]$. Notice that if we double each input symbol, then odd palindromes become even and thus, without loss of generality, we can concentrate on finding only the maximal radii of the even palindromes [6].

In the sequential setting, Manacher [33], and Knuth, Morris and Pratt [27] presented linear-time algorithms that find the initial palindromes (palindrome prefixes) of a string. Galil [22] and Slisenko [35] presented real-time algorithms on multi-tape Turing machines to find all initial palindromes.

In the parallel setting, Crochemore and Rytter [19] presented an algorithm that finds all palindromes in a string in $O(\log n)$ time using n processors and

$O(n^{1+\epsilon})$ space. Their algorithm assumes that the alphabet symbols are small integers. Breslauer and Galil [12] using an observation of Fischer and Paterson [21], described an algorithm that finds all initial palindromes in a string in $O(\log \log n)$ time and $n/\log \log n$ processors using equality tests.

Apostolico, Breslauer and Galil [6] gave an algorithm that can find all palindromes in a string using equality tests in $O(\log \log n)$ time and $n \log n / \log \log n$ processors. They also gave an optimal-speedup algorithm that finds all palindromes in a string over constant size alphabets in $O(\log n)$ time and $n/\log n$ processors, using the "four Russians technique". If the input alphabet has constant size then the number of processors used in their $O(\log \log n)$ time algorithm can be reduces to $n/\log \log n$ by applying string matching algorithm introduced in section 4.

Theorem 15. *There exists an algorithm that finds all even palindromes in a string $S[1..n]$ over a constant size alphabet in $O(\log \log n)$ time using $n/\log \log n$ processors and $O(n)$ space.*

6 Conclusions

The string-matching algorithm presented in this paper takes advantage of the bounded alphabet size to reduce the number of processor used. Since the lower bound of Breslauer and Galil [11, 12] does not hold if the alphabet has constant size, one can hope to design an optimal-speedup algorithms for several string problems, such as the string-matching, the square-detection and the palindrome-detection problems, that will achieve faster running times over constant size alphabets.

An other interesting open question remaining is whether there exists a fast optimal-speedup palindrome detection algorithm using only pairwise symbol comparisons.

References

1. A. Amir, G. Benson, and M. Farach. An alphabet-independent approach to two-dimensional pattern-matching. *SIAM J. Comput.*, 23(2):313–323, 1994.
2. A. Apostolico. On context constrained squares and repetitions in a string. *R.A.I.R.O. Informatique théorique*, 18(2):147–159, 1984.
3. A. Apostolico. Optimal Parallel Detection of Squares in Strings. *Algorithmica*, 8:285–319, 1992.
4. A. Apostolico and D. Breslauer. An Optimal $O(\log \log n)$ Time Parallel Algorithm for Detecting all Squares in a String. *SIAM J. Comput.*, to appear.
5. A. Apostolico, D. Breslauer, and Z. Galil. Optimal Parallel Algorithms for Periods, Palindromes and Squares. In *Proc. 19th International Colloquium on Automata, Languages, and Programming*, number 623 in Lecture Notes in Computer Science, pages 296–307. Springer-Verlag, Berlin, Germany, 1992.
6. A. Apostolico, D. Breslauer, and Z. Galil. Parallel Detection of all Palindromes in a String. *Theoret. Comput. Sci.*, to appear.

7. A. Apostolico and F.P. Preparata. Optimal off-line detection of repetitions in a string. *Theoret. Comput. Sci.*, 22:297–315, 1983.

8. V.L. Arlazarov, E.A. Dinic, M.A. Kronrod, and I.A. Faradzev. On economic construction of the transitive closure of a directed graph. *Soviet Math. Dokl.*, 11:1209–1210, 1970.

9. R.P. Brent. Evaluation of general arithmetic expressions. *J. Assoc. Comput. Mach.*, 21:201–206, 1974.

10. D. Breslauer and Z. Galil. An optimal $O(\log \log n)$ time parallel string matching algorithm. *SIAM J. Comput.*, 19(6):1051–1058, 1990.

11. D. Breslauer and Z. Galil. A Lower Bound for Parallel String Matching. *SIAM J. Comput.*, 21(5):856–862, 1992.

12. D. Breslauer and Z. Galil. Finding all Periods and Initial Palindromes of a String in Parallel. *Algorithmica*, to appear.

13. R. Cole, M. Crochemore, Z. Galil, L. Gąsieniec, R. Hariharan, S. Muthukrishnan, K. Park, and W. Rytter. Optimally fast parallel algorithms for preprocessing and pattern matching in one and two dimensions. In *Proc. 34th IEEE Symp. on Foundations of Computer Science*, pages 248–258, 1993.

14. M. Crochemore. An optimal algorithm for computing the repetitions in a word. *Inform. Process. Lett.*, 12(5):244–250, 1981.

15. M. Crochemore. Transducers and repetitions. *Theoret. Comput. Sci.*, 12:63–86, 1986.

16. M. Crochemore, Z. Galil, L. Gąsieniec, K. Park, and W. Rytter. Constant-Time Randomized Parallel String Matching. Manuscript, 1994.

17. M. Crochemore, L. Gąsieniec, R. Hariharan, S. Muthukrishnan, and W. Rytter. A Constant Time Optimal Parallel Algorithm for Two Dimensional Pattern Matching. Manuscript, 1993.

18. M. Crochemore and W. Rytter. Efficient parallel algorithms to test square-freeness and factorize strings. *Inform. Process. Lett.*, 38:57–60, 1991.

19. M. Crochemore and W. Rytter. Usefulness of the Karp-Miller-Rosenberg algorithm in parallel computations on strings and arrays. *Theoret. Comput. Sci.*, 88:59–82, 1991.

20. F.E. Fich, R.L. Ragde, and A. Wigderson. Relations between concurrent-write models of parallel computation. *SIAM J. Comput.*, 17(3):606–627, 1988.

21. M.J. Fischer and M.S. Paterson. String matching and other products. In R.M. Karp, editor, *Complexity of Computation*, pages 113–125. American Mathematical Society, Prividence, RI., 1974.

22. Z. Galil. Palindrome Recognition in Real Time by a Multitape Turing Machine. *J. Comput. System Sci.*, 16(2):140–157, 1978.

23. Z. Galil. Optimal parallel algorithms for string matching. *Inform. and Control*, 67:144–157, 1985.

24. Z. Galil. A Constant-Time Optimal Parallel String-Matching Algorithm. In *Proc. 24th ACM Symp. on Theory of Computing*, pages 69–76, 1992.

25. Z. Galil and K. Park. Truly Alphabet-Independent Two-Dimensional Pattern Matching. In *Proc. 33th IEEE Symp. on Foundations of Computer Science*, pages 247–256, 1992.

26. T. Goldberg and U. Zwick. Faster parallel string matching via larger deterministic samples. *J. Algorithms*, 16(2):295–308, 1994.

27. D.E. Knuth, J.H. Morris, and V.R. Pratt. Fast pattern matching in strings. *SIAM J. Comput.*, 6:322–350, 1977.

28. S.R. Kosaraju. Computation of Squares in a String. In *Proc. 5rd Symp. on Combinatorial Pattern Matching*, number 807 in Lecture Notes in Computer Science, pages 146–150. Springer-Verlag, Berlin, Germany, 1994.

29. R.E. Ladner and M.J. Fischer. Parallel prefix computation. *J. Assoc. Comput. Mach.*, 27(4):831–838, 1980.

30. R.C. Lyndon and M.P. Schützenberger. The equation $a^m = b^n c^p$ in a free group. *Michigan Math. J.*, 9:289–298, 1962.

31. G.M. Main and R.J. Lorentz. An $O(n \log n)$ algorithm for finding all repetitions in a string. *J. Algorithms*, 5:422–432, 1984.

32. G.M. Main and R.J. Lorentz. Linear time recognition of squarefree strings. In A. Apostolico and Z. Galil, editors, *Combinatorial Algorithms on Words*, volume 12 of *NATO ASI Series F*, pages 271–278. Springer-Verlag, Berlin, Germany, 1985.

33. G. Manacher. A new Linear-Time "On-Line" Algorithm for Finding the Smallest Initial Palindrome of a String. *J. Assoc. Comput. Mach.*, 22:346–351, 1975.

34. M.O. Rabin. Discovering Repetitions in Strings. In A. Apostolico and Z. Galil, editors, *Combinatorial Algorithms on Words*, volume 12 of *NATO ASI Series F*, pages 279–288. Springer-Verlag, Berlin, Germany, 1984.

35. A.O. Slisenko. Recognition of palindromes by multihead Turing machines. In V.P. Orverkov and N.A. Sonin, editors, *Problems in the Constructive Trend in Mathematics VI (Proceedings of the Steklov Institute of Mathematics, No. 129)*, pages 30–202. Academy of Sciences of the USSR, 1973. English Translation by R.H. Silverman, pp. 25–208, Amer. Math. Soc., Providence, RI, 1976.

36. A. Thue. Über unendliche zeichenreihen. *Norske Vid. Selsk. Skr. Mat. Nat. Kl. (Cristiania)*, (7):1–22, 1906.

37. A. Thue. Über die gegenseitige lage gleicher teile gewisser zeichenreihen. *Norske Vid. Selsk. Skr. Mat. Nat. Kl. (Cristiania)*, (1):1–67, 1912.

38. U. Vishkin. Optimal parallel pattern matching in strings. *Inform. and Control*, 67:91–113, 1985.

39. U. Vishkin. Deterministic sampling - a new technique for fast pattern matching. *SIAM J. Comput.*, 20(1):22–40, 1990.

Fast Approximate Matching using Suffix Trees

Archie L. Cobbs[*]

Computer Science Division, University of California Berkeley, Berkeley, CA 94720

Abstract. Let T be a text of length n and P a pattern of length m, both strings over a fixed finite alphabet Σ. We wish to find all approximate occurrences of P in T having weighted edit distance at most k from P: this is the approximate substring matching problem. We focus on the case in which T is fixed and preprocessed in linear time, while P and k vary over consecutive searches. We give an $O(mq + t_{occ})$ time and $O(q)$ space algorithm, where $q \leq n$ depends on the problem instance, and t_{occ} is the size of the output. The running time is proportional to the amount of matching, in the worst case as fast as standard dynamic programming. The algorithm uses the suffix tree representation of the text. The best previous algorithm requires $O(mq \log q + t_{occ})$ time and $O(mq)$ space.

1 Introduction

Advances in genetics and DNA sequence analysis have been particularly strong forces driving research in approximate pattern matching. In this paper we discuss a variant of the *approximate substring matching problem*, which is central in this area: given a text string T and pattern string P, the goal is to output all ending positions j of some substring P_j in T having edit distance at most k from P. Edit distance measures the minimum total cost of any sequence of weighted character insertions, deletions, and substitution operations that transform one string into the other. We are allowed linear time preprocessing of the text.

Let $m = |P|$ and $n = |T|$. Standard dynamic programming [9] solves the approximate string matching problem in $O(mn)$ time. Landau and Vishkin [5, 6] improve this to $O(kn)$ for the special case of unit cost edit operations (the *k differences problem*). Jokinen and Ukkonen [4, 10] were the first to treat the special case in which T can be preprocessed. This situation could arise, for example, if a researcher wants to compare newly sequenced DNA fragments against all of GenBank [1]. Meyers [8] describes another approach for this case.

Ukkonen [10] gives two algorithms for this problem, both using the suffix tree of T. The first runs in time $O(mq \log q + t_{occ})$ and space $O(mq)$; the second in time $O(m^2 q + t_{occ})$ and space $O(m^2 q)$. Here $t_{occ} \leq n$ is the size of the output, i.e., the total number of ending positions of approximate occurrences of P, and $q \leq n$ depends on the amount of matching between P and T. The value of q decreases as k decreases or the amount of repetition in the text increases. Both algorithms require $O(n)$ preprocessing to build the suffix tree.

[*] Supported by U.S. DOE Grant #DE-FG03-90ER60999.

In this paper, we give a similarly motivated algorithm that runs in time $O(mq + t_{occ})$, space $O(q)$, and retains the same $O(n)$ preprocessing time. Our algorithm thus runs asymptotically as fast or faster than standard dynamic programming, even if the preprocessing time is included. In certain cases substantial time can be saved over standard dynamic programming, since the resources required by the algorithm scale proportionally with the complexity of the particular problem instance. For example, when $k = 1$, $O(m^2 + t_{occ})$ time is required.

2 Definitions and Notation

Let Σ be a fixed, finite alphabet. If $w \in \Sigma^*$, the length of w is $|w|$ and the letters of w are $w = w_1 w_2 \cdots w_{|w|}$. ϵ denotes the empty string. Let $T, P \in \Sigma^*$, and let $n = |T|$ and $m = |P|$. $T_{j..j'}$ denotes the substring (subword) $T_j T_{j+1} \cdots T_{j'}$ of T. The concatenation of $w \in \Sigma^*$ and $a \in \Sigma$ is wa. A subword w *occurs* at j in T if $w = T_{j-|w|+1} \cdots T_j$; here j is the ending position. Num(w) is the number of occurrences of w in T, i.e., the number of distinct j such that w occurs at j. We invent the following term:

Definition 1. v is a *(proper) cosuffix* of w if w is a (proper) suffix of v.

Whereas any two suffixes of w must themselves satisfy a suffix relation, this is not true for cosuffixes.

An *edit operation* on a string w is the insertion, deletion, or substitution of a single character in w. Each edit operation is given a non-negative integer cost $d(\cdot, \cdot)$. If $a, b \in \Sigma$, $d(a, b)$ is the cost of substituting b for a, $d(\epsilon, a)$ is the cost of inserting a, and $d(a, \epsilon)$ is the cost of deleting a. $d(a, a) = 0$. If u, v are strings, $d(v, w)$ denotes the *weighted edit distance* between v and w, the minimum total cost of any sequence of edit operations transforming v into w. Let $k_{\max} \geq 0$ denote the maximum allowable edit distance for an approximate match.

Definition 2. Suppose for some subword $T_{j'..j}$ of T, $d(P, T_{j'..j}) \leq k_{\max}$. Then there is an *approximate occurrence of P in T at j*.

Our objective is to output all such positions j.

3 Suffix Trees

The $O(n)$ preprocessing is for generating the *suffix tree* [11, 7] of T. The algorithm requires the *suffix links* to be included with the suffix tree; they are generated automatically during construction. Here we refer to the suffix links of McCreight [7], which point towards the root of the tree.

For conceptual clarity, we refer in our discussion to the *suffix trie* of T. The suffix trie is just the suffix tree without compressed edges, but potentially quadratic in size. In an actual implementation the algorithm would use the suffix tree (or the *directed acyclic word graph* [2, 3]) to simulate the suffix trie; here no generality is lost. To simulate a node in the suffix trie, we just use a pointer to a suffix tree node plus an offset down an edge.

There is a one-to-one correspondence between nodes in the suffix trie and subwords of T. If p is a node and w is a subword, $p \sim w$ indicates this correspondence; p is the *locus* of w. Let the root node of the suffix trie be *root*. Then *root* $\sim \epsilon$; in general, $p \sim w$ iff p is the node reached by the path from *root* which spells out w. In this paper, we refer to a subword and its locus interchangeably. Abuse of notation such as "suffix of p" is well defined. Let $p \neq root$, and let p' be the suffix of p gotten by removing the first character from p. Then there is a *suffix link* from p to p'.

Each leaf of the suffix trie corresponds to a location j in T. The number of leaves in the subtree under node p equals Num(p). With a linear time depth first search over the suffix trie, we can compute Num(p) for each node p. We assume this has been done during preprocessing.

4 The Dynamic Programming Matrix

4.1 Approximate Substring Dynamic Programming

Standard dynamic programming for computing $d(P, T)$ generates an $m + 1$ by $n + 1$ matrix D such that $D(i, 0) = i$, $D(0, j) = j$, and entry $D(i, j)$ is computed using a simple recurrence:

$$D(i,j) = \min \begin{cases} D(i-1,j) + d(P_i, \epsilon) \\ D(i-1, j-1) + d(P_i, T_j) \\ D(i, j-1) + d(\epsilon, T_j) \end{cases}$$

Then $D(i, j) = d(P_{1..i}, T_{1..j})$ and $D(m, n) = d(P, T)$.

Solving the approximate substring problem involves a simple modification: setting the top row of D to all zeroes, yielding the *approximate substring dynamic programming matrix*. Now $D(i, j)$ is equal to the minimum cost of any alignment between $P_{1..i}$ and some *suffix* of $T_{1..j}$. The output consists of those j for which $D(m, j) \leq k_{\max}$. Entry (i, j) is *essential* if $D(i, j) \leq k_{\max}$. We can ignore non-essential entries of D since they have no effect on the output.

For example, let $T = atacatacatcat$, $P = agacatgc$, and $k_{\max} = 2$. The essential part of D is shown in Figure 1. Approximate matches end at columns 8 and 11. The number in the upper right of each square is $D(i, j)$. Originating from (i, j) is a path to the top row of D describing a minimum cost (less than $k_{\max}$) alignment between $P_{1..i}$ and some suffix of $T_{1..j}$.

4.2 Viable Subwords

For a particular (i, j) there may be several paths to the top row achieving minimum cost; each one corresponds to a different alignment between $P_{1..i}$ and some suffix of $T_{1..j}$. We chose vertical edges over diagonal edges over horizontal edges whenever equal cost choices exist. The result is the *canonical path* from (i, j).

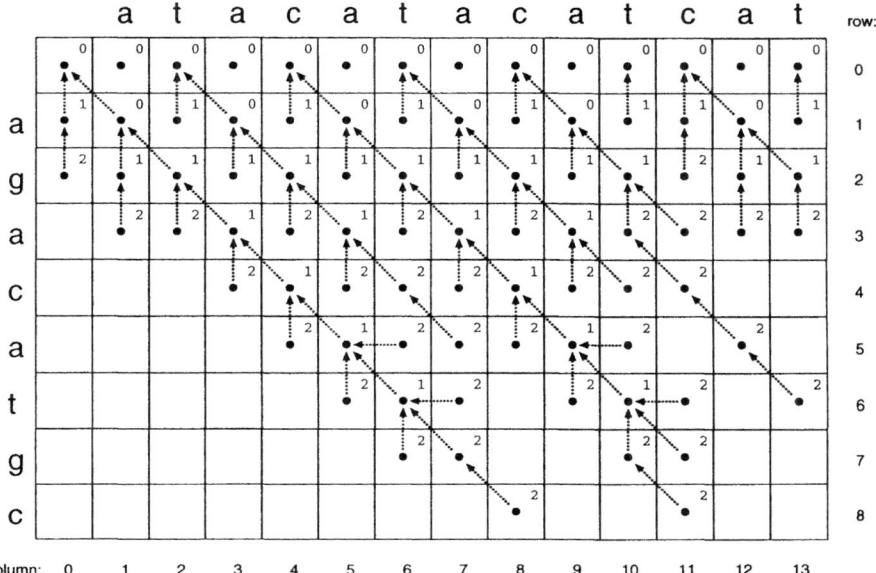

Fig. 1. Approximate substring dynamic programming matrix for $T = atacatacatcat$, $P = agacatgc$ with unit cost edit distances and maximum distance 2. Canonical paths are shown.

The canonical path choses the "rightmost" minimum alignment of $P_{1..i}$ with a suffix of $T_{1..j}$, and the suffix w implied by the canonical path is as short as possible. This suffix is the *viable subword for $D(i,j)$*, denoted $V(i,j)$.

For example, the viable subword at $D(3,4)$ in Figure 1 is ac and $d(aga, ac) =$ 2. Notice $d(aga, atac)$ is also 2, but $atac$ is not implied by the canonical path. If $D(i,j) > k_{max}$, then $V(i,j)$ does not exist; non-essential entries have no viable subword.

Observation 3. *Canonical paths cannot cross.*

4.3 Duplicated Work

The value of $D(i,j)$ can be computed knowing only the viable subword $V(i,j)$. It is just $d(P_{1..i}, V(i,j))$. Another way of stating this: $D(i,j)$ is completely determined by the *triangle* bounded by column j, the canonical path from (i,j), and the top row of the matrix. In turn, this triangle is determined by $V(i,j)$.

Consider entries $(7,6)$ and $(7,10)$ of Figure 1. In both cases the viable subword is $atacat$. Not only are $D(7,6)$ and $D(7,10)$ equal, but the corresponding triangles are isomorphic. In a sense, traditional dynamic programming performs redundant work by recomputing these values. As in [10], our algorithm saves time by computing "redundant" entries only once.

Definition 4. Let $0 \le i \le m$. $V_i = \{V(i,j) : 1 \le j \le n\}$ is the set of *viable subwords for row i*.

Clearly $|V_i| \le n$. In the example of Figure 1,

$$V_0 = \{\epsilon\},$$
$$V_1 = \{\epsilon, a\},$$
$$V_2 = \{\epsilon, a, ac, at\},$$
$$V_3 = \{a, at, ata, ac, aca, atc\},$$
$$V_4 = \{ata, atac, aca, acat, atc\},$$
$$V_5 = \{atac, ataca, atacat, acata, atca\},$$
$$V_6 = \{ataca, atacat, atacata, atacatc, atcat\},$$
$$V_7 = \{atacat, atacata, atacatc\},$$
$$V_8 = \{atacatac, atacatc\}.$$

Definition 5. If w is any subword of T the *score of w in row i*, denoted $D_i(w)$, is equal to $d(P_{1..i}, w)$.

If $w = V(i,j)$ then $D_i(w) = D(i,j)$. In any row, the algorithm computes one entry per viable subword, rather than one entry per column. So the more repetition in T, the more work is saved. When V_m is computed, we output all ending positions of subwords in V_m. In Figure 1, the output set is $\{8, 11\}$.

5   The Structure of V_i

The algorithm works on a row-by-row basis, computing V_i from V_{i-1}. In order to motivate the algorithm, we make a couple of observations to characterize V_i.

5.1   Extensions

Consider the canonical path from matrix entry (i,j): the edge out of (i,j) goes to $(i-1,j)$, $(i-1,j-1)$, or $(i,j-1)$. We describe $V(i,j)$ as *vertical, diagonal,* or *horizontal* respectively. In V_5 of Figure 1, *atac* is vertical; *ataca, acata* and *atca* are diagonal; and *atacat* is horizontal. Suppose $pa \in V_i$. If pa is vertical, diagonal, or horizontal, then $pa \in V_{i-1}$, $p \in V_{i-1}$, or $p \in V_i$, respectively. Therefore, the following is a suitable set of *candidates* for V_i:

- All p for which $p \in V_{i-1}$, plus
- All pa for which $p \in V_{i-1}$ and there is an edge labeled a out of p, plus
- All pa for which $p \in V_i$ and there is an edge labeled a out of p.

In fact, the algorithm generates candidates in exactly this way, by extending $p \in V_{i-1}$ vertically and diagonally, and $p \in V_i$ horizontally. Because of horizontal extensions, candidates must be generated in a sensible order, and the algorithm insures this.

5.2 Preemption

Each candidate is checked for viability. There are two ways in which a candidate can fail to be viable.

Suffix preemption. Suppose p' is a proper suffix of p and $D_i(p') \leq D_i(p)$. Then p cannot be viable; at every occurrence of p the canonical path could not imply p because a lower cost path (implying p') exists. This is *suffix preemption*—p is preempted from being in V_i by a lower scoring suffix.

For example, consider entry $(3,4)$ in Figure 1. Both *atac* and *ac* have edit distance 2 from the pattern prefix *aga*. Since it is shorter, *ac* preempts *atac* from being viable anywhere in row 3.

Observation 6. *Within the set V_i, taking a proper suffix implies a strictly worse (higher) score.*

Corollary 7. *Let j be an occurrence of p, and suppose $V(i,j)$ is a suffix of p. Then $V(i,j)$ is the longest suffix of p in V_i.*

Cosuffix preemption. On the other hand, suppose $q_1, q_2, \ldots, q_r \in V_i$ and each q_s is a proper cosuffix of p. Then from the above discussion, $D_i(q_s) < D_i(p)$. Consider an ending position j at which p appears in T. If some q_s also appears there, then $p \neq V(i,j)$ because q_s gives a strictly lower score. If this happens at *every* occurrence of p, then $p \notin V_i$. This is called *cosuffix preemption*—p is preempted from being viable by a set of cosuffixes.

For example, consider row 4 in Figure 1, where the pattern prefix is $P_{1..4} = agac$. No suffix of *ac* has a score equal or better than $d(ac, agac) = 2$, so *ac* is not suffix preempted. However, every occurrence of *ac* in T is as a suffix of *atac*, which has score 1. Therefore *ac* is cosuffix preempted in row 4, so $ac \notin V_4$.

We can characterize viable candidates with this lemma:

Lemma 8. *A candidate p is not viable for row i if and only if one or both of the following holds:*

- *For some occurrence j of p, $V(i,j)$ is a proper suffix of p (suffix preemption).*
- *For every occurrence j of p, $V(i,j)$ is a proper cosuffix of p (cosuffix preemption).*

5.3 Filling the Holes

If $D(i,j) > k_{\max}$ then $V(i,j)$ does not exist, leaving "holes" in the matrix, for example $(5, 11)$ of Figure 1. Define $Q(i,j)$ equal to $V(i',j)$ where $i' \leq i$ is the largest row such that $V(i',j)$ exists; then $Q(i,j)$ always exists. Let Q_i be the set of $Q(i,j)$ for all j. Q_i is V_i with the "holes" filled in. From the non-crossing of canonical paths (Observation 3), we have:

Observation 9. *Let $i_1 \leq i_2$. Then $Q(i_2, j)$ is a cosuffix of $Q(i_1, j)$.*

Lemma 10. *Let j be an occurrence of p, and suppose $Q(i,j)$ is a suffix of p. Then $Q(i,j)$ is the longest suffix of p in Q_i.*

Proof. Suppose not. Then some suffix p' of p, $p' \in Q_i$, lies properly between p and $Q(i,j)$. Then $p' = V(i_2, j')$ for some j' and $i_2 \leq i$. Since p' is a suffix of p, then $V(i_2, j)$ is a cosuffix of $p' \in V_{i_2}$ by Corollary 7. But $Q(i,j)$ is a cosuffix of $V(i_2, j)$ and therefore of p'—a contradiction. $\square$

Corollary 11. *$Q(i, j+1)$ is the longest suffix of $Q(i,j)a$ in Q_i, where $a = T_{j+1}$.*

Lemma 12. *Suppose $p \in V_{i_0}$. Then if p is a candidate in any round $i > i_0$, p is not suffix preempted in round i.*

Proof. Suppose j is an occurrence of p such that $p' = V(i,j)$ is a proper suffix of p. By Corollary 7, $V(i_0, j)$ is a cosuffix of p. But then the canonical paths to (i_0, j) and (i, j) cross, unless the (i, j) path passes through (i_0, j). But then $p' = p$, a contradiction. $\square$

Lemma 13. *Suppose p is cosuffix preempted in row i_1. Then p is not viable in any row $i > i_1$.*

Proof. Since p is cosuffix preempted in row i_1, at every occurrence j of p, $V(i,j) = Q(i,j)$ is a proper cosuffix of p. If p is a candidate in row $i > i_1$, then $V(i,j) = Q(i,j)$ exists for all occurrences j of p. By Observation 9, p is cosuffix preempted in row i as well. $\square$

Corollary 14. *Let p be a node and let I be the set of rows for which $p \in Q_i$. Then I is a continuous range of rows.*

6 Description of the Algorithm

In this section we describe a simplified version of the algorithm. In Section 7 we describe optimizations which bring down the time and space requirements.

6.1 Overview

The algorithm has m rounds. In the ith round, we generate V_i from V_{i-1} and the portion of V_i we've already generated. After computing V_m, we report the ending positions of all subwords in V_m. Initially we set $V_0 = \{\epsilon\}$ and $D_0(\epsilon) = 0$.

Round i proceeds as follows. Initially, V_i is empty. We generate candidates for V_i by extending members of V_{i-1} vertically and diagonally, and members of V_i horizontally. When a candidate p is generated it is checked for viability; specifically, we check for suffix preemption and cosuffix preemption according to Lemma 8. If p is viable, it is added to V_i. This continues until no more viable candidates can be generated.

The suffix tree of T is treated as a read-only object. A hash table is used to associate additional information with nodes of the (virtual) suffix trie. The space requirement of the algorithm is proportional to the size of our "working set" of suffix trie nodes. When an entry is no longer needed, it is flushed from the hash table (Section 7.3).

6.2 Generation

We generate candidates in *score-length order*, which means in order of increasing score $D_i(p)$, and among those candidates with the same score, in order of increasing length. This facilitates preemption checking and assures proper ordering of the horizontal extensions. We never generate nodes for which $D_i(p) > k_{\max}$.

This ordering is enforced by keeping V_{i-1} and V_i sorted into $k_{\max} + 1$ linked lists, where $p \in V_i$ is kept in the list corresponding to $D_i(p)$ (which is a non-negative integer). Each list is itself sorted by length. Then generating candidates in score length order is a matter of walking $2|\Sigma| + 1$ pointers through the lists, one pointer for each possible extension. Viable nodes are appended to the list corresponding to their score.

6.3 Suffix Preemption Checking

Suppose p is the next candidate. We check p for suffix preemption as follows. First, vertical extensions do not need to be checked (Lemma 12). We also ignore any candidates already in V_i, as the earlier extension would have a better score. Otherwise, pa is suffix preempted if and only if there is some proper suffix of pa already in V_i: this is a consequence of score-length ordering and Lemma 8. So to determine if pa is suffix preempted, we just follow suffix links until some $p'a \in V_i$ is found or we reach the root of the trie.

6.4 Cosuffix Preemption Checking

To check for cosuffix preemption, nodes are annotated with *cosuffix counters*. Let p be a node and let J be the set of occurrences of p in T. Note $|J| = \text{Num}(p)$. At the end of round $i - 1$, $\text{Cos}_{i-1}(p) \le |J|$ counts for how many occurrences j of p is $Q(i - 1, j)$ a cosuffix of p. At the beginning of round i we (implicitly) set $\text{Cos}_i(p) = \text{Cos}_{i-1}(p)$. By Observation 9, $\text{Cos}_{i-1}(p)$ is a lower bound on $\text{Cos}_i(p)$. Any difference will be a result of a *newly viable* node in V_i, i.e., some cosuffix q of p such that $q \in Q_i \setminus Q_{i-1}$. The cosuffix counter can be thought of as p's "coverage" by viable cosuffixes, and this coverage can only grow as i increases. So every time such a node q is added to V_i, we update the cosuffix counters for all suffixes of q (as described below).

Note that if p is a candidate in row i, then $V(i, j)$ exists for all $j \in J$, so $Q(i, j) = V(i, j)$ for all j. Therefore any candidate p is cosuffix preempted in round i if and only if $\text{Cos}_i(p) = \text{Num}(p)$. However, not all new members of V_i may have been generated at the time p is considered, so it is possible that $\text{Cos}_i(p)$ could read erroneously low. Fortunately, score-length ordering insures this will never happen: all $q \in V_i$ which are cosuffixes of p will have already been generated at the time p is generated and its cosuffix counter examined.

Cosuffix counters are updated as follows. Suppose we have just added q to V_i and $q \notin Q_{i-1}$. Then q's cosuffix counter has some value less than $\text{Num}(p)$. Let h be the difference. We then add h to the cosuffix counter of each proper suffix of p, up to and including the first suffix p' of p such that $p' \in Q_{i-1} \cap Q_i$. It is necessary and sufficient to stop at p', because every occurrence of p is also an occurrence of p', whose "coverage" has already been accounted for. Of course, the first time we encounter a node p, we set $Cos_i(p) = 0$. This method works (in one view) because of the tree-like structure of suffix links.

Cosuffix counters also allow us to identify members of Q_i:

Observation 15. $p \in Q_i$ *if and only if* $p \in V_i$ *or* $p \in Q_{i-1}$ *and at the end of round* i, $Cos_i(p) < Num(p)$.

6.5 Output

The output step is easy once V_m is known. For each $p \in V_m$ we output the positions associated with all leaves in the subtree under p by doing a depth-first search. We only do this for nodes for which no suffix p' is also in V_m; this guarantees each position j is output exactly once.

7 Optimizations

The essence of the algorithm is captured in the simplified version described above. This section optimizes the implementation to achieve the time and space bounds.

Definition 16. Let $q = Q(i, j)$. Define $Up_i(q) = Q(i - 1, j)$.

$Up_i(q)$ is well defined because the choice of j does not matter, since the triangles above occurrences of q are all isomorphic. Note also that if $q \in Q_{i-1}$ and $q \in Q_i$ then $Up_i(q) = Up_{i-1}(q)$ (Lemma 10).

Definition 17. Let $q \in Q_i$. $Next_i(q, a)$ is the longest (not necessarily proper) suffix of qa in Q_i.

The name comes from this corollary, which follows from Corollary 11:

Corollary 18. $Next_i(Q(i, j), T_{j+1}) = Q(i, j + 1)$.

Note that $Up_i(q)$ and $Next_i(q, a)$ are members of Q_i and always exist.

7.1 Cosuffix Counters

We sort cosuffix counter values according to following letter. That is, instead of a single cosuffix counter for p, we have $|\Sigma|$ cosuffix counters, $Cos_i^a(p)$, one for each possible extension pa of p. $Cos_i^a(p)$ counts for how many occurrences $j \in J$ is some cosuffix q_j of p, also occurring at j, already in Q_i, *and such that* $T_{j+1} = a$. Then

$$Cos_i(p) = \sum_{a \in \Sigma} Cos_i^a(p) \ .$$

Updating suffix counters now is just a matter of propagating $|\Sigma|$ separate values instead of one.

The point is that we can avoid extending p diagonally or horizontally to pa if $\text{Cos}_i^a(p) = \text{Num}(pa)$; any such extension will always be cosuffix preempted, namely by extensions $q_j a$ of the same q_j which account for $\text{Cos}_i^a(p)$. Moreover, the converse is true: suppose pa is cosuffix preempted by some set of nodes $q_j a$. Only diagonal or horizontal extensions can cosuffix preempt pa (otherwise there would be crossing of canonical paths). Therefore, $q_j \in V_{i-1}$ or $q_j \in V_i$ (and then $D_i(q_j) < D_i(pa)$), so $\text{Cos}_i^a(p)$ will have already been updated to reflect their presence. The net effect is that cosuffix preempted nodes are never even generated.

7.2 Suffix Preemption Checking

Instead of following suffix links all the way up the trie, we can use the structure of V_i to help make suffix preemption checking more efficient. As mentioned in Section 6.3, only diagonal and horizontal extensions need to be checked. In this section we assume pa is not cosuffix preempted.

Lemma 19. *Suppose pa is a diagonal candidate in row i which is not cosuffix preempted. If pa is suffix preempted, then $\text{Next}_{i-1}(p, a)$ preempts it.*

Proof. Let j be an occurrence of pa such that $V(i, j)$ is a proper suffix of pa (Lemma 8). Then $V(i, j)$ is the longest such suffix in V_i by Corollary 7. It suffices to show $V(i, j)$ equals $\text{Next}_{i-1}(p, a)$. There are three cases: $V(i, j)$ is either vertical, diagonal, or horizontal.

Since pa is diagonal, $p \in V_{i-1}$ and so $V(i - 1, j - 1)$ is equal to, or a cosuffix of, p. Then $V(i, j)$ cannot be horizontal, because the canonical paths to (i, j) and $(i - 1, j - 1)$ would cross. It cannot be diagonal, because then $V(i, j) = V(i - 1, j - 1)a$, contradicting that $V(i, j)$ is a proper suffix of pa.

Therefore $V(i, j)$ is vertical, in which case $V(i, j) = V(i - 1, j)$. Therefore $V(i-1, j)$ is a proper suffix of pa. By Corollary 7, $V(i-1, j)$ is the longest proper suffix of pa in V_{i-1}. Now $pa \notin V_{i-1}$, or else by Corollary 11 $V(i - 1, j) = pa$, contradicting that $V(i - 1, j)$ is a proper suffix of pa. Therefore $\text{Next}_{i-1}(p, a)$ is the longest *proper* suffix of pa in V_{i-1}, i.e. $V(i - 1, j)$, i.e. $V(i, j)$. $\qquad\square$

Lemma 20. *Suppose pa is a horizontal candidate in row i which is not cosuffix preempted. If pa is suffix preempted, then either $Up_i(p)a$ or $\text{Next}_{i-1}(p, a)$ preempts it.*

Proof. Since pa is not cosuffix preempted, $\text{Cos}_i^a(p) < \text{Num}(pa)$. Therefore there exists some occurrence $j - 1$ of p such that $Q(i, j - 1)$ is not a proper cosuffix of p and such that $T_j = a$. But then $Q(i, j - 1)$ is a suffix of p; since $p \in V_i$, $p = Q(i, j - 1) = V(i, j - 1)$ by Lemma 10.

Since pa is a candidate, $V(i, j)$ exists and $Q(i, j) = V(i, j)$. $V(i, j)$ is a proper suffix of pa (because $V(i, j - 1) = p$ and canonical paths don't cross), in fact the longest such suffix in V_i by Corollary 7. It suffices to show $V(i, j)$ equals either $\text{Up}_i(p)a$ or $\text{Next}_{i-1}(p, a)$. There are three cases: $V(i, j)$ is either vertical, diagonal, or horizontal.

$V(i, j)$ cannot be horizontal, because then $V(i, j) = V(i, j - 1)a = pa$, contradicting that $V(i, j)$ is a proper suffix of pa.

If $V(i, j)$ is diagonal, $V(i, j) = V(i - 1, j - 1)a = \text{Up}_i(Q(i, j - 1))a$. But $Q(i, j - 1) = p$, so $V(i, j) = \text{Up}_i(p)a$.

If $V(i, j)$ is vertical, the proof is exactly the same as in the proof of Lemma 19: $V(i, j) = \text{Next}_{i-1}(p, a)$. $\quad\square$

Corollary 21. *Let pa be a non-cosuffix preempted horizontal or diagonal candidate. To determine if pa is suffix preempted, it suffices to check $\text{Up}_i(p)a$ and $\text{Next}_{i-1}(p, a)$.*

Now we describe how to compute $\text{Up}_i(p)$ and $\text{Next}_i(p, a)$. Suppose p is a new member of V_i. Then $\text{Up}_i(p)$ is computed immediately simply by following suffix links until reaching the first $p' \in Q_{i-1}$. At the end of round i, for those $p \in Q_i$ which are not in V_i, it must be that $p \in Q_{i-1}$; then we just set $\text{Up}_i(p) = \text{Up}_{i-1}(p)$.

Since $\text{Next}_i(p, a)$ is not needed until round $i + 1$, we compute $\text{Next}_i(p, a)$ at the end of round i. Note that we only need to compute $\text{Next}_i(p, a)$ if pa is a candidate in round $i + 1$. Let $p \in Q_i$ and $a \in \Sigma$; we describe how to compute $\text{Next}_i(p, a)$. Take the a-edge out of p to get to pa. If pa is not in the suffix trie then it doesn't matter because pa will never be a candidate. Otherwise, we have three choices, given one assumption:

Lemma 22. *Let $p \in Q_i$ and suppose $\text{Cos}_i^a(p) < \text{Num}(pa)$. Then $\text{Next}_i(p, a)$ is equal to either pa, $\text{Up}_i(p)a$, or $\text{Next}_{i-1}(\text{Up}_i(p), a)$.*

Proof. Since $\text{Cos}_i^a(p) < \text{Num}(pa)$, there must exist some occurrence $j - 1$ of p such that $Q(i, j - 1)$ is not a proper cosuffix of p and such that $T_j = a$. Then since $p \in Q_i$, $p = Q(i, j - 1)$ by Lemma 10.

By Corollary 18, $\text{Next}_i(p, a) = Q(i, j)$. It suffices to show $Q(i, j)$ is either pa, $\text{Up}_i(p)a$, or $\text{Next}_{i-1}(\text{Up}_i(p), a)$. There are two cases: either $V(i, j)$ exists or not.

Suppose it doesn't exist. Then $Q(i, j) = Q(i - 1, j)$. So it suffices to show $Q(i - 1, j) = \text{Next}_{i-1}(\text{Up}_i(p), a)$. By Corollary 18, $Q(i - 1, j) = \text{Next}_{i-1}(Q(i - 1, j - 1), a)$. But $Q(i - 1, j - 1) = \text{Up}_i(Q(i, j - 1)) = \text{Up}_i(p)$.

Suppose $V(i, j)$ does exist. Then $Q(i, j) = V(i, j)$. $V(i, j)$ is either horizontal, diagonal, or vertical. If horizontal, then $V(i, j) = V(i, j - 1)a = pa$.

If diagonal, then $V(i, j) = V(i-1, j-1)a$. So it suffices to show $Q(i-1, j-1) = \text{Up}_i(p)$. But $p = Q(i, j - 1)$, so $\text{Up}_i(p) = Q(i - 1, j - 1)$.

If vertical, then $V(i, j) = V(i - 1, j) = Q(i - 1, j)$. So it suffices to show $Q(i - 1, j) = \text{Next}_{i-1}(\text{Up}_i(p), a)$. By Corollary 18, $Q(i - 1, j) = \text{Next}_{i-1}(Q(i - 1, j - 1), a)$. But $Q(i - 1, j - 1) = \text{Up}_i(Q(i, j - 1)) = \text{Up}_i(p)$. $\quad\square$

So assuming $\text{Cos}_i^a(p) < \text{Num}(pa)$, we can compute $\text{Next}_i(p, a)$ in constant time by taking the longest of the above choices in Q_i. If $\text{Cos}_i^a(p) = \text{Num}(pa)$, then we don't bother to compute $\text{Next}_i(p, a)$. This is acceptable because:

Lemma 23. *Suppose $Cos_i^a(p) = Num(pa)$. Then pa will not be a candidate in round $i + 1$.*

Proof. From Observation 9 the following is clear: if pa is cosuffix preempted in row i then pa is cosuffix preempted in row $i + 1$. So it follows that if pa is a candidate in round $i + 1$, pa is not cosuffix preempted in row i. Therefore for any candidate pa in round $i + 1$, $Cos_i^a(p) < Num(pa)$. □

7.3 Saving Space

The trie nodes are stored as references in a hash table. The simplest approach would be to keep a reference to every node ever visited by the algorithm in the hash table for the duration of the algorithm. However, this requires too much space. Ideally, we would like to keep only Q_{i-1} and Q_i in the hash table. This would insure that $Up_i(p)$ and $Next_{i-1}(p, a)$ were always accessible. Since the "lifetime" of a node is a continuos interval, we can just leave $p \in Q_i$ in the hash table. Then at the end of round i, we remove all nodes p for which $Cos_i(p) < Num(p)$. By Observation 15, this leaves exactly Q_i.

Branch nodes. The only problem with this idea is that the cosuffix counter of a node is expected to be known before that node is ever viable. We adjust for missing cosuffix counters by using *branch nodes*. Suppose $p, q \in Q_i$ and let r be the longest common suffix of p and q such that $|r| < |p|, |q|$ and no other node in Q_i lies between p and r or between q and r. Then r is a branch node and is included in the hash table. Branch nodes are exactly those nodes in any chain of suffix links where the cosuffix counter changes.

Consider the restriction of the trie of suffix links to just Q_i and the branch nodes. Leaf nodes are in Q_i. Internal nodes are either in Q_i or are branch nodes. An internal branch node not in Q_i always has at least two children. All nodes have at most $|\Sigma|$ children. Therefore $|\Sigma|$ child pointers are required at each node and there are less than $|Q_i|$ extra branch nodes.

So if $p \in Q_i \setminus Q_{i-1}$ is a new viable node, we update cosuffix counters by following suffix links until reaching $p' \in Q_{i-1} \cap Q_i$ or the next branch node, whichever comes first. We determine which child q of this node is also a cosuffix of p, if any; then the cosuffix counters for p are equal to the cosuffix counters for q, or zero if q does not exist. Then we update the branch pointers, possibly creating a new branch node.

8 Analysis

Q_m is Ukkonen's set of *k-approximate viable prefixes*. Let $q = |Q_m|$.

Lemma 24. $|Q_i| \leq q$.

Proof. It suffices to show that $Q(m, j) = Q(m, j')$ implies $Q(i, j) = Q(i, j')$. By Observation 9, since $i \leq m$ the triangle above $Q(m, j)$ contains the triangle above $Q(i, j)$. If $Q(m, j) = Q(m, j')$ these triangles are isomorphic, and so $Q(i, j) = Q(i, j')$ as well. □

Lemma 25. *The number of candidates generated during round i is $O(|\Sigma|q)$.*

Proof. In round i nodes are generated by extensions of viable nodes from V_{i-1} and V_i. Each $p \in V_{i-1}$ extends to at most one vertical candidate and at most $|\Sigma|$ diagonal ones. Each $p \in V_i$ extends to at most $|\Sigma|$ horizontal candidates. Therefore $O(|\Sigma|(|V_{i-1}| + |V_i|)) = O(|\Sigma|q)$ candidates are generated. □

Theorem 26. *The algorithm requires $O(|\Sigma|mq + t_{occ})$ time.*

Proof. List operations, pointer manipulation, and trie edge and link traversals are all constant time operations. Generating candidates and checking suffix and cosuffix preemption takes constant time per node. By Lemma 25, $O(|\Sigma|mq)$ total time is spent generating and checking candidates. Scanning the hash table after each round takes $O(|Q_{i-1}| + |Q_i|) = O(|\Sigma|q)$ time, or $O(|\Sigma|mq)$ for all rounds. Maintaining branch nodes costs $O(|\Sigma|)$ time per node visited. So the only operations possibly costing more than $O(|\Sigma|)$ time per node are following suffix links to compute to $\mathrm{Up}_i(p)$ and updating cosuffix counters. These are only done for $p \in Q_i$.

Fix some $Q(m, j)$ and look at all $Q(i, j)$ for $i \leq m$. When traversing suffix links from $Q(i, j)$ we never go past $\mathrm{Up}_i(Q(i, j)) = Q(i - 1, j)$, either when computing $\mathrm{Up}(Q(i, j))$ or updating cosuffix counters. So the total number of links traversed for these $Q(i, j)$ is $O(|Q(m, j)|) = O(m)$. Therefore the total time spent following suffix links for the whole algorithm is $O(|\Sigma|mq)$.

During the output phase, in order to determine if any suffix p' of p exists in V_m we simply follow suffix links. This takes $O(mq)$ total time. Non-leaf nodes in the suffix tree have out-degree at least two; therefore the depth first search takes time $O(t_{occ})$. □

Theorem 27. *The algorithm requires $O(|\Sigma|q)$ space.*

Proof. The lists and pointers require $O(|\Sigma|)$ space per node. Using the optimization of Section 7.3, the number of nodes stored in the hash table for round i is at most $O(|Q_{i-1}| + |Q_i|)$, or $O(q)$. Therefore $O(|\Sigma|q)$ total space is required. □

9 Conclusion

We have described an algorithm for solving the approximate substring matching problem. Over a fixed alphabet, the algorithm requires $O(n)$ preprocessing of T, and runs in time $O(mq + t_{occ})$ and space $O(q)$. The algorithm is implemented using simple data structures.

The algorithm will run faster than standard dynamic programming when the text contains a lot of repetition, k or m is small compared to n, or there is little matching between P and T. When $k = 1$ standard dynamic programming still requires $\Omega(n)$ time, whereas our algorithm runs in $O(m^2 + t_{occ})$ time. Indeed, for any fixed m or k the running time is independent of n, except for output size. If both are fixed, the running time is *constant* except for output size. Even when the text is not static the algorithm can be faster if the text contains repetitions; with a non-static text, $O(n + mq + t_{occ})$ time and $O(n)$ space is required. In general, the performance adapts gracefully to the particular problem instance.

References

1. D. Benson, D. J. Lipman, and J. Ostell. Genbank. *Nucl. Acids Res.*, 21(13):2963–2965, 1993.
2. A. Blumer, J. Blumer, A. Ehrenfeucht, D. Haussler, and R. McConnell. Building a complete inverted file for a set of text files in linear time. In *FOCS*, pages 349–358. ACM, January 1984.
3. M. T. Chen and Joel Seiferas. *Efficient and Elegant Subword Tree Construction*, pages 97–107. Springer-Verlag, Berlin, 1985.
4. P. Jokinen and E. Ukkonen. Two algorithms for approximate string matching in static texts. In *Proc. MFCS 1991*, volume 16, pages 240–248. Springer-Verlag, September 1991.
5. G. M. Landau and U. Vishkin. Fast string matching with k differences. *J. Comp. Sys. Sci.*, 37:63–78, 1988.
6. G. M. Landau and U. Vishkin. Fast parallel and serial approximate string matching. *J. Algorithms*, 10:157–169, 1989.
7. Edward M. McCreight. A space-economical suffix tree construction algorithm. *JACM*, 23(2):262–272, April 1976.
8. E. W. Myers. A sublinear algorithm for approximate keyword searching. *Algorithmica*, 12:345–374, 1994.
9. D. Sankoff and J. B. Kruskal, editors. *Time Warps, String Edits, and Macromolecules: The Theory and Practice of Sequence Comparison*. Addison-Wesley, 1983.
10. E. Ukkonen. Approximate matching over suffix trees. In *Proc. Combinatorial Pattern Matching 1993*, volume 4, pages 228–242. Springer-Verlag, June 1993.
11. P. Weiner. Linear pattern matching algorithms. In *14th Annual Symposium on Switching and Automata Theory*, pages 1–11. IEEE, 1973.

Common subsequences and supersequences and their expected length

Vlado Dančík[*]

Department of Computer Science, King's College London
London WC2R 2LS, England
e-mail: vlado@dcs.kcl.ac.uk

Abstract. Let $f(n, k, l)$ be the expected length of a longest common subsequence of l sequences of length n over an alphabet of size k. It is known that there are constants $\gamma_k^{(l)}$ such that $f(n, k, l) \to \gamma_k^{(l)} n$, we show that $\gamma_k^{(l)} = \Theta(k^{1/l-1})$. Bounds for the corresponding constants for the expected length of a shortest common supersequence are also presented.

1 Introduction and preliminaries

To find the expected length of a longest common subsequence of two sequences is a standard problem studied in the literature [7, 8]. In this paper we shall concentrate on the expected length of a longest common subsequence of several sequences. We show that this expected length for l sequences of length n over an alphabet of size k is $\Theta(\frac{n}{k^{1-1/l}})$ for $n >> k >> l$. We also consider a dual case, the expected length of a shortest common supersequence.

Let $\Sigma = \{0, 1, \ldots, k-1\}$ be a fixed alphabet of size k. Let Σ^* be the set of all strings over Σ. The length of string u is denoted by $|u|$. The set of all sequences of length n is denoted by Σ^n. We shall work over the set, $\Pi_n^l = (\Sigma^n)^l$, of all l-tuples of sequences of length n. We denote the set of all l-tuples of sequences by $\Pi^l = (\Sigma^*)^l$. The total length of the l-tuple $\mathbf{u} = (u_1, \ldots, u_l) \in \Pi^l$ is the sum of lengths of all sequences from the l-tuple, i.e. $|\mathbf{u}| = l(u_1, \ldots, u_l) = |u_1| + \cdots + |u_l|$. Concatenation of two l-tuples $\mathbf{u} = (u_1, \ldots, u_l)$ and $\mathbf{v} = (v_1, \ldots, v_l)$ is defined by $\mathrm{cat}(\mathbf{u}, \mathbf{v}) = (u_1 v_1, \ldots, u_l v_l)$. Concatenation of more than two l-tuples is defined analogously.

A sequence u is a *subsequence* of a sequence v if u can be obtained from v by a deletion of some symbols. A sequence v is a *common subsequence* of $\mathbf{u} = (u_1, \ldots, u_l) \in \Pi^l$ if for every $i = 1, \ldots, l$ sequence v is a subsequence of sequence u_i. A *longest common subsequence* is a common subsequence of maximal possible length. We shall denote the length of a longest common subsequence of sequences $\mathbf{u} = (u_1, \ldots, u_l)$ by $\mathbf{L}(\mathbf{u}) = \mathbf{L}(u_1, \ldots, u_l)$.

---

[*] Most of the work was done while author was a postgraduate student at Warwick University, England. Partially supported by the EPSRC grant GR/J 17844.

2 Longest common subsequences

The expected length $\mathbf{EL}_n^{(l)}$ of a longest common subsequence is given by

$$\mathbf{EL}_n^{(l)} = \frac{1}{k^{ln}} \sum_{\mathbf{u} \in \Pi_n^l} \mathbf{L}(\mathbf{u}).$$

It is known (Chvátal and Sankoff [2]) that for every $k \geq 2$ and every $l \geq 2$ there are constants

$$\gamma_k^{(l)} = \lim_{n \to \infty} \frac{\mathbf{EL}_n^{(l)}}{n} = \sup_n \frac{\mathbf{EL}_n^{(l)}}{n}.$$

To get lower bounds for the expected length we design an algorithm producing any common subsequence (not necessarily the longest one). The expected length of such a common subsequence will be a lower bound for $\mathbf{EL}_n^{(l)}$.

We can use the following algorithm to obtain lower bounds for $\gamma_k^{(l)}$ for sufficiently large k. We scan the first input tape until $k^{1-1/l}$ different symbols are found. Let K_1 be the set of these symbols. Then we scan the second tape until $k^{1-2/l}$ different symbols from K_1 are read. We denote the set of symbols from K_1 scanned on the second tape by K_2. We shall continue in this way. Suppose we have determined the set K_{i-1}. We scan the i-th tape until $k^{1-i/l}$ different symbols from K_{i-1} are found. These symbols form the set K_i. Having specified K_{l-1}, we search for some symbol $a \in K_{l-1}$ on the last tape. For convenience we set $K_0 = \Sigma$. Since

$$a \in K_{l-1} \subseteq K_{l-2} \subseteq \cdots \subseteq K_1,$$

we can put a into a common subsequence produced by the algorithm. Computing the average progress on each tape yields following lower bounds.

Theorem 1. *For every $l \geq 2$ we have*

$$\lim_{k \to \infty} k^{1-1/l} \gamma_k^{(l)} \geq 1.$$

Proof. Let us consider the i-th tape. We have constructed the set K_{i-1} of $k^{1-(i-1)/l}$ symbols. We are reading the i-th tape until $k^{1-i/l}$ symbols from K_{i-1} are found. Let $\mathbf{EK}_i$ be the expected number of symbols read to achieve this. Since $\lim_{n \to \infty} \frac{\mathbf{EL}_n}{n} \geq \frac{1}{\max\{\mathbf{EK}_i : 1 \leq i \leq l\}}$, to prove the theorem we have to show that for every i

$$\mathbf{EK}_i \leq k^{1-1/l}(1 + o(1)).$$

The expected waiting time to collect one of j symbols is $1 + (1 - \frac{i}{k}) + (1 - \frac{i}{k})^2 + \cdots = \frac{k}{j}$, therefore

$$\mathbf{EK}_i = \sum_{j=|K_{i-1}|-|K_i|+1}^{|K_{i-1}|} \frac{k}{j}.$$

There are $k^{1-i/l}$ summands and the largest of them is $\frac{k}{k^{1-(i-1)/l}-k^{1-i/l}+1} \leq \frac{k}{k^{1-(i-1)/l}-k^{1-i/l}}$, hence

$$\mathbf{EK}_i \leq k^{1-i/l}\frac{k}{k^{1-(i-1)/l}-k^{1-i/l}}$$

$$\leq k^{1-1/l}\frac{1}{1-k^{-1/l}}$$

$$\leq k^{1-1/l}(1+o(1)). \qquad \square$$

Now we show how to get upper bounds that match the lower bounds up to a constant factor. Let $\mathcal{F}(i,n)$ be the set of l-tuples from Π_n^l with a longest common subsequence of length i, i.e.

$$\mathcal{F}(i,n) = \{\mathbf{u} \in \Pi_n^l : \mathbf{L}(\mathbf{u}) = i\}.$$

The number of l-tuples in $\mathcal{F}(i,n)$ is denoted by $F(i,n)$. Let $H(i,n)$ be any upper bound for $F(i,n)$. If y is such that $\sum\limits_{i=\lceil yn\rceil}^{n} H(i,n)$ is very small compared with k^{ln}, then l-tuples with a longest common subsequence larger than yn do not contribute much to the average and therefore the expected length must be no greater than yn. This is the basic idea of the first upper bound for pairs of sequences given by Chvátal and Sankoff [2] and also the basic idea of all further upper bounds.

Lemma 1. *If y is such that $\sum\limits_{i=\lceil yn\rceil}^{n} H(i,n) = o(k^{ln})$ then $\gamma_k^{(l)} \leq y$.*

Proof. We note that

$$\mathbf{EL}_n^{(l)} = \frac{1}{k^{ln}}\sum_{i=0}^{n} iF(i,n), \quad \text{and} \quad \frac{1}{k^{ln}}\sum_{i=0}^{n} F(i,n) = 1.$$

Hence

$$\frac{\mathbf{EL}_n^{(l)}}{n} = \frac{1}{nk^{ln}}\left(\sum_{i=0}^{\lceil yn\rceil-1} iF(i,n) + \sum_{i=\lceil yn\rceil}^{n} iF(i,n)\right)$$

$$\leq \frac{1}{nk^{ln}}\left(yn\sum_{i=0}^{n} F(i,n) + n\sum_{i=\lceil yn\rceil}^{n} H(i,n)\right) = y + o(1) \qquad \square$$

An l-tuple of sequences $(u_1,\ldots,u_l) \in \Pi$ is a match if $u_1,\ldots,u_l$ have the same last symbol. A collation of order i is a sequence of i matches followed by an l-tuple. We say that collation $\mathbf{u}_1,\ldots,\mathbf{u}_{i+1}$ generates the l-tuple $\mathrm{cat}(\mathbf{u}_1,\ldots,\mathbf{u}_{i+1})$. For example collation $\begin{pmatrix}10\\110\\0\end{pmatrix}\begin{pmatrix}1\\01\\01\end{pmatrix}\begin{pmatrix}10\\0\\10\end{pmatrix}\begin{pmatrix}1\\\lambda\\10\end{pmatrix}$ of order 3 generates the triple $\begin{pmatrix}101101\\110010\\0011010\end{pmatrix}$. Let $H(i,m)$ be the number of all collations of order i generating an l-tuple of total length m. For every l-tuple $\mathbf{u}$ with a longest common subsequence of length i there is at least one collation of order i generating $\mathbf{u}$, therefore $F(i,n) \leq H(i,ln)$. To get upper bounds for $H(i,ln)$ we shall follow the framework given in [3].

Theorem 2. *Let* $h(z) = \frac{kz^l}{(1-kz)^l}$. *Let* z_0 *be such that* $h(z_0) < 1$. *Then*

$$\gamma_k^{(l)} \leq \frac{\log(k^l z_0^l)}{\log h(z_0)} . \tag{1}$$

Proof. To get an upper bound for $F(i,n)$ we have to count $H(i,m)$. We shall use generating functions to achieve this. There are k possibilities for a matching symbol and its contribution is kz^l. The contribution of every sequence, not counting the matching symbol, is $\frac{1}{1-kz}$, therefore the generating function for the number of matches is

$$h(z) = \frac{kz^l}{(1-kz)^l} .$$

The generating function for the number of all collations of order i is given by $s(z)(h(z))^i$, where $s(z)$ is independent of i. To bound coefficient A_m of a generating function $a(x) = \sum_{j=0}^{\infty} A_j x^j$, $A_j \geq 0$, we can use the following inequality

$$A_m \leq \inf \left\{ \frac{a(x)}{x^m} : a(x) \text{ converges} \right\} . \tag{2}$$

Let Z be a set of all $z \in (0,1)$ such that $h(z) < 1$. Now we can bound $H(i,n)$.

$$\sum_{i=\lceil yn \rceil}^{n} H(i,ln) \leq \sum_{i=\lceil yn \rceil}^{n} \inf_{z \in Z} \frac{s(z)(h(z))^i}{z^{ln}}$$

$$\leq \inf_{z \in Z} \sum_{i=\lceil yn \rceil}^{n} \frac{s(z)(h(z))^i}{z^{ln}}$$

$$\leq \inf_{z \in Z} ns(z) \left(\frac{(h(z))^y}{z^l} \right)^n$$

$$\leq ns(z_0) \left(\frac{(h(z_0))^y}{z_0^l} \right)^n .$$

For every $y > \frac{\log(k^l z_0^l)}{\log h(z_0)}$ we have $\frac{(h(z_0))^y}{z_0^l} < k^l$ and therefore $\sum_{i=\lceil yn \rceil}^{n} H(i,ln) = o(k^{ln})$. Using Lemma 1 we can conclude that $\gamma_k^{(l)} \leq y$. $\square$

For $l = 2$ the upper bound from Theorem 2 corresponds to the upper bound of Chvátal and Sankoff [2]. Now we can analyse the speed of convergence of the upper bounds (1).

Corollary 2. *For every $l \geq 2$ we have*

$$\lim_{k \to \infty} k^{1-1/l} \gamma_k^{(l)} \leq e .$$

Proof. Let $z_0 = \frac{k^{1-1/l}-e}{k^{2-1/l}}$, then $kz_0 = 1 - ek^{1/l-1}$. For $h(z_0)$ we have

$$h(z_0) = \left(\frac{k^{1/l-1}kz_0}{1 - kz_0} \right)^l = \left(\frac{k^{1/l-1}(1 - ek^{1/l-1})}{ek^{1/l-1}} \right)^l = \frac{(1 - ek^{1/l-1})^l}{e^l} .$$

From (1) we have

$$k^{1-1/l} \gamma_k^{(l)} \leq k^{1-1/l} \frac{\log(1 - ek^{1/l-1})}{\log(1 - ek^{1/l-1}) - \log e} .$$

Since $\lim_{x \to 0^+} \frac{\log(1+ax)}{x} = a \log e$, we get

$$\lim_{k \to \infty} k^{1-1/l} \gamma_k^{(l)} \leq \lim_{k \to \infty} \left(\frac{\log(1 - ek^{1/l-1})}{k^{1/l-1}} \cdot \frac{-1}{\log e - \log(1 - ek^{1/l-1})} \right)$$

$$\leq (-e \log e)(-1/\log e) = e . \qquad \square$$

In [9] Steele conjectures that $\gamma_k^{(l)} = (\gamma_k^{(2)})^{l-1}$ for $l > 2$. As we can see from Corollary 2 and Theorem 1 this is not the case. Sankoff and Mainville [8] conjecture that $\lim_{k \to \infty} \sqrt{k} \gamma_k^{(2)} = 2$. A natural extension of their conjecture is that $\lim_{k \to \infty} k^{1-1/l} \gamma_k^{(l)} = 2$. Jiang and Li [6] have considered the case when $l = n$. Using Kolmogorov complexity, they have shown that $|\mathbf{EL}_n^{(n)} - \frac{n}{k}| < n^{\frac{1}{2}+\varepsilon}$ for every $\varepsilon > 0$ and large enough n. For any constant l, their algorithm gives the lower bound $\gamma_k^{(l)} \geq \frac{1}{k}$. A slight modification of the proof of Corollary 2 gives us $\lim_{l \to \infty} k \gamma_k^{(l)} \leq e$.

3   Shortest common supersequences

In this section we shall consider a 'dual' notion, namely shortest common supersequences. Sequence u is a *supersequence* of v if v is a subsequence of u. We say, that $u \in \Sigma^*$ is a *common supersequence* of $v_1, \ldots, v_l$ if, for every $i = 1, \ldots, l$, sequence u is a supersequence of v_i. We say, that u is a *shortest common supersequence* if u is a common supersequence of minimal possible length. The length of a shortest common supersequence is denoted by $\mathbf{S}(v_1, \ldots, v_l)$. Creating shortest common supersequences is natural when merging sequences. This is useful for some types of compression [10] or for efficient planning [5]. Longest common subsequences and shortest common supersequences of two sequences are dual, and $\mathbf{S}(u, v) + \mathbf{L}(u, v) = |u| + |v|$. However, if we have more than two sequences and we know only their lengths and the length of their longest common subsequence then we are not able to compute the length of a shortest common supersequence.

We can define $\mathbf{ES}_n^{(l)}$ – the expected length of a shortest common supersequence of l sequences of length n analogously to $\mathbf{EL}_n^{(l)}$. Properties of $\mathbf{ES}_n^{(l)}$ are similar and the constants $\sigma_k^{(l)} = \lim\limits_{n \to \infty} \frac{\mathbf{ES}_n^{(l)}}{n}$ exist. For $l = 2$ we have $\sigma_k^{(2)} = 2 - \gamma_k^{(2)}$. To obtain bounds on $\sigma_k^{(l)}$, $l > 2$, we shall use methods similar to the methods from the previous section.

To obtain upper bounds for the expected length of a shortest common supersequence of l sequences we shall consider algorithms that produce a common supersequence of its input. The expected length of a common supersequence produced by the algorithm will be an upper bound for $\sigma_k^{(l)}$.

We can use a tournament-style algorithm to produce a common supersequence of l sequences $u_1, \ldots, u_l$. In the first round u_1 and u_2 produce a common supersequence $v_{1,1}$, while u_3 and u_4 produce a common supersequence $v_{1,2}$, and so on. In the second round we repeat the process with $v_{1,1}, \ldots, v_{1,\lceil l/2 \rceil}$. After $\lceil \log l \rceil$ steps we get a 'winner' $v_{\lceil \log l \rceil, 1}$ – a common supersequence of $u_1, \ldots, u_l$.

Timkovskij [11] has asked a question about how good an approximation is the tournament algorithm. Bradford and Jenkyns [1] have found an example of three sequences of length 12 over the alphabet of size nine such that no tournament-like algorithm can compute their shortest common supersequence.

We describe a different algorithm that is easy to analyse. Let $T^{(l)}(i) = \{u_1[i], \ldots, u_l[i]\} \subseteq \Sigma$ be the set of symbols that appear at the i-th positions. Let v_i, $i = 1, \ldots, n$ be a sequence of length $\mathbf{T}^{(l)}(i) = |T^{(l)}(i)|$ consisting of all symbols from $T^{(l)}(i)$. The algorithm in i-th step, $i = 1, \ldots, n$, finds $T^{(l)}(i)$ and appends v_i to the output. The sequence $v_1, \ldots, v_n$ produced by the algorithm is a common supersequence of $u_1, \ldots, u_l$ and $\mathbf{S}(u_1, \ldots, u_l) = \sum\limits_{i=1}^{n} \mathbf{T}^{(l)}(i)$. Since symbols $u_j[i]$, $j = 1, \ldots, l$, $i = 1, \ldots, n$ are independent, we have $\mathbf{ES}_n^{(l)} = n\mathbf{ET}^{(l)}$. The probability that $\mathbf{T}^{(l)}$ equals m is $\binom{k}{m} \sum\limits_{j=0}^{m} (-1)^j \binom{m}{j} \left(\frac{m+j}{k}\right)^l$ [4, Section IV.2]. Thus we get following upper bound.

Theorem 3. *For every $l \geq 3$ and $k \geq 2$ we have*

$$\sigma_k^{(l)} \leq \sum_{m=1}^{k} m \binom{k}{m} \sum_{j=0}^{m} (-1)^j \binom{m}{j} \left(\frac{m+j}{k}\right)^l.$$

To get lower bounds we shall use following lemma corresponding to Lemma 1.

Lemma 3. *Let $F(i, n) = |\{\mathbf{u} \in \Psi^n : \mathbf{S}(\mathbf{u}) = i\}|$ and let $H(i, n)$ be any upper bound for $F(i, n)$. If y is such that $\sum\limits_{i=0}^{\lfloor yn \rfloor} H(i, n) = o(k^{ln})$ then $y \leq \sigma_k^{(l)}$.*

For common supersequences the role of matches will be played by 'placements' and a 'distribution' will be the notion analogous to a collation.

An l-tuple $u_1, \ldots, u_l$ is a *placement* if $u_1, \ldots, u_l \in \{\lambda, a\}$ for some $a \in \Sigma$ and not all of $u_1, \ldots, u_l$ are empty sequences. A sequence of i placements will be called a *distribution* of order i. The distribution $\mathbf{d}_1, \ldots, \mathbf{d}_i$ *generates* l-tuple $\mathbf{u}$ if $\mathrm{cat}(\mathbf{d}_1, \ldots, \mathbf{d}_i) = \mathbf{u}$.

Let $\mathcal{D}(i)$ be the set of all distributions of order i. Let $\mathcal{G}(m)$ be the set of all distributions generating an l-tuple of total length m. Since for every l-tuple $\mathbf{u} \in \Psi^n$, there exists a distribution of order $\mathbf{S}(\mathbf{u})$ that generates $\mathbf{u}$, we have

$$F(i, n) \leq |\mathcal{G}(ln) \cap \mathcal{D}(i)| = G(i, n).$$

Every placement of total length m corresponds to a nonempty subset of $\{1, \ldots, l\}$ of size m, thus the generating function for the number of placements is $k((1+z)^l - 1)$. The generating function $\sum_{j=0}^{\infty} |\mathcal{G}(j) \cap \mathcal{D}(i)| x^j$ for the number of distributions in the $\mathcal{D}(i)$ then is $k^i ((1+z)^l - 1)^i$. From (2) for $z_0 > 2^{1/l} - 1$ and $y < \frac{l \log(z_0 k)}{\log(k(1+z_0)^l - k)}$ we have

$$\sum_{i=0}^{\lfloor yn \rfloor} G(i, n) \leq \sum_{i=0}^{\lfloor yn \rfloor} \inf_{z \in Z} \frac{k^i ((1+z)^l - 1)^i}{z^{ln}}$$

$$\leq \inf_{z \in Z} \sum_{i=0}^{\lfloor yn \rfloor} \frac{k^i ((1+z)^l - 1)^i}{z^{ln}}$$

$$\leq \sum_{i=0}^{\lfloor yn \rfloor} \frac{k^i ((1+z_0)^l - 1)^i}{z_0^{ln}}$$

$$\leq yn \left(\frac{k^y ((1+z_0)^l - 1)^y}{z_0^l} \right)^n = o(k^{ln}).$$

Thus we have just proved the following theorem.

Theorem 4. *For every $l \geq 2$ and $k \geq 2$ we have*

$$\sigma_k^{(l)} \geq \frac{l \log z_0 k}{\log k((1+z_0)^l - 1)},$$

where $z_0 \in \mathbb{R}$ is such that $z_0 > 2^{1/l} - 1$.

Upper and lower bounds for $k = 2, \ldots, 15$ and $l = 2, 3, 4, 5, 6, 8, 16$ obtained using Theorems 3 and 4 can be found in the table in Figure 1. Better bounds for $l = 2$ follow from bounds for longest common subsequences using duality properties. Unfortunately lower bounds from Theorem 4 do not appear to match upper bounds from Theorem 3.

Acknowledgement

I would like to thank Prof. M. Paterson for his help, many valuable comments and suggestions.

k	$l = 2$	$l = 3$	$l = 4$	$l = 5$	$l = 6$	$l = 8$	$l = 16$
2	1.50000	1.75000	1.87500	1.93750	1.96875	1.99219	1.99997
	1.09321	1.14253	1.17267	1.19287	1.20731	1.22656	1.25822
3	1.66667	2.11112	2.40741	2.60494	2.73663	2.88295	2.99544
	1.16377	1.25982	1.32366	1.36923	1.40336	1.45107	1.53569
4	1.75000	2.31250	2.73438	3.05079	3.28809	3.59955	3.95991
	1.21514	1.34913	1.44278	1.51244	1.56642	1.64476	1.79311
5	1.80000	2.44000	2.95200	3.36160	3.68928	4.16114	4.85927
	1.25447	1.41951	1.53883	1.63017	1.70270	1.81106	2.02791
6	1.83334	2.52778	3.10649	3.58874	3.99062	4.60460	5.67548
	1.28583	1.47679	1.61830	1.72893	1.81841	1.95512	2.24164
7	1.85715	2.59184	3.22158	3.76135	4.22402	4.96050	6.40578
	1.31162	1.52463	1.68549	1.81331	1.91818	2.08131	2.43661
8	1.87500	2.64063	3.31055	3.89673	4.40964	5.25113	7.05547
	1.33333	1.56541	1.74333	1.88656	2.00542	2.19301	2.61512
9	1.88889	2.67903	3.38135	4.00564	4.56057	5.49231	7.63290
	1.35196	1.60076	1.79387	1.95098	2.08260	2.29285	2.77923
10	1.90000	2.71000	3.43900	4.09510	4.68559	5.69533	8.14698
	1.36819	1.63183	1.83857	2.00828	2.15158	2.38282	2.93073
11	1.90910	2.73554	3.48686	4.16987	4.79079	5.86842	8.60608
	1.38251	1.65943	1.87851	2.05972	2.21377	2.46452	3.07115
12	1.91667	2.75695	3.52720	4.23327	4.88050	6.01764	9.01762
	1.39527	1.68418	1.91451	2.10627	2.27025	2.53919	3.20180
13	1.92308	2.77515	3.56168	4.28771	4.95788	6.14755	9.38799
	1.40675	1.70657	1.94721	2.14871	2.32189	2.60782	3.32378
14	1.92858	2.79082	3.59148	4.33494	5.02531	6.26162	9.72267
	1.41715	1.72696	1.97710	2.18762	2.36938	2.67122	3.43804
15	1.93334	2.80445	3.61749	4.37632	5.08457	6.36256	10.0263
	1.42664	1.74564	2.00458	2.22349	2.41326	2.73007	3.54540

Fig. 1. Upper and lower bounds for the expected length of a shortest common super-sequence of l sequences over an alphabet of size k.

References

1. James H. Bradford and T. A. Jenkyns. On the inadequacy of tournament algorithms for the n-scs problem. *Information Processing Letters*, 38(4):169–171, 1991.
2. Václav Chvátal and David Sankoff. Longest common subsequence of two random sequences. *Journal of Applied Probability*, 12:306–315, 1975.
3. Vlado Dančík and Mike Paterson. Upper bounds for the expected length of a longest common subsequence of two binary sequences. In P. Enjalbert, E. W. Mayr, and K.W.Wagner, editors, *11th Annual Symposium on Theoretical Aspects of Computer Science, Proceedings*, pages 669–678. Lecture Notes in Computer Science 775, Springer-Verlag, 1994.

4. William Feller. *An Introduction to Probability Theory and its Applications*, volume I. John Wiley & Sons, New York, third edition, 1968.
5. David E. Foulser, Ming Li, and Qiang Yang. Theory and algorithms for plan merging. *Artificial Intelligence*, 57:143–181, 1992.
6. Tao Jiang and Ming Li. On the approximation of shortest common supersequences and longest common subsequences. In S. Abiteboul and E. Shamir, editors, *Automata, Languages and Programming, Proceedings*, pages 191–202. Lecture Notes in Computer Science 820, Springer-Verlag, 1994.
7. Mike Paterson and Vlado Dančík. Longest common subsequences. In *19th International Symposium Mathematical Foundations of Computer Science, Proceedings*, pages 127–142. Lecture Notes in Computer Science 841, Springer-Verlag, 1994.
8. D. Sankoff and J. B. Kruskal. *Time Warps, String Edits, and Macromolecules: The theory and practice of sequence comparison*. Addison-Wesley, Reading, Mass, 1983.
9. J. Michael Steele. An Efron–Stein inequality for nonsymmetric statistics. *The Annals of Statistics*, 14(2):753–758, 1986.
10. James A. Storer. *Data Compression: Methods and Theory*. Computer Science Press, Rockville, Maryland, 1988.
11. V. G. Timkovskii. Complexity of common subsequence and supersequence problems and related problems. *Cybernetics*, 25:565–580, 1990.

Pattern Matching in Directed Graphs

Jianghai Fu

Department of Computer Science, University of Waterloo
Waterloo, ON, Canada, N2L 3G1
Email: jfu@neumann.uwaterloo.ca

Abstract. Pattern matching in directed graphs is a natural extension of pattern matching in trees and has many applications to different areas. In this paper, we study several pattern matching problems in ordered labeled directed graphs. For the rooted directed graph pattern matching problem, we present an efficient algorithm which runs in time and space $O(|E(P)| \times |V(T)| + |E(T)|)$, where $|E(P)|$, $|V(T)|$ and $|E(T)|$ are the number of edges in the pattern graph P, the number of nodes in the target graph T and the number of edges in the pattern graph T, respectively. It is by far the fastest algorithm for this problem. This algorithm can also solve the directed graph pattern matching problem without increasing time or space complexity. Our solution to this problem outperforms the best existing method by Katzenelson, Pinter and Schenfeld by a factor of at least $|V(P)|$. We also present an algorithm for the directed graph topological embedding problem which runs in time $O(|V(P)| \times |E(T)| + |E(P)|)$ and space $O(|V(P)| \times |V(T)| + |E(P)| + |E(T)|)$, where $|V(P)|$ is the number of nodes in the pattern graph P. To our knowledge, this algorithm is the first one for this problem.

1 Introduction

Pattern matching in trees has been successful in a number of application areas. However, because of the lack of mechanism in trees to express recursive structures, trees are not suitable for representing some complex objects in which explicit expressions of recursive relations are useful. As a consequence, there has been increasing demand in recent years that pattern matching be extended to more general graphs [2, 3, 5, 6, 7, 10, 11, 12].

A directed graph is a natural choice for expressing recursive relations. Pattern matching in directed graphs has been used by several research groups for different purposes in different areas. One example is Katzenelson, Pinter and Schenfeld's type-checking system [11] in which type expressions are represented by *type-graphs* which are ordered labeled *rooted directed graphs*, where a rooted directed graph is a directed graph in which there is a single node designated as the root from which there is at least one edge emanating. A key step in their system is to identify all equivalent subgraphs in a type-graph so that redundant parts in the type-graph can be eliminated. Using the technique of pattern matching, Katzenelson, Pinter and Schenfeld [11] describe a method which can identify all equivalent subgraphs in a type-graph G in time $O(|V(G)|^2 \times |E(G)|)$.

Pattern matching in directed graphs is also used in Holm's system, in which graph concepts are used in semantics descriptions of functional languages [10]. In Holm's system, recursively typed languages are represented by directed graphs. The task of pattern matching can be described as follows: the patterns describe a class of objects and are represented using directed graphs. When checking a recursive type, the recursive type value is matched against the patterns, i.e., the algorithm checks if the type value is an instance of a pattern. The matching process is implemented in a simple graph reduction machine which supports primitive graph operations and provides a base for language implementations.

Directed graphs are well suited for representing regular tree expressions [2]. Aiken and Murphy [2] discuss the implementation of regular tree expressions which are used in type inference and program analysis algorithms [3, 8]. The most commonly used operation among all fundamental operations in the implementation is the operation for testing inclusion relations which can be implemented using a pattern matching technique.

Pattern matching is a crucial component of term rewriting systems. Directed graphs are well suited for expressing the infinite terms with a finite number of distinct subterms in a *cyclic term graph rewriting* system [5, 6]. Motivated by the need for providing satisfactory interpretation for cyclic term graph rewriting, Corradini [5] discuss the extension of the classical theory of term rewriting systems to infinite and partial terms. A theoretical treatment of pattern matching in directed graphs has an important impact on the research of infinite term rewriting systems.

Although pattern matching in directed graphs has been used by individual research groups, a thorough study of the problems has not been carried out. This paper is a step towards an extensive investigation of various pattern matching problems in directed graphs and the development of techniques for these problems. Efficient solutions to these problems not only are of theoretical interest in their own right, but also improve the performance of many systems such as listed above.

In this paper, we first define in Section 2 the notation and terminology that are used throughout this paper and the problems with which we are concerned. Then in Section 3, we review the previous results for some problems. In Sections 4.1 and 4.2 we present an efficient algorithm for the ordered labeled rooted directed graph pattern matching problem which runs in time and space $O(|E(P)| \times |V(T)| + |E(T)|)$. Extension of this algorithm to the ordered labeled directed graph pattern matching problem is also discussed. In Section 5 we present an efficient algorithm for the ordered labeled directed graph topological embedding problem which runs in time $O(|V(P)| \times |E(T)| + |E(P)|)$ and space $O(|V(P)| \times |V(T)| + |E(P)| + |E(T)|)$. Our solution to the ordered labeled directed graph pattern matching problem is faster than the best method obtained by adapting Katzenelson, Pinter and Schenfeld's type equivalence testing algorithm [11] by a factor of at least $|V(P)|$. Section 6 gives the conclusion.

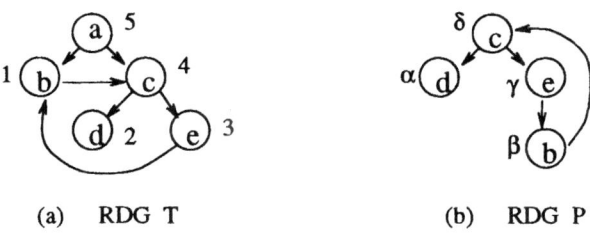

(a)   RDG T   (b)   RDG P

Figure 1:

2   Notation and Terminology

In the following description, we will use RDG as an abbreviation for "rooted directed graph". Figure 1(a) shows a RDG in which node 5 is the root. The following definitions pertain to a RDG G.

Definition 1 The *depth(x)* of a node x in G is the number of edges in the shortest path from the root of G to x. An edge (x_1, x_2) is an *ordinary edge* if *depth(x_1)* is less than *depth(x_2)*; an edge (x_1, x_2) is a *cross edge* if *depth(x_1)* equals *depth(x_2)*; an edge (x_1, x_2) is a *back edge* if *depth(x_1)* is greater than *depth(x_2)*.

In Figure 1(a), node 5 has depth 0, node 1 and node 4 have depth 1 and node 2 and node 3 have depth 2. Edges (5, 1), (5, 4), (4, 2) and (4, 3) are ordinary edges, edge (1, 4) is a cross edge and edge (3, 1) is a back edge.

In the following description we use the word "edge" to mean ordinary edge, back edge, or cross edge, unless otherwise specified.

Definition 2 Let x and y be two nodes in a RDG G. We say x is a *parent* of y if there is an edge pointing from x to y. Node y is a *child* of x if and only if x is a parent of y. We say x is an *ancestor* of y if there exist zero or more nodes $x_1, x_2, ..., x_q$ in G such that x is a parent of x_1, x_q is a parent of y, and x_i is a parent of x_{i+1} for $1 \le i < q$. Node y is a descendant of x if and only if x is an ancestor of y. Nodes x and y are *siblings* if they have the same depth and a common parent. A *leaf* of G is a node from which there is no ordinary edge emanating. A *subgraph* of G rooted at a node x in G is a RDG which contains all its descendants and the induced edges.

In Figure 1(a), node 4, node 3 and node 1 are parents of node 3, node 1 and node 4, respectively. Node 1 is an ancestor of node 4 and node 4 is an ancestor of node 1. Node 1 and node 4 are also siblings. The leaf nodes in RDG T are node 1, node 2 and node 3. The subgraph of RDG T rooted at node 4 is shown in Figure 1(b).

In this paper, we use Σ to denote an alphabet of labels, *root(G)* to denote the root of a RDG G, and *label(x)*, *subgraph(x)* and *child$_i$(x)* to denote the label of a node x, the subgraph rooted at x, and the ith child of x from the left, respectively. Unless otherwise specified, we use P and T to denote a pattern RDG and a target RDG, respectively, and use Greek letters to denote the nodes in P and integers or lower case English letters to denote the nodes in T.

An *ordered labeled* RDG or directed graph is a RDG or directed graph in which every node is associated with a label in an alphabet Σ and the left-to-right order of siblings is significant. The *unfolded RDG* of a RDG T, denoted $unfold(T)$, is the infinite tree obtained by unfolding T. The *rank* of a node x in $unfold(T)$ is the number of edges in $unfold(T)$ from $root(unfold(T))$ to x. Let T^k denote the tree which consists of all nodes of $unfold(T)$ whose ranks are less than or equal to k, and the induced edges. Let $subtree(x)$ denote the subtree rooted at node x. The following is the definition of tree pattern matching between P^k and T^k:

Definition 3 P^k *matches* T^k if i) $label(root(P^k)) = label(root(T^k))$, ii) $degree(root(P^k)) = 0$ or $degree(root(P^k)) = degree(root(T^k))$ and iii) $subtree(child_i(root(P^k)))$ matches $subtree(child_i(root(T^k)))$ for each i, $1 \leq i \leq degree(root(P^k))$.

We define the *match* between ordered labeled RDGs in terms of unfolded RDGs:

Definition 4 A RDG P *matches* a subgraph S of a RDG T if for every integer $k \geq 0$, P^k matches S^k.

As an example, RDG P matches $subgraph(4)$ of RDG T in Figure 1. The following defines the *topological embedding* between ordered labeled RDGs. We first define the topological embedding between unfolded RDGs, then use the definition to define the topological embedding between RDGs.

Definition 5 P^k is *topologically embeddable* in T^j if i) $label(root(P^k)) = label(root(T^j))$, ii) $degree(root(P^k)) = 0$ or $degree(root(P^k)) = degree(root(T^j))$ and iii) let d_i be $child_i(root(T^j))$ or a descendant of $child_i(root(T^j))$. $subtree(child_i(root(P^k)))$ is topologically embeddable in $subtree(d_i)$ for each i, $1 \leq i \leq degree(root(P^k))$.

Definition 6 A RDG P is *topologically embeddable* in a subgraph S of a RDG T if for every integer $k \geq 0$, there exists an integer $j \geq k$ such that P^k is topologically embeddable in S^j.

The following are the definitions of the problems we consider in this paper:

Definition 7 The *root preserving RDG pattern matching problem* is to determine whether or not P matches T. The *RDG pattern matching problem* is to find all subgraphs of T which P matches. The *directed graph pattern matching problem* is to find all subgraphs of T which a subgraph of P matches for every subgraph of P. The *directed graph topological embedding problem* is to find all subgraphs of T in which a subgraph of P is topologically embeddable for each subgraph of P.

3 Previous Work

For ordered RDGs, two algorithms can be adapted to solve the root preserving RDG pattern matching problem: one is the unification algorithm by Aho, Sethi and Ullman [1] and the other is the type equivalence testing algorithm by Katzenelson, Pinter and Schenfeld [11]. The two algorithms are based on the same idea which can be described as follows when adapted to perform matching:

Function $matching(\alpha, x)$
  If α has been matched against x then return MATCH;
  If $label(\alpha) \neq label(x)$ or $degree(\alpha) \neq degree(x)$ then return NOT-MATCH;
  For i from 1 to $degree(\alpha)$ do
    If $matching(child_i(\alpha), child_i(x)) =$ NOT-MATCH then return NOT-MATCH;
  Return MATCH;

Given P and T, we apply the function to the roots of P and T and report that P matches T if the function returns MATCH. Since there is a finite number of pairs of nodes, the function always terminates. A $|V(P)| \times |V(T)|$ matrix can be used to store all pairs that have been processed so that, given a pair of nodes, one can check in constant time whether or not they have been processed to avoid processing the same pair of nodes more than once. The time and space complexities of the above algorithms are $O(|E(P)| + |E(T)|)$ and $O(|V(P)| \times |V(T)| + |E(P)| + |E(T)|)$, respectively. Katzenelson, Pinter and Schenfeld [11] also consider the problem of finding all equivalent subgraphs in an ordered labeled RDG G, which is essentially the same problem as the directed graph pattern matching problem. Their solution runs in time $O(|V(G)|^2 \times |E(G)|)$ which is $O(|V(P)| \times |V(T)| \times (|E(P)| + |E(T)|))$ when it is applied to solving the directed graph pattern matching problem. Note that we use Katzenelson, Pinter and Schenfeld's algorithm [11] for comparison because we are not aware of any better existing algorithm for this problem than an adaptation of their algorithm. This algorithm is not the main subject of their paper [11] and the efficiency of this algorithm may not be their main concern.

The two algorithms for the root preserving RDG pattern matching problem can be extended to the ordered labeled RDG pattern matching problem by comparing P with every subgraph of T. The time complexity will be $O((|E(P)| + |E(T)|) \times |V(T)|)$. There is no existing algorithm for the ordered directed graph topological embedding problem.

The unordered RDG pattern matching problem is different from the ordered RDG pattern matching problem in that one can change the left-to-right order of the children of any pair of nodes when checking for the matches between the children of the pair. The problem is NP-complete [4]; as a result, the directed graph pattern matching problem and the directed graph topological embedding problem are NP-hard.

4 Pattern Matching

4.1 A Bottom-up Algorithm

In this section we describe an algorithm for the ordered labeled RDG pattern matching problem which adopts the bottom-up approach. It works as follows: traverse P in postorder, where the postorder is obtained by ignoring the back edges and cross edges, and for each node α in P we match $subgraph(\alpha)$ against all subgraphs of T in postorder. Unlike in tree pattern matching [9], the bottom-up approach faces the difficulty of dealing with the cyclic dependencies in RDGs. In Figure 1, it is easy to tell whether or not $subgraph(\alpha)$ matches $subgraph(2)$, but it is not so clear whether or not $subgraph(\beta)$ matches $subgraph(1)$ before we have determined whether or not $subgraph(\delta)$ matches $subgraph(4)$. However, we cannot determine whether or not $subgraph(\delta)$ matches $subgraph(4)$ without knowing whether or not $subgraph(\beta)$ matches $subgraph(1)$. We need to find some way to break the dependency cycle.

Before discussing how to break the dependency cycle, we define some notation which is useful for the description of the algorithm. A pair (β, y) is a *condition of matching*, denoted $cm(\beta, y)$, on which another pair (α, x) depends if $subgraph(\alpha)$ matches $subgraph(x)$ only if $subgraph(\beta)$ matches $subgraph(y)$. A $cm(\beta, y)$ is *determined* if $subgraph(\beta)$ has been matched against $subgraph(y)$ during the bottom-up matching process, or *undetermined* otherwise. A $subgraph(\alpha)$ *conditionally matches* $subgraph(x)$ if i) $label(\alpha) = label(x)$, ii) $degree(\alpha) = degree(x)$, iii) there exist some undetermined conditions on which (α, x) depends and iv) there exists no i such that it has been determined that $subgraph(child_i(\alpha))$ does not match $subgraph(child_i(x))$. A $cm(\beta, y)$ is *confirmed* if it has been determined that $subgraph(\beta)$ matches or conditionally matches $subgraph(y)$. A $cm(\beta, y)$ is *violated* if it has been determined that $subgraph(\beta)$ does not match $subgraph(y)$.

The basic idea of how to break the dependency cycle can be described as follows: when we cannot determine whether or not $subgraph(\alpha)$ matches $subgraph(x)$ because of cyclic dependency, we conclude that $subgraph(\alpha)$ conditionally matches $subgraph(x)$ and discover the undetermined conditions on which (α, x) depends. These conditions will be examined in later computation and we will by then determine whether or not $subgraph(\alpha)$ matches $subgraph(x)$. If all conditions on which (α, x) depends are confirmed, we conclude that $subgraph(\alpha)$ matches $subgraph(x)$; if one of the conditions on which (α, x) depends is violated, we conclude that $subgraph(\alpha)$ does not match $subgraph(x)$. In Figure 1, when matching $subgraph(\beta)$ against $subgraph(1)$, we determine that $subgraph(\beta)$ conditionally matches $subgraph(1)$ under $cm(\delta, 4)$. Since P and T are processed in a bottom-up fashion, we will eventually match $subgraph(\delta)$ against $subgraph(4)$ and will confirm $cm(\delta, 4)$. Then we conclude that $subgraph(\beta)$ and $subgraph(\delta)$ match $subgraph(1)$ and $subgraph(4)$, respectively. Suppose $label(3)$ were not e. We again determine that $subgraph(\beta)$ conditionally matches $subgraph(1)$ under the same condition as above. However, the later computation shows that $subgraph(\gamma)$ does not match $subgraph(3)$ and thus $subgraph(\delta)$ does not match $subgraph(4)$, i.e., $cm(\delta, 4)$ is violated. Then we conclude that $subgraph(\beta)$, $subgraph(\gamma)$ and $subgraph(\delta)$ do not match $subgraph(1)$, $subgraph(3)$ and $subgraph(4)$, respectively.

The following lemma proves that the above idea is correct:

Lemma 1 *For any pair of nodes α in P and x in T, $subgraph(\alpha)$ matches $subgraph(x)$ if and only if i) $label(\alpha) = label(x)$, ii) $degree(\alpha) = degree(x)$ and iii) all conditions on which (α, x) depends are confirmed during the bottom-up computation.*

Proof Outline. It is obvious that if $subgraph(\alpha)$ matches $subgraph(x)$, then all above i), ii) and iii) are true. By induction on k, we can prove that if the above i), ii) and iii) are true then for any $k \geq 0$, $subgraph(\alpha)^k$ matches $subgraph(x)^k$. □

A $|V(P)| \times |V(T)|$ matrix M is used to store the intermediate information about the matches between the subgraphs of P and T. Each entry $M[\alpha, x]$ may have one of the four values, recording whether or not $subgraph(\alpha)$ matches or conditionally matches $subgraph(x)$: UNKNOWN (we have not matched $subgraph(\alpha)$ against $subgraph(x)$ yet), MATCH ($subgraph(\alpha)$ matches $subgraph(x)$), NOT-MATCH ($subgraph(\alpha)$ does not match $subgraph(x)$) and COND-MATCH ($subgraph(\alpha)$ conditionally matches $subgraph(x)$). We say an entry $M[\alpha, x]$ depends on a condition $cm(\beta, y)$ if (α, x) depends on $cm(\beta, y)$.

To keep track of the conditions and the pairs depending on these conditions, for each entry $M[\alpha, x]$ whose value is COND-MATCH, we use an abstract data type *cond-list* to store all undetermined conditions on which (α, x) depends, and an abstract data type *dep-list* to store every pair (δ, z) such that $M[\delta, z]$ is COND-MATCH and depends on the conditions on which (α, x) depends. The pairs in a dep-list indicate the entries of M whose values need to be changed to MATCH or NOT-MATCH in later computation.

Cond-lists and dep-lists can be computed as follows: for any entry $M[\alpha, x]$, its cond-list consists of all $cm(child_i(\alpha), child_i(x))$ such that $M[child_i(\alpha), child_i(x)]$ = UNKNOWN and all conditions except $cm(\alpha, x)$ in $M[child_i(\alpha), child_i(x)]$'s cond-list for $1 \leq i \leq degree(\alpha)$. Its dep-list consists of (α, x) and all elements from $M[child_i(\alpha), child_i(x)]$'s dep-list for $1 \leq i \leq degree(\alpha)$. Once $M[\alpha, x]$'s cond-list and dep-list have been computed, the cond-list and dep-list of $M[child_i(\alpha), child_i(x)]$ for $1 \leq i \leq degree(\alpha)$ can be removed. It is clear that each $cm(\beta, y)$ can be checked only once since $subgraph(\beta)$ is matched against $subgraph(y)$ only once during the whole course of bottom-up computation. There are cases where $M[\beta, y]$ is computed but $cm(\beta, y)$ is in the cond-list of another entry $M[\gamma, w]$. This means that $M[\gamma, w]$ depends on the conditions on which (β, y) depends. When this happens, we need to merge $M[\beta, y]$'s cond-list and dep-list into $M[\gamma, w]$'s cond-list and dep-list, respectively, and remove (β, y) from $M[\gamma, w]$'s cond-list.

We illustrate how the algorithm works using Figure 2. After node β is processed, $M[\beta, 3]$ is assigned COND-MATCH and has $cm(\gamma, 6)$ in its cond-list and $(\beta, 3)$ in its dep-list, $M[\beta, 5]$ is assigned COND-MATCH and has $cm(\gamma, 4)$ in its

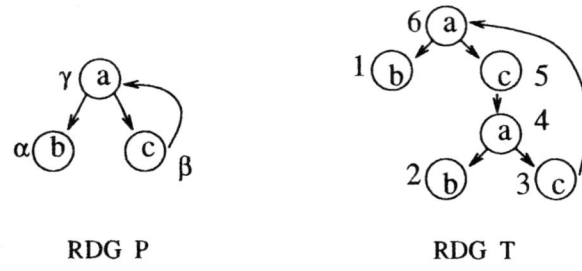

RDG P   RDG T

Figure 2:

cond-list and $(\beta, 5)$ in its dep-list. Then $M[\gamma, 4]$ is determined to be COND-MATCH and has $cm(\gamma, 6)$ in its cond-list and $(\beta, 3)$ and $(\gamma, 4)$ in its dep-list. Since $cm(\gamma, 4)$ is in $M[\beta, 5]$'s cond-list, we merge $cm(\gamma, 6)$ into $M[\beta, 5]$'s cond-list and remove $cm(\gamma, 4)$ from $M[\beta, 5]$'s cond-list, and merge $(\beta, 3)$ and $(\gamma, 4)$ into $M[\beta, 5]$'s dep-list. Then $subgraph(\gamma)$ is matched against $subgraph(6)$. Since $label(\gamma) = label(6)$, $degree(\gamma) = degree(6)$ and all conditions on which $(\gamma, 6)$ depends have been confirmed, we change the values of all entries of M in $M(\gamma, 6)$'s dep-list, which consists of $(\beta, 3)$, $(\gamma, 4)$, $(\beta, 5)$ and $(\gamma, 6)$, to MATCH.

If we use linked lists to implement the cond-lists and dep-lists, the time complexity of this algorithm is quite high. The next section discusses a more efficient implementation.

4.2   An Efficient Implementation of the Algorithm

In the algorithm described in the above section, we use cond-lists and dep-lists to keep track of the undetermined conditions and all entries of M which have been assigned COND-MATCH. The values of the entries of M indicated by the pairs in dep-lists will in later computation be changed to either MATCH or NOT-MATCH according to Lemma 1. However, for any pair (α, x) in a dep-list, we do not have to change $M[\alpha, x]$ from COND-MATCH to MATCH or NOT-MATCH right after we have determined that $subgraph(\alpha)$ matches or does not match $subgraph(x)$. Consider an algorithm which consists of two steps: a bottom-up traversal step in which M is filled in with MATCH, NOT-MATCH and COND-MATCH, and a propagation step in which NOT-MATCH is propagated from each entry $M[\beta, y]$ which has value NOT-MATCH to every entry $M[\alpha, x]$ depending on $cm(\beta, y)$. After the propagation step, each entry which has value MATCH or COND-MATCH signals a match. We will prove this idea later.

To efficiently propagate NOT-MATCH, we associate with each entry a *prop-list* to record the entries to which NOT-MATCH is propagated when needed: (α, x) is in $M[\beta, y]$'s prop-list if and only if $M[\alpha, x] =$ COND-MATCH and α and x are the parents of β and y, respectively. Cond-lists and dep-lists are no longer needed since we do not need to keep track of the conditions and the pairs depending on the conditions.

Now we give the algorithm which consists of a function *match*, a procedure *propagate* and the main body. Function *match* takes as input a node α in P and a node x in T and determines whether or not *subgraph*(α) matches or conditionally matches *subgraph*(x), given that the information about the matches between the subgraphs rooted at the children of α and the subgraphs rooted at the children of x is kept in M. The value returned from Function *match* may be MATCH, COND-MATCH, or NOT-MATCH. Procedure *propagate* takes as input a node α in P and a node x in T and propagate NOT-MATCH to the entries of M depending on $cm(\alpha, x)$. To make sure that each pair of nodes α and x are passed to Procedure *propagate* only once, we put a mark on $M[\alpha, x]$ when they are passed to Procedure *propagate*.

Pattern Matching Algorithm (P, T)
  Create a $|V(P)| \times |V(T)|$ matrix M and initialize all entries of M to UNKNOWN;
  For each node α in P in postorder do
    For each node x in T in postorder do
      $M[\alpha, x] = match(\alpha, x)$;
  For each unmarked entry $M[\alpha, x]$ which has value NOT-MATCH do
    Mark $M[\alpha, x]$ and $propagate(\alpha, x)$;
  Output all $(root(P), x)$ s.t. $M[root(P), x]$ is MATCH or COND-MATCH;

Function $match(\alpha, x)$
  If $label(\alpha) = label(x)$, $degree(\alpha) = 0$ or $degree(\alpha) = degree(x)$, and for each k
    $M[child_k(\alpha), child_k(x)] = $ MATCH then $M[\alpha, x] = $ MATCH;
  Else if $label(\alpha) \neq label(x)$, $degree(\alpha) \neq degree(x)$ or there exists k s.t.
    $M[child_k(\alpha), child_k(x)] = $ NOT-MATCH then return NOT-MATCH;
  Else
    For each k put (α, x) in $M[child_k(\alpha), child_k(x)]$'s prop-list;
    Return COND-MATCH;

Procedure $propagate(\alpha, x)$
  For each (β, y) in $M[\alpha, x]$'s prop-list s.t. $M[\beta, y]$ is not marked do
    Assign NOT-MATCH to $M[\beta, y]$, mark $M[\beta, y]$ and $propagate(\beta, y)$;
  Remove $M[\alpha, x]$'s prop-list;

Lemma 2 *For any pair of nodes α and x, $M[\alpha, x]$ has value NOT-MATCH after the propagation step if and only if subgraph(α) does not match subgraph(x).*

Proof. It is obvious that if $M[\alpha, x] = $ NOT-MATCH after the propagation step then *subgraph*(α) does not match *subgraph*(x).

If *subgraph*(α) does not match *subgraph*(x), then at least one of the following is true: i) $label(\alpha) \neq label(x)$, ii) $degree(\alpha) \neq degree(x)$ or iii) there exists a $cm(\beta, y)$ on which (α, x) depends such that *subgraph*(β) does not match *subgraph*(y). During the bottom-up traversal step, if the above i) or ii) is true, then $M[\alpha, x]$ is assigned NOT-MATCH; otherwise if the above iii) is true, then $M[\alpha, x]$ may be assigned COND-MATCH. If $M[\alpha, x]$ is assigned COND-MATCH, then there must exist $cm(\beta, y)$ such that $M[\beta, y] = $ NOT-MATCH and (α, x) depends on $cm(\beta, y)$, because otherwise $subgraph(\alpha)$ would match $subgraph(x)$ according to Lemma 1. Furthermore, there must exist (α_1, x_1), (α_2, x_2), ..., (α_n, x_n) such that (α, x) is in $M[\alpha_1, x_1]$'s prop-list, (α_n, x_n) is in $M[\beta, y]$'s prop-list and (α_i, x_i) is in $M[\alpha_{i+1}, x_{i+1}]$'s prop-list for $1 \le i \le n - 1$. Therefore, NOT-MATCH will be propagated from $M[\beta, y]$ to $M[\alpha, x]$ in Procedure *propagate*. $\square$

Function *match* takes at most $O((\sum_{\alpha \in V(P)} degree(\alpha)) \times |V(T)|) = O(|E(P)| \times |V(T)|)$ steps during the whole course of computation. It is clear that the total number of elements in all prop-lists is bounded by $|E(P)| \times |V(T)|$. Since each element in any prop-list is visited at most once during the propagation step, the propagation takes at most $O(|E(P)| \times |V(T)|)$ steps in total. The postorder of P and T can be computed in time $|E(P)| + |E(T)|$. Therefore, the time complexity of the algorithm is $O(|E(P)| \times |V(T)| + |E(T)|)$. The space cost of our algorithm is clearly $O(|E(P)| \times |V(T)| + |E(T)|)$. As T is usually larger than P, this algorithm is faster than the one obtained by extending Aho, Sethi and Ullman's algorithm [1] or Katzenelson, Pinter and Schenfeld's algorithm [11] to ordered labeled RDG pattern matching.

Our algorithm establishes not only the match between P and every subgraph of T, but also the match between every subgraph of P and every subgraph of T. This means that it solves not only the ordered labeled RDG pattern matching problem, but also the ordered labeled directed graph pattern matching problem at the same time. Compared with Katzenelson, Pinter and Schenfeld's method for this problem [11], our solution is faster than theirs by a factor of at least $|V(P)|$.

5 Topological Embedding

The ordered labeled directed graph topological embedding problem can also be solved using the bottom-up approach. Let $DAG(G)$ denote the directed acyclic graph obtained by replacing each strongly connected component in a directed graph G with a single node and attaching to the single node all edges attaching to the nodes in the strongly connected component. We use SCC as an abbreviation for "strongly connected component" from now on. The SCCs in a directed graph can be identified using Tarjan's linear time algorithm [13]. We associate with each node x in $DAG(G)$ a unique number i, $1 \le i \le |V(DAG(G))|$, such that i is greater than all numbers associated with the descendants of x. These numbers ensure that if $DAG(G)$ is traversed in the increasing order of these numbers, then a node is visited only after all its children have been visited. Each node i in $DAG(G)$ corresponds to one or more nodes in G. We use $org(i)$ to denote the node or nodes in G represented by node i in $DAG(G)$ throughout the description.

Our algorithm traverses $DAG(P)$ and $DAG(T)$ in the increasing order of the numbers associated with the nodes. During the traversal, it compares each pair of nodes i in $DAG(P)$ and j in $DAG(T)$ and determines whether or not the subgraph or subgraphs of P represented by node i are topologically embeddable in the subgraph or subgraphs represented by node j. If node i or j corresponds to a SCC, we may need to deal with cyclic dependency and we will describe how to deal with it later. We say that a subgraph P_1 of P is *desc-embeddable* in a subgraph T_1 of T if P_1 is topologically embeddable in a subgraph of T_1. A $|V(P)| \times |V(T)|$ matrix M is used to record whether or not the subgraphs of P are topologically embeddable or desc-embeddable in the subgraphs of T. Each entry $M[\alpha, x]$ may have one of four values: UNKNOWN (α has not been processed against x yet), EMBED ($subgraph(\alpha)$ is topologically embeddable in $subgraph(x)$), DESC-EMBED ($subgraph(\alpha)$ is desc-embeddable but not topologically embeddable in $subgraph(x)$) and NOT-DESC-EMBED ($subgraph(\alpha)$ is not desc-embeddable in $subgraph(x)$)). A procedure *cycleembed* is used to handle the situation when we need to deal with cyclic dependency.

Topological Embedding Algorithm (P, T)
  Compute $DAG(P)$ and $DAG(T)$ and number the nodes;
  Create a $|V(P)| \times |V(T)|$ matrix M and initialize each entry to UNKNOWN;
  For i from 1 to $|V(DAG(P))|$ do
    For j from 1 to $|V(DAG(T))|$ do
      If j corresponds to a SCC in T then $cycleembed(i, j)$;
      Else for each node α in $org(i)$ do $dagembed(\alpha, org(j))$;
  Output all (α, x) s.t. $M[\alpha, x] = $ EMBED;

Procedure $dagembed(\alpha, x)$
  If $label(\alpha) = label(x)$, $degree(\alpha) = 0$ or $degree(\alpha) = degree(x)$, and for each k
    $M[child_k(\alpha), child_k(x)]$ is EMBED or DESC-EMBED then $M[\alpha, x] = $ EMBED;
  Else if there exists a child c of x s.t. $M[\alpha, c]$ is DESC-EMBED or EMBED then
    $M[\alpha, x] = $ DESC-EMBED;
  Else $M[\alpha, x] = $ NOT-DESC-EMBED;

Before describing Procedure *cycleembed*, we define some notation. A pair (β, y) is a *condition of embedding*, denoted $ce(\beta, y)$, on which another pair (α, x) depends if i) $subgraph(\alpha)$ is topologically embeddable in $subgraph(x)$ only if $subgraph(\beta)$ is desc-embeddable in $subgraph(y)$ and ii) α and x are the parents of β and y, respectively. A $ce(\beta, y)$ is *confirmed* if it has been determined that $subgraph(\beta)$ is desc-embeddable in $subgraph(y)$. A $ce(\beta, y)$ is *violated* if it has been determined that $subgraph(\beta)$ is not desc-embeddable in $subgraph(y)$. A $subgraph(\alpha)$ is *conditionally topologically embeddable* in $subgraph(x)$ if i) $label(\alpha) = label(x)$, ii) $degree(\alpha) = degree(x)$, iii) there exists some condition on which (α, x) depends that is not confirmed and iv) no condition on which (α, x) depends is violated.

For each input pair of nodes i and j, Procedure *cycleembed* traverses $org(i)$ and $org(j)$ in postorder, where the postorder is obtained by ignoring the back edges and cross edges. If for each node α in $org(i)$, there exists a node x in $org(j)$ such that $subgraph(\alpha)$ is conditionally topologically embeddable or desc-embeddable in $subgraph(x)$, then for all i in $org(i)$ and j in $org(j)$ such that $subgraph(\beta)$ is conditionally topologically embeddable in $subgraph(y)$, we conclude that $subgraph(\beta)$ is topologically embeddable in $subgraph(y)$; otherwise we conclude that $subgraph(\alpha)$ is not desc-embeddable in any subgraph rooted at a node in $org(j)$. The following lemmas prove the idea.

Lemma 3 *Let α be a node in $org(i)$ and x be a node in $org(j)$. If $subgraph(\alpha)$ is conditionally topologically embeddable in $subgraph(x)$ and for each node β in $org(i)$, there exists a node y in $subgraph(x)$ such that $subgraph(\beta)$ is topologically embeddable or conditionally topologically embeddable in $subgraph(y)$, then $subgraph(\alpha)$ is topologically embeddable in $subgraph(x)$.*

Proof Outline. By induction on k, we can prove that for any $k \geq 0$, there exists j such that $subgraph(\alpha)^k$ is topologically embeddable in $subgraph(x)^j$. $\qquad\Box$

Lemma 4 *For each node α in $org(i)$, if $subgraph(\alpha)$ is not desc-embeddable in a subgraph S of T, then there exists no node β in $org(i)$ such that $subgraph(\beta)$ is desc-embeddable in S.*

Proof. Suppose that there were a node β in $org(i)$ such that $subgraph(\beta)$ is desc-embeddable in S. Then there would exist a subgraph S_1 of S such that $subgraph(\beta)$ is topologically embeddable in S_1. Since α is a descendant of β, there would exist a subgraph S_2 of S_1 such that $subgraph(\alpha)$ is topologically embeddable in S_2. This contradicts the fact that $subgraph(\alpha)$ is not desc-embeddable in S. $\qquad\Box$

Procedure *cycleembed* uses a $|org(i)| \times |org(j)|$ matrix *EM* to record the intermediate results, where $|org(i)|$ and $|org(j)|$ are the numbers of nodes in $org(i)$ and $org(j)$, respectively. Each entry $EM[\alpha, x]$ may have one of the four values: UNKNOWN ($subgraph(\alpha)$ has not been processed against $subgraph(x)$ yet), EMBED ($subgraph(\alpha)$ is topologically embeddable in $subgraph(x)$), NOT-EMBED ($subgraph(\alpha)$ is not topologically embeddable in $subgraph(x)$) and COND-EMBED ($subgraph(\alpha)$ is conditionally topologically embeddable in $subgraph(x)$). Procedure *cycleembed* also uses a size $|org(i)|$ array *DE* to record whether or not the nodes in $org(i)$ are desc-embeddable in any node in $org(j)$: $DE[\alpha] = 1$ if there exists a node x in $org(j)$ such that $subgraph(\alpha)$ is conditionally topologically embeddable in $subgraph(\alpha)$ or there exists a child y of any node in $org(j)$ such that $subgraph(\alpha)$ is desc-embeddable in $subgraph(y)$; $DE[\alpha] = 0$ otherwise. A function *embed* is used to determine the values of the entries of *EM*.

```
Procedure cycleembed(i, j)
    Create matrix EM and array DE and initialize each entry of EM to UNKNOWN
        and each entry of DE to 0;
    For each node α in org(i) in postorder do
        For each node x in org(j) in postorder do
            EM[α, x] = embed(α, x, i, j);
            If EM[α, x] is EMBED or COND-EMBED or there exists k s.t. M[α, child_k(x)]
                is EMBED or DESC-EMBED then DE[α] = 1;
    If there exists β s.t. DE[β] = 0 then for each α in org(i) and x in org(j) do
        M[α, x] = NOT-DESC-EMBED;
    Else for each α in org(i) and x in org(j) do
```

If $EM[\alpha, x]$ is EMBED or COND-EMBED then $M[\alpha, x] =$ EMBED;
Else $M[\alpha, x] =$ DESC-EMBED;
Remove matrix EM and array DE;

Function $embed(\alpha, x, i, j)$
  If $label(\alpha) = label(x)$, $degree(\alpha) = 0$ or $degree(\alpha) = degree(x)$, and for each k
    $M[child_k(\alpha), child_k(x)]$ is EMBED or DESC-EMBED then return EMBED;
  Else if $label(\alpha) \neq label(x)$, $degree(\alpha) \neq degree(x)$ or there exists k s.t.
    $M[child_k(\alpha), child_k(x)] =$ NOT-DESC-EMBED then return NOT-EMBED;
  Else return COND-EMBED;

It takes $O(|E(P)| + |E(T)|)$ steps to compute $DAG(P)$ and $DAG(T)$ and number the nodes in them. Let A be the set of nodes which are in the SCCs in T and B be the set of nodes which are not in any SCC. Procedure $dagembed$ takes at most $O(|V(P)| \times (\sum_{x \in B}(degree(x))))$ steps during the whole course of computation. Procedure $cycleembed$ and Function $embed$ take at most $O(|V(P)| \times (\sum_{x \in A}(degree(x))))$ steps during the whole course of computation. Therefore, the time complexity of this algorithm is $O(|V(P)| \times (\sum_{x \in B}(degree(x)))) + O(|V(P)| \times (\sum_{x \in A}(degree(x)))) + O(|E(P)| + |E(T)|) = O(|V(P)| \times |E(T)| + |E(P)|)$. The space cost of this algorithm is clearly $O(|V(P)| \times |V(T)| + |E(P)| + |E(T)|)$.

6 Conclusion

We have presented efficient algorithms for the RDG pattern matching problem, the directed graph pattern matching problem and the directed graph topological embedding problem. All of these algorithms use the bottom-up approach. It would be interesting to see whether or not the running times of these algorithms can be improved, although dramatical improvements seem unlikely. It would also be interesting to see whether or not the bottom-up approach can be used to solve other embedding problems in directed graphs.

Acknowledgment

The author wishes to thank Naomi Nishimura who read the early versions of the manuscript and provided many valuable comments.

References

[1] A.V. Aho, R. Sethi, and J. D. Ullman. *Compilers - Principles, Techniques, and Tools*, chapter 6.7. Addison Wesley, 1986.

[2] Alexander Aiken and Brian R. Murphy. Implementing regular tree expressions. In *Proceedings of 5th ACM Conference on Functional Programming Languages and Computer Architecture*, pages 427–447, 1991.

[3] Alexander Aiken and Brian R. Murphy. Static type inference in a dynamically typed language. In *Proceedings of Eighteenth Annual ACM Symposium on Principles of Programming Languages*, pages 279–290, 1991.

[4] S. A. Cook. The complexity of theorem-proving procedures. In *Proceedings of the 3rd Annual Symposium on the Theory of Computing*, pages 151–158, 1971.

[5] Andrea Corradini. Term rewriting in CT_Σ. In *Proceedings of International Joint Conference on Theory and Practice of Software Development*, pages 468–484, 1993.

[6] Nachum Dershowitz and Stéphane Kaplan. Rewrite, rewrite, rewrite, rewrite, rewrite ... In *Proceedings of Sixteenth Annual Symposium on Principles of Programming Languages*, pages 250–259, 1989.

[7] J. R. W. Glauert, J. R. Kennaway, and M. R. Sleep. Dactl: An experimental graph rewriting language. In *Proceedings of Fourth International Workshop on Graph Grammars and Their Application to Computer Science*, pages 378–395, 1990.

[8] N. Heintze and J. Jaffar. A finite presentation theorem for approximating logic programs. In *Proceedings of Seventeenth Annual ACM Symposium on Principles of Programming Languages*, pages 197–209, 1990.

[9] Christoph M. Hoffmann and Michael J. O'Donnell. Pattern matching in trees. *Journal of ACM*, 29(1):68–95, 1982.

[10] Kristoffer H. Holm. Graph matching in operational semantics and typing. In *Proceedings of Colloquium on Trees in Algebra and Programming*, pages 191–205, 1990.

[11] Jacob Katzenelson, Shlomit S. Pinter, and Eugen Schenfeld. Type matching, type-graphs, and the Schanuel conjecture. *ACM Transactions on Programming Languages and Systems*, 14(4):574–588, Oct. 1992.

[12] Jan Willem Klop. Term rewriting systems. Technical Report CS-R9073, Free University, Department of Mathematics and Computer Science, 1990.

[13] R. Tarjan. Depth first search and linear graph algorithms. *SIAM Journal on Computing*, 1(2):146–160, 1972.

Constant-space string matching with smaller number of comparisons: sequential sampling

Leszek Gąsieniec Wojciech Plandowski * Wojciech Rytter **

Instytut Informatyki, Uniwersytet Warszawski, Warszawa,
{lechu,wojtekpl,rytter}@mimuw.edu.pl

Abstract. A new string-matching algorithm working in constant space and linear time is presented. It is based on a powerful idea of sampling, originally introduced in parallel computations. The algorithm uses a sample S which consists of two positions inside the pattern P. First the positions of the sample S are tested *against* the corresponding positions of the text T, then a version of Knuth-Morris-Pratt algorithm is applied. This gives the simplest known string-matching algorithm which works in constant space and linear time and which does not use any linear order of the alphabet. A refined version of the algorithm gives the fastest (in the sense of number of comparisons) known algorithm for string-matching in constant space. It makes $(1 + \varepsilon)n + O(\frac{n}{m})$ symbol comparisons. This improves substantially the result of [3], where a $(\frac{3}{2} + \varepsilon)n$ comparisons constant space algorithm was designed. Additionally, our preprocessing to this algorithm makes only $(1 + \varepsilon)m + O(\frac{1}{\varepsilon})$ comparisons and uses constant space.

1 Introduction

Assume we are given two strings: a pattern P of length m and a text T of length n. The *string-matching problem* consists in finding all occurrences of P in T. The algorithms solving this problem with linear cost and (simultaneously) constant space are the most interesting and usually the most sofisticated. By the cost of computations we mean the number of symbol comparisons between pattern and text symbols. The first algorithm which used constant amount of additional memory was given by Galil and Seiferas in [14]. Later Crochemore and Perrin in [9] have shown how to achieve $2n$ comparisons algorithm preserving small amount of memory. An alternative algorithm was presented by Gąsieniec, Plandowski and Rytter in [15]. The first small space algorithm which beats the bound of $2n$ comparisons was presented by Breslauer in [3]. He designed $(\frac{3}{2} + \varepsilon)n$ comparisons, constant space algorithm. In this paper we present constant space and $(1 + \varepsilon)n + O(\frac{n}{m})$ comparisons algorithm. For constant space algorithms this bound seems to be tight due to the last results of Breslauer [2]. We introduce also a very simple constant space algorithm with $2n$ comparisons called *sequential sampling*.

---
* Supported partially by the KBN grant.
** Supported partially by the DFG grant.

All strings considered in the paper are built over a general alphabet Σ (without any restriction). Given a word $w \in \Sigma^*$, by $w[i]$ we mean the symbol at the ith position of the word w. Notice that positions in a word are enumerated from 1. We say that the word w has *a period per* if and only if $w[i] = w[i + per]$ for all positions $1 \leq i \leq |w| - per$. The shortest period of w is called *the period* of w. If the period *per* $\leq |w|/2$, then the word w is called *periodic*, otherwise w is *nonperiodic*.

Lemma 1. *If the word w has two periods c, d and $c + d \leq |w|$ then w has also a period $gcd(c, d)$, where gcd stands for the greatest common divisor.*

Denote by w^- the word w without its last symbol, then Lemma 1 implies:

Observation 1 *Given periodic word w with the period per, let v be the prefix of w of size per.*
Then the word vv^- is nonperiodic.

The main novel idea behind our algorithm is a use of a two-point sample S of the pattern P. The samples were extensively used in parallel computations, see [18]. Here the sample is used in a sequential setting, another application of sample in sequential computations appears in [7]. Assume that a nonperiodic pattern P has a periodic prefix. Denote by π the longest periodic prefix of P. Let $q - 1$ be the length of π, let *per* be the size of the shortest period of π and let $p = q - per$. Define the set $S = \{ p, q \}$ as the *sample S* of the prefix $P[1..q]$. Introduce the predicate:

$$MatchSample(i, S) = (T[i + p] = P[p] \text{ and } T[i + q] = P[q])$$

Observation 2
If $MatchSample(i, S)$ then no occurrence of the pattern starts at any position of T in $[i + 2 \ldots i + p]$.

The observation implies that if the pattern matches the text at the positions of the sample, then the next *safe* shift is at least p. For example if $P = aaaaaaab$ then $S = \{7, 8\}$. In this case if $MatchSample(i, S)$ then the next shift is at least 7.

Our algorithm is also based on Knuth Morris Pratt (KMP in short) algorithm, see [16].

```
ALGORITHM KMP;
i:= 0; j:=0;
while i ≤ n − m do
begin
    j := max{k : T[i + j + 1 . . . i + k] = P[j + 1 . . . k] or k = 0};
    if j = m then report match at i + 1;
    i:= i + max(1, j − FT[j]);
    j := FT[j];
end
```

The KMP algorithm makes *exact* shifts, using the *failure table FT*, we refer to [11] and [16] for the definition of the failure table. We shall make approximate shifts and use a partial failure table.

2 A simple algorithm performing 2n comparisons: sequential sampling

Recall that we call the word w periodic if it has a period *per* $\leq \frac{1}{2}|w|$. We consider first the case when none of the prefixes of P is periodic. For this case we have a very simple algorithm for text-searching.

> ALGORITHM *Simple_Text_Searching*;
> { none of the prefixes is periodic }
> $i:= 0$;
> **while** $i \leq n - m$ **do**
> **begin**
> $j := \max\{k \; : \; T[i+1\ldots i+k] = P[1\ldots k] \; or \; k = 0\}$;
>   **if** $j = m$ **then** report match at $i + 1$;
> $i:= i + \lceil\frac{i+1}{2}\rceil$;
> **end**

During every stage (iteration of outer while loop) of the algorithm *Simple_Text_Searching* the total work (number of comparisons) is equal to $j + 1$. Since every prefix is nonperiodic we can make shift of size $\lceil\frac{i+1}{2}\rceil$. Notice that $2\lceil\frac{i+1}{2}\rceil \geq j+1$ and we get the algorithm which performs at most $2n$ comparisons. The second case is when P is nonperiodic and there is a periodic prefix of P.

> ALGORITHM *SequentialSampling*;
> {the case when P is nonperiodic and has a sample $S = \{p, q\}$ }
> $i:= 0$;
> **while** $i \leq n - m$ **do**
> **begin**
>   { first test positions of the sample {p, q} in P}
>   **if** not $MatchSample(i, S)$ **then** $i := i + 1$ **else**
>   **begin**
> $j := \max\{k \; : \; T[i+1\ldots i+k] = P[1\ldots k] \; or \; k = 0\}$;
>   **if** $j = m$ **then** report match at $i + 1$;
>   **if** $j < q - 1$ **then** $i:= i + p$ **else** $i:= i + \lceil\frac{i+1}{2}\rceil$;
>   **end**

Remark We assume that when we compute $j := \max\{k \; : \; T[i+1\ldots i+k] = P[1\ldots k] \; or \; k = 0\}$ then the positions $T[i+p]$ and $T[i+q]$ are not tested, since we have already tested them when computing $MatchSample(i, S)$.

In this case the algorithm starts with the test if pattern symbols associated with positions of the sample S are the same as corresponding symbols in the text. Then we try to match full occurence of prefix $P[1..q]$ without inspecting $P[p]$ and $P[q]$ for the second time. In case a mismatch is found we make shift of size p. Otherwise we start to match longer nonperiodic prefixes and use the approach from the previous algorithm.

We give the name *sequential sampling* to the whole algorithm consisting of subalgorithms for three cases:

(A) the simple case, when the algorithm *Simple_Text_Searching* is applied;

(B) the (main) case of nonperiodic pattern having a sample, when the algorithm *SequentialSampling* is applied;

(C) the case of periodic patterns, which is reduced to one of the cases (A) or (B) in the proof of the theorem below.

Theorem 2. *The algorithm sequential sampling performs at most $2n$ symbol comparisons and uses a constant additional space.*

Proof.

We prove the theorem in three stages:

A: P is nonperiodic and all its prefixes are nonperiodic (the algorithm *Simple_Text_searching* is applied)

The text searching is done as follows inductively. Assume that we start to recognize an occurence of the pattern P at position i in text T. We compare, one by one, pattern and text symbols. Assume that first j comparisons were positive (i.e. $P[1..j] = T[i..i+j-1]$) and $(j+1)$th one was negative ($P[j+1] \neq T[i+j]$). In that case we can make shift of length $s = \lceil \frac{i+1}{2} \rceil$, since prefix $P[1..j]$ is nonperiodic. The work of $j+1$ comparisons is amortized by shift s, since $2 \cdot s \geq j+1$.

B: P is nonperiodic and $P[1..q-1]$ is the longest periodic prefix of P (the algorithm *SequentialSampling* is applied).

Let *per* be the period of $P[1...q-1]$. We have that $P[k] = P[k+per]$ for all $1 \leq k \leq q - per - 1$. Since prefix $P[1..q]$ is nonperiodic we know that $P[q-per] \neq P[q]$. Let $p = q-per$ and recall that the pair $S = \{p, q\}$ is the sample of P. Negative tests are amortized by immediate shifts, i.e. two comparisons are amortized by shift of length one. In case of positive match of the sample S, we start to test the full match of $P[1..q]$, omitting recognized earlier symbols of the sample. Symbols from S don't belong to period of the prefix $P[1..q-1]$, so if a mismatch between text and prefix is found we can make shift of length $s = p$. Hence the total work is not greater than $q-1$ and $p \geq \frac{q-1}{2}$ (prefix $P[1..q-1]$ is periodic) so $q \leq 2 \cdot s$ and we get proper amortization. In case the whole prefix $P[1..q]$ was matched all longer prefixes are nonperiodic and we come to case **A**.

C: Pattern P is periodic with a period *per*.

Recall that the pattern P is periodic with the period *per* iff $per \leq \frac{|P|}{2}$, i.e. $P = v^k v'$ where v is the prefix of P of length *per*, $k \geq 2$ and v' is a prefix of v. We know from observation 1 that the prefix vv^- is nonperiodic. Our algorithm starts with searching for prefix vv^-. One of the cases (A) or (B) is applied. Since the word vv^- is nonperiodic the work to find all its occurrences is amortized by the shifts according to nonperiodic case parts (A and B). If the word vv^- is found then we start to match next symbols of the pattern P. In case of any later mismatch the work equals to $2 \cdot per + k$, for some $0 \le k \le m - 2 \cdot per + 1$, but the shift is $s = per + k$, since matched prefix has the period per. Also $2 \cdot s \ge 2 \cdot per + k$ so we get proper amortization in this case, too. Note that in the case the whole pattern is found, the shift equals per and amortize, one to one, comparisons involved in testing first per symbols of the occurrence of P. Since we can remember pattern prefix of size $m - per$, all further comparisons are amortized by one of the shifts explained above.

3 Partial Algorithm KMP

Let FT be the *failure table* of the pattern. The α-*part* of FT is the table:

$$FT_\alpha[j] = \begin{cases} FT[j] \text{ for } FT[j] > \alpha j \\ 0 \qquad FT[j] \le \alpha j \end{cases} \tag{1}$$

We say that a table X can be *represented in constant space* if using a pre-computed information of a constant size we can compute each value of $X[j]$ in constant time, for any index j.

Lemma 3. *Assume $0 < \alpha$, $\sigma < 1$ are constants. Then the α-part of the failure table corresponding to prefixes longer than σm can be represented in constant space.*

Proof. A technical proof is omitted in this version.

The algorithm is designed as a function *Partial_KMP* which starts from the position s in the text and assumes that length r prefix of the pattern matches the text at this position. Then it works in the same way as KMP until the values of the failure table is not available or KMP wants to use the value for a pattern prefix shorter than αr. The main point is that when this function stops the *shift* in the algorithm KMP would be *large* enough to amortize symbol comparisons.

Clearly, the algorithm reports correctly all *matches* that start in the interval $[s..s']$, where s' is the value returned by *Partial_KMP*.

Note, that the algorithm uses the table only for prefixes not shorter than αr.

function *PartiaLKMP(α, s, r)*;
$i := s - 1$; $j := r$; { initially $T[i+1 \ldots i+r] = P[1 \ldots r]$ }
while $i \leq n - m$ **do**
begin $j := \max\{k : T[i+j+1 \ldots i+k] = P[j+1 \ldots k]$ or $k = 0\}$;
    if $j = m$ **then** report match at i;
    if $j \leq \alpha r$ **then** { the matched prefix is shorter than αr }
        return i and STOP
    if $FT_\alpha[j] = 0$ **then** { shift is at least $j - \alpha j$}
        return $\max(i, i+j-\alpha j)$ and STOP
    else begin { $FT_\alpha[j] > \alpha j$ }
        $i := i + (j - FT_\alpha[j])$;
        $j := FT_\alpha[j]$;
    end
end
return i;

Our next lemma states formally that the shift which is made by the function *Partial_Match* is large enough with respect to the number of comparisons.

Lemma 4. *Assume the function* Partial_KMP(α,s,r) *returns the value* s'. *Then it can be modified to work with* $\frac{n'}{(1-\alpha)} - r + O(\frac{n'}{r})$ *comparisons for* $n' = s' - s$.

Proof. The algorithm makes shifts ($j - FT_\alpha[j]$). Denote by t the number of *matches*. Since at the beginning of the algorithm the length r pattern prefix matches the text the number of matched symbols is $t + r$. It can be $\alpha(t + r)$ matched symbols at the end not amortized by shifts. Thus the number n' of matched symbols amortized by shifts is not less than $(1 - \alpha)(t + r)$, i.e. $t \geq \frac{n'}{(1-\alpha)} - r$. For each mismatch we make a shift. If all shifts are large enough (larger than $r' = \frac{1}{2}(1 - \alpha)r$) then the number of mismatches is $O(\frac{n'}{r})$. We have to modify slightly the algorithm if the shift is smaller than r'. In this case a prefix of P, of size at least $(1 - \alpha)r$, has a period of size at most $\frac{1}{2}(1 - \alpha)r$. In this moment instead of continuing the algorithm we compute the maximal continuation of such period in the text. This gives us later the shift which is large enough. Then we resume the algorithm given above. We omit the details.

4 Reducing the number of comparisons

In the latter we need to redefine the notion of periodicity. We say that the word w is periodic (or c-periodic) if it has a period $q \leq c|w|$. Observe that each text is 1-periodic.

As previously we search for the pattern P of length m in the text T of length n. The algorithm which is described in this section depends on two constants δ, c such that $\delta > 0$, $0 < c < 1$. Our algorithm makes at most $(1 + \varepsilon(\delta, c))n + O(\frac{n}{m})$ comparisons and $\lim_{\delta \to 0, c \to 1} \varepsilon(\delta, c) = 0$.

Denote by *OnLine* one of the known online algorithms for string-matching with small number of comparisons (i.e. $n + \mathcal{O}((n-m)/m)$) and $\mathcal{O}(m)$ space, e.g. the algorithm from [5]. Such algorithm finds at the position i in a text the first from the left occurrence of a pattern making at most $(i+m) + \mathcal{O}(\frac{i}{m})$ comparisons. We shall apply the algorithm *OnLine* to *very short* patterns.

Define another constant M to be the least number K such that the algorithm *OnLine* finds in a given text the first (from the left) occurrence i of length K pattern with at most $(1 + \delta)i+$K comparisons. Note, that if δ tends to 0 then M tends to infinity. The pattern (or its prefix) is *very short* iff its length does not exceed M.

Denote by $PREF(k)$ the set of the pattern prefixes longer than k. The behavior of our algorithm depends on the type of the pattern.

Case 1. The pattern is *very short* $(m \leq \text{M})$.
Apply algorithm *OnLine*.

Case 2. One of the prefixes in $PREF(\text{M})$ is c-periodic with the period shorter than the shortest period of the whole pattern.

Let *pref* be the longest prefix of P with the period *per* shorter than $c|pref|$ which is not a period of P. Take the positions $q = |pref| + 1$, $p = q - per$ as the sample S. The algorithm in this case is a modification of the algorithm *SequentialSampling*. We replace the predicate *MatchSample* by the function *ModifiedMatchSample(i, S)* which works as follows. If $P[q-1] = P[q]$ then test the position p of the sample first and then the position q, otherwise do it in the reverse order. Return 0 if the sample occurs and the number of made comparisons otherwise. Since $P[q-1] = P[p-1]$ if the first symbol compared by the function matches a symbol in the text, the next shift is at least 2.

ALGORITHM Economic_SequentialSampling;
  { P is nonperiodic and p, q give the *sample S* }
  $i := 1$;
  while $i \leq n - m$ **do**
  begin $k := ModifiedMatchSample(i, S)$;
      $i := i + k$;
      if $k = 0$ **then begin** $j := \max\{k : P[1..k] = T[i+1..i+k] \text{ or } k = 0\}$;
                 if $j <= p$ **then** $i := i+p$
                 else $i := Partial\_KMP(1 - c, i, j)$;
             end
  end

Lemma 5. *Assume the pattern satisfies the conditions in Case 2 and $S = \{p, q\}$ is the sample of this pattern. Then the $(1 - c)$-part of the failure table for the prefixes in $PREF((1 - c)p)$ may be represented in constant space.*

Proof. The prefixes in $PREF(p)$ are either non c-periodic or c-periodic with the period being the shortest period *sh_per* of the pattern. All c-periodic ones are longer than those which are not c-periodic. The values of the $(1 - c)$-part of the failure table for non c-periodic prefixes equal 0 and for length k c-periodic prefixes with the period *sh_per* equal $k - sh\_per$. Hence, having *sh_per* we are able to retrieve those values in constant time. On the other hand since $q < p + per \leq p + c \cdot q$ we have $q < \frac{p}{(1-c)}$ and, by Lemma 3, the $(1 - c)$-part of the failure table for the prefixes longer than $(1 - c)p$ and shorter than q can be encoded in constant space. This completes the proof.

Lemma 6. *The algorithm Economic_SequentialSampling makes at most $n/c + \mathcal{O}(n/M)$ symbol comparisons and may be implemented in constant space.*

Proof. The fact that the algorithm may be implemented in constant space is a consequence of Lemma 5. Let $shift$ be the shift of the pattern over the text which is made by execution of the statements inside the while-loop. Let $comp$ be the number of symbol comparisons made during this execution of the statements. If the sample does not match the text then $shift = comp$. If the sample matches the text and $j \leq p$ then $shift = q \geq comp$ and again the shift amortizes the number of comparisons. In the case $j > p$, the function $Partial\_KMP$ is called, and, by Lemma 4, the number of comparisons in this loop is at most $shift/c + \mathcal{O}(shift/j)$. Since the sum of all shifts does not exceed n and $j > p > M$ the result follows.

Case 3. Each prefix in $PREF(M)$ is either not c-periodic or its shortest period is the period of the pattern.

Denote by $OnLine(i, s)$ the function which starts the algorithm $OnLine$ from the ith position in T and finds the first to the left occurrence of length s pattern prefix.

$$ALGORITHM \ \ Economic\_KMP;$$
$$i := 1;$$
$$\textbf{while } i \leq n - m \textbf{ do}$$
$$\textbf{begin } i := OnLine(i, M)$$
$$i := Partial\_KMP(1 - c, i, M));$$
$$\textbf{end}$$

The proof of our next lemma is similar to the proof of Lemma 5.

Lemma 7. *Assume the pattern satisfies the requirements of Case 3. Then the $(1 - c)$-part of the failure table for the prefixes longer than $(1 - c)M$ may be represented in constant space.*

The upper bound for the number of symbol comparisons in $Economic\_KMP$ is a consequence of such upper bounds in algorithms $Online$ and $Partial\_KMP$.

Lemma 8. *The algorithm Economic_KMP makes at most $(\max\{1+\delta, 1/c\})n + \mathcal{O}(n/M)$ symbol comparisons and may be implemented in constant space.*

The theorem below is a direct consequence of Lemmas 6 and 8 and the fact that if δ tends to 0 then M tends to infinity.

Theorem 9.
For any constant $\varepsilon > 0$ there is a linear time constant space string-matching algorithm which makes $(1 + \varepsilon)n + O(\frac{n}{m})$ symbol comparisons.

5   Pattern Preprocessing

Our preprocessing phase works in an *on-line* left-to-right way (similarly as the searching phase), we scan the pattern and preprocess the actually seen prefix P' of P together with several sufficiently long prefixes of P'.

As an auxiliary operation we use the searching phase (described before) to search for a suitable prefix of P'. The occurrences of this prefix determine the continuation of periodicity of the scanned prefix in the extended new prefix and helps to compute next portion of preprocessing data .

The data computed in the preprocessing for a given pattern P depends on the type of the pattern.

- If the shortest period of P is also the shortest period of all pattern prefixes of length at least M, then preprocessing finds data generated by the preprocessing phase of the algorithm *Online* for the pattern prefix of length M and the values of the failure table $FT_{1-c}[j]$ for $(1-c)\text{M} \leq j \leq m$.
- If some prefix of P of length at least M is c-periodic with a period which is not a period of the pattern then the preprocessing phase finds the values of the failure table $FT_{1-c}[j]$ for $(1-c)p \leq j \leq m$ and the sample $\{p.q\}$ where $q - p$ is the length of the shortest period of the prefix $w = P[1..q-1]$ and w is the longest c-periodic prefix of P with a period shorter than the period of the pattern. Observe that $(1-c)p \geq (1-c)(1-c)q \geq (1-c)^2\text{M}$.

Denote by *Preprocess*$[i..j]$ all the data needed in the searching for the prefixes $P[1..k]$ for $i \leq k \leq j$. Denote *Preprocess*$[i] = $ *Preprocess*$[i..i]$. Hence the aim of the preprocessing phase is to find only *Preprocess*$[m]$. However the preprocessing algorithm needs itself also data related to *Preprocess*$[i..j]$, where j is the last scanned position of the preprocessed part of the pattern.

Lemma 10. *Let $0 < \sigma \leq 1$. The data in Preprocess$[\sigma j..j]$ for some $j \leq m$ can be computed in constant space with no symbol comparisons if Preprocess$[\text{M}]$, the values $FT_{1-c}[\sigma j..j]$ and the values of FT_{1-c} being in Preprocess$[\sigma j]$ are given. There is a compressed representation of the needed values in FT_{1-c} such that the above can be done in constant time.*

An interesting feature of the KMP algorithm is that its searching phase can be used in the preprocessing and the failure table can be computed as a side effect of searching pattern inside pattern. The entries of the table are computed before they are used.

```
ALGORITHM  KMP_Preprocessing;
  i:= 1; j:=0;
  while i ≤ m do
  begin
      j_old:=j;
      j := max{k  :  P[i + j + 1...i + k] = P[j + 1...k] or k = 0};
      for k ∈ [i + j_old + 1..i + j] do
            FT[k]:= k-i
      i:= i + max(1, j − FT[j]);
      j := FT[j];
  end
```

In a similar way we modify the function *Partial_KMP* to obtain the function *Modified_Partial_KMP*. *Partial_KMP* is a version of the KMP algorithm with "approximate" failure table (stored in a constant space). We show how *Partial_KMP* can compute the values of "approximate" failure table before using it.

```
function  Modified_Partial_KMP(α, s, r);
    i:= s − 1; j:=r; t := r;{ initially P[i + 1...i + r] = P[1...r]}
    while i ≤ m do
    begin j_old:=j;
            j := max{k  :  P[i + j + 1...i + k] = P[j + 1...k] or k = 0};
            for k ∈ [i + j_old + 1..i + j] do
                    FT_α[k]:= if i > (1 − α)k then 0 else k-i
            if FT_α[i + j] ≠ 0 then {P[1..i + j] is α-periodic}
            begin   t := α(i + j);
                    forget about values in FT_α that are not
                    in Preprocess[α(i + j)..i + j];
            end
            if j ≤ αt then { the matched prefix is shorter than αt }
                    return i and STOP
            if FT_α[j] = 0 then { shift is at least j − αj}
                    return max(i, i + j − αj) and STOP
            else begin { FT_α[j] > αj }
                    i:= i + ( j − FT_α[j]);
                    j := FT_α[j];
                    end
    end
    return i;
```

The above function works similarly as the function *Partial_KMP(α,s,r)* and calculates as a side effect all values in the table $FT_α[s + r..i]$ where i the value returned by the function. It can happen that all calculated values cannot be represented in constant space. Therefore, from time to time the function forgets some calculated values. This in turn force us to do one more change in the algorithm. We introduce new variable t which says when to stop. The algorithm stops when it needs a value of the failure table for non c-periodic prefix or a prefix shorter than αt where t starts from r and grows during execution of the algorithm. The complexity analysis of the function *Modified_Partial_KMP* is similar as the one for the function *Partial_KMP*.

In the following we need one property of the function *Modified_Partial_KMP*.

Observation 3 *Suppose that* Modified_Partial_KMP(α, s, r) *returned value* i. *Then* $FT_i \leq j - i$ *for* $j \geq i$.

Now we are ready to describe our preprocessing phase. During the execution of the algorithm we compute all values of the table FT_{1-c} for prefixes longer than $(1-c)^2 M$. As in the function *Modified_Partial_KMP* from time to time we have to forget some calculated values.

The values in FT_{1-c} are nonzero for c-periodic pattern prefixes. Hence we start from searching for the shortest c-periodic prefix of P which is longer than $k = (1-c)^2 M$. Suppose that its shortest period is *per*. Then there is an occurrence of length $(1-c)k$ pattern prefix at position *per* in the pattern. So we start our algorithm from searching this prefix in the pattern. As we find it we call the function *Modified_Partial_KMP*. Suppose the returned value is i. Then all values of the table FT_{1-c} that are needed to compute $Preprocess[(1-c)i..i]$ are calculated and the symbol comparisons are amortized by shifts.

Now we try to find the shortest c-periodic pattern prefix longer than i. At this moment by Observation 3 its shortest period is at least i. As previously we search for the pattern prefix of length $(1-c)i$ and run *Modified_Partial_KMP*. We repeat this process until we reach the end of the pattern. Then on the basis of the computed part of the table FT_{1-c} we compute $Preprocess[m]$.

The main part of the preprocessing algorithm looks as follows.

ALGORITHM Preprocessing;
{ preprocessing the whole pattern corresponds to $Preprocess[m]$}
$i:=1$; $d:=(1-c)^3 M$;
while $i + d \leq m$ **do**
    begin { $FT_{1-c}[j] \leq j - i$ for $j \geq i$}
        $i_{old} := i$;
        compute $Preprocess[d]$ using available part of FT_{1-c};
        $i:=$ the first to the right of i occurrence of length d
            prefix of P (use our searching phase);
        for $i_{old} \leq k \leq i + d - 1$ **do** $FT_{1-c}[k] := 0$
        $i:= Modified\_Partial\_KMP(1-c, i, d)$;
        $d:=(1-c)i$;
    end
for $i + 1 \leq k \leq m$ **do** $FT_{1-c}[k] := 0$
compute $Preprocess[m]$ using available part of FT_{1-c}

The upper bound for the number of comparisons in the following theorem is a consequence of such bounds for our searching phase and for the function *Partial_KMP*. The details of the proof are omitted.

Theorem 11. *For any constant ε there is string-matching algorithm which runs in $O(n)$ time and makes $(1+\varepsilon)n+O(\frac{n}{m})$ comparisons and the preprocessing phase for it runs in $O(m)$ time and makes $(1+\varepsilon)m + O(\frac{1}{\varepsilon})$ comparisons.*

References

1. A.V. Aho, Algorithms for Finding Patterns in Strings. In *Handbook of Theoretical Computer Science*, p. 257–300. Elsevier Science Publishers B. V., Amsterdam, The Netherlands, 1990.
2. D. Breslauer, private communication.
3. D. Breslauer, Saving Comparisons in the Crochemore–Perrin String Matching Algorithm. In Proc. of *1st European Symp. on Algorithms*, p. 61–72, 1993.
4. D. Breslauer and Z. Galil, Efficient Comparison Based String Matching. *J. Complexity* 9(3), p. 339–365, 1993.
5. R. Cole, R. Hariharan, Tighter bounds on the exact complexity of string matching. In Proc. of *33rd Annual Symp. on Foundations of Comp. Sci.*, p. 600–609, 1992.
6. L. Colussi, Correctness and efficiency of string matching algorithms. *Inform. and Control*, 95, p. 225–251, 1991.
7. M.Crochemore, L. Gąsieniec, W. Plandowski and W. Rytter, Two-dimensional pattern matching in small time and space, in Proc. of *STACS'95*, Springer-Verlag, 1995.
8. M. Crochemore, String-matching on ordered alphabets. *Theoret. Comput. Sci.*, 92, p. 33–47, 1992.
9. M. Crochemore and D. Perrin, Two-way string-matching. *J. Assoc. Comput. Mach.*, 38(3), p. 651–675, 1991.
10. M. Crochemore and W. Rytter, Periodic Prefixes in Texts. In Proc. of *Sequences'91 Workshop "Sequences II: Methods in Communication, Security and Computer Science"*, p. 153–165, Springer–Verlag, 1993.
11. M. Crochemore and W. Rytter, Text algorithms, *Oxford University Press*, New York, 1994
12. Z. Galil and R. Giancarlo, On the exact complexity of string matching: lower bounds. *SIAM J. Comput.*, 20(6), p. 1008–1020, 1991.
13. Z. Galil and R. Giancarlo, The exact complexity of string matching: upper bounds. *SIAM J. Comput.*, 21(3), p. 407–437, 1992.
14. Z. Galil and J. Seiferas, Time-space-optimal string matching. *J. Comput. System Sci.*, 26, p. 280–294, 1983.
15. L. Gąsieniec, W. Plandowski and W. Rytter, The zooming method: a recursive approach to time-space efficient string-matching. *Theoret. Comput. Sci.*, to appear.
16. D.E. Knuth, J.H. Morris and V.R. Pratt, Fast pattern matching in strings. *SIAM J. Comput.*, 6, p. 322–350, 1977.
17. M. Lothaire, Combinatorics on Words. Addison-Wesley, Reading, MA., U.S.A., 1983.
18. U.Vishkin, Deterministic sampling - a new technique for fast pattern matching, *SIAM J. Comput.*, 20, p. 22–40, 1991.

Multi-Dimensional Pattern Matching with Dimensional Wildcards

Raffaele Giancarlo[1]* and Roberto Grossi[2]**

[1] Dipartimento di Matematica e Applicazioni, Università di Palermo
via Archirafi 34, I-90123 Palermo, Italy
[2] Dipartimento di Sistemi e Informatica, Università di Firenze
via Lombroso 6/17, I-50134 Firenze, Italy

Abstract. We introduce a new multi-dimensional pattern matching problem, which is a natural generalization of the on-line search in string matching. We are given a text matrix $A[1:n_1, \cdots, 1:n_d]$ of size $N = n_1 \times n_2 \times \cdots \times n_d$, which we may preprocess. Then, we are given, on-line, an r-dimensional pattern matrix $B[1:m_1, \ldots, 1:m_r]$ of size $M = m_1 \times m_2 \times \cdots \times m_r$, with $1 \leq r \leq d$. We would like to know whether $B^* = B^*[*, 1:m_1, *, \cdots, 1:m_r, *]$ occurs in A, where $*$ is a *dimensional wildcard* such that B^* is *any* d-dimensional matrix having size $1 \times \cdots \times m_1 \times \cdots 1 \times m_r \times \cdots 1$ and containing the same elements as B. Notice that there might be $\binom{d}{r} \leq 2^d$ occurrences of B^* for each position of A. We give CRCW-PRAM algorithms for preprocessing A in $O(d \log N)$ time with $N^2/n_{\max}$ processors, where $n_{\max} = \max\{n_1, \ldots, n_d\}$. The on-line search for B^* can be done in $O(d \log M)$ time and optimal $O(dM)$ work.

1 Introduction

In its simplest and most classic version, *multi-dimensional pattern matching* is the following problem. We are given an $N = n_1 \times n_2 \cdots n_d$ *text* matrix A and an $M = m_1 \times m_2 \times m_d$ *pattern* matrix B and we want to know whether B occurs as a submatrix of A. The study of such a problem has both theoretical and practical motivations. We refer the interested reader to [5] for a discussion along with many pointers to the literature. Another aspect of multi-dimensional pattern matching is the gathering of statistical information about the text A, for example: which one is the largest submatrix of B appearing at least twice in it? Algorithms solving the multi-dimensional pattern matching are roughly in two complementary classes (see [5, 6] for the references):

- *Pattern Matching Algorithms — No Statistics about Text*: They preprocess the pattern B first (preprocessing step) and then they check whether B occurs into the text A (matching step).

---

* The work of this author is partially supported by MURST Grant "Algoritmi, Modelli di Calcolo e Strutture Informative". This work was done while the author was with AT&T Bell Laboratories, Murray Hill, NJ. U.S.A. . Email: raffaele@altair.math.unipa.it

** Work supported by MURST of Italy, partially done while the author was visiting AT&T Bell Laboratories. Email: grossi@di.unipi.it

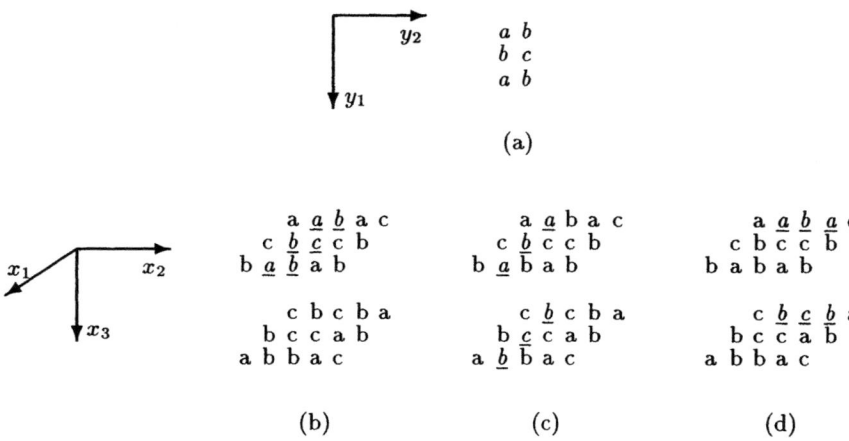

Fig. 1. (a) A 2-dimensional matrix of shape 3×2. (b)–(d) Three copies of the same 3-dimensional matrix $A = A[1{:}3, 1{:}5, 1{:}2]$ of shape $3 \times 5 \times 2$, illustrating the $\binom{3}{2}$ types of occurrences (shown in underlined italic) of the matrix (a) in position $(1, 2, 1)$.

- *Pattern Matching Algorithms — Statistics about Text:* They preprocess the text A to build an index data structure, which is the analog of the suffix tree for a string [9]. The index must support a wide variety of queries, taking time independent of the size N of A. The most basic query is *occurrence*(B), check whether B occurs in A. Giancarlo [3] showed an $\Omega(N^2/n_{\max})$ lower bound, $n_{\max} = \max\{n_1, \ldots, n_d\}$, on the time needed to build a wide class of such indexes (the proof is for $d = 2$, but it generalizes easily to the case $d > 2$).

Statement of the Problem. The pattern matching algorithms found in literature assume that A and B are *both* d-dimensional matrices. They were not originally designed to check whether an r-dimensional pattern occurs as a submatrix of a d-dimensional text, with $1 \leq r \leq d$. That is, they do not account for the degrees of freedom that one has in higher dimensions. We provide here an index $\mathcal{H}_A$ which does that, being efficiently built and queried in optimal time by PRAM algorithms. Clearly, we also obtain the first efficient RAM algorithms for the construction and query of $\mathcal{H}_A$.

Let $A = A[1{:}n_1, 1{:}n_2, \ldots, 1{:}n_d]$ and $B = B[1{:}m_1, 1{:}m_2, \ldots, 1{:}m_r]$ be respectively the text and the pattern, with $1 \leq r \leq d$. We say that A has *dimension* d, *size* $N = n_1 n_2 \cdots n_d$, and *shape* $n_1 \times n_2 \times \cdots \times n_d$. The same terminology applies to any other matrix. The size of B is denoted by $M = m_1 m_2 \cdots m_r$. Both A and B have entries drawn from an ordered alphabet Σ.

We say that B occurs as an r-dimensional submatrix in position $\mathbf{x}_d = (x_1, x_2, \ldots, x_d)$ of A if, and only if, there exist r distinct increasing integers $i_1, \ldots, i_r \in [1, d]$ such that $B[\mathbf{y}_r] = A[\hat{\mathbf{x}}_d]$ for all $\mathbf{y}_r = (y_1, \ldots, y_r) \in [1, m_1] \times \cdots \times [1, m_r]$, where $\hat{\mathbf{x}}_d$ is as $\mathbf{x}_d$ except that its i_k-th component is $x_{i_k} + y_k - 1$ instead of x_{i_k}, for all $k \in [1, r]$ (clearly, $x_{i_k} + m_k - 1 \leq n_{i_k}$). An intuitive example is shown in Fig. 1. The occurrences of B in A can be interpreted as the occurrences of a d-dimensional matrix $B^* = B^*[*, 1{:}m_1, *, \cdots, 1{:}m_r, *]$ in A, where $*$ is a *dimensional wildcard* such that B^* is *any* d-dimensional matrix having shape $1 \times \cdots \times m_1 \times \cdots 1 \times m_r \times \cdots 1$ and the same elements as B. Therefore we say that B^* *occurs in* A. Notice that there might be $\delta = \binom{d}{r} \leq 2^d$ distinct occurrences of B^* for each position of A.

We are interested in the following. Preprocessing step: build an index representing all submatrices of A. On-line queries: given an r-dimensional matrix B, with $r \in [1, d]$, check whether there exists at least one position x_d of A in which B^* occurs. The queries can be extended to ask statistics about A. For example, what is the largest B such that B^* occurs at least twice in A?

Notice that our problem requires a solution falling into the second class of pattern matching algorithms defined earlier, and simply checking whether B^* occurs in a position of A must take $\Omega(dM)$ work since M d-tuples should be read. Moreover, there is some analogy with string pattern matching with "don't care" symbols introduced by Fischer and Paterson [2]. Intuitively, here $*$ is a dimensional wildcard "matching any dimension," whereas the don't care symbol in [2] is a wildcard matching any character.

Our Solution. We work on an Arbitrary Concurrent Read Concurrent Write (CRCW) PRAM model of computation (e.g., see [7]). The preprocessing step of A requires $O(d \log N)$ time with $N^2/n_{\max}$ processors; the on-line query for B^* takes $O(d \log M)$ time with optimal $O(dM)$ work. In Section 3 we show that an easy solution through standard algorithmic techniques would obtain non-optimal $O(\delta dM)$ query work (with the same preprocessing bounds as ours), thus depending linearly on the size N of the text A. Indeed, δ might be as large as $O(N/n_{\max})$ (for the non-degenerate case in which $n_i > 1$ for $1 \leq i \leq d$).

Based upon new combinatorial and algorithmic techniques (Section 4), we propose a more sophisticated solution that allows us to achieve an exponential speed-up in the query work, from $O(\delta dM)$ to $O(dM)$, at no expense for the preprocessing work and time. Its key component is the introduction of a new data structure, the *heterogeneous index* $\mathcal{H}_A$, defined on special pieces of A called *heterogeneous blocks*. Since $\mathcal{H}_A$ satisfies the definition of index for a matrix proposed in [3], the parallel algorithm building $\mathcal{H}_A$ is a poly-logarithmic factor far from optimal work. Using $\mathcal{H}_A$ in our query, we can show that the task of checking whether B^* occurs in A is reduced to checking whether at most *two* suitably chosen matrices, instead of δ ones, appear into $\mathcal{H}_A$. We remark that it remains an open problem to devise more efficient indices for the representation of all submatrices of A. However, the speed-up obtained here can be regarded as a fundamental contribution of this paper. Indeed, it is not difficult to imagine situations in which there is plenty of time to preprocess the text, but the queries must be answered quickly.

Both the algorithms for the construction of $\mathcal{H}_A$ and for its query hinge on generalizations to heterogeneous blocks of the naming technique of Karp *et al.* [8] and of the algorithms for the parallel construction of the suffix tree by Apostolico *et al.* [1]. However an efficient implementation of the naming technique for heterogeneous blocks and the proof of its correctness turn out to be challenging tasks.

2 Preliminaries

A tuple $(x_1, x_2, \ldots, x_k)$ of $k > 0$ integers is denoted by $\mathbf{x}_k$, where $\mathbf{x}_0$ denotes $\mathbf{1}_1$ and $\mathbf{1}_k$ is used for the tuple $(1, 1, \ldots, 1)$ of k 1s. We use $c\mathbf{x}_k$, $\mathbf{x}_k + \mathbf{y}_h$ and $(\mathbf{x}_k, \mathbf{y}_h)$ with the obvious meaning. We make $\mathcal{E}[\mathbf{x}_k]$ denote all tuples $\hat{\mathbf{x}}_h$ extending $\mathbf{x}_k$ with arbitrary 1s, where $h \geq k$. That is, the sequence $\hat{x}_1, \hat{x}_2, \ldots, \hat{x}_h$ is obtained from $x_1, x_2, \ldots, x_k$ by interleaving it with a sequence of 1s (if $k = 0$ then $\mathcal{E}[\mathbf{x}_0] = \{\mathbf{1}_s : s \geq 1\}$). For example, $(1, 3, 1, 1, 4, 1)$ and $(1, 1, 3, 1, 1, 1, 4)$ are in $\mathcal{E}[(3, 4)]$. The shorthand $[a, b]$ for an interval is extended to the Cartesian product $[\mathbf{a}_k, \mathbf{b}_k] = [a_1, b_1] \times [a_2, b_2] \times \cdots \times [a_k, b_k]$. Setting $k = d$, we use $A[\mathbf{a}_d, \mathbf{b}_d]$ to denote the submatrix $A[a_1 : b_1, a_2 : b_2, \ldots, a_d : b_d]$ of a d-dimensional matrix A. The work for accessing $A[\mathbf{x}_d]$ is $O(d)$, and one memory cell suffices to store compactly $\mathbf{x}_d$ with $O(d)$ work. When $\mathbf{x}_d \notin [\mathbf{1}_d, \mathbf{n}_d]$, we assume implicitly that $A[\mathbf{x}_d]$ contains a *distinct* instance of a special endmarker $\$ \notin \Sigma$. For the sake of presentation, let $n_d = n_{\max} = \max\{n_k : k \in [1, d]\}$ and $d > 1$. Our techniques work also for $n_d \neq n_{\max}$ or $d = 1$.

A somewhat "relaxed" definition of trie on sequences that are drawn from a *parametric* alphabet Γ and satisfy a generic equivalence relation $\equiv$ is given below. We require that $\alpha[1 : n] \equiv \beta[1 : n]$ if and only if $a[i] \equiv b[i]$ for each $i \in [1, n]$. Obviously, two equivalent sequences might have different characters in homologous positions. We use the relaxed notion that α is prefix, suffix or subsequence of β, meaning that α is equivalent ($\equiv$) to a prefix, suffix or contiguous subsequence of β, respectively.

Given a set Y of sequences $y_1, y_2, \ldots, y_s \in \Gamma^*$ such that no y_i is prefix of any other y_j (for $1 \leq i, j \leq s$ and $i \neq j$), a *parametric compacted trie* CT built on Y under $\equiv$ satisfies: **(C0)** No unary nodes are allowed, except possibly for the root. **(C1)** Each arc is labeled with a nonempty subsequence $y_i[j : k] \in \Gamma^*$, for some $i \in [1, s]$ and $j, k \in [1, |y_i|]$. **(C2)** Sibling arcs are labeled with sequences from Γ^* that start with nonequivalent characters under $\equiv$. **(C3)** For each sequence $y_i \in Y$, with $i \in [1, s]$, there is exactly one leaf v such that the concatenation, say γ, of the labels along the path connecting the root to v satisfies $\gamma \equiv y_i$ (not necessarily $\gamma = y_i$). That leaf v is labeled with y_i. Moreover, if $y_i[j : k]$ labels an arc e, then e must be along the path from the root to v.

We adopt the same definition for *locus* and *extended locus* in parametric compacted tries as in suffix trees [9] (which can be seen as a special case with $\equiv$ being $=$). One can show that, for fixed $\equiv$, a nonempty sequence $\gamma \in \Gamma^*$ is prefix of at least one sequence in Y if, and only if, γ has extended locus v in CT.

The elegant CRCW-PRAM algorithms of Apostolico *et al.* [1, 7] can be easily adapted to be used as "black boxes" to build and query a parametric compacted trie CT:

- *Refine*: The construction of CT is done with Algorithm *Refine*, which has three main *inputs* for given Γ and $\equiv$. (i) A set Y of $s \geq 1$ non-equivalent sequences $y_1, y_2, \ldots, y_s \in \Gamma^*$, assuming w.l.o.g. that they have the same length n, and $\log n$ is an integer (if not, we pad the sequences in Y with instances of a new character not in Γ). (ii) A $Q \times Q$ non-initialized matrix BB, called *Bulletin Board*, for a parameter $Q \geq \max(s, n)$ to be specified. (iii) A function *Name* on which all the computation hinges: Its argument is a subsequence $y_i[j: j + 2^r - 1]$ of length a power of two (with $i \in [1, s]$, $j \in [1, n]$ and $j + 2^r - 1 \leq n$ for some $r \geq 0$), and it returns an integer in $[1, Q]$ called the *name* of $y_i[j: j + 2^r - 1]$. The property underlying *Name* is that two sequences of the *same* length are constrained to have the same name if, and only if, they are equivalent under $\equiv$. The *output* of *Refine* is a sequence of trees, called *refinement trees* $D^{(r)}$, for $r = \log n, \log n - 1, \ldots, 0$. The final tree $D^{(0)}$ is isomorphic to CT. Algorithm *Refine* requires $O(\log n)$ time with s processors on an Arbitrary CRCW PRAM, with the proviso that each call to *Name* is executed in $O(1)$ work.

- *ExLocus:* The query of CT with a sequence $\gamma[1: m] \in \Gamma^*$, in order to find out whether or not γ is prefix of any sequence in Y, is done with Algorithm *ExLocus*. Let $r_0, r_1, \ldots, r_f$, with $f \leq \lfloor \log m \rfloor$, be the decreasing sequence of integers such that $m = \sum_{j=0}^{f} 2^{r_j}$. Consider the partition of γ into f contiguous sequences $\gamma_0 \gamma_1 \ldots \gamma_f = \gamma$. Each γ_j has length 2^{r_j}, with $0 \leq j \leq f$, and is given by $\gamma[\sigma_{j-1} + 1: \sigma_j]$, where $\sigma_{-1} = 0$ and $\sigma_j = 2^{r_0} + \cdots + 2^{r_j}$. The *inputs* to *ExLocus* are (i) all the refinement trees $D^{(r)}$ produced by *Refine*, and (ii) the *consistent* names η_j of the sequences γ_j, for $j \in [0, f]$. The name η_j of γ_j is said to be *consistent* if, and only if, $\eta_j = Name(\beta)$ for every sequence β such that $\gamma_j \equiv \beta$, with β subsequence of sequences in Y. The *output* of *ExLocus* is the extended locus (if any) of γ in CT, implying that γ is prefix of some sequence in Y. Algorithm *ExLocus* requires $O(\log m)$ time and work on an Arbitrary CRCW PRAM, assuming to have the consistent names η_j.

3 Sources, Embeddings, and the "Raw" Index $\mathcal{R}_A$

Given $A = A[1_d, \mathbf{n}_d]$ and $B = B[1_r, \mathbf{m}_r]$, with $d \geq r \geq 1$, we present now an equivalent definition of the problem of searching for B^* in A that allows us to use the "black boxes" *Refine* and *ExLocus*. We need the related notions of source and embedding of a matrix to express the condition by which two matrices have the same elements, independently of their shape and dimension.

An s-dimensional matrix $D[1_s, \mathbf{q}_s]$ is called *the source* of a t-dimensional matrix $C[1_t, \boldsymbol{\ell}_t]$, with $t \geq s \geq 1$, if and only if (1) either $\boldsymbol{\ell}_t = 1_t$ and thereby $s = 1$ and $\mathbf{q}_s = 1_1$, or $\boldsymbol{\ell}_t \neq 1_t$ and thereby $\boldsymbol{\ell}_t \in \mathcal{E}[\mathbf{q}_s]$ with $q_i > 1$ for all $i \in [1, s]$; (2) D contains the same elements as C. That is, for each $\mathbf{y}_s \in [1_s, \mathbf{q}_s]$, we have $D[\mathbf{y}_s] = C[\hat{\mathbf{y}}_t]$, where $\hat{\mathbf{y}}_t \in \mathcal{E}[\mathbf{y}_s]$ has its i-th component enforced to be 1 whenever $\ell_i = 1$, for $i \in [1, t]$.

We say that C is a t-dimensional *embedding* of source matrix D. We also distinguish between *type-I* embeddings when $\ell_t = 1$, and *type-II* otherwise (i.e., $\ell_t = q_s$). As an example, matrix $D[1:3, 1:2]$ in Fig. 1a is the source of matrix $A[1:3, 2:2, 1:2] = C$ in Fig. 1c, and C is a 3-dimensional embedding of D, of type-II. Notice that there is only *one* source D for a matrix C. In contrast, except for the trivial case $\ell_t = 1_t$, there are $\binom{t}{s}$ *different* t-dimensional embeddings.

Fact 1. *Given* $B = B[1_r, \mathbf{m}_r]$, B^* *occurs in position* $\mathbf{x}_d$ *of* A *if, and only if, the source of* B *has at least one* d-dimensional embedding occurring in $\mathbf{x}_d$.

Following Fact 1, we discuss only the case in which $m_i > 1$ for $1 \leq i \leq r$. That is, B is its own source.

The "Raw" Index. We present now a natural generalization to d-dimensional matrices of notions introduced in previous papers [3, 4]. The purpose is to define a "raw" index $\mathcal{R}_A$ and give a first attempt in solving our problem with standard techniques. This intermediate step is then used to define the more powerful index $\mathcal{H}_A$ in Section 4.

Consider the alphabet Σ, and fix $\ell_{d-1} \in [1_{d-1}, \mathbf{n}_{d-1}]$. The new alphabet $\Sigma^{\ell_{d-1}}$ contains ℓ_{d-1}-*blocks* of *shape* ℓ_{d-1}, where each ℓ_{d-1}-block (or, simply block) represents a d-dimensional matrix of shape $\ell_1 \times \ell_2 \times \cdots \times \ell_{d-1} \times 1$ having entries from Σ. When $\ell_{d-1} = 1_{d-1}$, we adopt the convention that $\Sigma^{\ell_{d-1}} = \Sigma$, i.e., each character in Σ is seen as a d-dimensional matrix of shape $1 \times \cdots \times 1$. Two ℓ_{d-1}-blocks α and β are equivalent (shortly, $\equiv_B$) if and only if they represent two equal d-dimensional matrices, of shape $\ell_1 \times \cdots \times \ell_{d-1} \times 1$. The concatenation of ℓ_{d-1}-blocks is analogous to the one of characters of Σ, producing ℓ_{d-1}-*block strings*. There is a one-to-one correspondence between ℓ_{d-1}-block strings of length ℓ and d-dimensional matrices of shape $\ell_1 \times \cdots \times \ell_{d-1} \times \ell$.

The definition of compacted tries for ℓ_{d-1}-block strings under $\equiv_B$ easily follows from the one of parametric compacted tries given in Section 2, by setting $\Gamma = \Sigma^{\ell_{d-1}}$ and letting $\equiv$ be $\equiv_B$. Thus the we can define the *raw index* $\mathcal{R}_A$ as the forest of compacted tries $CT^{\ell_{d-1}}$, where ℓ_{d-1} ranges in $[1_{d-1}, \mathbf{n}_{d-1}]$ and $CT^{\ell_{d-1}}$ is defined as follows. For each $\mathbf{x}_{d-1} \in [1_{d-1}, \mathbf{n}_{d-1}]$ and $i \leq j$, let $M(\mathbf{x}_{d-1}, i, j)$ be the ℓ_{d-1}-block string of length $j - i + 1$ representing the submatrix $A[\mathbf{y}_d, \mathbf{z}_d]$, where $\mathbf{y}_d = (\mathbf{x}_{d-1}, i) \in [1_d, \mathbf{n}_d]$ and $\mathbf{z}_d = (\mathbf{x}_{d-1} + \ell_{d-1} - 1_{d-1}, j) \in [\mathbf{y}_d, 2\mathbf{n}_d]$. Let $Y^{\ell_{d-1}}$ be the set of ℓ_{d-1}-block suffixes $M(\mathbf{x}_{d-1}, x_d, x_d + n_d)$, where $\mathbf{x}_d$ ranges in $[1_d, \mathbf{n}_d]$. Notice that $Y^{\ell_{d-1}}$ contains $N = n_1 \cdots n_d$ distinct block strings of the same length $n_d + 1$ (they all end with \$s by our assumption that $A[\mathbf{x}_d] = \$$ for $\mathbf{x}_d \notin [1_d, \mathbf{n}_d]$). Then, $CT^{\ell_{d-1}}$ is the compacted trie over $\Sigma^{\ell_{d-1}}$ representing the block strings of $Y^{\ell_{d-1}}$ under $\equiv_B$. Notice that there are N/n_d such tries in $\mathcal{R}_A$, and thus $\mathcal{R}_A$ can be stored into $O(N^2/n_d)$ space. Moreover, every submatrix of A is represented by a path in one compacted trie of $\mathcal{R}_A$:

Lemma 2. *Let* C *be a matrix of shape* $\ell_1 \times \ell_2 \times \cdots \times \ell_d$, *with entries defined over the alphabet* Σ, *and let* c *be its* ℓ_{d-1}-*block string. Then* C *is a submatrix of* A *if, and only if,* c *has extended locus in the compacted trie* $CT^{\ell_{d-1}}$ *of* $\mathcal{R}_A$.

We remark that $\mathcal{R}_A$ can be built in $O(dN^2/n_d \log N)$ work by using Algorithm *Refine* mentioned in Section 2.

We could use $\mathcal{R}_A$ to search for B^* by considering all d-dimensional embeddings of B, implicitly represented by B^* (Fact 1). For each embedding of B, we could search for the extended locus of its corresponding block string in the appropriate trie in $\mathcal{R}_A$ by Lemma 2. We point out (leaving the details to the reader) that such a strategy would need $\delta = \binom{d}{r} = O(N/n_d)$ queries, each one taking at least $O(dM)$ work. It would be better if we had an index storing all the submatrices of A in terms of their sources. Then, one would check whether the source of B, i.e., B itself, is equal to any of the sources stored in the index. As shown below, we can efficiently do it.

4   The "Heterogeneous" Index $\mathcal{H}_A$

To point out the intuitive idea behind the new data structure, consider the matrices $A[1:3, 2:2, 1:2]$ (shown in italic in Fig. 1c) and $A[1:1, 2:4, 1:2]$ (shown in italic in Fig. 1d). For the sake of discussion, assume that $d = 4$ and thus they are part of a 4-dimensional matrix, say $\widehat{A}$. With such a convention, they have shape $3 \times 1 \times 2 \times 1$ and $1 \times 3 \times 2 \times 1$ and are represented by two distinct blocks $\hat{\alpha}$ and $\hat{\beta}$ of shape $(3, 1, 2)$ and $(1, 3, 2)$, respectively. In $\mathcal{R}_{\widehat{A}}$, the blocks $\hat{\alpha}$ and $\hat{\beta}$ would appear on paths of two *distinct* compacted tries, $CT^{(3,1,2)}$ and $CT^{(1,3,2)}$, respectively. However, those two matrices have the same 3×2 source matrix, shown in Fig. 1a. Therefore $\hat{\alpha}$ and $\hat{\beta}$ contain the same elements and should appear on the same path of some suitably defined compacted trie. Below we formalize such a concept, letting $\mathcal{M} = \mathcal{M}(\mathbf{n}_{d-1})$ be the set of tuples $\mathbf{m}_j \in [2, n_{i_1}] \times [2, n_{i_2}] \times \cdots \times [2, n_{i_j}]$ such that $1 \leq i_1 \leq \cdots \leq i_j \leq d$ for some $j < d$ and all components in $\mathbf{m}_j$ are greater than one. We add to $\mathcal{M}$ the special tuple $\mathbf{m}_0 = \mathbf{1}_1$, which is the only one containing 1s.

Given $\mathbf{m}_j \in \mathcal{M}$, the alphabet $\Sigma[\mathbf{m}_j] = \cup_{\boldsymbol{\ell}_{d-1} \in [\mathbf{1}_{d-1}, \mathbf{n}_{d-1}] \cap \mathcal{E}[\mathbf{m}_j]} \Sigma^{\boldsymbol{\ell}_{d-1}}$ contains the *heterogeneous* $[\mathbf{m}_j]$-*blocks* (or, *h-blocks*). They are the $\boldsymbol{\ell}_{d-1}$-blocks from the alphabets introduced in Section 3, representing d-dimensional matrices whose sources are *always* of shape $m_1 \times m_2 \times \ldots \times m_j$. That condition is expressed by taking $\boldsymbol{\ell}_{d-1} \in [\mathbf{1}_{d-1}, \mathbf{n}_{d-1}] \cap \mathcal{E}[\mathbf{m}_j]$. (Notice that $\Sigma[\mathbf{m}_0] = \Sigma$, since $\boldsymbol{\ell}_{d-1} = \mathbf{1}_{d-1}$.) The rationale behind such a definition is that, if we pick any two blocks from $\Sigma[\mathbf{m}_j]$ and ignore the 1s in their shapes, we make the matrices represented by such two blocks become of the same shape, which is exactly the one of their source matrices. Given two h-blocks $\alpha, \beta \in \Sigma[\mathbf{m}_j]$, we say that α is equivalent to β (shortly, $\alpha \equiv_H \beta$) if and only if the d-dimensional matrices they represent as blocks have the *same source*.

The operation of concatenation of h-blocks is analogous to the one for strings, producing $[\mathbf{m}_j]$-*block strings*. Two $[\mathbf{m}_j]$-block strings $c_1[1:n]$ and $c_2[1:n]$ are equivalent (shortly, $c_1 \equiv_H c_2$) if and only if $c_1[i] \equiv_H c_2[i]$ for each $i \in [1, n]$. We remark that there is no one-to-one correspondence between matrices and h-block strings. However, for any two block strings α, β representing two matrices Z_1, Z_2, we prove in [5] that $\alpha \equiv_H \beta$ when α, β are seen as h-block strings if, and only if, Z_1 and Z_2 have the same source. Therefore, block strings have been introduced for representing matrices, whereas h-block strings are introduced for labeling compacted tries.

The definition of *compacted trie* for h-block strings under $\equiv_H$ easily follows from the one of parametric compacted trie given in Section 2, by setting $\Gamma = \Sigma[\mathbf{m}_j]$ and letting $\equiv$ be $\equiv_H$. Thus for a fixed $\mathbf{m}_j \in \mathcal{M}$, take $Y[\mathbf{m}_j] = \cup_{\ell_{d-1} \in [1_{d-1}, \mathbf{n}_{d-1}] \cap \mathcal{E}[\mathbf{m}_j]} Y^{\ell_{d-1}}$ and let $CT[\mathbf{m}_j]$ be the heterogeneous compacted trie for $Y[\mathbf{m}_j]$ under $\equiv_H$ (such a trie is over the alphabet $\Sigma[\mathbf{m}_j]$). The sequence of labels found along a path in $CT[\mathbf{m}_j]$ represents now a sequence of matrices of various shapes. That motivates the use of the term "heterogeneous," since it is not clear how to concatenate those heterogeneous matrices. Indeed, we may think of $CT[\mathbf{m}_j]$ as obtained by coalescing the compacted tries $CT^{\ell_{d-1}}$ together, for all $\ell_{d-1} \in [1_{d-1}, \mathbf{n}_{d-1}] \cap \mathcal{E}[\mathbf{m}_j]$, under the new relation $\equiv_H$. However, we prove in [5] that every sequence of heterogeneous arcs' labels encountered along a path from the root always corresponds to some submatrix of A via $\equiv_H$!

We define the *heterogeneous index* $\mathcal{H}_A$ for a matrix A as the forest of the heterogeneous compacted tries $CT[\mathbf{m}_j]$, for all $\mathbf{m}_j \in \mathcal{M}$. In [5], we show that $\mathcal{H}_A$ can be stored in $O(N^2/n_d)$ space.

Now, our problem of checking whether B^* occurs in A can be solved by querying $\mathcal{H}_A$ at most *twice* rather than $\delta = \binom{d}{r}$ times, as stated in Theorem 3. Assume $\mathbf{m}_r \in \mathcal{M} - \{\mathbf{m}_0\}$ (the case $\mathbf{m}_r = \mathbf{m}_0$ is easy), and consider any two d-dimensional embeddings $C_1[1_d, \mathbf{a}_d]$ and $C_2[1_d, \mathbf{b}_d]$ of $B[1_r, \mathbf{m}_r]$ such that C_1 is of type-I (such an embedding will exist only when $r < d$) and C_2 is of type-II (such an embedding exists since $\mathbf{m}_r \neq \mathbf{m}_0$). Notice that $\mathbf{a}_d, \mathbf{b}_d \in \mathcal{E}[\mathbf{m}_r]$ by definition of source. Let γ_1 be the $\mathbf{a}_{d-1}$-block representing C_1, seen also as an $[\mathbf{m}_r]$-block. Similarly for C_2, define the $\mathbf{b}_{d-1}$-block string γ_2, seen also as an $[\mathbf{m}_{r-1}]$-block string of length $m_r > 1$.

Theorem 3. *With the aforementioned conventions, B^* occurs in A if, and only if, either γ_2 has extended locus in $CT[\mathbf{m}_{r-1}]$, or γ_1 is the initial block labeling an arc that departs from the root of $CT[\mathbf{m}_r]$ (when $r < d$).*

5 Parallel Construction of Index $\mathcal{H}_A$ — An Outline

Building $\mathcal{H}_A$ needs N^2/n_d processors, indexed from 1 to N^2/n_d and partitioned into N/n_d *groups*. Each group, $G^{\ell_{d-1}}$, contains N processors and corresponds to a distinct choice of ℓ_{d-1} into the range $[1_{d-1}, \mathbf{n}_{d-1}]$. The algorithm follows the two-phase scheme in [1], called *naming* and *refining* (we assume w.l.o.g. that $n_d + 1$ is a power of two).

Naming requires $O(\log N)$ time to assign integers to some suitably chosen set $\mathcal{B}$ of h-block strings of length a power of two, obtained from $Y[\mathbf{m}_j]$ for all $\mathbf{m}_j \in \mathcal{M}$. Processing matrix A has the main goal to allow constant time calls to a function $Name_H : \mathcal{B} \longrightarrow [1, N^2/n_d]$ satisfying Property 4 (which is a restatement of the constraints given for the function $Name$ in Section 2).

Property 4. *For any two h-block strings $\alpha, \beta \in \mathcal{B}$, we have $Name_H(\alpha) = Name_H(\beta)$ and $|\alpha| = |\beta|$ if, and only if, $\alpha \equiv_H \beta$.*

Refining takes $O(d \log N)$ time and $\sum_{\mathbf{m}_j \in \mathcal{M}} |Y[\mathbf{m}_j]| = N^2/n_d$ processors (note that the size $|Y[\mathbf{m}_j]|$ depends on $\mathbf{m}_j$). It uses $Name_H$ to execute a different instance of Algorithm *Refine* for each $CT[\mathbf{m}_j] \in \mathcal{H}_A$ (see Section 2). All those instances share a new Bulletin Board BB, which is of size $Q \times Q$, where $Q = N^2/n_d$. For a given $CT[\mathbf{m}_j]$, the input parameters $\{\Gamma, \equiv, Y, s, n\}$ of *Refine* are $\{\Sigma[\mathbf{m}_j], \equiv_H, Y[\mathbf{m}_j], |Y[\mathbf{m}_j]|, n_d + 1\}$. The processors are clustered together, processors in $G^{\ell_{d-1}}$ being in the cluster for $CT[\mathbf{m}_j]$, for all $\ell_{d-1} \in \mathcal{E}[\mathbf{m}_j]$. Clusters are created using BB on the sequence of integers given by each ℓ_{d-1}.

Thus refining is a rather standard application of ideas in [1] and notions in Section 4. However the computation of function $Name_H$ is not a straightforward extension of ideas in [8].

6 Computation of Function $Name_H$

Define $\mathcal{B}(\ell_{d-1})$ as the set of the ℓ_{d-1}-block strings $M(\mathbf{x}_{d-1}, x_d, x_d + 2^g - 1)$, for all $\mathbf{x}_d \in [\mathbf{1}_d, \mathbf{n}_d]$ and $g \in [0, \log(n_d + 1)]$ such that $x_d + 2^g - 1 \leq 2n_d$ (see Section 3). Thus, $\mathcal{B} = \cup_{\ell_{d-1}} \mathcal{B}(\ell_{d-1})$ contains the h-block substrings of length a power of two taken from $\cup_{\mathbf{m}_j \in \mathcal{M}} Y[\mathbf{m}_j]$, and $Name_H : \mathcal{B} \longrightarrow [1, N^2/n_d]$.

The set $\mathcal{B}$ is partitioned into $\mathcal{B}_1 = \cup_{\ell_{d-1}} \mathcal{B}_1(\ell_{d-1})$ and $\mathcal{B}_2 = \cup_{\ell_{d-1}} \mathcal{B}_2(\ell_{d-1})$, where $\mathcal{B}_1(\ell_{d-1})$ contains the ℓ_{d-1}-blocks $M(\mathbf{x}_{d-1}, x_d, x_d)$ and $\mathcal{B}_2(\ell_{d-1})$ contains the ℓ_{d-1}-block strings $M(\mathbf{x}_{d-1}, x_d, x_d + 2^g - 1)$ of length 2^g ($1 \leq g \leq \log(n_d+1)$). The names of the blocks in $\mathcal{B}_1$ are used to compute the names of the block strings in $\mathcal{B}_2$, which have length at least two. Thus we can "enforce" Property 4 independently on $\mathcal{B}_1$ and $\mathcal{B}_2$.

Due to our convention that $A[\mathbf{x}_d] = \$$ when $\mathbf{x}_d \notin [\mathbf{1}_d, \mathbf{n}_d]$, we need the auxiliary set $\mathcal{B}' = \cup_{\ell_{d-1}} \mathcal{B}'(\ell_{d-1})$, where $\mathcal{B}'(\ell_{d-1})$ contains all ℓ_{d-1}-block strings $M(\mathbf{x}_{d-1}, x_d, x_d + 2^g - 1)$ for which $\mathbf{x}_d \notin [\mathbf{1}_d, \mathbf{n}_d]$ and $g \geq 0$. Since $\mathcal{B}' \cap \mathcal{B} = \emptyset$, their name is assigned *a-priori* as the distinct instance of $\$ = A[\mathbf{x}_d]$, which is coded as a distinct integer in $[N^2/n_d + 1, Q]$, where $Q = O(N^2/n_d)$. This way $Name_H$ obviously satisfies Property 4, when its domain is $\mathcal{B} \cup \mathcal{B}'$, its range is $[1, Q]$, and all names of the block strings in $\mathcal{B}$ are in $[1, N^2/n_d]$.

Therefore, we focus our attention on $\mathcal{B}$, introducing the notion of **canonical decomposition** for a block string $\alpha \in \mathcal{B}(\ell_{d-1})$.

- **Case $\alpha \in \mathcal{B}_1(\ell_{d-1})$** : If $\ell_{d-1} = \mathbf{1}_{d-1}$ then α is a character from Σ and hence its own canonical decomposition. Otherwise, assume that ℓ_k is the rightmost component of ℓ_{d-1} strictly greater than one (i.e., $\ell_{k+1} = \cdots = \ell_{d-1} = 1$). Let 2^q be the largest power of two such that $2^q < \ell_k$, and ℓ'_{d-1} be the same as ℓ_{d-1}, except that $\ell'_k = 2^q$. Now, consider the matrix, say $C[1:\ell_1, \ldots, 1:\ell_{d-1}, 1]$, corresponding to α. Take C_1 as the submatrix obtained from C by reducing the k-th dimension of C from "$1:\ell_k$" to "$1:2^q$" and, similarly, C_2 as the submatrix obtained by reducing the k-th dimension of C to "$\ell_k - 2^q:\ell_k$." Then the blocks $\alpha_1 \in \mathcal{B}_1(\ell'_{d-1})$ and $\alpha_2 \in \mathcal{B}_1(\ell'_{d-1}) \cup \mathcal{B}'(\ell'_{d-1})$ corresponding to C_1 and C_2, respectively, form the canonical decomposition of α. (Notice that, if $\alpha \in \Sigma[\mathbf{m}_j]$, then both α_1 and α_2 are either in $\Sigma[\mathbf{m}_{j-1}]$ or in $\Sigma[\mathbf{m}'_j]$, where $\mathbf{m}'_j = (\mathbf{m}_{j-1}, 2^q)$.)

- **Case** $\alpha \in \mathcal{B}_2(\ell_{d-1})$: The length of α is $2^g > 1$, for some $g \in [1, \log(n_d+1)]$. This case is managed as in [8] for strings. The canonical decomposition of α is given by its block prefix α_1 and its block suffix α_2, both of length 2^{g-1}. Note that $\alpha_1 \in \mathcal{B}(\ell_{d-1})$ and $\alpha_2 \in \mathcal{B}(\ell_{d-1}) \cup \mathcal{B}'(\ell_{d-1})$ (i.e., if $\alpha \in (\Sigma[\mathbf{m}_j])^*$ then $\alpha_1, \alpha_2 \in (\Sigma[\mathbf{m}_j])^*$).

Lemma 5. *Consider two block strings* $\alpha \in \mathcal{B}(\ell_{d-1})$ *and* $\beta \in \mathcal{B}(\hat{\ell}_{d-1})$, *where* $\ell_{d-1}, \hat{\ell}_{d-1} \in \mathcal{E}[\mathbf{m}_j] - \{\mathbf{1}_{d-1}\}$. *Let* (α_1, α_2) *and* (β_1, β_2) *be their canonical decompositions, respectively. Then* $\alpha \equiv_H \beta$ *if and only if* $\alpha_i \equiv_H \beta_i$ *for* $i = 1, 2$.

The canonical decomposition of the block string α induces a **decomposition tree** T_α, which is a complete binary tree having the internal nodes labeled with integer tuples from $\mathcal{M}$, and the leaves labeled with characters from Σ. Its depth is $O(\log S)$, where S is the size of the matrix corresponding to α. The tree T_α is defined recursively as follows. If $\alpha \in \Sigma$, then T_α is composed of only one node labeled with the character α. If $\alpha \in \Sigma[\mathbf{m}_j]$ for some $\mathbf{m}_j \in \mathcal{M}$, then the root of T_α is labeled with the tuple $\mathbf{m}_j$, and its left and right children are the roots of T_{α_1} and T_{α_2} respectively, where (α_1, α_2) is the canonical decomposition of α.

There are two remarks on the block strings associated with the internal nodes of T_α at depth ω from the root, assuming that α has shape ℓ_{d-1} and length 2^c for some $c \geq 0$. **(a)** If $0 \leq \omega < c$ then those block strings have all the same shape ℓ_{d-1} and the same length $2^{c-\omega}$. **(b)** If $\omega \geq c$ then they are all blocks having the same shape, $\hat{\ell}_{d-1} = (\ell_{h-1}, m, \mathbf{1}_{d-h-1})$, for some $h \in [1, d-1]$ such that $\ell_h > 1$ and $m = \min(2^g, \ell_h)$ for some g in $[1, \lceil \log \ell_h \rceil]$.

We say that two decomposition trees T_α and T_β are equivalent (shortly, $T_\alpha \sim T_\beta$) if, and only if, they have the same shape and their homologous nodes contain the same label.

Lemma 6. *Consider two block strings* $\alpha, \beta \in \mathcal{B}$. *Let* T_α *and* T_β *be their decomposition trees, respectively. Then* $\alpha \equiv_H \beta$ *if, and only if,* $T_\alpha \sim T_\beta$ *and* $|\alpha| = |\beta|$.

Lemma 6 reduces computing $Name_H$ to the easier task of assigning equal integers to equivalent trees in the forest $\mathcal{T} = \{T_\alpha : \alpha \in \mathcal{B}\}$. Note that the set $\mathcal{T}$ is closed under the canonical decomposition, i.e., $T_\alpha \in \mathcal{T}$ implies $T_{\alpha_1}, T_{\alpha_2} \in \mathcal{T}$, where (α_1, α_2) is the canonical decomposition of α. Or put in other words, any subtree of $T_\alpha \in \mathcal{T}$ belongs to $\mathcal{T}$. From now on, we make no difference between the name of α and the name of T_α.

Sketch of the computation. The computation is based upon a bottom-up visit of the decomposition trees in $\mathcal{T}$, although they are not explicitly built, but introduced only for the sake of presentation and analysis of our algorithms. During the visit, the name for T_α is assigned by using those for T_{α_1} and T_{α_2}, where (α_1, α_2) is the canonical decomposition of α. Essential is a set of non-initialized $Q \times Q$ Bulletin Boards BB_g^h (for $h \in [1, d]$ and $g \in [0, \log(n_d + 1)]$) to assign equal integers to equal pairs of names (the overall space can be significantly reduced as shown in [1]). The Bulletin Boards are unusually employed here as an *asynchronous* mechanism to exchange "global information" among "active" groups of (synchronous) processors in the different rounds of the computation. Moreover, a careful scheduling of the processors is required to ensure that the names for $T_{\alpha_1}, T_{\alpha_2}$ can be correctly retrieved in some memory location.

Consider the generic group $G^{\ell_{d-1}}$. Each of the N processors in $G^{\ell_{d-1}}$ is assigned to exactly one position $x_d \in [1_d, n_d]$ of A. We use N/n_d two-dimensional arrays $ID^{\ell_{d-1}}[g, x_d]$ (where $\ell_{d-1} \in [1_{d-1}, n_{d-1}]$, $x_d \in [1_d, n_d]$ and $0 \le g \le \log(n_d + 1)$), with the convention that $ID^{\ell_{d-1}}[g, x_d]$ contains the unique integer $A[x_d] \in [N^2/n_d + 1, Q]$ whenever $x_d \notin [1_d, n_d]$.

The computation consists of $d + 1$ rounds. Rounds $k = 0, \ldots, d - 1$ process the blocks in $\mathcal{B}_1$, and the last round $k = d$ processes the block strings in $\mathcal{B}_2$.

Round $k = 0$ assigns names to the characters in the leaves of $\mathcal{T}$, storing the name of $A[x_d]$ into $ID^{1_{d-1}}[0, x_d]$. All groups are idle, except for $G^{1_{d-1}}$, which becomes idle for the remainder of the rounds. So, let us assume $\ell_{d-1} \neq 1_{d-1}$. Each $G^{\ell_{d-1}}$ initializes $ID^{\ell_{d-1}}[0, x_d] = ID^{1_{d-1}}[0, x_d]$ for all $x_d \in [1_d, n_d]$.

Round k, for $1 \le k < d$, is divided into steps $g = 1, 2, \ldots, \lceil \log n_k \rceil$, which are executed only by the processors in the active groups. Group $G^{\ell_{d-1}}$ is *active* in round k when the k-th component, ℓ_k, in ℓ_{d-1} is strictly greater than 1. During step g of round k, the active group $G^{\ell_{d-1}}$ assigns names to the $\hat{\ell}_{d-1}$-blocks $\alpha = M(x_{d-1}, x_d, x_d)$ of shape $\hat{\ell}_{d-1} = (\ell_{k-1}, \min(2^g, \ell_k), 1_{d-k-1})$, according to remark (b) on T_α. Those names are then stored into $ID^{\ell_{d-1}}[s, x_d]$, where $s = g \bmod \lceil \log \ell_k \rceil$. Consider the canonical decomposition (α_1, α_2) of α.

Lemma 7. *At round k and step g of active group $G^{\ell_{d-1}}$, take y_d as x_d except that its kth component is $y_k = x_k + \min(2^g, \ell_k) - 2^q$, where q is the largest integer such that $2^q < \min(2^g, \ell_k)$. The names of T_{α_1} and T_{α_2} needed for assigning name to T_α can be found into $ID^{\ell_{d-1}}[q, x_d]$ and $ID^{\ell_{d-1}}[q, y_d]$, respectively.*

At the end of round $d - 1$, each group $G^{\ell_{d-1}}$ has computed the names of the blocks in $\mathcal{B}_1(\ell_{d-1})$ and has stored them into $ID^{\ell_{d-1}}[0, x_d]$ for all $x_d \in [1_d, n_d]$.

The last round $k = d$ is divided into steps $g = 1, 2, \ldots, \log(n_d + 1)$ and resembles the doubling technique by Karp *et al.* [8]. During step $g = 1, \ldots, \log(n_d + 1)$, all groups are active and $G^{\ell_{d-1}}$ computes the names of block strings in $\mathcal{B}_2(\ell_{d-1})$ of length 2^g, storing them into $ID^{\ell_{d-1}}[g, x_d]$ for all x_d. So, at the end of all rounds, the arrays ID have become a tabular implementation of function $Name_H$. More technical discussion is given in [5].

Theorem 8. *The text preprocessing of an $N = n_1 \times n_2 \times \cdots \times n_d$ matrix A can be correctly performed in $O(d \log N)$ time with N^2/n_d processors.*

7 Query of the Pattern

To find whether B^* occurs in A, it suffices to search only for the two embeddings C_1 and C_2 of B indicated by Theorem 3. Here we discuss only the case for C_2.

Let γ_2 be its $\mathbf{b}_{d-1}$-block string, where $\mathbf{b}_{d-1} \in \mathcal{E}[\mathbf{m}_{r-1}]$. Also this computation is organized in two phases: *naming* and *searching*.

Naming is still the major issue, since searching is done through a call to Algorithm *ExLocus*. We partition γ_2 as in Section 2 (with $m = m_r$), producing the set $\mathcal{P} = \{\beta_0, \beta_1, \ldots, \beta_f\}$ of block strings, with $f \leq \lfloor \log m_r \rfloor$. Then we extend the range of function $Name_H$ to the block strings in $\mathcal{P}$, ensuring consistency under Property 4. Indeed, each block string $\beta \in \mathcal{P}$, of length 2^c for some c, is assigned a name using the decomposition tree T_β (the definition of canonical decomposition and decomposition tree are analogous to the ones in Section 6, and remarks (a)–(b) still hold). To ensure consistent names, processors read from the BB's information computed during the preprocessing (they can detect garbage entries). If a new name must be used, then B^* cannot occur in A.

The decomposition tree T_β, for each $\beta \in \mathcal{P}$, has $V_\beta = O(m_1 \cdots m_{r-1} 2^c)$ vertices and $O(\log V_\beta) = O(\log M)$ height, for some $c \geq 0$. Consider the tree T obtained by connecting the roots of the trees T_β for all $\beta \in \mathcal{P}$ to a new root. It has still $O(\log M)$ height and $\Sigma_{\beta \in \mathcal{P}} V_\beta = O(M)$ nodes. The computation of names becomes a standard parallel tree computation on T (e.g., see [7]) and it can be done in $O(d \log M)$ time with $M/\log M$ processors. Finally, the execution of *ExLocus* takes $O(d \log M)$ time and work.

Theorem 9. *Having preprocessed the text A, whether a pattern B occurs as B^* in A can be checked on-line in $O(d \log M)$ time with $M/\log M$ processors.*

References

1. A. Apostolico, C. Iliopoulos, G. Landau, B. Schieber, and U. Vishkin. Parallel construction of a suffix tree with applications. *Algorithmica*, 3:347–365, 1988.
2. M.J. Fischer and M.S. Paterson. String matching and other products. In R.M. Karp, editor, *Complexity of Computation*, 113–125, Providence, RI., 1974. SIAM-AMS, American Mathematical Society.
3. R. Giancarlo. An index data structure for matrices, with applications to fast two-dimensional pattern matching. In *Proc. WADS*, 337–348. Springer-Verlag, 1993.
4. R. Giancarlo and R. Grossi. Parallel construction and query of suffix trees for two-dimensional matrices. In *Proc. 5th SPAA*, 86–97. ACM, 1993.
5. R. Giancarlo and R. Grossi, Multi-dimensional pattern matching with dimensional wildcards: data structures and optimal on-line search algorithms. AT&T Bell Labs. Technical Memorandum TM-11272-941003-18, 1994.
6. R. Giancarlo and R. Grossi, On the construction of classes of suffix trees for square matrices: algorithms and applications. In *Proc. 22nd ICALP*. LNCS, 1995.
7. J. JáJá. *An Introduction to Parallel Algorithms*. Addison-Wesley, Reading, MA., 1992.
8. R. Karp, R. Miller, and A. Rosenberg. Rapid identification of repeated patterns in strings, arrays and trees. In *Proc. 4th STOC*, 125–136. ACM, 1972.
9. E.M. McCreight. A space economical suffix tree construction algorithm. *J. of ACM*, 23:262–272, 1976.

Minimizing Phylogenetic Number to find Good Evolutionary Trees

Leslie Ann Goldberg[1], Paul W. Goldberg[1], Cynthia A. Phillips[1], Elizabeth Sweedyk[2] and Tandy Warnow[3]

[1] Sandia National Laboratories, MS 1110, P.O. Box 5800, Albuquerque, NM 87185, U.S.A.

[2] 593 Soda Hall, Dept. of Computer Science, UC Berkeley, Berkeley, CA 94720, U.S.A.

[3] Dept. of Computer and Information Science, University of Pennsylvania. Philadelphia, PA 19104, U.S.A.

Abstract. Inferring phylogenetic trees is a fundamental problem in computational-biology. We present a new objective criterion, the *phylogenetic number*, for evaluating evolutionary trees for species defined by biomolecular sequences or other qualitative characters. The phylogenetic number of a tree T is the maximum number of times that any given character state arises in T. By contrast, the classical *parsimony* criterion measures the total number of times that different character states arise in T. We consider the following related problems: finding the tree with minimum phylogenetic number, and computing the phylogenetic number of a given topology in which only the leaves are labeled by species. When the number of states is bounded (as is the case for biomolecular sequence characters), we can solve the second problem in polynomial time. We can also compute a fixed-topology 2-phylogeny (when one exists) for an arbitrary number of states. This algorithm can be used to further distinguish trees that are equal under parsimony. We also consider a number of other related problems.

1 Introduction

The problem of evolutionary tree construction involves taking a given set of species, and constructing a tree which describes the evolutionary history of that set of species. We would expect a pair of species to be close together in the tree if they are closely related. Numerous variants of this general problem have been studied, the variants arising from the differing kinds of information that may be assumed to be available concerning the species.

In *character-based phylogeny*, the scenario is the following. A character c is a function from the species set S to some set R_c of *states*. For example, the character vertebrate-invertebrate has two states, so we can choose $R_c = \{0, 1\}$ and we can define c so that $c(s) = 0$ for every species s that is a vertebrate and $c(s) = 1$ for every species s that is an invertebrate. As another example, we could define a character c based on average life-span. In this case R_c might be a set of ranges such as $R_c = \{0\text{--}10\text{ years}, 10\text{--}20\text{ years}, 20\text{--}60\text{ years}, \text{more than } 60\text{ years}\}$. Then the function c could be defined to map each species s to the range containing its average life-span. We can think of a sequence of k characters $c_1, \ldots, c_k$ as mapping each species s in the species set to a vector $(c_1(s), \ldots, c_k(s))$ in $R_{c_1} \times \cdots \times R_{c_k}$. The species sets that we will consider will have the property that for any two distinct species, s and s', that are in a species set, $(c_1(s), \ldots, c_k(s)) \neq (c_1(s'), \ldots, c_k(s'))$. Thus, we will be able to identify each species s with a vector $(c_1(s), \ldots, c_k(s))$ in $R_{c_1} \times \cdots \times R_{c_k}$. Furthermore, we will think of the set $R_{c_1} \times \cdots \times R_{c_k}$ as being the set containing all *possible* species, including those in S.

The inputs to the phylogeny construction problem are the species set S (we will use n to denote the size of S) and a sequence of characters, $c_1, \ldots, c_k$. We will let r_{c_j} denote $|R_{c_j}|$, and r denote $\max_j r_{c_j}$. A *phylogenetic tree* for the input is a node-labeled tree in which every node of the tree is labeled with a vector in $R_{c_1} \times \cdots \times R_{c_k}$, and each species in S is the label of some node of the tree†. Thus, each character c_j can be extended to a function from the set of vertices of T to R_{c_j}.

We can think of a species as a string of length k over the alphabet $\{1, \ldots, r\}$. A phylogeny is a way of expressing similarity amongst a *set* of strings rather than expressing similarity between pairs of strings. Strings which have strong similarities (as measured by matches in many locations) are located closer to each other in the tree than those that are more disparate. The output tree is the *pattern* of similarity amongst the entire set of input strings.

The measure of fitness of a phylogenetic tree is computed in one of several ways. Classically, there have been essentially two measures of fitness. The first is *parsimony*, in which a tree is sought such that the total weight of the tree is minimized, where the weight of the tree is the sum of the edge-weights, and the weight of an edge is the number of characters on which the species at its endpoints disagree. The other criterion is called *compatibility*, in which a tree is sought to maximize the number of characters that are *convex* on the tree. A character c_j is convex on a tree T if for each state $i \in R_{c_j}$, $c_j^{-1}(i)$ defines a connected component of T.

---

† A phylogenetic tree for the input $S, c_1, \ldots, c_k$ is sometimes defined to be a node-labeled tree in which every node of the tree is labeled with a vector in $R_{c_1} \times \cdots \times R_{c_k}$, and each species in S is the label of some *leaf* of the tree. It is clear that every tree satisfying this alternative definition also satisfies our definition above. The alternative definition is equivalent to ours in the sense that we can convert a tree T satisfying our definition into a tree T' satisfying the alternative definition by adding extra leaves. Under all reasonable measures of fitness for phylogenetic trees, T and T' will have the same measure of fitness.

Both criteria, compatibility and parsimony, result in NP-hard optimization problems [4,5]. An ideal tree is one in which all characters are compatible (i.e., all characters are convex on the tree). Such a tree is optimal under parsimony and compatibility criteria and is called a *perfect phylogeny*. The question of whether a perfect phylogeny exists for a given input is NP-Complete [3, 14].

In this paper, we consider an alternative measure of fitness which combines some of the properties of compatibility and parsimony. We will say that a phylogenetic tree T for an input consisting of a species set S and a sequence of characters $c_1, \ldots, c_k$ is an *ℓ-phylogeny* if, for every character c_j and every state $i \in R_{c_j}$, the set of vertices $c_j^{-1}(i)$ form at most ℓ connected components in T. (A 1-phylogeny is the same as a perfect phylogeny). The *ℓ-phylogeny problem* is the problem of determining whether an input has an ℓ-phylogeny. The *phylogenetic number* of an input is the minimum ℓ such that the input has an ℓ-phylogeny. The *phylogenetic number problem* is the problem of determining the phylogenetic number of an input.

The ℓ-phylogeny problem and the phylogenetic number problem both have fixed-topology versions which are defined as follows. The input is a species set S, a sequence of characters $c_1, \ldots, c_k$, and a tree T in which internal nodes are unlabeled and each leaf is labeled with a species $s \in S$. Each species $s \in S$ is the label of exactly one leaf of T. A phylogenetic tree for the input is formed by taking T and labeling the internal nodes of T with vectors in $R_{c_1} \times \cdots \times R_{c_k}$. The *fixed-topology ℓ-phylogeny problem* is the problem of determining whether the input has an ℓ-phylogeny. The fixed-topology phylogenetic number problem is defined analogously.

The ℓ-phylogeny problem and the phylogenetic number problem also have restricted versions in which new ancestral species may not be added. The restricted versions are defined as follows. The input is a species set S and a sequence of characters $c_1, \ldots, c_k$. A *restricted phylogenetic tree* for the input is a node-labeled tree in which every node of the tree is labeled with a vector in S, and each species in S is the label of some node of the tree. The *restricted ℓ-phylogeny problem* is the problem of determining whether the input has a restricted ℓ-phylogeny. The restricted phylogenetic number problem is defined analogously.

The ℓ-phylogeny problem can be generalized as follows. Fix positive integers $r, \ell_1, \ldots, \ell_r$. Suppose that $S, c_1, \ldots, c_k$ is a phylogeny input such that $\max_j r_{c_j} \leq r$. An *$(\ell_1, \ldots, \ell_r)$-phylogeny* for an input is defined to be a phylogenetic tree for the input such that, for each character c_j and each integer $i \leq |R_{c_j}|$, the set of vertices that are mapped to the ith state in R_{c_j} by c_j forms at most ℓ_i connected components in T. The *$(\ell_1, \ldots, \ell_r)$-phylogeny problem* is the problem of determining whether an input has an $(\ell_1, \ldots, \ell_r)$-phylogeny. A generalized version of the restricted ℓ-phylogeny problem is defined analogously.

1.1 Summary of Results and Outline of Paper:

The 1-phylogeny problem is also known as the perfect phylogeny problem. It was shown to be NP-hard by Bodlaender, Fellows, Warnow, and (independently) Steel [3, 14]. The hardness of 1-phylogeny implies that the phylogenetic number problem is NP-hard. In Section 2 of this paper we show that for any fixed $\ell > 1$ the ℓ-phylogeny problem is also NP-hard.

Having shown that the ℓ-phylogeny problem is NP-hard, we consider in Section 3 the fixed topology ℓ-phylogeny problem. It is known that the fixed-topology 1-phylogeny problem can be solved in polynomial time [7]. We show that the fixed-topology 2-phylogeny problem can also be solved in polynomial time and that the fixed-topology ℓ-phylogeny problem is NP-hard for fixed $\ell > 2$. (We show that the fixed-topology ℓ-phylogeny problem is NP-hard for fixed $\ell > 2$ even when the input is guaranteed to have an $\ell + 1$-phylogeny and the degree of the topology is restricted to be at most 3.)

In Section 4 we consider the restricted ℓ-phylogeny problem. We show that there is a polynomial-time algorithm for the restricted 1-phylogeny problem, but the restricted ℓ-phylogeny problem is NP-hard for fixed $\ell \geq 2$.

Although the 1-phylogeny problem is NP-hard, it can be solved in polynomial time if the number, n, of species is fixed, or the number, k, of characters is fixed [12, 1], or the quantity $r = \max_j r_{c_j}$ is fixed [2, 11]. A full analysis of fixed parameter ℓ-phylogeny problems is outside the scope of this paper. However, we observe that all of the phylogeny problems can be solved in polynomial time (by brute force) if n is fixed. In Section 5 we use interesting combinatorial techniques to show that for $k = 2$ the phylogenetic number problem can be solved in polynomial time. The complexity of the ℓ-phylogeny problem remains open for fixed $\ell > 1$ and fixed $k > 2$. The difficulty of fixed-topology phylogeny problems does not change if k is fixed. In Section 6 we show that the fixed-topology phylogenetic number problem can be solved in polynomial time for fixed r. On a related note, we show that if r is fixed, there is a polynomial-delay algorithm for listing fixed-topology ℓ-phylogenies. We also show that for fixed $r \geq 2$ and fixed $\ell \geq 3$ the restricted ℓ-phylogeny problem is NP-hard. (This result follows from a more general result. Namely, we show that the restricted (ℓ_1, ℓ_2)-phylogeny problem is NP-hard for fixed $\ell_1 \geq 2$ and $\ell_2 \geq 2$ as long as one of ℓ_1, ℓ_2 is greater than 2.)

Finally, in section 7 we offer some concluding remarks and present some open problems.

1.2 Preliminary Facts

The following fact is used in some of the proofs and in the restricted 1-phylogeny algorithm.

Fact 1: *If an input S, $c_1, \ldots, c_k$ has an ℓ-phylogeny then it has an ℓ-phylogeny in which:*

1. *Each leaf has a label from S.*

2. *Each species is the label of at most one node.*

3. *Every node whose label is not in S has degree at least 3.*

4. *There are at most $\max(0, n - 2)$ nodes with labels that are not in S.*

Proof: It is easy to see that conditions 1–3 can be satisfied. (One can convert an ℓ-phylogeny into one that satisfies conditions 1–3 by removing leaves with labels that are not in S, combining branches of the tree to accomplish condition 2, and then "splicing out" the appropriate degree 2 nodes to accomplish condition 3.) To prove that condition 4 can also be satisfied, suppose that T is an ℓ-phylogeny for the input that satisfies conditions 1–3 and contains at least one node, w, with a label that is not in S. Let T' be the tree obtained from T by splicing out any nodes of degree 2. (Condition 3 guarantees that no node with a label outside of S is spliced out in this process.) Consider T' to be rooted at w. It is easy to see that we can add one or more new internal nodes to T' to obtain a complete binary tree T'' which is rooted at w and has the same leaves as T' †. Conditions 1 and 2 imply that T, and therefore T' and T'', have at most n leaves. Since T'' has at most n leaves, it has at most $n - 1$ internal nodes. Therefore, T' has at most $n - 2$ internal nodes, and T has at most $n - 2$ nodes with labels that are not in S. □

Fact 1 implies that if an input has an ℓ-phylogeny then it has a polynomial-sized ℓ-phylogeny.

---

† To see how to construct T'', let the "level" of a vertex denote its distance from the root. Start with level 0 of T' and proceed through the levels of the tree in increasing order. Consider each vertex v on each level. If v has children $x_1, \ldots, x_j$ with $j > 2$ remove the edges $(v, x_2), \ldots, (v, x_j)$ and add a new node y which is a child of v and the parent of nodes $x_2, \ldots, x_j$. Note that at least one new internal node is added in the process, as w has at least three children in T'.

2 The Hardness of ℓ-Phylogeny

In this section we show that for any fixed $\ell > 1$, the ℓ-phylogeny problem is NP-hard. Our reduction is from the 1-phylogeny problem, which was shown to be NP-hard by Bodlaender, Fellows, Warnow, and (independently) Steel [3, 14].

We define the *weight* of an edge (v_1, v_2) in a phylogeny to be the number of characters c_j such that $c_j(v_1) \neq c_j(v_2)$. That is, the weight of (v_1, v_2) is the number of characters on which the species labeling v_1 and v_2 disagree. We define the *weight* of a phylogeny to be the sum of the weights of its edges. We start with the following observation.

Observation 2: Let $S, c_1, \ldots, c_k$ be any input to the ℓ-phylogeny problem and let r denote $\max_j r_{c_j}$. Any ℓ-phylogeny for this input has weight at most $k(\ell r - 1)$.

We will use the following lemma.

Lemma 3: For every integer ℓ there is an input $I_\ell = S, c_1, \ldots, c_{2\ell}$ in which $|S| = 2\ell^3 - 2\ell + 1$ and $R_{c_j} = \{0, \ldots, \ell - 1\}$ for $1 \leq j \leq 2\ell$ such that

1. For every state i in the range $0 \leq i < \ell$, the species $i^{2\ell}$ is in S.

2. I_ℓ has an ℓ-phylogeny

3. In any ℓ-phylogeny for I_ℓ the subgraph induced by all of the nodes with any given label is connected.

4. In any ℓ-phylogeny for I_ℓ all of the nodes are labeled by species in S. (That is, no new species are introduced.)

5. In any ℓ-phylogeny for I_ℓ the path between the species $i^{2\ell}$ and $j^{2\ell}$ for $i \neq j$ passes through at least $2\ell - 1$ distinct species.

Example: The Input I_3

The species set S of input I_3 consists of 49 species. The values of the six characters on these species are defined as in figure 1:

000000 100000 110000 111000 211000 221000 222000
       000100 000110 000111 000211 000221 000222
010000
010010
110010
111010
111110
111111 211111 212111 212121 012121 010121 010101
       121111 121211 121212 101212 101012 101010
112111
122111
222111
222211
222221
222222 022222 002222 002202 102202 112202 112212
       220222 220022 220020 221020 221120 221121

Fig. 1. The Input I_3

The Input I_3 has a 3-phylogeny in which species are connected along the rows and along the first column. Each node i^6 is also connected to the row beneath it. By Observation 2 any 3-phylogeny for I_3 has weight at most 48. However, 48 edges with positive weight are needed just to hook up the 49 species in S into a tree. We conclude that any 3-phylogeny for I_3 consists of 48 edges with weight 1 plus possibly some edges with weight 0. Thus, the subgraph induced by all of the nodes with any given label forms a single connected component. Furthermore, no new species are introduced. Finally, since i^6 and j^6 differ in 6 characters, any path between them in any 3-phylogeny for I_3 passes through at least 5 distinct species.

Construction of $I_\ell = S, c_1, \ldots, c_{2\ell}$:

For $1 \leq j \leq 2\ell$ we set $R_{c_j} = \{0, \ldots, \ell - 1\}$. For each state i in the range $0 \leq i < \ell$ we put the species $i^{2\ell}$ into S. The other species in S will be the species in the following phylogeny:

For each state i in the range $0 \leq i < \ell$ we will choose a unique partition P_i of the 2ℓ characters into two sets of size ℓ. (In the construction of I_3 above we used $P_0 = \{0,1,2\}, \{3,4,5\}, P_1 = \{0,2,4\}, \{1,3,5\}$, and $P_2 = \{0,1,4\}, \{2,3,5\}$.)

We will use each of the parts of the partition P_i to form a "row" of species which will be connected to the species $i^{2\ell}$. To construct each row, consider the ordered list $c_{i_1}, \ldots, c_{i_\ell}$ consisting of the characters in the appropriate part of the partition. From the species $i^{2\ell}$ form a new species by changing the state of character c_{i_1} to $(i+1) \bmod \ell$. Then form a new species by changing the state of character c_{i_2} to $(i+1) \bmod \ell$. Continue on until the state of character c_{i_ℓ} is changed to $(i+1) \bmod \ell$. Then change the state of character c_{i_1} to $(i+2) \bmod$ ℓ and continue on in this manner until finally the state of character c_{i_ℓ} is changed to $(i+(\ell-1)) \bmod \ell$.

Finally, we will add species to connect the species $i^{2\ell}$ to the species $(i+1)^{2\ell}$ in the vertical *spine* (for i in the range $0 \leq i < \ell-1$). Let c_λ be the second character in the first part of the partition corresponding to i and construct a new species from $i^{2\ell}$ by changing the state of character c_λ to $i+1$. Next, let c'_λ be the first character such that c_λ and c'_λ are in different parts of i's partition and c_λ and c'_λ are in different parts of $(i+1)$'s partition. Construct a new species by changing the state of character c'_λ to $i+1$. Now, construct $2\ell-3$ more species by considering each remaining character in turn and changing it from state i to state $i+1$.

Proof of Lemma 3: By construction, S contains the species $i^{2\ell}$ for every state i in the range $0 \leq i < \ell$. To see that the phylogeny constructed above is indeed an ℓ-phylogeny for I_ℓ note that for each state i and for each state $j \neq i$ a character c_λ only has state i in one of the two rows connected to $j^{2\ell}$ and the species with c_λ in state i are connected in this row. Furthermore, there is a single connected component with character c_λ in state i in the rows connected to $i^{2\ell}$ and this connected component contains all species on the vertical spine with character c_λ in state i. We now wish to show that all of the species introduced in the construction are distinct. Suppose that instead two species s_1 and s_2 have identical labels. Note that, by construction, s_1 and s_2 could not be of the form $i^{2\ell}$. Furthermore, they could not be on the same horizontal row and they could not both be on the vertical spine. There are three cases to consider:

1. s_1 and s_2 are on different rows, both of which are attached to $i^{2\ell}$.

   In this case s_1 has state i for all of the characters in one part of the partition P_i and s_2 has state i for all of the characters in the other part of the partition P_i so it must be the case that $s_1 = s_2 = i^{2\ell}$ which is a contradiction.

2. s_1 is on a horizontal row connected to $i^{2\ell}$ and s_2 is on a horizontal row connected to $j^{2\ell}$ for some $j \neq i$.

   In this case s_1 has state i for all of the characters in some part of the partition P_i so s_2 must have character i for all of the characters in that part of the partition P_i and character j on all other characters. But then the partition P_j is the same as the partition P_i, which is not true by construction.

3. s_1 is on the vertical spine between $i^{2\ell}$ and $(i+1)^{2\ell}$ and s_2 is on a horizontal row.

   By construction s_2 must be on a row attached to i or on a row attached to $i+1$. However, the choice of c_λ and c'_λ ensures that s_2 cannot be on either of these rows.

Now that we know that the species are distinct, we count them. There are ℓ species of the form $i^{2\ell}$. Each of the 2ℓ horizontal rows has $\ell(\ell-1)$ species. Finally, there are $(\ell-1)(2\ell-1)$ additional species on the vertical spine. We conclude that S has $2\ell^3 - 2\ell + 1$ distinct species. By Observation 2, any ℓ-phylogeny for I_ℓ has weight at most $2\ell(\ell^2 - 1) = 2\ell^3 - 2\ell$. However, $2\ell^3 - 2\ell$ edges with positive weight are needed just to hook up the $2\ell^3 - 2\ell + 1$ species in S into a tree. We conclude that any ℓ-phylogeny for I_ℓ consists of $2\ell^3 - 2\ell$ edges with weight 1 plus possibly some edges with weight 0. Thus, the subgraph induced by all of the nodes with any given label forms a single connected component. Furthermore, no new species are introduced. Finally, since $i^{2\ell}$ and $j^{2\ell}$ differ in 2ℓ characters, any path between them in any ℓ-phylogeny for I_ℓ passes through at least $2\ell - 1$ distinct species. $\square$

We will use Lemma 3 to prove the following theorem

Theorem 4: *For any fixed $\ell > 1$ the ℓ-phylogeny problem is NP-hard.*

Proof: The reduction is from the 1-phylogeny problem. Let S, $c_1, \ldots, c_k$ be an input to the 1-phylogeny problem such that $R_{c_j} = \{0, \ldots, r - 1\}$ for $1 \leq j \leq k$. Let S', $c'_1, \ldots, c'_{2\ell}$ be an input to the ℓ-phylogeny problem satisfying the conditions in Lemma 3. Let $S^* = \{ sr^k \mid s \in S' \}$. For each i in the range $0 \leq i < \ell$ let $S_i = \{ i^{2\ell}y \mid y \in S \}$. Let $S'' = S^* \cup \bigcup_{0 < i < \ell} S_i$. Let I be the input to the ℓ-phylogeny problem with species set S'' and characters $c'_1, \ldots, c'_{2\ell}, c_1, \ldots, c_k$. (Note that in input I the range of c_j has been extended from R_{c_j} to $R_{c_j} \cup \{r\}$.)

$\rightarrow$ Suppose that T is a 1-phylogeny for $S, c_1, \ldots, c_k$. For each i in the range $0 \leq i < \ell$ let T_i be a copy of T in which each label y has been changed to $i^{2\ell}y$. (T_i is a 1-phylogeny for $S_i, c'_1, \ldots, c'_{2\ell}, c_1, \ldots, c_k$.) Let T^* be an ℓ-phylogeny for $S^*, c'_1, \ldots, c'_{2\ell}, c_1, \ldots, c_k$. (Part 2 of Lemma 3 guarantees that T^* exists.) Now for each i in the range $0 \leq i < \ell$ connect an arbitrary node in T_i to the node $i^{2\ell}r^k$ in T^*. (The construction, together with Part 1 of Lemma 3 guarantees that there is a vertex of T^* labeled $i^{2\ell}r^k$.) It is easy to see that the resulting tree is an ℓ-phylogeny for I.

$\leftarrow$ Suppose that T is an ℓ-phylogeny for I. If we restrict our attention to characters $c'_1, \ldots, c'_{2\ell}$, we still have an ℓ-phylogeny. Therefore, by Part 3 of Lemma 3, the subgraph induced by all of the species which have some particular set of states for characters $c'_1, \ldots, c'_{2\ell}$ is connected. We will use the notation T_i to refer to the induced subtree of T containing those species that have state i for characters $c'_1, \ldots, c'_{2\ell}$.

We claim that for any j in the range $1 \leq j \leq k$ any path in T between a node $t_i \in T_i$ and a node $t_h \in T_h$ (for $h \neq i$) contains some species s with $c_j(s) = r$. Clearly, this claim implies that T_0 is a 1-phylogeny for $S_0, c'_1, \ldots, c'_{2\ell}, c_1, \ldots, c_k$. Hence, $S, c_1, \ldots, c_k$ has a 1-phylogeny.

To prove the claim note that by Part 5 of Lemma 3 the path between T_i and T_h passes through $2\ell - 1$ nodes $v_1, \ldots, v_{2\ell-1}$, no two of which agree on all of characters $c'_1, \ldots, c'_{2\ell}$. By construction and by Part 1 of Lemma 3, S'' contains the species $i^{2\ell}r^k$ and by Part 3 of Lemma 3 it is part of T_i. Similarly, S'' contains the species $h^{2\ell}r^k$ and it is part of T_h. Furthermore, (by construction and by Part 4 of Lemma 3), for each node v_m, S'' contains a species v'_m that agrees with v_m on characters $c'_1, \ldots, c'_{2\ell}$ and has characters $c_1, \ldots, c_k$ in state r. By Part 3 of Lemma 3 v'_m is in the connected subgraph of T induced by species which agree with v_m on characters $c'_1, \ldots, c'_{2\ell}$. Now suppose that none of $v_1, \ldots, v_{2\ell-1}$ has character c_j in state r. Then the sub-graph of T induced by those nodes that have character c_j in state r has $2\ell + 1$ connected components, which contradicts the fact that T is an ℓ-phylogeny. □

3   The Fixed-Topology ℓ-Phylogeny Problem

It is known that the fixed-topology 1-phylogeny problem can be solved in polynomial time [7]. In Subsection 3.1, we show that the fixed-topology 2-phylogeny problem can also be solved in polynomial time. In Subsection 3.2 we show that the fixed-topology ℓ-phylogeny problem is NP-hard for fixed $\ell > 2$. (We show that the fixed-topology ℓ-phylogeny problem is NP-hard for fixed $\ell > 2$ even when the input is guaranteed to have an $\ell + 1$-phylogeny and the degree of the topology is restricted to be at most 3.)

3.1   The Fixed-Topology 2-Phylogeny Problem

In this subsection, we show that the fixed-topology 2-phylogeny problem can be solved in polynomial time. The algorithm runs in time $O(nrk)$ where n is the number of species, r is the maximum number of states in any character, and k is the number of characters. If a 2-phylogeny exists, then our algorithm computes a labeling that achieves a 2-phylogeny.

Since the topology is fixed, the characters are independent and can be handled one at a time. We will now show how to compute the labels for a single character in time $O(nr)$, where in this case r is the number of states for this character. The overall bound then follows.

Although the input tree is unrooted, for this algorithm, we root this tree from an arbitrary internal node. The choice of root does not affect the existence of a 2-phylogeny, but it may affect the labeling.

Let T be the input tree with leaves labeled by states $1, 2, \ldots, r$. Consider a single state i and let T_i be the subtree of tree T consisting of all the leaves labeled i and the unique set of paths connecting this set of leaves. For state i to have a single connected component in tree T, every node in T_i must be labeled i. For state i to have at most two connected components, every node in tree T_i with degree greater than 2 must be labeled i (otherwise state i would be split into at least 3 components). We call such nodes *branch points* of tree T_i. The branch points and the leaves already labeled i are the *forced points* of tree T_i. At most one path of degree-2 nodes between two forced points can be labeled something other than i.

We begin by computing T_i for $i = 1, \ldots, r$. Each branch point of T_i is labeled as such, each path between two forced points is given a unique label, and each degree-2 node in T_i is labeled with its path label. Note that the root of tree T_i need not be a branch point. If each node of tree T is given a length-r vector, then information for all r trees T_i can be stored on top of each other. For example, node v could be a branch point for tree T_i (ith slot of the vector indicates branch point), on the lth path for tree T_j (the jth slot of the vector has the number l), and not in tree T_h (the hth slot is null). We can compute all r trees in time $O(nr)$ using depth-first search.

The first phase of the algorithm (the forced phase) computes all forced labels. For each tree T_i, each branch point of T_i is labeled i and a pointer to the node is placed into a queue. If at any time we try to label a node that is already labeled with something else, then we stop and report that there is no 2-phylogeny for this topology.

Now all path conflicts have to be settled for the labeled nodes. We remove the first node from the queue. Suppose it is node v and it is labeled i. If this node is also in path l of tree T_j for some $j \neq i$, then tree T_j must give up path l. Once path l is broken, then in order to achieve 2 connected components for state j, every other path in tree T_j must be labeled j. We traverse tree T_j, clearing path l (setting slot j to null for all nodes on path l of tree T_j) and labeling all other nodes j. If we attempt to label a node that is already labeled, then we stop. There can be no 2-phylogeny. Otherwise, the newly-labeled nodes are added to the queue. We do this for all paths that go through node v, then clear path conflicts on all the other nodes in the queue. Because each node can be labeled, enqueued, dequeued, and processed at most once, and each tree can be traversed at most once, this phase can be completed in time $O(nr)$.

The final phase completes the labeling of the tree. If we succeed in emptying the queue without encountering a fatal conflict, it is still possible that some nodes remain unlabeled. We show that there is always a 2-phylogeny. Let trees T_i and T_j be left undetermined by the forced phase of the algorithm. If the intersection of these two trees is empty, there is no conflict between them. Otherwise, the intersection is connected* and contains exactly one path from each tree†. Furthermore, the root of one of the trees (possibly both) is in the intersection‡. Suppose that the root of T_i is contained in $T_i \cap T_j$. Then tree T_i gives up the path through its root (if both roots are contained in $T_i \cap T_j$, one of the trees chosen arbitrarily will give up the path through its root). By the structure of the intersection, this clears the conflict between tree T_i and T_j. We can solve all conflicts between pairs of trees in a similar manner. Since each tree

---

* If two nodes v_1 and v_2 are both in T_i and both in T_j, then every node on the unique path in T between v_1 and v_2 must also be in both trees.

† If the intersection contained pieces of two paths from tree T_i, then it must contain a branch point for tree T_i and therefore tree T_j would have been forced to relinquish a path and left completely determined by the forced phase.

‡ Consider a node in the intersection. If its parent in T is in the intersection, move up to it. Continue until some parent is no longer in the intersection. That node is the root of at least one of T_i and T_j

was not forced to give up a path in the forced phase of the algorithm (otherwise it would have been fully determined then), it is free to give up one path in this phase. Each tree will give up at most one path, namely the one through its root. Therefore, all conflicts are resolved and we have a 2-phylogeny. This phase of the algorithm can be implemented in $O(nr)$ time by processing each remaining tree in order (determining whether it must relinquish the path through its root, and claiming all other paths).

Thus we have shown how to compute the labelings of the internal nodes of the input tree T in time $O(nr)$ per character for an overall time of $O(nrk)$. Thus, we have proved the following theorem.

Theorem 5: *The fixed-topology 2-phylogeny problem can be solved in polynomial time.*

3.2 The Fixed-Topology ℓ-Phylogeny Problem for $\ell > 2$

In this subsection we prove the following theorem.

Theorem 6: *The fixed-topology ℓ-phylogeny problem is NP-hard for fixed $\ell > 2$.*

Proof: The proof is by reduction from 3SAT. Let $\ell > 2$ be fixed. Suppose that we are given an input to 3SAT. We will show how to construct a one-character input S, c, T to the fixed-topology ℓ-phylogeny problem such that the phylogeny input has an ℓ-phylogeny if and only if the input to 3SAT is satisfiable.

The species set S, the set R_c of states, and the character c are constructed as follows. For each of the n variables, x, in the satisfiability input we have states s_x and $s_{\overline{x}}$ and species $s_{(x,1)}, \ldots, s_{(x,\ell+1)}$ and $s_{(\overline{x},1)}, \ldots, s_{(\overline{x},\ell+1)}$ where $c(s_{(x,j)}) = s_x$ and $c(s_{(\overline{x},j)}) = s_{\overline{x}}$. For each of the m clauses, C, in the satisfiability input we have state s_C and species $s_{(C,1)}, \ldots, s_{(C,\ell+1)}$ where $c(s_{(C,j)}) = s_C$. For the ith occurrence of the literal x in the satisfiability input, we have state s_{x_i} and species $s_{(x_i,1)}, \ldots, s_{(x_i,\ell+1)}$ where $c(s_{(x_i,j)}) = s_{x_i}$. Similarly, for the ith occurance of the literal $\overline{x}$ in the satisfiability input, we have state $s_{\overline{x}_i}$ and species $s_{(\overline{x}_i,1)}, \ldots, s_{(\overline{x}_i,\ell+1)}$ where $c(s_{(\overline{x}_i,j)}) = s_{\overline{x}_i}$. Let N denote $n(2\ell - 3) + m(4\ell - 11)$. For each h in the range $1 \leq h < N$ we have a state s'_h and species $s'_{(h,1)}, \ldots, s'_{(h,\ell+1)}$ where $c(s'_{(h,j)}) = s'_h$.

We will show how to construct a tree T in which internal nodes are unlabeled and each leaf is labeled with a species in S. Each species in S will be the label of exactly one leaf of T. To construct T we will first construct trees $T_1, \ldots, T_N$. Finally, we will hook T_i to T_{i+1} for $1 \leq i < N$

We start by showing how to hook tree T_i to tree T_{i+1}. Let t_i be an internal node in T_i of degree at most 2 and let t_{i+1} be an internal node in T_{i+1} of degree at most 2 (it will be clear from the construction that such small degree internal nodes exist in T_i and T_{i+1}). Connect t_i and t_{i+1} with a chain of $\ell + 1$ new internal nodes. Finally, give each of the internal nodes in the chain a leaf and label the new leaves with the species $s'_{(i,1)}, \ldots, s'_{(i,\ell+1)}$. For example, if $\ell = 3$ then connect t_i and t_{i+1} as in figure 2:

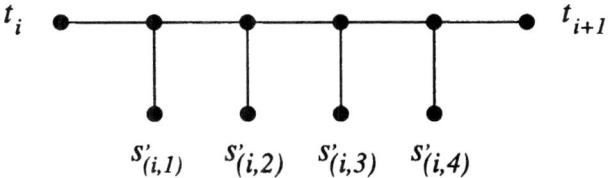

Fig. 2. Example for $\ell = 3$

Note that in any ℓ-phylogeny for the input, at least one of the internal nodes in the chain will be labeled with a species s such that $c(s) = s'_i$. Since we have now used all $\ell + 1$ species s with $c(s) = s'_i$, neither T_i nor T_{i+1} contains a leaf s such that $c(s) = s'_i$. Therefore when T_i is hooked to T_{i+1} as above, any leaves $\ell_i \in T_i$ and $\ell_{i+1} \in T_{i+1}$ with $c(\ell_i) = c(\ell_{i+1})$ are in different connected components in the subgraph induced by $c^{-1}(c(\ell_i))$.

We next show how to construct the trees $T_1, \ldots, T_N$. Trees $T_1, \ldots, T_{N-n-m}$ will each consist of a single internal node connected to a single leaf. In particular, we will construct one such tree for each of the following species: for each variable x, species $s_{(x,1)}, \ldots, s_{(x,\ell-2)}$ and $s_{(\overline{x},1)}, \ldots, s_{(\overline{x},\ell-2)}$; for each clause C, species $s_{(C,1)}, \ldots, s_{(C,\ell-3)}$; for the ith occurance of the literal x, species $s_{(x_i,1)}, \ldots, s_{(x_i,\ell-3)}$; for the ith occurance of the literal $\overline{x}$, species $s_{(\overline{x}_i,1)}, \ldots, s_{(\overline{x}_i,\ell-3)}$.

Trees $T_{N-n-m+1}, \ldots, T_{N-m}$ will be used for truth-setting. For each variable x in the satisfiability input we will construct a tree as follows. Suppose that the literal x appears i times in the satisfiability input and that the literal $\overline{x}$ appears j times in the satisfiability input. Construct a tree consisting of a chain of $2i + 2j + 6$ internal nodes. Each internal node will have one leaf, and the species at the leaves will be (in order): first, $s_{(x,\ell-1)}$; then, $s_{(x_1,\ell-2)}, s_{(x_1,\ell-1)}, s_{(x_2,\ell-2)}, s_{(x_2,\ell-1)}, \ldots, s_{(x_i,\ell-2)}, s_{(x_i,\ell-1)}$; then $s_{(x,\ell)}, s_{(\overline{x},\ell-1)}$, $s_{(x,\ell+1)}, s_{(\overline{x},\ell)}$; then $s_{(\overline{x}_1,\ell-2)}, s_{(\overline{x}_1,\ell-1)}, \ldots, s_{(\overline{x}_j,\ell-2)}, s_{(\overline{x}_j,\ell-1)}$; finally, $s_{(\overline{x},\ell+1)}$. For example, if $\ell = 3$, $i = 1$, and $j = 2$ construct a tree as in figure 3:

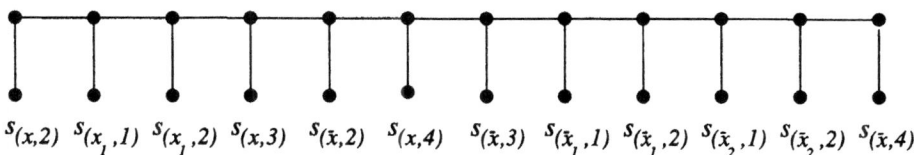

Fig. 3. Example for $\ell = 3$, $i = 1$, $j = 2$

Because we have already introduced single-leaf trees for the species $s_{(x,1)}, \ldots,$ $s_{(x,\ell-2)}$ and $s_{(\overline{x},1)}, \ldots, s_{(\overline{x},\ell-2)}$, we observe that in any ℓ-phylogeny, the truth-setting tree for variable x must have at most 2 connected components for each of the states s_x and $s_{\overline{x}}$. We will say that an ℓ-phylogeny sets the satisfiability variable x to "true" if and only if the leaves $s_{(x,\ell)}$ and $s_{(x,\ell+1)}$ are in the same connected component for state s_x. If the variable x is set to "true" then the leaf $s_{(x,\ell-1)}$ can be in a different connected component for state s_x. Therefore, for $1 \leq h \leq i$, state s_{x_h} can form a single connected component in the truth-setting tree for x. Otherwise, state s_{x_h} must have two connected components in the truth-setting tree for x. Similarly, if x is set to "false" then leaves $s_{(\overline{x},\ell-1)}$ and $s_{(\overline{x},\ell)}$ can be in the same connected component for state $s_{\overline{x}}$ and leaf $s_{(\overline{x},\ell+1)}$ can be in a different connected component. Therefore, for $1 \leq h \leq j$, state $s_{\overline{x}_h}$ can form a single connected component in the truth-setting tree for x. Otherwise, state $s_{\overline{x}_h}$ must have two connected components in the truth-setting tree for x.

Trees $T_{N-m+1}, \ldots, T_N$ will be used for clause-checking. For each clause $C = x_i \vee \overline{y}_j \vee z_k$ in the satisfiability input we will construct a tree consisting of a chain of 10 internal nodes. Each internal node will have one leaf, and the species at the leaves will be (in order): $s_{(C,\ell-2)}, s_{(x_i,\ell)}, s_{(x_i,\ell+1)}, s_{(C,\ell-1)}, s_{(\overline{y}_j,\ell)}, s_{(\overline{y}_j,\ell+1)}, s_{(C,\ell)},$ $s_{(z_k,\ell)}, s_{(z_k,\ell+1)}, s_{(C,\ell+1)}$. For example, if $\ell = 3$, construct a tree as in figure 4:

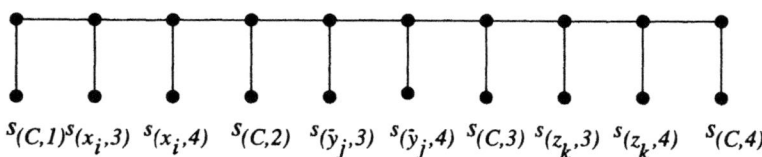

$s_{(C,1)} \, s_{(x_i,3)} \quad s_{(x_i,4)} \quad s_{(C,2)} \quad s_{(\overline{y}_j,3)} \quad s_{(\overline{y}_j,4)} \quad s_{(C,3)} \, s_{(z_k,3)} \quad s_{(z_k,4)} \quad s_{(C,4)}$

Fig. 4. Example for $\ell = 3$

Because we have already introduced single-leaf trees for the species $s_{(C,1)}, \ldots,$ $s_{(C,\ell-3)}$, we observe that in any ℓ-phylogeny, the clause-checking component for clause C must have at most 3 connected components for the state s_C. This is possible if one of the literals in the clause has been set to "true" by the truth checking component and not otherwise. The correctness of the reduction follows. □

The input to the fixed-topology ℓ-phylogeny problem that is constructed in the proof of Theorem 6 had two notable features. First, (because there are only $\ell + 1$ species with each state), the input is guaranteed to have an $\ell + 1$-phylogeny. Second, the degree of the tree T is at most 3. Therefore, the fixed-topology ℓ-phylogeny problem is NP-hard for fixed $\ell > 2$ even when the input is guaranteed to have an $\ell + 1$-phylogeny and the degree of the topology is restricted to be at most 3.

4 The Restricted ℓ-Phylogeny Problem

In this section we show that there is a polynomial time algorithm for the restricted 1-phylogeny problem. We then show that the restricted ℓ-phylogeny problem is NP-hard for fixed $\ell \geq 2$.

We start by describing the algorithm for solving the restricted 1-phylogeny problem. Suppose that $S, c_1, \ldots, c_k$ is an input to the restricted 1-phylogeny problem. If the input has a restricted 1-phylogeny, it has one in which each species in S is the label of exactly one node (if not, combine branches).

We define the *weight* of an edge (v_1, v_2) in a phylogeny to be the number of characters c_j such that $c_j(v_1) \neq c_j(v_2)$. That is, the weight of (v_1, v_2) is the number of characters on which the species labeling v_1 and v_2 disagree. We define the *weight* of a phylogeny to be the sum of the weights of its edges.

Let G denote the complete graph with vertex set S. We seek a spanning tree T of G in which, for every character c_j and every state $i \in R_{c_j}$, the set of vertices $c_j^{-1}(i)$ form a connected component in T. Let the *weight* of an edge (s, s') in G be the number of characters c_j such that $c_j(s) \neq c_j(s')$. It is easy to see that a spanning tree of G is a 1-phylogeny for the input if and only if its weight is $\sum_{j=1}^{k}(r_{c_j} - 1)$. Therefore, the restricted 1-phylogeny problem reduces to the minimum weight spanning tree problem, which can be solved in polynomial time [13]. We have proved the following theorem

Theorem 7: *The restricted 1-phylogeny problem can be solved in polynomial time.*

In the remainder of this section, we prove the following theorem.

Theorem 8: *The restricted ℓ-phylogeny problem is NP-hard for fixed $\ell \geq 2$.*

Proof: The reduction is from the ℓ-*consecutive ones problem*, which is defined as follows:

INSTANCE: A (0,1)-matrix M.

QUESTION: Can the rows of M be permuted in such a way that for each column in the resulting matrix, there are at most ℓ sequences of consecutive ones.

The ℓ-consecutive ones problem is known to be solvable in polynomial time for $\ell = 1$ [8]. However, it is NP-complete for fixed $\ell > 1$ [9].

Let ℓ be a positive integer that is greater than or equal to 2. Suppose that we are given an input M to the ℓ-consecutive ones problem with n rows and m columns. (We will assume that $n \geq 3\ell$.) We will show how to construct an input $S, c_1, \ldots, c_{m+\binom{n}{\ell-1}}$ to the restricted ℓ-phylogeny problem such that the phylogeny input has a restricted ℓ-phylogeny if and only if the rows of M can be permuted in such a way that for each column in the resulting matrix there are at most ℓ sequences of consecutive ones.

The phylogeny input is constructed as follows. Let M' be a matrix derived from M by replacing the zeroes in each column of M with integers in the range $2, ..., n+1$ in such a way that each column of M' has at most one occurance of each integer in the range $2, ..., n+1$. The species set S will have n species — one for each row of M'. For j in the range $1 \leq j \leq m$ character c_j will map the species corresponding to row r to the entry in column j of row r of M'. We will define the remaining $\binom{n}{\ell-1}$ characters as follows. For j in the range $1, \ldots, \binom{n}{\ell-1}$ we will have $R_{c_{j+m}} = \{0, 1\}$. We will let S_j denote the jth size-$(\ell-1)$ subset of S and we will set $c_{j+m}(s) = 1$ for $s \in S_j$ and $c_{j+m}(s) = 0$ for $s \notin S_j$.

$\rightarrow$ Suppose that T is a restricted ℓ-phylogeny for $S, c_1, \ldots, c_{m+\binom{n}{\ell-1}}$. Using Fact 1, we can assume that each species in S is the label of exactly one node in T. Let $V = \{v_1, \ldots, v_{\ell-1}\}$ be any set of $\ell-1$ vertices of T and let j be the integer such that the species labeling the vertices in V correspond to the set S_j. Observe that the graph obtained by removing the vertices in V from T has at most ℓ connected components (otherwise, the set of vertices $c_{j+m}^{-1}(0)$ form more than ℓ connected components in T, so T is not an ℓ-phylogeny). We will show that every node in T has degree at most 2. Suppose instead that T has a vertex, v_1, of degree greater than or equal to 3. We will show that there are $\ell-2$ other vertices, $v_2, \ldots, v_{\ell-1}$ such that the graph obtained by removing the vertices in $V = \{v_1, \ldots, v_{\ell-1}\}$ from T has at least $\ell+1$ connected components. This will be a contradiction, so we will conclude that every node in T has degree at most 2. To show that $v_2, \ldots, v_{\ell-1}$ exist, note that the subgraph of T formed by removing vertex v_1 has at least 3 connected components. Furthermore, if any subgraph T' of T that is formed by removing up to $\ell-1$ vertices has fewer than $\ell+1$ connected components, it is possible to remove a vertex so as to increase the number of connected components†. Let v_2 be a vertex such that removing v_2 from $T - v_1$ increases the number of connected components. Similarly, let v_3 be a vertex such that removing v_3 from $T - \{v_1, v_2\}$ increases the number of connected components. Continuing this process we identify $v_2, \ldots, v_{\ell-1}$. We have now shown that T is a path. It follows that we can arrange the rows of M in the order that the corresponding species occur on path T and that, in such an arrangement, each column has at most ℓ sequences of consecutive ones.

$\leftarrow$ Suppose that $p = \{p_1, ..., p_n\}$ is a permutation of $\{1, ..., n\}$ such that when the rows of M are permuted according to p each column has at most ℓ sequences of consecutive ones. Let T be a path consisting of the species in S, arranged according to permutation p. Then T is a restricted ℓ-phylogeny for $S, c_1, \ldots, c_{m+\binom{n}{\ell-1}}$. □

---

† To see this, note that (since $n \geq 3\ell$) T' has some connected component with more than 2 vertices.

5 Two-Character Phylogeny

In this section we show that for $k = 2$ the phylogenetic number problem can be solved in polynomial time. We start by proving the following fact.

Fact 9: *If a phylogeny input S, c_1, c_2 has an ℓ-phylogeny then it has a restricted ℓ-phylogeny T in which each species in S is the label of exactly one node and for each character $j \in \{1, 2\}$ and each state $i \in R_{c_j}$, at most one of the connected components in the subgraph of T induced by the set of vertices $c_j^{-1}(i)$ has more than one vertex.*

Proof: Suppose that T' is a an ℓ-phylogeny for S, c_1, c_2. We start by showing that S, c_1, c_2 has a restricted ℓ-phlogeny in which each species in S is the label of exactly one node. We can assume that each species is the label of at most one node of T' (if not, combine branches). Now, suppose that a species $s \notin S$ is the label of some node of T'. We can assume that this node, v, is an internal node of T' (otherwise delete it). Let U_1 be the set of neighbors u of v such that $c_1(u) = c_1(v)$. Let U_2 be the the the set of neighbors u of v such that $c_2(u) = c_2(v)$. Note that $U_1 \cap U_2 = \emptyset$ since s is the only species that can label a node in $U_1 \cap U_2$ and v is the only node with label s. Let U_3 be the set of neighbors of v that are not in U_1 or U_2. We can form a new ℓ-phylogeny for S, c_1, c_2 by deleting node v, connecting the vertices in U_1 in a path, connecting the vertices in U_2 in a path, connecting the vertices in U_3 in a path, and connecting some node from U_1 to some node in U_2 and some node from U_2 to some node from U_3.

We have now shown that S, c_1, c_2 has a restricted ℓ-phylogeny in which each species in S is the label of exactly one node. Let T be such an ℓ-phylogeny. Suppose that for character $j \in \{1, 2\}$ and state $i \in R_{c_j}$, C and C' are two non-singleton connected components in the subgraph of T induced by the set of vertices $c_j^{-1}(i)$. Let $c \in C$ and $c' \in C'$ be vertices such that the path connecting c to c' in T does not include any other vertices in C or C'. (Note that c and c' are uniquely defined.) For every neighbor v of c in C note that the path between v and c' passes through vertex c. Remove the edge (v, c) from T and add the edge (v, c'). Note that the resulting tree is an ℓ-phylogeny for S, c_1, c_2. (To see this, note that since the species labeling v is different from the species labeling c, the character other than character j disagrees on v and c.) □

In this section, we will represent the phylogeny input S, c_1, c_2 as a bipartite graph. One set of vertices in the graph will be the set R_{c_1} and the other set of vertices in the graph will be the set R_{c_2}. For $i \in R_{c_1}$ and $j \in R_{c_2}$ the edge (i, j) will be present in the graph if and only if S contains a species s such that $c_1(s) = i$ and $c_2(s) = j$. We will use the notation $d(u)$ to denote the degree of a vertex u in this graph. We will define a *special ℓ-coloring* of the graph to be a coloring of the edges with the colors white, blue, red, and purple such that each vertex i in R_{c_1} has $\max(0, d(i) - \ell + 1)$ of its neighboring edges colored either red or purple and the rest of its neighboring edges colored either white or blue and each vertex j in R_{c_2} has $\max(0, d(j) - \ell + 1)$ of its neighboring edges colored either blue or purple and the rest of its neighboring edges colored either white or red. (Intuitively, think of each edge as starting out white. Then each vertex i in R_{c_1} adds red color to $\max(0, d(i) - \ell + 1)$ of its neighboring edges and each vertex j in R_{c_2} adds blue color to $\max(0, d(j) - \ell + 1)$ of its neighboring edges. Edges that get colored both red and blue in this process become purple.) We will prove the following lemma.

Lemma 10: *A phylogeny input S, c_1, c_2 has an ℓ-phylogeny if and only if the corresponding bipartite graph has a special ℓ-coloring with no purple cycle.*

Proof: First, suppose that the input S, c_1, c_2 has an ℓ-phylogeny. By Fact 9 it has a restricted ℓ-phylogeny T in which each species in S is the label of exactly one node and for each character $h \in \{1, 2\}$ and each state $i \in R_{c_h}$, at most one of the connected components in the subgraph of T induced by the set of vertices $c_h^{-1}(i)$ has more than one vertex. Construct a special ℓ-coloring as follows. For each vertex $i \in R_{c_1}$ let C_i be the largest connected component in the subgraph of T induced by the set of vertices $c_1^{-1}(i)$. Arbitrarily choose $\max(0, d(i) - \ell + 1)$ of the vertices in C_i and add red color to the corresponding edges in the graph. For each vertex $j \in R_{c_2}$ let C_j be the largest connected component in the subgraph of T induced by the set of vertices $c_2^{-1}(j)$. Arbitrarily choose $\max(0, d(j) - \ell + 1)$ of the vertices in C_j and add blue color to the corresponding edges in the graph. We will now argue that the special ℓ-colored graph has no purple cycle. Suppose instead that the special ℓ-colored graph has a purple cycle consisting of the edges $(i_1, j_1), (i_2, j_1), (i_2, j_2), \ldots, (i_m, j_m), (i_1, j_m)$. Then, by construction, there is a path in T between the species (i_1, j_1) and the species (i_2, j_1) which is contained in C_{j_1}. Similarly, there is a path in T between the species (i_2, j_1) and the species (i_2, j_2) which is contained in C_{i_2}. These paths intersect exactly at the species (i_2, j_1). Continuing in this manner, we construct a cycle in T, which contradicts the fact that T is a phylogeny.

Next, suppose that the graph has a special ℓ-coloring with no purple cycle. Construct an ℓ-phylogeny T as follows. The nodes of T are the species in S. For each vertex $i \in R_{c_1}$ let C_i be the set of species in $c_1^{-1}(i)$ such that the corresponding edges in the graph have red color. Add a path to T which traverses the nodes in C_i. All of the species on this path have the same state in character 1. Also, these species correspond to red edges in the special ℓ-coloring. For the purpose of the proof, we will think of the corresponding nodes in the path as having red color. For each vertex $j \in R_{c_2}$ let C_j be the set of species in $c_2^{-1}(j)$ such that the corresponding edges in the graph have blue color. Add a path to T which traverses the nodes in C_j. All of the species on this path have the same state in character 2. Also, these species correspond to blue edges in the special ℓ-coloring. For the purpose of the proof, we will think of the corresponding nodes in the path as having blue color. We will now argue that T has no cycle. Suppose instead that T has a cycle. Note by construction that every edge in the cycle either fixes character 1 or fixes character 2 (but not both). For example, the cycle might look like $(i_1, j_1), (i_1, j_2), (i_1, j_3), (i_2, j_3), (i_2, j_4), (i_2, j_5), (i_1, j_5)$. Let $(x_1, y_1), \ldots, (x_m, y_m)$ be the sequence of nodes that we get when we traverse the nodes in the cycle in order, skipping any node such that the edge into the node fixes the same character as the edge out of the node. (For the above example, we get the sequence $(i_1, j_1), (i_1, j_3), (i_2, j_3), (i_2, j_5), (i_1, j_5)$.) Each species (x_a, y_a) is colored purple in T, so each edge (x_a, y_a) is colored purple in the graph. (To see that species (x_a, y_a) is colored purple in T, note that it is part of a path fixing the state of character 1 (hence, red color is added). It is also part of a path fixing the state of character 2 (hence, blue color is added).) Finally, we observe that the edges (x_a, y_a) form a cycle in the graph, which contradicts the fact that the graph has no purple cycle. We conclude that T has no cycle. If T is disconnected, we arbitrarily add edges making it into a tree. □

We now present a polynomial-time algorithm that takes as input an integer ℓ and a bipartite graph G and determines whether G has a special ℓ-coloring with no purple cycle. The algorithm proceeds by considering a sequence of special ℓ-colored graphs $G_0, G_1, \ldots$. Graph G_0 is an arbitrary special ℓ-coloring of the graph G. For $t \geq 1$, G_t is constructed by modifying the coloring in G_{t-1}. We will use the notation $\mathcal{E}(G_t)$ to denote the set of edges that are contained in some purple cycle in G_t. When the algorithm considers the graph G_{t-1} it will either produce a graph G_t such that $\mathcal{E}(G_t) \subset \mathcal{E}(G_{t-1})$ or it will terminate with the answer "no". If the algorithm ever produces a graph G_t such that $\mathcal{E}(G_t) = \emptyset$ it will terminate with the answer "yes".

We now show how to construct the graph G_t from G_{t-1} (or to terminate with the answer "no"). Fix an edge $e \in \mathcal{E}(G_{t-1})$. The procedure will consider a sequence of special ℓ-colored graphs $G'_0 = G_{t-1}, G'_1, G'_2, \ldots$. For each graph G'_j in the sequence, e will be a member of $\mathcal{E}(G'_j)$. For each graph G'_j, let $P(G'_j)$ be the graph that is obtained by considering all of the purple edges in G'_j (and no other edges) and let $(U_e(G'_j), V_e(G'_j))$ be the vertices of the connected component in $P(G'_j)$ that contains e. Let $M_e(G'_j)$ be the set of edges in the connected component in $P(G'_j)$ that contains e. To transform G'_j into G'_{j+1} the algorithm may make one e-*move* in which it either selects a vertex $u \in U_e(G'_j)$ and transfers the red color from one edge adjacent to u to another edge adjacent to u that does not already have red color or the procedure selects a vertex $v \in V_e(G'_j)$ and transfers the blue color from one edge adjacent to v to another edge adjacent to v that does not already have blue color. The move is *legal* if and only if $\mathcal{E}(G'_{j+1}) \subseteq \mathcal{E}(G'_j)$. Such a move is called a *finishing* move if $\mathcal{E}(G'_{j+1}) \subset \mathcal{E}(G'_j)$. It is called an e-*continuing* move if it is not finishing, but $M_e(G'_{j+1}) \subset M_e(G'_j)$. When it considers the special ℓ-colored graph G'_j, the algorithm checks every possible e-move. If it finds a legal e-move, it constructs G'_{j+1} by making this move. If the move is finishing, then the procedure returns the graph $G_t = G'_{j+1}$. If the move is not finishing, but it is e-continuing, the procedure now considers the graph G'_{j+1}. (Note that in this case $\mathcal{E}(G'_{j+1}) = \mathcal{E}(G'_j)$ so $e \in \mathcal{E}(G'_j)$.) If there are no legal e-moves that are finishing or e-continuing, the algorithm terminates with the answer "no". Note that at most $|M_e(G'_0)|$ continuing moves can be made, so the procedure terminates in polynomial time.

The correctness of the algorithm follows from the following lemma.

Lemma 11: *If a bipartite graph G has a special ℓ-coloring with no purple cycle and H is a special ℓ-coloring of G with $e \in \mathcal{E}(H)$ then there is a legal e-move from H that is either finishing or e-continuing.*

Proof: Let G_e be the subgraph of G induced by $U_e(H) \cup V_e(H)$ and let S_e denote the set of edges in G_e. We wish to compute an upper bound for $|S_e|$. To do so, let $d'(w)$ denote the degree of vertex w in graph G_e. Since G has a special ℓ-coloring with no purple cycle, G_e has a special ℓ-coloring with no purple cycle. Let H'_e be such a special ℓ-coloring of G_e. The number of edges with red color added in H'_e is at least $\sum_{u \in U_e(H)} (d'(u) - \ell + 1)$. The number of edges with blue color added is at least $\sum_{v \in V_e(H)} (d'(v) - \ell + 1)$. The number of purple edges (which have both red color and blue color) is at most $|U_e(H)| + |V_e(H)| - 1$. Hence,

$$|S_e| \geq \sum_{u \in U_e(H)} (d'(u) - \ell + 1) + \sum_{v \in V_e(H)} (d'(v) - \ell + 1) - (|U_e(H)| + |V_e(H)| - 1)$$

and therefore $|S_e| < \ell(|U_e(H)| + |V_e(H)|)$.

Now consider H. Let S_1 be the set of edges that are adjacent to vertices in U_e and do not have red color. Let S_2 be the set of edges that are adjacent to vertices in V_e and do not have blue color. Suppose that some edge e' is in $S_1 \cap S_2$. Let e'' be a purple edge that is adjacent to e'. Clearly, the e-move that transfers color from e'' to e' is legal. Suppose that it is not finishing and let H' be the graph obtained from H by making this move. Then $M_e(H') \subseteq M_e(H) - \{e''\}$. Hence, the move is e-continuing.

Suppose instead that $S_1 \cap S_2 = \emptyset$. Every vertex $w \in U_e(H) \cup V_e(H)$ has $d(w) \geq \ell$. (If $d(w) < \ell$ then w will not add color to its neighboring edges in any special ℓ-coloring of G so w will not be in the connected component containing e in $P(H)$.) Therefore $|S_1| \geq \sum_{u \in U_e(H)} (\ell - 1)$ and $|S_2| \geq \sum_{v \in V_e(H)} (\ell - 1)$. Let S_3 be the set of purple edges with endpoints in $U_e(H) \cup V_e(H)$. Note that $|S_3| \geq |U_e| + |V_e|$. S_3 is disjoint from S_1 and S_2 so $|S_1 \cup S_2 \cup S_3| \geq \ell(|U_e(H)| + |V_e(H)|)$. We conclude that some edge in S_1 or S_2 must have an endpoint outside of G_e.

Without loss of generality, assume that there is an edge $e' \in S_1$ that has endpoint $u \in U_e(H)$ and its other endpoint, v, outside of $V_e(H)$. There are two cases. Suppose that u is contained in a purple cycle in H. Let (u, w) be an edge in such a cycle. Consider the e-move that transfers color from (u, w) to e'. This move is legal. (Since v is not in $V_e(H)$ no purple cycles are created by the move.) Let H' be the graph obtained from H by making this move. $\mathcal{E}(H') \subseteq \mathcal{E}(H) - \{(u, w)\}$, so the move is finishing. Suppose instead that u is not contained in a purple cycle in H. Let (u, w) be the first edge on the unique path from u to e in $P(H)$. Consider the legal e-move that transfers color from (u, w) to e'. Suppose that it is not finishing and let H' be the graph obtained from H by making this move. Then $M_e(H') \subseteq M_e(H) - \{(u, w)\}$. Hence, the move is e-continuing. $\square$

In Lemma 10 we showed that a phylogeny input S, c_1, c_2 has an ℓ-phylogeny if and only if the corresponding bipartite graph has a special ℓ-coloring with no purple cycle. We then described a polynomial time algorithm that takes as input an integer ℓ and a bipartite graph G and determines whether G has a special ℓ-coloring with no purple cycle. Hence, we have shown that there is a polynomial time algorithm that takes input ℓ and a phylogeny input S, c_1, c_2 and determines whether the phylogeny input has an ℓ-phylogeny. (In fact, our algorithm constructs an ℓ-phylogeny if one exists.) Using binary search (or even linear search) on ℓ, we obtain a polynomial-time algorithm that takes as input a phylogeny input S, c_1, c_2 and determines the phylogenetic number of the input. Hence, we have proved the following theorem.

Theorem 12: *The phylogenetic number problem can be solved in polynomial time for $k = 2$.*

Unfortunately, Fact 9 no longer holds if we add a third character c_3. Hence, our approach does not solve the phylogenetic number problem (or even the ℓ-phylogeny problem) for fixed $k \geq 3$. (To see that Fact 9 does not hold for $k > 2$, consider the 3-species 3-character input $\{100, 010, 001\}$. One can construct a 1-phylogeny for this input by attaching each species to the new species 000. However, the input does not have a restricted 1-phylogeny.)

6 Phylogeny With a Fixed Number of States

In subsection 6.1 we show that the fixed-topology phylogenetic number problem can be solved in polynomial time for fixed r. On a related note, we show that if r is fixed, there is a polynomial-delay algorithm for listing fixed-topology ℓ-phylogenies. In subsection 6.2 we show that for fixed $r \geq 2$ and fixed $\ell \geq 3$ the restricted ℓ-phylogeny problem is NP-hard. (This result follows from a more general result. Namely, we show that the restricted (ℓ_1, ℓ_2)-phylogeny problem is NP-hard for fixed $\ell_1 \geq 2$ and $\ell_2 \geq 2$ as long as one of ℓ_1, ℓ_2 is greater than 2.)

6.1 Fixed-Topology Phylogeny with a Fixed Number of States

In this subsection we prove the following theorem.

Theorem 13: *The fixed-topology phylogenetic number problem can be solved in polynomial time for fixed r.*

It suffices to consider each character independently. We are given an input tree T with each of its n leaves labeled by a state in the range $\{1, \ldots, r\}$. We wish to label the internal nodes of T to construct a phylogeny with the smallest possible phylogenetic number. We root the tree at an arbitrary node, constructing the child and parent pointers. The choice of root will not affect the phylogenetic number of the tree.

For a given character, this problem can be solved by a two-pass algorithm: once up the tree and once down. In the upward phase, for each node v, and for each vector in the set

$$\{ (i, \ell_1, \ldots, \ell_r) \mid (1 \leq i \leq r) \text{ and } 1 \leq \ell_j \leq n \text{ for } 1 \leq j \leq r \}$$

we construct, if possible, a labeling of the nodes in the subtree rooted at v such that v is labeled with state i and, in the subtree rooted at v, the subgraph induced by nodes labeled j has exactly ℓ_j connected components. We call such a labeling a *configuration* of the subtree rooted at v, or a configuration of v for short. If there are no leaves in this subtree labeled j for some $j \in \{1, \ldots, r\}$, then we have $\ell_j = 0$ for all configurations (there are no connected components labeled j in the subtree rooted at v).

There are $O(rn^r)$ possible configurations for the subtree rooted at any node, with one possible configuration for each leaf. Once the possible configurations have been constructed for the children of a node, we can construct the possible configurations for the parent by combining configurations of the children incrementally. Consider the first two children v_1 and v_2 of parent node v. For each pairing of a configuration for v_1 with a configuration for v_2, we construct r configurations for the subtree consisting of parent node v and the subtrees rooted at children v_1 and v_2, one configuration for each possible labeling of the parent v. If node v is labeled i, and the configurations of v_1 and v_2 are represented by the vectors $(i_1, \ell_{11}, \ell_{12}, \ldots, \ell_{1r})$ and $(i_2, \ell_{21}, \ell_{22}, \ldots, \ell_{2r})$ respectively, then the resulting configuration is $(i, \ell_1, \ell_2, \ldots, \ell_r)$ where $\ell_j = \ell_{1j} + \ell_{2j}$ for all $j \neq i$, and $\ell_i = \ell_{1i} + \ell_{2i} + 1 - m$, where $m \in \{0, 1, 2\}$ is the number of children (considering only v_1 and v_2) which are labeled i. That is, the number of components of state j is the sum of the number of components in each child for most states. The only state that can differ is the state with which node v is labeled (i). In this case, if neither v_1 nor v_2 is labeled i, then we create a new component of state i (the node v) in addition to the components present in the children. If exactly one child is labeled i, then the label of node v becomes part of that component. If both v_1 and v_2 are labeled i, then one component of state i from each child can merge through node v, and the number of components in the combination is one fewer than the sum.

Whenever a new possible configuration is achieved through a combination of configurations in the two children, it is recorded along with pointers to the configurations of v_1 and v_2 that achieve this phylogenetic configuration. Although there are $r^2 n^{2r}$ ways to pair up the configurations of two children, there can be at most rn^r configurations for the parent. If a configuration is achieved multiple ways, we only remember one way.

After computing the $O(rn^r)$ configurations for the subtree consisting of node v with the subtrees rooted at v_1 and v_2 (call this tree T'), we now add child v_3. The compuation is almost the same as before. Let possible configurations for T' and the subtree rooted at v_3 be represented by vectors $(i, \ell'_1, \ell'_2, \ldots, \ell'_r)$ and $(j, \ell_{31}, \ell_{32}, \ldots, \ell_{3r})$ respectively. Then the combined configuration is $(i, \ell_1, \ell_2, \ldots, \ell_r)$ where $\ell_k = \ell'_k + \ell_{3k}$ for all k, unless $i = j$. In this case, we have $\ell_i = \ell_i' + \ell_{3i} - 1$ because one component of state i from the subtree rooted at v_3 can connect to components of state i from the other children through the parent v.

Each child of node v is added in this way until we have computed the $O(rn^r)$ possible configurations for the entire subtree rooted at node v. We continue up the tree until we have computed all possible configurations for the root. This computation takes $O(r^2 n^{2r+1})$ time. We then pick a possible configuration with the minimum phylogenetic number and go down the tree generating labels by following the pointers to the subconfigurations that achieve the optimal configuration.

The above algorithm makes it clear that if r is fixed, there is a polynomial-delay algorithm for listing fixed-topology ℓ-phylogenies.

6.2 Restricted Phylogeny with a Fixed Number of States

In this subsection we show that for fixed $r \geq 2$ and fixed $\ell \geq 3$ the restricted ℓ-phylogeny problem is NP-hard.

We start by proving the following more general theorem.

Theorem 14: *The restricted (ℓ_1, ℓ_2)-phylogeny problem is NP-hard for fixed $\ell_1 \geq 2$ and $\ell_2 \geq 2$ as long as one of ℓ_1, ℓ_2 is greater than 2.*

Proof: Without loss of generality, assume that $\ell_1 \geq \ell_2$. The reduction is from the 2-consecutive ones problem.

Let M be the matrix in the input to the 2-consecutive ones problem. Let n' denote the number of rows of M and m denote the number of columns of M. (We will assume that $n' \geq 3\ell_2$.) We will show how to construct an input to the restricted (ℓ_1, ℓ_2)-phylogeny problem such that the phylogeny input has a restricted (ℓ_1, ℓ_2)-phylogeny if and only if the rows of M can be permuted in such a way that for each column in the resulting matrix there are at most 2 sequences of consecutive ones.

The phylogeny input is constructed as follows. Let M' be a matrix derived from M by adding $2(\ell_2 - 2)$ rows to the bottom of M. The entries in the $(n' + i)$th row are equal to 0 for odd $i > 0$ and are equal to one for even $i > 0$. Let n denote $n' + 2(\ell_2 - 2)$. Note that M' has n rows. The species set $S = \{s_1, \ldots, s_n\}$ will have n species. Species s_i will correspond to row i of M'. Let k_1 denote $\binom{n}{\ell_2 - 1}$. Let k_2 denote $\binom{n'}{\ell_2 - 1}$. Let k_3 denote $\max(0, n - n' - 1)$. Let k denote $m + k_1 + k_2 + k_2 k_3$. The input to the phylogeny problem will be $S, c_1, \ldots, c_k$. The characters $c_1, \ldots, c_k$ will be defined as follows:

1. (Characters that describe M') For j in the range $1 \leq j \leq m$ character c_j will map species s_i to the entry in column j of row i of M'.

2. (Characters that make every phylogeny a path) For j in the range $1 \leq j \leq k_1$ let S_j denote the jth size-$(\ell_2 - 1)$ subset of S. We set $c_{m+j}(s) = 0$ for $s \in S_j$ and $c_{m+j}(s) = 1$ for $s \notin S_j$.

3. (Characters that place s_n at one end of the path) For j in the range $1 \leq j \leq k_2$ let S'_j denote the jth size-$(\ell_2 - 1)$ subset of $\{s_1, \ldots, s_{n'}\}$. We set $c_{m+k_1+j}(s) = 0$ for $s \in S'_j$ and $c_{m+k_1+j}(s_n) = 0$ and $c_{m+k_1+j}(s) = 1$ for every other species s.

4. (Characters that place $s_{n'+1}, \ldots, s_n$ consecutively at the end of the path) For j in the range $1 \leq j \leq k_2$ and i in the range $1 \leq i \leq k_3$ let m' denote $m + k_1 + k_2 + (i-1)k_2 + j$. We set $c_{m'}(s_r) = 0$ for $s_r \in S'_j$ and $c_{m'}(s_r) = 1$ for $s_r \in \{s_1, \ldots, s_{n'}\} - S'_j$. Furthermore, we set $c_{m'}(s_{n'+1}) = \cdots = c_{m'}(s_{n-i-1}) = 1$ and we set $c_{m'}(s_{n-i}) = \cdots = c_{m'}(s_n) = 0$.

$\rightarrow$ Suppose that T is a restricted (ℓ_1, ℓ_2)-phylogeny for $S, c_1, \ldots, c_k$. Using Fact 1, we can assume that each species in S is the label of exactly one node in T. Following the proof of Theorem 8, we can show that every node in T has degree at most 2. That is, T is a path. If $n = n'$ (i.e., $\ell_2 = 2$) then it follows that we can arrange the rows of M in the order that the species occur in path T and that, in such an arrangement, each column has at most 2 sequences of consecutive ones. Suppose instead that $n > n'$. We will now show that the node labeled s_n has degree 1. Suppose instead that it has degree 2. We argue as in the proof of Theorem 8 that there is a size-$(\ell_2 - 1)$ set $S' \subseteq \{s_1, \ldots, s_{n'}\}$ such that if s_n and the species in S' are removed from T, the resulting subgraph has at least $\ell_2 + 1$ connected components. Let j be the integer such that $S' = S'_j$. Then the set of vertices $c_{m+k_1+j}^{-1}(1)$ form more than ℓ_2 connected components in T, which is a contradiction. We conclude that the node labeled s_n is an endpoint of the path. For i in the range $1 \leq i \leq k_3$ we will now argue that the node labeled s_{n-i} is adjacent to a node with a label in $\{s_{n-i+1}, \ldots, s_n\}$. Suppose that this is not the case. We argue as in the proof of Theorem 8 that there is a size-$(\ell_2 - 1)$ set $S' \subseteq \{s_1, \ldots, s_{n'}\}$ such that if the species in $S' \cup \{s_{n-i}, \ldots, s_n\}$ are removed from T then the resulting subgraph has at least $\ell_2 + 1$ connected components. Let j be the integer such that $S' = S'_j$. Then the set of vertices $c_{m+k_1+k_2+(i-1)k_2+j}^{-1}(1)$ form more than ℓ_2 connected components in T, which is a contradiction. We conclude that T is a path consisting of the species in $\{s_1, \ldots, s_{n'}\}$ (in some order) followed by $s_{n'+1}, \ldots, s_n$. It follows that we can arrange the rows of M in the order that the species occur in path T and that, in such an arrangement, each column has at most 2 sequences of consecutive ones.

$\leftarrow$ Suppose that $p = \{p_1, \ldots, p'_n\}$ is a permutation of $\{1, \ldots, n'\}$ such that when the rows of M are permuted according to p each column has at most 2 sequences of consecutive ones. If $\ell_2 = 2$ then let T be the path consisting of the species in $\{s_1, \ldots, s_{n'}\}$, arranged according to p. T is a restricted $(3,2)$-phylogeny for $S, c_1, \ldots, c_k$. Hence, T is a restricted (ℓ_1, ℓ_2)-phylogeny for $S, c_1, \ldots, c_k$. Suppose instead that $\ell_2 > 2$. Let T be a path consisting of the species in $\{s_1, \ldots, s_{n'}\}$, arranged according to permutation p, followed by $s_{n'+1}, \ldots, s_n$. Then T is a restricted (ℓ_2, ℓ_2)-phylogeny for $S, c_1, \ldots, c_k$. Hence, T is a restricted (ℓ_1, ℓ_2)-phylogeny for $S, c_1, \ldots, c_k$. $\square$

Note that Theorem 14 has the following corollary.

Corollary 15: *For fixed $r \geq 2$ and fixed $\ell \geq 3$ the restricted ℓ-phylogeny problem is NP-hard.*

7 Conclusions

In this section we present some open problems. There are several restrictions of the parameters which yield problems for which the complexity is still open. Recall that k is the number of characters, r is the maximum number of states for any character, and ℓ is the phylogenetic number. It is unknown whether the following restricted versions of the ℓ-phylogeny problem can be solved by polynomial-time algorithms:

1. Finding an ℓ-phylogeny where the number k of characters is a constant greater than 2 (for $\ell > 1$),

2. Finding an ℓ-phylogeny where the number r of states per character is a constant.

3. For the case where $r = 2$, determining whether an input has a $(1, 2)$-phylogeny or a $(2, 2)$ phylogeny. Recall that for $r = 2$, the problem of finding a $(1, 1)$-phylogeny is in $\mathcal{P}$, but finding a $(2, 3)$-phylogeny is $\mathcal{NP}$-complete.

This paper also leaves open the problems of randomly generating phylogenies with constraints upon their phylogenetic number and approximation algorithms for the $\mathcal{NP}$-complete versions of the ℓ-phylogeny problem. In particular, suppose that there exists a perfect phylogeny. For what ℓ can we find an ℓ-phylogeny in polynomial time (with ℓ possibly a function of k and r)?

8 Acknowledgements

The work of Leslie Ann Goldberg, Paul Goldberg and Cynthia Phillips was supported by the U.S. Department of Energy under contract DE-AC04-76AL85000. Elizabeth Sweedyk was supported by the California Legislative Grant. The work of Tandy Warnow was supported by a National Science Foundation Young Investigator Award under contract CCR-9457800, and by the U.S. Department of Energy under contract DE-AC04-76AL85000.

References

1. R. Agarwala and D. Fernandez-Baca, "Fast and Simple Algorithms for Perfect Phylogeny and Triangulating Colored Graphs", DIMACS TR#94-51, 1994.

2. R. Agarwala, D. Fernández-Baca, "A Polynomial-Time Algorithm for the Perfect Phylogeny Problem when the Number of Character States is Fixed", procs. of the 34th annual Symposium on Foundations of Computer Science, 1993.

3. H. Bodlaender, M. Fellows, T. Warnow, "Two Strikes Against Perfect Phylogeny", *procs. of the 19th International Congress on Automata, Languages and Programming (ICALP)*, pp. 273-287, Springer-Verlag Lecture Notes in Computer Science, 1992.

4. W.H.E. Day, "Computationally difficult parsimony problems in phylogenetic systematics," *Journal of theoretical biology*, 103: 429-438.

5. W.H.E. Day and D. Sankoff, "Computational complexity of inferring phylogenies by compatibility", *Systematic Zoology*, 35(2): 224-229, 1986.

7. Wm. Fitch, "Toward defining the course of evolution: minimum change for a specified tree topology", *Syst. Zool.*, 20:406-416, 1971.

8. D.R. Fulkerson, D.A. Gross, "Incidence matrices and interval graphs", Pacific J. Math., 15 (3), 1965.

9. P.W. Goldberg, M.C. Golumbic, H. Kaplan, R. Shamir, "Four Strikes Against Physical Mapping of DNA", Tech. Rept. 287/93, Tel Aviv University, 1993.

10. S. Kannan and T. Warnow, "Inferring Evolutionary History from DNA Sequences", *SIAM J. on Computing*, Vol. 23, No. 4, August 1994.

11. S. Kannan and T. Warnow, "A fast algorithm for the computation and enumeration of perfect phylogenies", to appear, *ACM/SIAM Symposium on Discrete Algorithms*, 1995.

12. F.R. McMorris, T. Warnow, T. Wimer, "Triangulating Colored Graphs", *SIAM J. on Discrete Mathematics*, Vol. 7, No. 2, pp. 296-306, 1994.

13. R.C. Prim, "Shortest Connection Networks and Some Generalisations", Bell System Tech. J., **36** 1389-1401, 1957.

14. M.A. Steel, "The complexity of reconstructing trees from qualitative characters and subtrees", *Journal of Classification*, **9** 91-116, 1992.

15. T. Warnow, "Efficient Algorithms for The Character Compatibility Problem", *New Zealand Journal of Botany*, Vol. 31, (1993), pp. 239-248.

Making the Shortest-Paths Approach to Sum-of-Pairs Multiple Sequence Alignment More Space Efficient in Practice (Extended Abstract)

Sandeep K. Gupta[1*], John D. Kececioglu[2**], Alejandro A. Schäffer[3***]

[1] Department of Computer Science, Rice University, Houston, TX 77005–1892
[2] Department of Computer Science, The University of Georgia, Athens, GA 30602–7404
[3] Department of Computer Science, Rice University, Houston, TX 77005–1892

Abstract

The MSA program, written and distributed in 1989, is one of the few existing programs that attempts to find optimal alignments of multiple protein or DNA sequences. MSA implements a branch-and-bound technique on a variant of Dijkstra's shortest paths algorithm to prune the basic dynamic programming graph. We have made substantial improvements in the time and space usage of MSA. On some runs, we achieve an order of magnitude reduction in space usage and a significant multiplicative factor speedup in running time. To explain these improvements, we give a much more detailed description of MSA than has been previously available.

1 Introduction

Alignment of multiple DNA and protein sequences is a fundamental problem in molecular biology. A variety of combinatorial definitions of the alignment problem are used in practice. Finding an optimal multiple sequence alignment seems to require time and space exponential in the number of sequences for all definitions, and is provably hard for a few [14, 5, 12, 19]. Nevertheless, it is often useful to align a small number of sequences, that is greater than two, optimally.

One of the existing programs that attempts to construct an optimal alignment, for some definition of the problem, is MSA [6, 13]. MSA (version 1.0) was completed and distributed in 1989; most of the implementation of the resource-intensive part was performed by the second author of this paper. Reference [13] gives a short overview of the program, but no detailed description of the implementation has ever been published. The description in Sections 2, 3, and 4 below of the shortest paths algorithm is derived from notes written and circulated by the second author [11].

---

* Present address: Laboratory for Computer Science, Massachusetts Institute of Technology, 545 Technology Square, Cambridge, MA 02139; skgupta@pdos.lcs.mit.edu
** Some of the work of this author was carried out at the Department of Computer Science of the University of California at Davis; kece@cs.uga.edu
*** schaffer@cs.rice.edu

Some other programs that attempt to compute a global multiple sequence alignment include: AMULT [4, 3], DFALIGN [9], MULTAL [17, 18], TULLA [16], CLUSTAL V [10], and MWT [12]. Of these methods, MWT is the only one other than MSA that attempts to compute an optimal alignment based on some well-defined score. It is important to point out that in practice, the MSA program rarely produces a provably optimal alignment, even on successful termination; the precise reasons for lack of optimality are discussed at length in the full paper. Methods other than MSA and MWT tend to use much less space, mainly because they do not search for an optimal alignment. Two surveys of these programs, and other issues in multiple sequence alignment are [7, 15].

The stimulus for this paper is a successful attempt to reduce the space usage of MSA by significant multiplicative factors. On some runs, we achieve an order of magnitude reduction in space usage and a significant multiplicative improvement in running time. The amount of improvement varies widely with the data set. In practice, the space usage of the original MSA is a more severe constraint than its time usage. The space reduction we have achieved makes many runs that were not feasible with the original MSA now feasible for the first time, and speeds up many runs that before were marginally feasible but would thrash extensively while allocating virtual memory.

The resource-intensive part of MSA is an implementation of a complex variant of Dijkstra's single-source shortest-paths algorithm [8]. The version of multiple sequence alignment that MSA solves can be formulated as a single-source, single-destination shortest-paths problem on a mesh-shaped dynamic programming graph. The dimension of the mesh is the number of sequences, and the number of vertices is the product of the sequence lengths.

To save space (even in the original version) the dynamic programming graph is generated "on-the-fly", so that only subgraphs are stored in memory at any given time. The essential idea in our space usage improvements is to prove algorithmic invariants regarding when edges and vertices of the graph first need to be created, and when they can be safely destroyed. To further improve the running time we carefully recoded a few subroutines that consumed the bulk of the running time, as detected through runtime profiling of the code.

Though MSA will never be able to compete with heuristic methods that can align dozens of sequences of triple-digit lengths, the new version is significantly more powerful than the original. The source code is currently available by anonymous ftp to softlib.cs.rice.edu. Contact the third author by electronic mail (schaffer@cs.rice.edu) for details.

The paper is organized as follows. Section 2 defines the version of multiple sequence alignment that MSA solves, and the basic mathematical method it uses. Section 3 describes in more detail how MSA implements this method in the absence of gap penalties. Section 4 explains how gap penalties are incorporated. Section 5 describes our space and time improvements. Section 6 compares the two versions of MSA with measurements on real data sets of protein sequences.

2   Definitions and the Carrillo-Lipman heuristic

In this section we define the version of multiple sequence alignment solved by MSA
and review the method of Carrillo and Lipman that motivated the development
of MSA. Let $S_1, \ldots, S_K$, be the input sequences and assume that K is at least 2.
Let Σ be the input alphabet; we assume that Σ does not contain the character
'-', so that a dash can be used to denote a gap in the alignment.

A multiple sequence alignment A is a rectangular character array on the
alphabet $\Sigma' := \Sigma \cup \{-\}$ that satisfies the following conditions:

1. There are exactly K rows.
2. Ignoring dashes, row I is precisely the string S_I.

Each multiple sequence alignment induces a pairwise alignment $A_{i,j}$ on the pair
of sequences S_i, S_j. The alignment is obtained by simply copying rows i and j
of A, with the proviso that columns with a '-' in both rows are ignored.

Figure 1 shows an example multiple alignment computed by MSA using the
default parameter settings. It has four rows, one per sequence. In this case each
input sequence has 30 characters, and the output alignment has 33 columns. In
general, the input sequences can be of varying lengths. The four sequences in
this figure are the first 30 characters of each of the four globin sequences HAHU,
HBHU, MYHU, IGLOB from the globin data set in [15].

```
--VLSPADKTNVKAAWGKVGAHAGEYGAEALE-
-VHLTPEEKSAVTALWGKVNVD--EVGGEALGR
--GLSDGEWQLVLNVWGKVEADIPGHGQEVLI-
MKFFAVLALCIVGAIASPLTADEASLVQSS---
```

Fig. 1. An alignment of prefixes of four globin sequences

In the simplest cost model there is a cost function sub : $\Sigma' \times \Sigma' \to \mathbf{N}$, such
that $\mathrm{sub}(a, b)$ is the cost of substituting a b in the second sequence for an a in
the first sequence; also, $\mathrm{sub}(-, b)$ is the cost for columns where the first sequence
has a gap and the second has a b, and $\mathrm{sub}(a, -)$ is the cost for columns where
the first sequence has an a and the second has a $-$. Function sub should be
symmetric, and $\mathrm{sub}(-, -)$ should have value 0. The cost of a pairwise alignment
$A_{i,j}$ induced in a rectangular array A of width w is

$$c(A_{i,j}) := \sum_{1 \leq k \leq w} \mathrm{sub}(A[i][k], A[j][k]).$$

A crucial point is that the cost of the pairwise alignment $A_{i,j}$ induced by A will
always be at least as great as the cost of an optimal alignment of S_i, S_j alone.
MSA attempts to minimize pairwise alignment costs.

By default, MSA supports a more sophisticated cost model where gaps that span several consecutive columns cost less than the sum of sub applied individually to each column. In particular, computing the substitution cost of a column depends on whether the previous column has a − or not. We use the notation $c(A_{i,j})$ for all cost models, and rely on the context to specify the cost model.

The basic *sum of pairs* multiple sequence alignment problem is to minimize the pairwise sum

$$c(A) := \sum_{i<j} c(A_{i,j})$$

over all alignments A. MSA can solve sum of pairs multiple sequence alignment, if the user supplies the appropriate flags. By default, however, the MSA program solves a variation where each pairwise alignment cost is multiplied by a weight, denoted by $\text{scale}(S_i, S_j)$, with weights chosen from an evolutionary tree of the input sequences. A method described in [2] is used to compute the weights. The weights are useful to diminish representational bias caused by having more sequences from some species and fewer sequences from others.

Although in principle MSA can solve problems to optimality, it often does not; furthermore, in most cases where it arrives at a solution that is optimal, it does not actually explore enough of the solution space to *guarantee* that the output alignment is optimal. The reasons for lack of optimality are discussed at length in the full paper.

As Carrillo and Lipman [6] realized, to solve the sum-of-pairs alignment problem, not all alignments A have to be considered. Define $d(S_1, S_2, \ldots, S_K) := \min_A c(A)$. Let

$$L := \sum_{i<j} d(S_i, S_j) \cdot \text{scale}(S_i, S_j)$$

be the sum-of-pairs lower bound. Let U be an upper bound on $d(S_1, \ldots, S_K)$, and let $\mathcal{A}$ be an alignment of minimum cost. Then for any distinct p and q,

$$U - L \geq \sum_{i<j} \left(c(\mathcal{A}_{i,j}) - d(S_i, S_j) \right) \geq c(\mathcal{A}_{p,q}) - d(S_p, S_q),$$

since for all i, j, $c(\mathcal{A}_{i,j}) \geq d(S_i, S_j)$. Rearranging, we derive the Carrillo and Lipman bound

$$c(\mathcal{A}_{i,j}) \leq d(S_i, S_j) + U - L. \tag{1}$$

Note that $\Omega(K^2)$ terms are thrown away to arrive at (1); while it is possible for the bound to achieve equality, for K equally dissimilar sequences $c(\mathcal{A}_{i,j}) - d(S_i, S_j)$ will be overestimated by a factor $O(K^2)$. In the implementation, MSA applies (1) only during an initial preprocessing phase.

In the classic dynamic programming algorithm for sequence alignment, a directed acyclic graph is constructed in which source-to-sink paths of weight C correspond to alignments of cost C. For multiple sequence alignment, this graph is conveniently represented with vertices embedded at integer coordinates in K-space and sequences labeling the coordinate axes. A path P starting at the source vertex $s = \langle 0, \ldots, 0 \rangle$ that ends at vertex $t = \langle n_1, \ldots, n_k \rangle$ represents an alignment of the K prefixes $S_i[1 \ldots n_i]$ for $1 \le i \le K$. There is a one-to-one correspondence between path edges and alignment columns. In the alignment of Fig. 1 the first two edges are

$$\langle 0,0,0,0 \rangle \to \langle 0,0,0,1 \rangle \to \langle 0,1,0,2 \rangle$$

For column c, if we do not have a dash in row i, then the cth edge in the path advances the coordinate in dimension i by 1; a dash in row i means the coordinate in dimension i stays the same.

Inequality (1) restricts consideration to those paths whose projection onto the planes defined by all pairs S_i, S_j has weight at most $d(S_i, S_j) + U - L$. MSA 1.0 also prunes vertices by lower-bounding costs from the current vertex to the sink. If the minimum cost to reach v from the source plus a lower bound on the cost to reach the sink from v is higher than U, v can be pruned. We found that not performing the lower-bound pruning eliminates one field in a record, saving enough space that our new version actually runs faster without the pruning. For each pair of sequences S_i, S_j, MSA computes the standard two-dimensional dynamic programming graph $D_{i,j}$ and the cost of an optimal alignment passing through each vertex. However, MSA considers only paths that pass through vertices whose two dimensional projections are near the optimal two-dimensional alignment paths for each pair of dimensions. Details of how "near" can be defined by the user are given in the full paper.

3 Basic implementation without gap penalties

We design an algorithm that incorporates inequality (1) and prunes vertices by lower-bounding costs, where pairwise alignments are scored as the sum of single-symbol insertion, deletion, and substitution costs; this algorithm is later extended for an affine gap cost function. Input consists of sequences $S_1, S_2, \ldots, S_K$ of lengths $N_1, N_2, \ldots, N_K$, and the user-selected difference $\delta := U - L$. We will write N for $\max_i N_i$.

The algorithm is naturally organized in two stages. In the first stage, $O(K^2)$ pairwise sequence comparisons are performed to determine, for each plane defined by sequence pair S_i and S_j, the set of points $\langle p, q \rangle$ on a path from $\langle 0,0 \rangle$ to $\langle N_i, N_j \rangle$ of weight at most $d(S_i, S_j) + \epsilon_{i,j}$, where $\epsilon_{i,j} \le \delta$ is a user-specifiable parameter. Let $F_{i,j}$ be the set of such points $\langle p, q \rangle$ on the face formed by S_i and S_j. At the conclusion of this stage, the $F_{i,j}$ are determined, and the lower bound $L = \sum_{i<j} d(S_i, S_j)$ and upper bound $U = L + \delta$ can be computed. The first stage is neither time- nor space-intensive, so we give it superficial treatment.

In the second stage, an alignment corresponding to a shortest source to sink path is computed; it is guaranteed that the projections of the path lie wholly within the $F_{i,j}$. Formally the weighted directed acyclic graph $G = \langle V, E \rangle$ has vertex set

$$V := [0, N_1] \times [0, N_2] \times \cdots \times [0, N_K]$$

and edge set

$$E := \left\{ v \to w \,\|\, w - v \in \{0,1\}^K - \{0\}^K \right\},$$

where we use the standard componentwise subtraction of vectors. For charac-
ter a, define $a^1 := a$ and $a^0 := '-'$. Using this notation, edge $\langle v_1, \cdots, v_K \rangle \xrightarrow{e} \langle w_1, \cdots, w_K \rangle$ has cost

$$\text{cost}(e) = \sum_{i<j} \text{sub}\left((S_i[w_i])^{w_i-v_i}, (S_j[w_j])^{w_j-v_j} \right).$$

Let $R_{i,j}$ be the region in K-space of points $\langle p_1, p_2, \ldots, p_K \rangle$ such that $\langle p_i, p_j \rangle \in F_{i,j}$, and define $R := \bigcap_{i<j} R_{i,j}$. Then the paths of interest lie within $G \setminus R$, the subgraph of G induced on the vertex subset R.

There are some subtleties to the shortest-path algorithm, as vertices and edges are generated dynamically in a graph with a mesh vertex structure. We say a vertex or edge "exists" when the algorithm has allocated space for it and has not freed that space. Consider the computation when it examines the edges leaving a vertex v with coordinates given by point p. Adjacent vertices are those with coordinates q such that $q - p \in \{0,1\}^K - \{0\}^K$. Having generated the coordinates of point q for an adjacent vertex, we must find the vertex w corresponding to point q if it exists, or create w if it does not exist. (Vertex w may already exist if another path to point q has already been explored.) We can accomplish this with a trie defined on point coordinates of existing vertices. In the trie, coordinates of a point q spell a path from the root to a leaf, with the leaf pointing to the corresponding vertex w. Leaves all have depth K. Each level corresponds to a dimension.

This algorithm is formalized in Fig. 2. Function PRIORITY() creates an empty priority queue; INSERT(x, k, q) inserts item x with key k into priority queue q; EXTRACT(q) returns and removes the item of minimum key in priority queue q; DECREASE(x, k, q) decreases to k the key of x in q. Function TRIE() creates an empty trie; INSTALL(x, s, t) installs item x with string s into trie t; LOOKUP(s, t) returns the item with string s in trie t, or Λ if none exists. In addition, two new data types are used. Type **point** is a K-tuple of integers; a point variable p has named fields $p.1, p.2, \cdots, p.K$, and is created by $p \leftarrow$ POINT($p_1, p_2, \cdots, p_K$). Type **vertex** has two named fields: a distance D, and a point P; statement $v \leftarrow$ VERTEX(d, p) creates a vertex v with $v.D = d$ and $v.P = p$. The priority queue of vertices is ranked by distance, while the trie of vertices is constructed over the integer sequences forming points. Shortest paths are computed using the distance field; adjacent vertices are generated and accessed using the point field.

Function TRACEBACK(v) returns an alignment by tracing backward through the graph from vertex v. (By examining the distance and point fields of preceding vertices, a shortest path can be reconstructed.)

One optimization incorporated in the algorithm of Fig. 2 is that edges leaving some vertices may be ignored. When a vertex v is removed from the priority queue, $v.D$ is the length of a shortest path from s to v. (It is assumed that all insertion, deletion, and substitution costs—and hence all edge weights—are positive.) A lower bound on the length of a path from v to t is $\sum_{i<j} C_r[v.P.i, v.P.j]$, whose terms are computed in Stage 1. If this quantity plus $v.D$ exceeds U, then

algorithm MSA(sequence $S_1, S_2, \cdots, S_K$; **number** δ)
  trie T ; **priority queue** Q ; **vertex** s, t, v, w ; **point** p, q
  (Stage 1.)
  $L \leftarrow 0$
  for $I \leftarrow 1$ **to** $K - 1$ **do**
    for $J \leftarrow I + 1$ **to** K **do**
      Let $a_1 a_2 \cdots a_n = S_I$ and $b_1 b_2 \cdots b_m = S_J$.
      for $i \leftarrow 1$ **to** n **do**
        for $j \leftarrow 1$ **to** m **do**
          $C_{f,I,J}[i,j] := d(a_1 \cdots a_i, b_1 \cdots b_j)$
          $C_{r,I,J}[i,j] := d(a_{i+1} \cdots a_n, b_{j+1} \cdots b_m).$
        od
      od
      for $i \leftarrow 1$ **to** n **do**
        $F_{I,J}[i] \leftarrow \{j \parallel C_{f,I,J}[i,j] + C_{r,I,J}[i,j] \leq C_{f,I,J}[n,m] + \epsilon_{i,j}\}$
      od
      $L \leftarrow L + C_{f,I,J}[n,m]$
    od
  od
  $U \leftarrow L + \delta$
  (Stage 2.)
  $s \leftarrow$ VERTEX$(0,$ POINT$(0, 0, \cdots, 0))$; $t \leftarrow$ VERTEX$(\infty,$ POINT$(N_1, N_2, \cdots, N_K))$
  $T \leftarrow$ TRIE$()$; INSTALL$(s, s.P, T)$; INSTALL$(t, t.P, T)$
  $Q \leftarrow$ PRIORITY$()$; INSERT$(s, s.D, Q)$
  while $\neg$ EMPTY(Q) **do**
    $v \leftarrow$ EXTRACT(Q) ; $p \leftarrow v.P$
    if $v = t$ **then**
      output TRACEBACK(t) ; **return**
    fi
    if $v.D + \sum_{i<j}(\text{scale}(S_i, S_j) \cdot C_{r,I,J}[p.i, p.j]) \leq U$ **then**
      for all $q \leftarrow$ POINT$(q_1, q_2, \cdots, q_K)$
        s.t. $(q - p \in \{0,1\}^K - \{0\}^K$ **and for all** $i < j, q_j \in F_{i,j}[q_i])$ **do**
        $w \leftarrow$ LOOKUP(q, T)
        if $w = \Lambda$ **then**
          $w \leftarrow$ VERTEX(∞, q) ; INSERT$(w, w.D, Q)$; INSTALL$(w, w.P, T)$
        fi
        if $v.D + \text{cost}(p \xrightarrow{e} q) < w.D$ **then**
          $w.D \leftarrow v.D + \text{cost}(e)$; DECREASE$(w, w.D, Q)$
        fi
      od
    fi
  od
  output There does not exist a multiple sequence alignment with cost at most U.
end

Fig. 2. Algorithm without gap costs.

the length of any path from s to t through v exceeds U, and edges leaving v do not need to be examined.

We now give some more details about the concrete implementation of the data structures. Making the realistic assumption that edge costs are integers, we can implement a discrete priority queue with buckets for the possible path lengths, the number of buckets being bounded by U. Then INSERT() and DECREASE() are constant-time operations, and all EXTRACT() operations take $O(U + |R|)$ total time.

Since the trie is constructed over integer sequences, the children of a node are conveniently indexed through an array of pointers. Consider accessing the vertex in the trie associated with point $\langle p_1, p_2, \cdots, p_K \rangle$ in R. After having traversed the trie on prefix $p_1 p_2 \cdots p_{j-1}$, the next coordinate p_j must be in the set $\bigcap_{i<j} F_{i,j}[p_i]$. Let m and n be the minimum and maximum integers in this set. Then pointers to child nodes, and an encoding of the intersection, can be represented in a vector of length $O(n - m)$. These vectors tend to be short, and once allocated, never need change in length as new vertices are installed. Generation of adjacent vertices is often short-circuited by "falling off" the trie. To minimize space consumption, the P vertex field is not represented as a K-tuple of integers; $v.P$ points to the trie leaf associated with vertex v, and coordinates are recovered from trie node labels by following father links to the root.

4 Adding gap penalties

In this section we explain how to modify the shortest-paths algorithm to acco-modate affine gap costs in the alignment scoring function.

When costs are determined character-by-character without special treatment of multiple-character gaps, the shortest paths in the graph representing least-cost alignments satisfy an important property, often called the "principle of optimality". Specifically, any least-cost path from the source s to a vertex w, ending in edge $v \rightarrow w$, must take a least-cost subpath from s to v. This justifies computing the distance from s to w, denoted $w.D$, by

$$w.D \leftarrow \min_v \{v.D + \text{cost}(v \rightarrow w)\},$$

which the algorithm of the previous section does implicitly.

With affine gap costs, in which the cost of a multi-character gap in an in-duced pairwise alignment is a per-character penalty times the length of the gap, plus a gap startup penalty, for initiating a multi-character gap, the principle of optimality does not apply in the above form. It may be preferable to reach w through a more costly path from s to v that ends with a gap already initiated, so that following with edge $v \rightarrow w$ does not incur additional gap startup cost.

In general, gap startup penalties under the sum-of-pairs scoring function cannot be determined column-by-column. Determining whether a column starts a gap in a pairwise alignment may require knowing the gapping structure of an arbitrary number of previous columns. MSA compromises, and uses the scheme of

```
algorithm COST(edge f, edge e)
    point p, q, r ; integer Δₑ[1 ... K], Δf[1 ... K] ; character subchar[1 ... K]
    p ←e.tail.P
    q ←f.tail.P
    r ←f.head.P
    for I ←1 to K do
        Δₑ[I] ←q[I] − p[I]
        Δf[I] ←r[I] − q[I]
        if Δf[I] = 1 then
            subchar[I] ←S[I][r[I]]
        else
            subchar[I] ←DASH()
        fi
    od
    totalcost ←0
    for I ←2 to K do
        for J ←1 to I − 1 do
            if DoNotPenalizeTerminalGaps
                and (q[I] = 0 or q[I] = N[I] or q[J] = 0 or q[J] = N[J]) then
                totalcost ←totalcost + scale(Sⱼ, Sᵢ) * sub(subchar[J], subchar[I])
            else
                totalcost ←totalcost + scale(Sⱼ, Sᵢ) * (sub(subchar[J], subchar[I])+
                GAPPENALTY(Δₑ[I], Δₑ[J], Δf[I], Δf[J]))
            fi
        od
    od
    return totalcost
end
```

Fig. 3. Edge cost algorithm.

Altschul [1] for counting gap startup events. In this scheme, gap startup events for a given column are determined solely from the preceding column in the alignment. The number of events counted by this scheme is never less than the true number of events under the sum-of-pairs measure; hence the lower bounds used by the shortest-path algorithm for pruning remain valid.

Thus, with affine gap penalties, the cost of an edge f in a path is a function of f and the preceding edge e on the path. The way edge costs are computed in MSA 1.0 is shown in Fig. 3. Suppose that $e = u \rightarrow v$ and $f = v \rightarrow w$. Let the points in K-dimensional space represented by vertices u, v, w be p, q, r respectively. We compute the cost of edge f under the assumption that it is preceded by edge e. From points p and q, we can determine which sequences already have a dash from edge e: the S_I for which $q[I] = p[I]$. Otherwise $q[I] - p[I] = 1$, in which case S_I contains a real character in the column corresponding to edge e. The difference $q[I] - p[I]$ is stored in Δ_e; similarly we compute Δ_f for edge f.

The double loop on I, J at the bottom of the cost algorithm computes the sum of pairwise contributions from edge f. When there are no gaps in the sequences, the pairwise contribution depends only on $\text{sub}(S_I[r[I]], S_J[r[J]])$. When there are gaps in the sequences, the GAPPENALTY term, which implements the scheme of Altschul [1], assigns different costs depending on whether following edge e by edge f is counted as beginning a new gap or continuing an old gap.

algorithm MSA(**sequence** $S_1, S_2, \cdots, S_K$; **number** δ)
    trie T ; **priority queue** Q ; **vertex** s, t, v, w ; **point** p, q, r ; **edge list** E ;
    edge e, f
    (Stage 2.)
    $s \leftarrow$VERTEX(POINT$(0, 0, \cdots, 0)$) ; $t \leftarrow$VERTEX(POINT$(N_1, N_2, \cdots, N_K)$)
    $T \leftarrow$TRIE() ; INSTALL$(s, s.P, T)$; INSTALL$(t, t.P, T)$
    $e \leftarrow$EDGE$(NULL, s, 0)$
    $E \leftarrow$EDGELIST() ; INCLUDE(e, E)
    $Q \leftarrow$PRIORITY() ; INSERT$(e, e.D, Q)$
    while $\neg$ EMPTY(Q) **do**
        $e \leftarrow$EXTRACT(Q) ; $v \leftarrow e.head$
        $q \leftarrow v.P$; $p \leftarrow e.tail.P$
        if $v = t$ **then**
            output TRACEBACK(t) ; **return**
        fi
        if $e.D + \sum_{i<j} \text{scale}(S_i, S_j) \cdot C_r[q.i, q.j]) \leq U$ **then**
            for all $r \leftarrow$POINT$(r_1, r_2, \cdots, r_K)$
                s.t. $(r - q \in \{0, 1\}^K - \{0\}^K$ **and for all** $i < j$ $r_j \in F_{i,j}[r_i])$ **do**
                $w \leftarrow$LOOKUP(r, T)
                if $w = NULL$ **then**
                    $w \leftarrow$VERTEX(r) ; INSTALL$(w, w.P, T)$
                fi
                $f \leftarrow$FIND(q, r, E)
                if $f = \Gamma$ **then**
                    $f \leftarrow$EDGE(v, w, ∞) ; INCLUDE(f, E) ; INSERT$(f, f.D, Q)$
                fi
                if $e.D + $COST$(f, e) < f.D$ **then**
                    $f.D \leftarrow f.D + $COST$(f, e)$; DECREASE$(f, f.D, Q)$
                fi
            od
        fi
    od
    output There does not exist a multiple sequence alignment with cost at most U.
end

Fig. 4. Modified Stage 2 for affine gap costs.

Figure 4 shows the modifications to the alignment algorithm for affine gap costs. Stage 1 does not change significantly, so we show only the new Stage 2. It is natural to modify Stage 2 so that it is an edge-based version of Dijkstra's algorithm. Instead of computing the shortest path starting at s and ending at each vertex, we compute the shortest path starting at s and ending at each edge. To make things work completely in terms of edges, a dummy source node $NULL$ is introduced, and an edge from $NULL$ to s is created to start the algorithm. Now the D distance fields are associated with edges and $e.D$ is the current minimum cost of a path starting at s and ending with edge e.

Looking at the data structures, the priority queue Q now stores edges. The trie still stores vertices to allow lookups by coordinate values. EDGELIST() creates an empty list of edges. The call EDGE(v, w, d) makes an edge from vertex v to vertex w with distance label d. INCLUDE(e, E) adds edge e to edge list E. FIND(q, r, E) returns the edge with tail q and head r in EDGELIST E if there is one, or returns Γ if such an edge is not on the list.

5   Improvements

In this section we describe the algorithmic changes that we implemented to improve the performance of MSA.

By profiling MSA with the function-level profiler **gprof** we learned two important facts:

1. The running time is dominated by three functions:
   - COST, which computes the cost of an edge e followed by an edge f,
   - ADJACENT, which generates the next vertex adjacent to the tail of the edge just extracted from the heap, and
   - EDGE, which searches through a small list of edges for an edge between two endpoints, and if the edge is not found, creates a record for a new edge and initializes it. Our measurements showed that the vast majority of the time over all calls to EDGE was spent searching for edges that already existed.
2. The bulk of the space is taken by records that store edges.

Starting here we naturally set out to investigate exactly when edges and vertices are needed. Later we recoded the cost computation to save more time.

We discovered that a major problem with edges is that they were stored in lists at the incoming vertex; i.e., the edge $v \rightarrow w$ would be stored in a list at vertex w. This choice had two virtues: it allowed for easy backtracking from source to sink to report the final alignment, and it was the easiest way to check whether an edge already existed or needed to be created. However, it also has two significant problems: it requires many list searches (which do not take $O(1)$ time) to find records for existing edges, and it precludes several opportunities for deleting edges and vertices. To see the first problem consider what happens each time v is extracted from the priority queue. We will want to find all the edges $v \rightarrow w$, but each such edge is in an arbitrary location on a list at w, and we need to search the list to find it. The second problem is more subtle, and we will explain it later at length.

If we instead store the list of outgoing edges $v \to w$ at v, we can scan the list in $O(1)$ time per edge and we never need to regenerate an edge. We changed the implementation, so that edges are stored in a list at the tail vertex. If the list of outgoing edges is NULL, then the vertex has no outgoing edges.

To provide for easy backtracking, we added a BACKTRACK field to each edge, which is its preceding edge in some optimal path. This makes backtracking very simple and fast, although it costs one extra pointer per edge. Each edge also has a reference count, which keeps track of how many BACKTRACK fields point to it. This reference count is very helpful in detecting when edges and vertices are no longer needed.

With these changes, we can now point out four facts about the creation of vertices and edges. We modified the code to take advantage of the opportunities to delete edges and vertices suggested by these facts. More details are given in the full paper.

Fact 1. Once an edge e is extracted from the priority queue, $e.D$ will never again decrease.

Fact 2. It is correct to delete (and never recreate) an edge $e = v \to w$ with $w \neq t$, that has already been extracted from the priority queue, for either of the following two reasons:

1. w has no outgoing edges (can be detected in $O(1)$ time by empty outgoing list), or
2. all edges of the form $w \to x$ have an edge other than e as backtrack edge (detected in $O(1)$ time by reference count on e).

Fact 3. If the list of outgoing edges for v ever becomes empty, after v has been visited, then all edges into v can be deleted.

The test in Fact 3 is extremely useful in deleting edges, and helps explain why it is a good idea to store the edges at their tail.

Fact 4. If vertex v has no outgoing edges or incoming edges after some edge outgoing from v is extracted, then v can never occur on an optimal path. In this case, v can be deleted, and need not be recreated.

Next we turn to some improvements that reduce the time spent computing COST. The first improvement we made was to inline COST by moving its code inside the MSA function that has the basic loop for Dijkstra's algorithm. An obvious, but minor advantage of inlining is that the time needed for the actual function call is saved.

A much more significant advantage of inlining COST is that we can use invariants that apply across consecutive calls. Recalling the main loop from the previous section, suppose e has just been extracted from the queue. Let the endpoints of e have coordinate values p and q. We then find each point r that can follow q and call COST with points p, q, r as arguments. The key points are that there will usually be multiple values of r for one p, q pair, but much of the COST computation depends only on p and q. After inlining COST, we identified those pieces of the cost computation that depended only on p, q and pulled them out of the loop on r.

The second collection of improvements involved the fundamental comparison of D labels in Dijkstra's algorithm. Recall that if we just extracted $e = p \rightarrow q$ (using coordinates) from the heap and $f = q \rightarrow r$ is a succeeding edge, we compare whether $e.D + \text{cost}(e, f) < f.D$. Our experiments with the basic-block-level profiler tcov made it clear that the test fails very often.

Our experiments suggested that we should see if the test would fail before all of the costs in cost are calculated. That is, it may be the case that $e.D + extra \geq f.D$, where $extra < \text{cost}(e, f)$. As a start, we tried with $extra = 0$, avoiding the cost computations entirely. Surprisingly this showed a measurable speedup. Then we noticed that the COST computation now inlined had a double loop on i, and j, where each loop iteration adds to the cost. A natural place to put a more refined partial test is after each iteration of the outer loop on i, using $extra$ as the pairwise sum of costs for all $j < i$.

Finally, we noticed from a tcov profile that there is a different way to arrange the double loop on i and j, which might allow for more early exits with a sufficiently large vale of $extra$. In the original formulation the first iteration on i did 1 value of j, the second iteration on i did 2 values of j and so on, until the last iteration did $K - 1$ values of j. We reversed the loop on i, so that the first iteration did $K - 1$ values, the second $K - 2$ and so on.

In response to a question from a referee, we tried deleting the pruning of edges by lower-bound to sink vertex mentioned in Section 2. Deleting the lower-bound to sink test eliminates one field from a record type of which there are many instances. The resulting space-savings also caused an improvement in running time, although this may be architecture-dependent.

6   Results

In this section, we compare the performance of MSA 1.0 with our improved version on real data sets. We ran all the tests on a Sun SparcStation 10 running SunOS 4.1.3 with 128 Mbytes of RAM. We used the cc C compiler with the -O flag for optimization. To measure time we took the sum of the user and system times reported by the time command. To measure space usage we ran the top command while MSA was running and noted the peak amount of space usage that it reported.

We used data sets extracted from the data sets presented by McClure, Vasi, and Fitch [15]. They compare 4 data sets consisting of 12 globin sequences, 12 kinase sequences, 12 protease sequences, and 12 RH sequences respectively. We extracted subsets of these 12-element sets that would give rise to MSA runs of moderate time and space usage.

In all runs, we used MSA with the default parameter settings, except that we made the upper bound on cost large enough for MSA to find some alignment. As the default settings consider restricted $F_{I,J}$ regions, the alignments produced are not provably optimal.

The data sets we used, in the nomenclature of [15], are:

- Globins A: HUMA (human hemoglobin, α chain), HBHU (human hemoglobin β chain), MYHU (human myoglobin), IGLOB (insect hemoglobin from *Chironomus thummi*), GPUGNI (nonlegume hemoglobin from swamp oak), GGZLB (bacteria, *Viteoscilla sp.*), GPYL (legume, yellow lupine).
- Globins B: HAOR (duckbill platypus, α-chain hemoglobin), HADK (duck, α-chain hemoglobin), HBOR (duckbill platypus, β-chain hemoglobin), HBDK (duck, β-chain hemoglobin), MYHU, MYOR (duckbill platypus, myoglobin), IGLOB, GPYL, GPUGNI, GGZLB.
- Proteases A: HIV-I (human immunodeficiency virus, Type I), SRV-I (simian retrovirus, type I), MoMLV (Moloney murine leukemia virus), 17.6 (retrotransposon from *Drosophila melanogaster*), PEPH (human pepsin sequence).
- Proteases B: PEPH, MoMLV, CaMV (cauliflower mosaic virus), Copia (retrotransposon from *Drosophila melanogaster*).
- Kinases A: CD28 (*S. cerevisiae*), MLCK (rat skeletal muscle), PSKH (hela cell), CAPK (bovine cardiac muscle), WEE1 (dual specificity kinase from *S. pombe*).
- Kinases B: CD28, MLCK, PSKH, CAPK, WEE1, CSRC (chicken oncogenic protein).
- Kinases C: CD28, CAPK, WEE1, PDGMR (mouse, PDGF receptor).
- RH A: HTLV-II (human T-cell leukemia virus), RSV (Rous sarcoma virus), MoMLV (Moloney murine leukemia virus), 17.6 (retrotransposon from *Drosophila melanogaster*), HBV (human hepatitis B virus, ayw strain).

Table 1 shows the relative time and space usage of MSA 1.0 (old) and MSA 2.0 (new) on these data sets. In the case of Kinases C, we had to stop the old MSA run after it had run overnight and used more than 300 Mbytes of memory. In the case of RH A, the old MSA run thrashed extensively, and therefore, the actual wall-clock time from start to finish was 44 hours (much larger than the cpu time usage reported in the table) compared to approximately 1 hour for the new version. We note that the running time seems to depend more on how similar the sequences are, rather than on how many sequences there are. For example, Proteases B has only 4 sequences, but takes much longer than Proteases A, which has 5 sequences.

Removing the pruning of edges by lower-bound to sink improved the time and space further. For example, it improved Globins A to 135 s/4.3 Mbytes, Proteases B to 7 min/5.2 Mbytes, and Kinases A to 8 min/6.0Mbytes.

The last column of Table 1 measures how many motifs MSA correctly aligned out of those given by McClure et al. [15] for the sequences. Each motif is a block of consecutive alignment columns, ranging from one to five columns, that a good alignment should contain, based on biochemical and crystallographic considerations. For each data set, we report the sum over the motifs of the fraction of

Table 1. Comparison of resource usage for two versions of MSA. Number of sequences in each data set is in parentheses.

# Data Set	Old Time	New Time	Old Space	New Space	Correct Motifs
Globins A (7)	429 s	157 s	15.1 Mbytes	4.8 Mbytes	4.86/5
Globins B (10)	385 s	130 s	11.3 Mbytes	6.1 Mbytes	5.00/5
Proteases A (5)	79 s	37 s	5.7 Mbytes	3.0 Mbytes	2.80/3
Proteases B (4)	24 min	9 min	81 Mbytes	5.8 Mbytes	0.50/3
Kinases A (5)	35 min	10 min	67 Mbytes	6.6 Mbytes	8.00/8
Kinases B (6)	stopped	118 m	> 300 Mbytes	25 Mbytes	8.00/8
Kinases C (4)	530 s	210 s	35 Mbytes	4.7 Mbytes	6.75/8
RH A (4)	394 min	68 min	380 Mbytes	31 Mbytes	2.60/4

sequences correctly aligned, followed by the number of motifs in the data set. For example on Globins B, with 10 sequences, all five motifs were correctly aligned. On Globins A, with 7 sequences, MSA aligned four of the five motifs correctly, and in the fifth motif, 6 of the 7 sequences were correctly aligned. Generally, on data sets with more sequences, more motifs were correctly aligned. Since the motifs are defined based on sets of 12 sequences, the motifs may not be as easy to detect in small subsets of those 12.

Acknowledgments

The first and third authors thank Dr. Sandhya Dwarkadas for useful discussions about MSA and Prof. Willy Zwaenepoel for his encouragement. The second author thanks Prof. Marcella McClure for providing her data sets from [15]. Thanks to the referees for useful suggestions. Research of the first author was supported by grants from the National Science Foundation and the Texas Advanced Technology Program. Research of the second author was supported by a Department of Energy Human Genome Distinguished Postdoctoral Fellowship. Research of the third author was supported by a contract from IBM and a grant from the National Institutes of Health.

References

1. S. F. Altschul. Gap costs for multiple sequence alignment. *J. Theor. Biol.*, 138:297–309, 1989.

2. S. F. Altschul, Raymond J. Carroll, and David J. Lipman. Weights for data related by a tree. *J. Molecular Biology*, 207:647–653, 1989.

3. G. J. Barton and M. J. E. Sternberg. Evaluation and improvements in the automatic alignment of protein sequences. *J. Mol. Biol.*, 198:327–337, 1987.

4. G. J. Barton and M. J. E. Sternberg. A strategy for the rapid multiple alignment of protein sequences. *Protein Engineering*, 1:89–94, 1987.

5. H. Bodlaender, R. G. Downey, M. R. Fellows, and H. T. Wareham. The parameterized complexity of sequence alignment and consensus. In *Proc. of the 5th Symp. on Combinatorial Pattern Matching, Lecture Notes Comp. Sci. 807*, pages 15–30, 1994.

6. H. Carrillo and D. Lipman. The multiple sequence alignment problem in biology. *SIAM J. Appl. Math.*, 48:1073–1082, 1988.

7. S. C. Chan, A. K. C. Wong, and D. K. Y. Chiu. A survey of multiple sequence comparison methods. *Bulletin of Mathematical Biology*, 54:563–598, 1992.

8. E. W. Dijkstra. A note on two problems in connexion with graphs. *Numerische Mathematik*, 1:269–271, 1959.

9. D. Feng and R. Doolittle. Progressive sequence alignment as a prerequisite to correct phylogenetic trees. *J. Molecular Evol.*, 25:351–360, 1987.

10. D. G. Higgins, A. J. Bleasby, and R. Fuchs. Clustal v: improved software for multiple sequence alignment. *CABIOS*, 8:189–191, 1992.

11. J. Kececioglu. Notes on an approach of Carrillo and Lipman to minimum sum of pairs multiple sequence alignment. Unpublished notes, 1989.

12. J. Kececioglu. The maximum weight trace problem in multiple sequence alignment. In *Proc. of the 4th Symp. on Combinatorial Pattern Matching, Springer-Verlag Lecture Notes in Comp. Sci. 684*, pages 106–119, 1993.

13. D. J. Lipman, S. F. Altschul, and J. D. Kececioglu. A tool for multiple sequence alignment. *Proc. Natl. Acad. Sci. USA.*, 86:4412–4415, 1989.

14. D. Maier. The complexity of some problems on subsequences and supersequences. *J. ACM*, 25:322–336, 1978.

15. M. A. McClure, T. K. Vasi, and W. M. Fitch. Comparative analysis of multiple protein-sequence alignment methods. *Mol. Biol. Evol.*, 11:571–592, 1994.

16. S. Subbiah and S. C. Harrison. A method for multiple sequence alignment with gaps. *J. Mol. Biol.*, 209:539–548, 1989.

17. W. R. Taylor. Multiple sequence alignment by a pairwise algorithm. *CABIOS*, 3:81–87, 1987.

18. W. R. Taylor. A flexible method to align large numbers of biological sequences. *Journal of Molecular Evolution*, 28:161–169, 1988.

19. L. Wang and T. Jiang. On the complexity of multiple sequence alignment. *J. Computational Biology*, 1:337–348, 1994.

An efficient algorithm for developing topologically valid matchings*

Liz Hanks, Ron K. Cytron, and Will Gillett

Washington University Box 1045
Department of Computer Science
St. Louis, Missouri 63130, USA

Abstract. We examine a problem that arises in physical DNA mapping, namely determining what common DNA is represented in two maps. We first present an example illustrating the properties of DNA mapping, and present some biological background supporting our approach. We present a new graph structure, called the $\mathcal{Z}$-graph, that takes advantage of structure that develops during the mapping process, thus catalyzing the discovery of all maximum, topologically valid matchings. We describe an algorithm based on this structure and present experimental data supporting its improved performance as compared with a naive approach.

1 Introduction

Unraveling the information content of different species' genomes, for instance the human genome containing some 3 billion nucleotides, remains one of the grand challenges of molecular biology. Since the time required to obtain results depends on both time spent obtaining laboratory data as well as analyzing the results of those experiments, the magnitude of this effort mandates significant automation in both areas. We focus our attention on the analysis required for *high-resolution restriction-fragment mapping.*

In this paper, we view a *genome* as a sequence of unique *genomic fragments.* Biologically, a *clone* is manufactured by copying a contiguous excerpt of the genome. We therefore view a clone as a subsequence of a genome. When a clone is *fingerprinted*, the excerpted fragments are released; while the order of these *clonal fragments* as they occurred in the genome is lost, the fingerprinting method determines the *approximate* length of each such fragment.

Given this abstraction, we define *high-resolution restriction-fragment mapping*, or just *mapping* for short, as a process in which fragment-length information from the fingerprinting of a set of overlapping clones is used to reconstruct the original genome. Naturally, there may not be enough information to reconstruct a total ordering of fragments, so successively more refined partial orderings are used to approximate the desired result.

Consider the small hypothetical genome and random clone extractions shown below:

* This work was supported by the James S. McDonnell Foundation under Grant 87–24 and NIH under grant R01 HG00180.

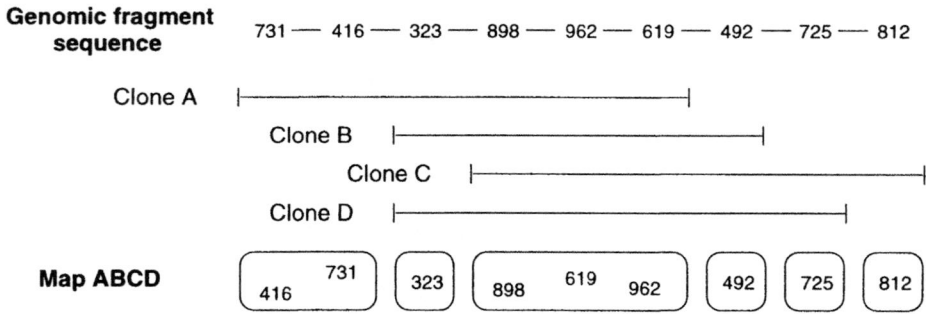

Genomic fragment sequence

731 — 416 — 323 — 898 — 962 — 619 — 492 — 725 — 812

If we were to map these clones' fragments, given perfect knowledge about the correspondence between clonal and genomic fragments, map *ABCD* shown above would be produced. The structure of this map shows *groups*, represented by boxes, whose order is known. Each group contains fragments whose order is not known.

While clone fingerprinting can supply more information than we retain, a "sequence of groups" is a simple form of expressing partial orderings of interest. Any such structure is referred to as a *map*. In actuality, a map has two possible orientations, and any mapping activity must continuously consider them both. We assume for simplicity throughout this paper that the orientation of all maps is fixed. We also assume that all groups and all maps are fragment-disjoint.

A reoccurring fundamental concept is that of "contiguity of fragments". Even though the order of a clone's fragments is unknown after fingerprinting, its fragments must be contiguous in the genome. A clone's fragments must therefore remain contiguous in any map that contains them. A simple corollary is that fragments common to multiple clones must also be contiguous in a map.

During a realistic mapping activity, no clear indication of fragment correspondence is supplied; the task at hand is to construct the most compact map possible given clone fingerprints. Although contiguity of a clone's fragments is still the guiding principle, algorithmic activities must include recognition of identifiable fragments, including the uncertainty caused by measurement error, while assembling maps that comply with topological constraints.

Continuing our example, below is shown the only maximum matching between the fragments in maps *A* and *B*, along with the resultant map *AB*. It is assumed that fragments in distinct maps whose lengths are within 3% of each other are *identifiable*, that is, could correspond to the same genomic fragment. Maps *A* and *B* each contain one single group that corresponds to the set of fragments released during the fingerprinting of clones *A* and *B*. Map *AB* includes all the information from maps *A* and *B*, and from the matching between their fragments.

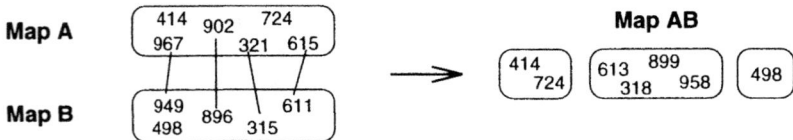

Fragments participating in a realistic matching must be contiguous in the resulting map, and to maintain fragment contiguity, those that do not participate in the matching must be relegated to opposite sides of the common overlap region. Note that the lengths of the clonal fragments differ from their genomic fragment lengths because of measurement error and that the length of each fragment in a map is the average of all lengths of clonal fragments that contributed to its creation. Maps AB and C can be merged using the unique maximum matching shown below:

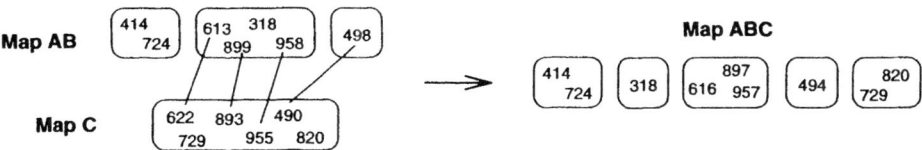

In essence, the process of mapping constitutes an ongoing activity in which the length of an emerging map is extended and the partial ordering of its fragments is refined. Note that, in the last mapping step, although the fragments of lengths 724 and 729 are identifiable, their pairing within a matching would violate topological constraints. This exemplifies the distinction between a *matching* in the usual graph-theoretic sense and a *topologically valid matching* that respects contiguity. As discussed in Section 7, many such topologically invalid matchings can occur in practice.

In attempting to map ABC and D together, two maximum matchings are possible, one leading to a map consistent with the genome and the other not. These two matchings involve pairing fragment 725 of clone D with either fragment 724 or fragment 729 of map ABC. In general, there is no reason to prefer one matching over the other, so this mapping step does not succeed because of *ambiguity*. Perhaps later in the mapping process, after maps undergo further refinement, the ambiguity will disappear.

Conceptually, mapping can be accomplished via a three-phase match/check/merge process. During the match phase, all potential matchings of interest between the two maps are constructed. During the check phase, each matching is tested for various forms of validity. During the merge phase, a composite map is produced for each validated matching. The TopoMatch algorithm presented in this paper pushes some of this validation, namely that of topological validity, into the match phase, thus greatly reducing the number of matchings that must be checked for overall validity. Section 7 presents some modest experiments that measure the benefits of this approach. Since the objective is to construct maps that are as compact as possible, only valid matchings of maximum cardinality pass into the merge phase. All such matchings must be constructed to facilitate detection and resolution of ambiguity.

Finally we make the following observations:

- An approach based on first finding all maximum matchings and then checking for validity can fail to find any matchings, where valid matchings exist with lesser cardinality.

– Any approach determined to find *all* valid matchings is necessarily exponential in its worst case, since one can construct maps that contain an exponential number of such matchings.

Section 2 presents the background needed to understand the significance of high-resolution restriction-fragment mapping. It need not be read in order to understand the rest of the paper. Section 3 defines the relevant discrete structures used in this paper. Section 4 describes topological validity in greater detail. Section 5 describes the graph structure upon which the TopoMatch algorithm is based. Section 6 describes the TopoMatch algorithm, and Section 7 presents empirical results on its performance.

2 Biological Background

In this section, some biological foundations are presented along with a comparison of the high-resolution restriction-fragment method described in this paper to other mapping methods. Using language structure as analogy, a genome is a sentence, and the fragments are words. Thus, a genome is a sequence of words. Each fragment is in turn a sequence of letters, or *nucleotides*, drawn from a 4-letter alphabet. The activity of *sequencing* a genome concerns discovering the order of letters in the genome sentence. Among other things, these letters code for amino acids, which constitute the building blocks of proteins. Thus, knowledge of the sequence of a genome enables discovery of the proteins producible by the host organism.

Sequencing technology is currently limited to direct analysis of relatively short sequences, approximately 500 nucleotides. With no major sequencing breakthrough in sight, this limitation suggests a hierarchical *subcloning* strategy: segments of DNA are copied and randomly subdivided into shorter overlapping segments, known as *clones*. If

1. the overlap relationship between the clones at each level of the hierarchy can be determined, and
2. sequence reads can be obtained at the bottom level,

then in principle the aggregate genomic sequence can be constructed.

The *fingerprint* of a clone is often used to determine the overlap relationship. There are many forms of such fingerprint, one of which is the restriction-fragment fingerprint which is based on the complete digestion of a clone by a restriction enzyme. A *restriction enzyme* is a protein that can cleave DNA at each *restriction site*: a short fixed subsequence of nucleotides intrinsic to the specific enzyme. Upon *complete digestion*, restriction-fragments whose lengths correspond to adjacent occurrences of these sites are released. These restriction-fragments can be separated by length using a process known as *electrophoresis*; their lengths can be then then be approximated. The digestion process eliminates all knowledge about the order of fragments in the clone, and the overall activity produces a set of lengths along with multiplicities corresponding to fragments released during the digestion.

At the lower levels of the hierarchy, two approaches have been used to determine clone overlap: contig building and high-resolution restriction-fragment mapping. Each approach can be fueled by the type of fragment-length fingerprint assumed in this paper, although contig-building data usually does not reflect fragment multiplicities. The difference is that contig building uses such data to find similarity among clones, with the goal of determining the order in which the clones occurred in the underlying genome, whereas high-resolution mapping has the goal of determining the positions of the restriction-fragments in the underlying genome. In contig building:

1. The fingerprints of all clones are pairwise compared to estimate apparent overlap, based on the *number* of identifiable fragments.
2. Based on overlap estimates, a simple polynomial-order algorithm then estimates the order of the clones in the underlying genome, in essence using similarities between the unstructured clone fingerprints as a coarse estimate of the probability of overlap and "joining" sequences of clones whose ends have the highest apparent probability of clone-clone overlap.

While this method is computationally simple, it is error-prone, primarily because no advantage is taken of *topological* information that develops when clones are positioned via fragment identification. Errors are especially likely when the genome features complex repeat structures.

On the other hand, high-resolution mapping joins maps only when:

1. fragments in the maps, that are composites of the fragments in the clones, are identifiable, and
2. identified fragments can be arranged contiguously in each map to comply with topological constraints present in the maps.

This exhaustive accounting for all fragments allows determination of not only the order of the clones, but also the *position* of the fragments and clones in the underlying genome. A much larger degree of confidence can be associated with the outcome of high-resolution mapping. This increased attention to detail, especially when high automation is desired, can significantly decrease the human effort required to obtain reliable maps. However, these advantages are achieved at the expense of higher computational and algorithmic complexity, and the requirement for precise laboratory data.

While this paper assumes restriction-fragment fingerprinting, other fingerprints are possible. For example, specific short sequences of nucleotides, called *oligo nucleotides*, can probe for occurrence of their sequences by DNA-DNA hybridization experiments. A large suite of oligos can be independently hybridized against a library of clones. As a result, each clone is in principal fingerprinted by the set of oligos whose sequence is contained therein. The raw laboratory data consist of intensity measurements from the radioactively or fluorescently tagged oligos, whose interpretation determines the apparent occurrence of the oligos. Unfortunately, the fundamental nature of such hybridization experiments admits a continuous range of intensities for which an occurrence/non-occurrence threshold must be set. Thus, relatively high percentages of false positives and false negatives can occur, and it is difficult even to estimate the multiplicity of oligo occurrences within a clone. Thus, because of the inaccuracy of the data, oligo fingerprints are not suitable for high-resolution mapping although they can be useful for contig building.

In summary, high-resolution restriction-fragment mapping has advantages over other mapping techniques which make it especially desirable if a high degree of automation is an important objective:

1. There is a relatively well understood model for interpreting the electrophoretic patterns, which constitute the raw laboratory data [4]; thus, there is the potential to automate this phase of the data analysis to obtain complete and accurate data, including precise fragment sizing with accurate multiplicities. Reliable data of this kind motivates a methodology based on the absolute accounting of *all* fragments present in the fingerprints; such is the methodology of the algorithm presented in this paper.
2. The precision of clone placement obtained by requiring topological constraints to be met while constructing the restriction-fragment maps—as compared with simpler methods that only order clones—significantly reduces the occurrence of false positives, especially when applied in the broader context of multiple-complete-digestion mapping [5, 7]; thus, many fewer errors need be addressed by human intervention.
3. The construction of restriction-fragment maps supplies significant information for consistency verification during the subsequent sequencing activity. Specifically, the positions of restriction sites are known and can be verified as the underlying sequence unfolds; thus, certain kinds of errors in the maps or sequence can be detected automatically.

Because the mapping problem is known to be NP-hard [11, 9], a *greedy* approach is often taken at the top level to reduce computational complexity. The algorithm presented in this paper assumes that two maps have been selected; the algorithm then determines matchings between the maps' fragments that satisfy certain identifiable and topological conditions. The problem of finding all relevant matchings is exponential in nature. If only one relevant matching is found, then the maps are merged into one; otherwise, the situation is *ambiguous* and no merge is performed. Ambiguity is prevalent even where the underlying DNA is not particularly repetitive. If successful, the merged map replaces the two contributors; it is potentially longer and more refined than its contributors.

The successful deployment of the overall approach requires the construction of *all* relevant matchings. This stems from the need to detect and resolve ambiguity, especially in the context in which it usually occurs, that of multiple-complete-digestion mapping. In this paradigm multiple digestions of the same clones, using different restriction enzymes, are employed simultaneously. Matchings are constructed in each restriction domain. Apparent ambiguity within a single domain may be resolvable at a higher level, where the regions spanned by the matchings can be compared across all domains to determine which are consistently placed.

Thus, all relevant matchings in each domain must be constructed to allow such resolution at a higher level.

A precursor to the technique presented here was used to map the 16 chromosomes of the 15 million nucleotide genome of *S. cerevisiae* by the Olson laboratory [8, 10]. Because of the simplicity of their original algorithmic approach, in which topology was not taken into account during the preliminary construction of the matchings:

1. The combinatorially-explosive nature restricted the size of clone that could be effectively handled.
2. Ad hoc heuristics were developed and deployed to combat computational complexity and apparent unmappability.
3. Significant human intervention was needed to detect and repair errors.

The mapping activity required 8 years of effort, the bottleneck being map assembly, not data collection.

The techniques presented here are particularly effective for mapping at the *cosmid* ($\sim$ 35–45 thousand nucleotides) level, which is currently the major cloning unit used for sequencing, and the small-insert clone ($\sim$ 1.5–5 thousand nucleotides) level, which is the cloning unit from which individual sequence reads are extracted.

3 Notation and Preliminary Definitions

The length of a sequence S is denoted $|S|$. For $1 \leq i \leq |S|$, the ith element of S is denoted S_i.

A slight variation of E. W. Dijkstra's notation for variable-binding constructs is used [3, 12]. Universal quantification is denoted by $(\forall l : D : P)$, where l is a list of bound variables, predicate D determines the domain over which the bound variables range, and predicate P need only be defined for values that satisfy D. Existential quantification is denoted in a similar way, using the symbol $\exists$, as is generalized union, using $\cup$. The quantifier for set construction is implicit in the braces. For instance, the left elements of a set E of pairs, denoted $Left(E)$ is defined to be the set

$$\{ left, right : \langle left, right \rangle \in E : left \}$$

Similarly, $Right(E)$ denotes E's right elements.

Definition 1. *A relation on a set A is a subset of $A \times A$. We use the symbol $\sqsubseteq$ for partial orderings: relations which are reflexive, antisymmetric, and transitive, and $\preceq$ for total orderings: partial orderings in which any two elements are comparable. The anti-reflexive companions of orderings $\sqsubseteq$ and $\preceq$ will be denoted $\sqsubset$ and $\prec$, respectively, and when defining a particular partial or total ordering, it can be notationally convenient to define its anti-reflexive companion instead. If $\sqsubseteq$ is a partial ordering on A, then $\langle A, \sqsubseteq \rangle$ is a partially ordered set, or poset. If $\preceq$ is a total ordering on A, then $\langle A, \preceq \rangle$ is a chain. The minimum element of a chain C is denoted $Min(C)$ and the maximum element is denoted $Max(C)$.*

Definition 2. *A poset* $\langle A, \sqsubseteq' \rangle$ *is said to be a superposet of poset* $\langle A, \sqsubseteq \rangle$, *denoted* $\langle A, \sqsubseteq' \rangle \supseteq \langle A, \sqsubseteq \rangle$ *when* $\sqsubseteq' \supseteq \sqsubseteq$.

Definition 3. *A* group *is a non-empty set of fragments, and a* map *is a non-empty sequence of fragment-disjoint groups. The set of groups of map* S *is denoted* $Groups(S)$; *its set of fragments is denoted* $Frags(S)$. *For fragment* f *in* $Frags(S)$, $Group(f, S)$ *denotes the group that* f *belongs to in* S.

Definition 4. *We are provided with* Id, *a relation on fragments. Fragments* f *and* f' *are said to be* identifiable, *denoted* $Id(f, f')$ *when they are included in this relation. It is common for* Id *to relate fragments whose lengths are considered equal within measurement error, and is usually not a transitive relation. Groups* g *and* g' *are said to be identifiable, denoted* $Id(g, g')$ *when*

$$(\exists f, f' : f \in g \wedge f' \in g' : Id(f, f')).$$

For maps S *and* T, *the pairs of identifiable groups, denoted* $GroupIds(S, T)$ *is the set* $\{ g, g' : g \in Groups(S) \wedge g' \in Groups(T) \wedge Id(g, g') : \langle g, g' \rangle \}$.

Definition 5. *The total ordering of groups in map* S *is denoted* $\preceq_S$, *and is defined by*

$$S_i \preceq_S S_j \Leftrightarrow i \leq j$$

for all $1 \leq i \leq |S|$, *and* $1 \leq j \leq |T|$. *The partial ordering of fragments in map* S *denoted* $\sqsubseteq_S$, *is defined by*

$$f \sqsubseteq_S f' \Leftrightarrow Group(f, S) \prec_S Group(f', S)$$

for all f, f' *in* $Frags(S)$. *The chain* $\langle Groups(S), \preceq_S \rangle$ *is denoted* $GroupChain(S)$, *and the poset* $\langle Frags(S), \sqsubseteq_S \rangle$ *is denoted* $FragPoset(S)$.

Definition 6. *Given two distinct maps* S *and* T, *when a set* E *is a subset of* $Frags(S) \times Frags(T)$, *the structure* $\langle Frags(S), Frags(T), E \rangle$ *will be referred to as a* graph. *This is a special form of the traditional bipartite graph in which a partitioning of the vertices is known a priori. A* matching *in the graph* $\langle Frags(S), Frags(T), E \rangle$ *is any fragment-disjoint subset of* E. *When* E *is a subset of* $Groups(S) \times Groups(T)$, *the structure* $\langle S, T, E \rangle$ *will be referred to as an* ordered graph.

4 Topological validity

4.1 Definitions

There are three concepts central to topological validity: orderability, contiguity, and positionability. Each concept is defined below, and then topological validity is defined using these new concepts.

Definition 7. *Given two chains $\langle A, \preceq \rangle$ and $\langle B, \preceq' \rangle$, a subset E of $A \times B$ is said to be* orderable, *denoted $Orderable(E, \langle A, \preceq \rangle, \langle B, \preceq' \rangle)$ when no two elements of E cross each other, that is, when*

$$(\forall \, a, a', b, b' : (a, b) \in E \wedge (a', b') \in E : a \prec a' \Rightarrow b \preceq' b').$$

Definition 8. *Given a chain $\langle A, \preceq \rangle$, a subset A' of A is said to be* contiguous, *denoted $Contiguous(A', \langle A, \preceq \rangle)$ when*

$$(\forall \, a, b, c : a \in A' \wedge b \in A \wedge c \in A' : a \preceq b \preceq c \Rightarrow b \in A').$$

In other words, there are no elements of $A - A'$ that are surrounded by elements of A'. A subset E of $A \times B$ is said to be contiguous *in the two chains $\langle A, \preceq \rangle$ and $\langle B, \preceq' \rangle$, denoted $Contiguous(E, \langle A, \preceq \rangle, \langle B, \preceq' \rangle)$ when*

$$Contiguous(Left(E), \langle A, \preceq \rangle) \wedge Contiguous(Right(E), \langle B, \preceq' \rangle).$$

Definition 9. *Given two chains $\langle A, \preceq \rangle$ and $\langle B, \preceq' \rangle$, a subset E of $A \times B$ is said to be* positionable, *denoted $Positionable(E, \langle A, \preceq \rangle, \langle B, \preceq' \rangle)$ when it uses the first element in one chain, and the last element in one chain, that is, when*

$$(Min(\langle A, \preceq \rangle) \in Left(E) \vee Min(\langle B, \preceq' \rangle) \in Right(E)) \wedge$$
$$(Max(\langle A, \preceq \rangle) \in Left(E) \vee Max(\langle B, \preceq' \rangle) \in Right(E)).$$

Definition 10. *Consider a matching M in the graph*

$$\langle Frags(S), Frags(T), Frags(S) \times Frags(T) \rangle,$$

where S and T are maps. This matching is topologically valid, *denoted $Topo\text{-}Valid(M, S, T)$, if there exist chains $C = \langle Frags(S), \preceq \rangle$ and $C' = \langle Frags(T), \preceq' \rangle$ such that all of the following conditions are satisfied:*

- $(\forall f, f' : \langle f, f' \rangle \in M : Id(f, f'))$: *each edge in M joins identifiable fragments.*
- $C \supseteq FragPoset(S) \wedge C' \supseteq FragPoset(T)$: *$C$ and C' are superposets of S and T's fragment orderings.*
- $Orderable(M, C, C')$: *the edges in M do not cross.*
- $Contiguous(M, C, C')$: *the matched fragments are contiguous in C and C'.*
- $Positionable(M, C, C')$: *there are no unmatched fragments on the same side of S and T.*

4.2 Examples

Given the matching shown below, any pair of valid total orderings of S and T's fragments will lead to crossed edges. Contiguity and positionability can be satisfied in several ways, but orderability is always violated.

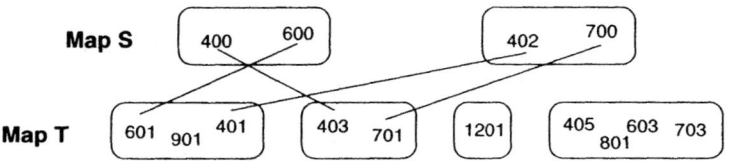

In the matching shown below, any valid total ordering of the fragments in S will satisfy contiguity, but all valid orderings of T's fragments will result in the unmatched fragment of length 1201 being surrounded by matched fragments.

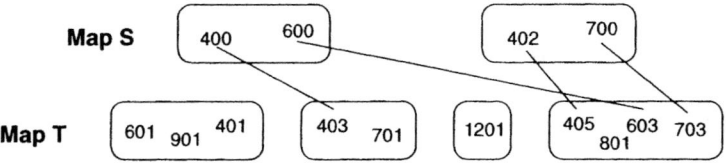

In the matching shown below, the positioning condition cannot be satisfied: the minimum element of any valid chain of S or T's fragments is unmatched.

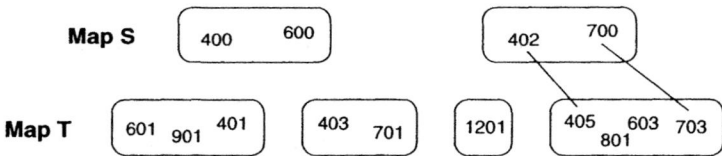

5 The $\mathcal{Z}$-graph

Associated with each set of fragment pairs that satisfies orderability, contiguity, and positionability, there is a set of group pairs that also satisfies these three conditions. Let S and T be two maps, and let M be a matching between their fragments such that $TopoValid(M, S, T)$ holds. Consider the following subset E of $Groups(S) \times Groups(T)$:

$$\{ f, f' : \langle f, f' \rangle \in M : \langle Group(f, S), Group(f', T) \rangle \}.$$

The following predicates follow directly from the contiguity, orderability, and positionability of M:

- $Contiguous(E, GroupChain(S), GroupChain(T))$
- $Orderable(E, GroupChain(S), GroupChain(T))$
- $Positionable(E, GroupChain(S), GroupChain(T))$.

Definition 11. The ordered graph $\langle S, T, E \rangle$ is said to be a $\mathcal{Z}$-graph if it satisfies the three above conditions.

For example, consider the following matching of the fragments of maps U and V:

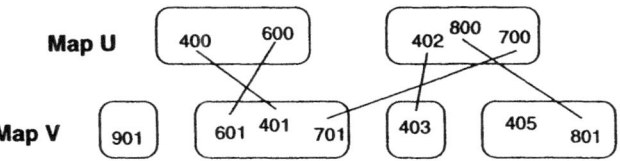

The chains $400-600-700-402-800$ in $Frags(U)$ and $901-401-601-701-403-801-405$ in $Frags(V)$ satisfy all the necessary conditions for this matching to be topologically valid. The corresponding set of group edges is depicted below, and satisfies contiguity, orderability and positionability.

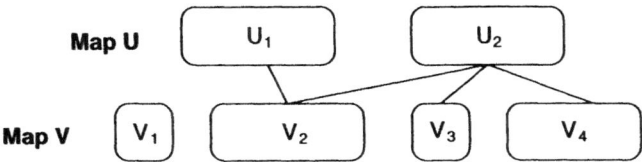

6 The TopoMatch algorithm

Since each topologically valid matching has an associated $\mathcal{Z}$-graph, we can find all maximum topologically valid matchings by first finding all maximal $\mathcal{Z}$-graphs, and then for each $\mathcal{Z}$-graph, finding all matchings in which the only fragments that are allowed to pair are those whose groups are joined by an edge in the $\mathcal{Z}$-graph. These matchings will satisfy orderability, but must be checked for contiguity and positionability.

6.1 Definitions

Before presenting our algorithm, we will introduce two related concepts.

Definition 12. *Given a poset $\langle A, \sqsubseteq \rangle$, and an element a in A, the successors of a, denoted $Successors(a, \langle A, \sqsubseteq \rangle)$ is the set*

$$\{ c : a \sqsubset c \wedge \neg(\exists b : a \sqsubset b : b \sqsubset c) : c \}$$

Definition 13. *Given two chains $\langle A, \preceq \rangle$ and $\langle B, \preceq' \rangle$, and a subset E of $A \times B$, the edge ordering of E, denoted $EdgeOrdering(E, \langle A, \preceq \rangle, \langle B, \preceq' \rangle)$, is the following partial ordering:*

$$\{ a, a', b, b' : \langle a, b \rangle \in E \wedge \langle a', b' \rangle \in E \wedge a \preceq a' \wedge b \preceq' b' : \langle \langle a, b \rangle, \langle a', b' \rangle \rangle \}$$

Note that when $Orderable(E, \langle A, \preceq \rangle, \langle B, \preceq' \rangle)$, $EdgeOrdering(E, \langle A, \preceq \rangle, \langle B, \preceq' \rangle)$ is a total ordering.

Before finding the $\mathcal{Z}$-graphs for two maps, we will create an ordered graph in which any two groups from different maps that are identifiable, are joined by an edge. The edge ordering in this graph, as defined above, can then be used to repeatedly find successor edges that can safely be added to a set of $\mathcal{Z}$-graphs under construction. The edge orderings in the $\mathcal{Z}$-graphs under construction will be total orderings, so their maximum elements are well defined, and are in fact the edges whose successors are of interest.

6.2 Pseudocode

The algorithm $FindTopoMatchings(S,T)$ is shown in Figure 1. We begin by

Function $FindTopoMatchings(S,T)$
    $answer \leftarrow \phi$
    $size \leftarrow 0$
    $zgraphs \leftarrow FindZgraphs(S,T)$ $\Leftarrow$ ⬚1
    foreach $(Z \in zgraphs)$ **do** $\Leftarrow$ ⬚2
        $matchings \leftarrow FindZgraphMatchings(Z)$ $\Leftarrow$ ⬚3
        foreach $(M \in matchings)$ **do**
            if $(IsValid(M,S,T))$ **then** $\Leftarrow$ ⬚4
                if $(|M| > size)$ **then** $\Leftarrow$ ⬚5
                    $answer \leftarrow \{ M \}$
                    $size \leftarrow |M|$
                else if $(|M| = size)$ **then**
                    $answer \leftarrow answer \cup \{ M \}$
                fi
            fi
        od
    od
    return $(answer)$
end

Fig. 1.

finding all maximal $\mathcal{Z}$-graphs in step ⬚1. In step ⬚3, all matchings between identifiable fragments whose groups are joined by an edge in a $\mathcal{Z}$-graph are found; this is an algorithmic abstraction: for brevity we omit the case analysis that allows finding only maximum matchings. The function $IsValid$ called in step ⬚4 determines if there exist valid total orderings of S and T's fragments that satisfy contiguity and positionability. Since the elements of $matchings$ aren't necessarily all of the same size, step ⬚5 updates $answer$ to contain all valid $maximum$ matchings created up to that point, and no others.

    As can be seen in Definition 9, there are four ways in which a $\mathcal{Z}$-graph can satisfy positionability. $FindZgraphs(S,T)$, shown in Figure 2, finds the $\mathcal{Z}$-graphs for each one of these ways.

    Assuming g_{first} is the minimum element of $GroupChain(S)$ or of $GroupChain(T)$, and g_{last} is the maximum element of one of those two chains, $FindBoundedZgraphs(\langle S,T,E_G\rangle, g_{first}, g_{last})$, shown in Figure 3, will find all maximal $\mathcal{Z}$-graphs whose minimum edge is incident to g_{first}, and whose maximum edge is incident to g_{last}. We start in step ⬚7 by finding $graphs$, a set of singleton-edge ordered graphs. Initially, each edge in each graph is incident to g_{first}. The following invariants are maintained in this algorithm, for each element $\langle S,T,E_Z \rangle$ in $graphs$:

Function $FindZgraphs(S,T)$
   $G \leftarrow \langle S, T, GroupIds(S,T) \rangle$ $\Leftarrow \boxed{6}$

$$\textbf{return} \left(\begin{array}{l} (\cup g, g' : g \in \{ S_1, T_1 \} \wedge g' \in \{ S_{|S|}, T_{|T|} \} \\ \quad : FindBoundedZgraphs(G, g, g')) \end{array} \right)$$

end

Fig. 2.

Function $FindBoundedZgraphs(\langle S, T, E_G \rangle, g_{first}, g_{last})$
   $answer \leftarrow \phi$
   $graphs \leftarrow FindSingletonGraphs(\langle S, T, E_G \rangle, g_{first})$ $\Leftarrow \boxed{7}$
   while $(graphs \neq \phi)$ **do**
      foreach $(\langle S, T, E_Z \rangle \in graphs)$ **do**
         $graphs \leftarrow graphs - \{ \langle S, T, E_Z \rangle \}$
         $\langle S_i, T_j \rangle \leftarrow Max(EdgeOrdering(E_Z, S, T))$
         $successors \leftarrow Successors(\langle S_i, T_j \rangle, EdgeOrdering(E_G, S, T))$
         if $(successors = \phi \wedge Incident(\langle S_i, T_j \rangle, g_{last}))$ **then**
            $answer \leftarrow answer \cup \{ \langle S, T, E_Z \rangle \}$
         fi
         foreach $(\langle S_{i'}, T_{j'} \rangle \in successors)$ **do**
            if $(i' \leq i + 1 \wedge j' \leq j + 1)$ **then**
               $graphs \leftarrow graphs \cup \{ \langle S, T, E_Z \cup \{ \langle S_{i'}, T_{j'} \rangle \} \rangle \}$ $\Leftarrow \boxed{8}$
            fi
         od
      od
   od
   return $(answer)$
end

Fig. 3.

1. $Contiguous(E_Z, GroupChain(S), GroupChain(T))$
2. $Orderable(E_Z, GroupChain(S), GroupChain(T))$
3. $g_{first} \in Left(E_Z) \vee g_{first} \in Right(E_Z)$

Initially these three invariants are easily met by the singleton-edge graphs in *graphs*. During each iteration, only edges that maintain contiguity and orderability are added to a graph: any successor of the maximum edge of a graph can be added to it without violating orderability, and only successors that do not skip any groups are added in step $\boxed{8}$ so that contiguity is not violated. When the maximum edge of a graph in *graphs* has no successors, it is guaranteed to be a $\mathcal{Z}$-graph if its maximum edge is incident to g_{last}: positionability is satisfied, and the graph is added to *answer*. This maximum edge is well defined since the edge ordering of all graphs constructed is a total ordering.

Function *FindSingletonGraphs*($\langle S, T, E_G \rangle, g$)

$\quad E_Z \leftarrow \{\, i, j : \langle g_{first}, T_{j+1} \rangle \notin E_G \wedge \langle S_{i+1}, g_{first} \rangle \notin E_G : \langle S_i, T_j \rangle \,\}$

$\quad$**return** $(\{\, e : e \in E_Z : \langle S, T, \{\, e \,\} \rangle \,\})$

end

Fig. 4.

The algorithm for *FindSingletonGraphs*($\langle S, T, E_G \rangle, g$) shown in Figure 4 finds the set E_Z of all edges incident to g that have no immediate successors in E_G incident to g. For $1 \leq i \leq |S|$ and $1 \leq j < |S|$, edge $\langle S_i, T_j \rangle$ has one *immediate successor* that is incident to S_i: $\langle S_i, T_{j+1} \rangle$. A similar successor that is incident to T_j exists, if $i < |S|$. If an edge and its immediate successor are in E_G, it is useless to explore $\mathcal{Z}$-graphs that start with the second edge, since they will all be subgraphs of the $\mathcal{Z}$-graphs that start with the first edge. Note that no two edges in E_Z can belong to the same $\mathcal{Z}$-graph, and all $\mathcal{Z}$-graphs that can be created from S and T must start with some edge in E_Z. Once E_Z is created, a singleton-edge graph will be created for each edge in E_Z.

6.3 Example

The algorithm's behavior is now illustrated using the maps X and Y depicted below along with their identifiable fragments.

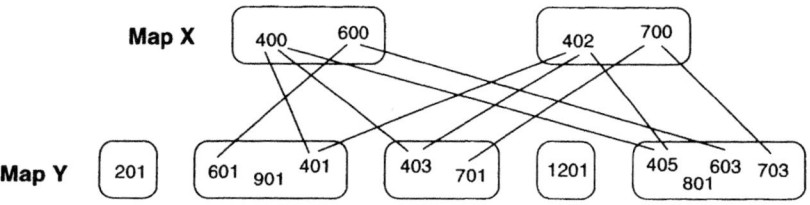

- Step $\boxed{1}$ of *FindTopoMatchings*(X, Y) begins by calling *FindZgraphs*(X, Y) to find all maximal $\mathcal{Z}$-graphs
- In step $\boxed{6}$, we construct the graph G containing pairs of identifiable groups as depicted below.

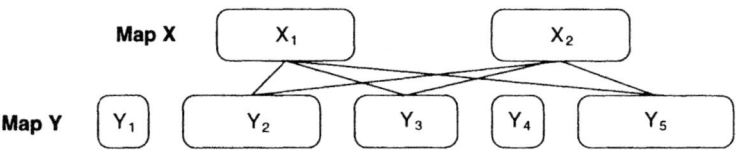

- *FindBoundedZgraphs* will then be called four times, but two of the calls will not result in any $\mathcal{Z}$-graphs, since Y_1 is not identifiable with any other group.
- In step $\boxed{7}$ of *FindBoundedZgraphs*($\langle X, Y, GroupIds(X, Y) \rangle, X_1, X_2$) we begin by finding the initial singleton-edge graphs. Group X_1 is incident to three groups in Y: Y_2, Y_3, and Y_5, as illustrated below.

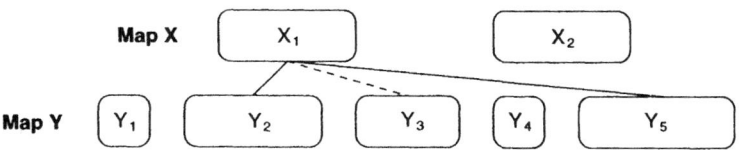

Since the dashed edge is an immediate successor of an edge incident to X_1, it will not be a part of the initial set of graphs. Each of the other two edges will be used to create a graph.

— We depict the development of *graphs* by a tree, in which paths show how one graph is constructed by adding a successor edge to another. There are a total of three maximal graphs that are developed from the two starting graphs.

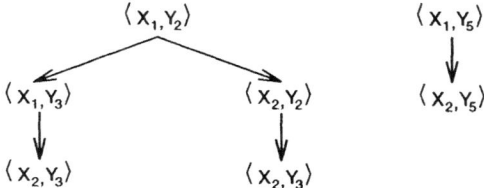

— Only maximal graphs that have an edge incident to group X_2, the last group in X, will be added to *answer*. In this case, all three graphs meet this condition.

— The three $\mathcal{Z}$-graphs shown below are left.

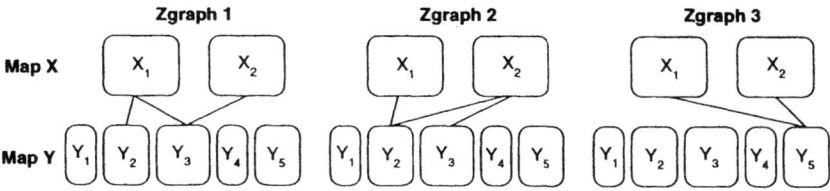

— In the other relevant call to *FindBoundedZgraphs*, we have $g_{first} = X_1$ and $g_{last} = Y_5$. The development of *graphs* is the same as in the previous call, since g_{first} has not changed, but only graphs that include group Y_5 in some edge will be added to *answer*. Only one graph meets this requirement, and Zgraph 3, depicted above, is returned.

— *FindZgraphs* returns the three $\mathcal{Z}$-graphs shown above, since no new ones are found in the remaining calls to *FindBoundedZgraphs*.

— Back in step $\boxed{3}$ of *FindTopoMatchings*, *FindZgraphMatchings* is called for each of the three $\mathcal{Z}$-graphs. The first two $\mathcal{Z}$-graphs shown above will lead us to find the same topologically valid matching, labeled Matching 1 below. So, after each of the first two iterations of step $\boxed{2}$, *size* will be 4, and *answer* will contain one matching. From the third $\mathcal{Z}$-graph the other two matchings illustrated below are found.

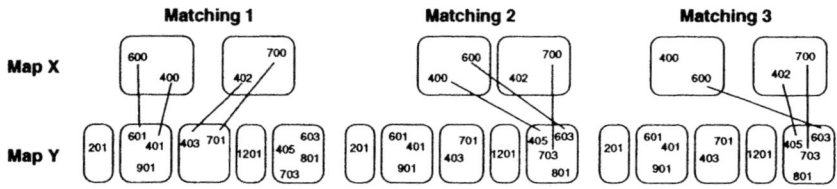

- Matching 2 passes the validity check in step ④, but its size is too small. Matching 3 is not valid: any chain of map X's fragments will either have the unmatched fragment of length 400 as a minimum element, in which case positionability is violated, or will have it surrounded by matched fragments, in which case contiguity is violated. So neither Matching 2 nor Matching 3 will be added to *answer*.
- The set of matchings returned contains only Matching 1.

Note that since one single relevant matching is left, maps X and Y, along with Matching 1 can be used to create the new map depicted below.

7   Experiments and Results

In this section, we present the results of two modest sets of experiments that measure the efficiency of the TopoMatch algorithm as compared with a straightforward approach that we call NaiveMatch:

1. All maximum matchings are constructed.
2. Each such matching is checked for validity.

As mentioned in Section 1, this approach is inadequate since it may miss valid matchings of submaximum cardinality. Actually, our implementation of Naive-Match contains some heuristics [5], such as "windowing" and "scrunching", that boost its performance as well as its chances of finding topologically valid matchings.

Each experiment consisted of executing the following steps:

1. Simulate the creation of randomly overlapping clones from a DNA strand.
2. Simulate the fingerprinting of the clones created.
3. Map the clones together using NaiveMatch.
4. Map the clones together using TopoMatch.

The DNA strands used in the first set of experiments were part of the *C. elegans* worm genome. In the second set, the TCR-β region of the human genome was used. The simulation of the fingerprinting activity included inducing normal random error to the fragments, in order to mimic the measurement errors produced in a laboratory. In each experiment 50 clones were created and mapped together using either NaiveMatch or TopoMatch. This resulted in several thousand invocations of both algorithms for each experiment.

The figure below shows the results of 12 experiments in each DNA realm. Each bar represents one experiment. The height of each bar reflects the *speedup* of TopoMatch, computed as the ratio of the NaiveMatch time to the TopoMatch time. Note that a logarithmic scale was used.

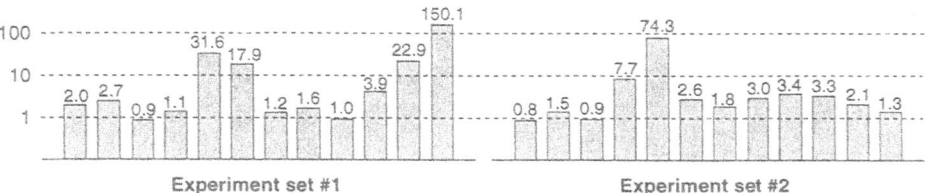

Even though the number of experiments performed was small, hints of near worst-case behavior surfaced, causing NaiveMatch to slow the mapping activity down unreasonably. In the experiment which resulted in a speedup of over 150, there was a pair of maps that caused NaiveMatch to spend hours producing and sifting through 138,240 invalid matchings. A valid matching was found using TopoMatch in a few seconds.

We also note that, in addition to the performance improvements attributable to TopoMatch, the algorithm is also responsible for producing maps with *fewer errors*. Since TopoMatch considers significantly fewer topologically infeasible maps, its notion of maximum matchings is more precise than NaiveMatch's notion.

8 Conclusion

We have described a new approach to finding topologically valid matchings for high-resolution restriction-fragment mapping. The algorithm has been implemented, and its improved performance measured in a realistic setting. Based on these results and on our own experience, TopoMatch is a very good replacement for NaiveMatch.

Not only does the new algorithm allow mapping to proceed efficiently, even when clones contain a fairly large number of fragments, but it has also been extended to compute *almost topologically valid matchings* in an efficient manner. The discovery of matchings with slight topological mistakes is essential to the process of fixing errors in maps [1, 2, 6]. This process is computationally infeasible when a naive matching algorithm is used.

Acknowledgements

The authors thank Maynard Olson for seminal discussions and guidance on the problems and issues in DNA mapping.

References

1. James Daues and Will Gillett. DNA mapping algorithms: Fragment splitting and combining. Technical report, Washington University, Department of Computer Science, 1991. Report WUCS–91–50.

2. James Daues and Will Gillett. DNA mapping algorithms: Fragment matching mistake detection and correction. Technical report, Washington University, Department of Computer Science, 1993. Report WUCS–93–50.

3. Edsger W. Dijkstra. *A Method of Programming*. Addison-Wesley, 1988.

4. Heather A. Dury, Philip Green, Bridgid K. McCauley, Maynard V. Olson, David G. Politte, and Jr. Lewis J. Thomas. Spatial normalization of one-dimensional electrophoretic gel images. *Genetics*, 8:119–126, 1990.

5. Will Gillett. DNA mapping algorithms: Strategies for single restriction enzyme and multiple restriction enzyme mapping. Technical report, Washington University, Department of Computer Science, 1992. Report WUCS–92–29.

6. Will Gillett, Jim Daues, Liz Hanks, and Rob Capra. Fragment collapsing and splitting while assembling high-resolution restriction maps. *Journal of Computational Biology*, 1995. In press.

7. Will Gillett, Liz Hanks, and Maynard Olson. Assembling high-resolution restriction maps using multiple complete digestions of a redundant set of clones. *Genetics*, 1995. Submitted.

8. Maynard V. Olson, James E. Dutchik, Madge Y. Graham, Garret M. Brodeur, Cynthia Helms, Mark Frank, Mia MacCollin, Robert Scheinman, and Thomas Frank. Random-clone strategy for genomic restriciton mapping in yeast. *Proceedings of the National Academy of Science (Genetics)*, 83:7826–7830, October 1986.

9. Gwangsoo Rhee. *Computational Models for DNA Restriction Mapping*. PhD thesis, Washington University in St. Louis, 1990.

10. Linda Riles, James E. Dutchik, Amara Baktha, Brigid K. McCauley, Edward C. Thayer, Mary P. Leckie, Valerie V. Braden, Julie E. Depke, and Maynard V. Olson. Physical maps of the six smallest chromosomes of *Saccharomyces* cerevisiae at a resolution of 2.6 kilobase pairs. *Genetics*, 134:81–150, May 1993.

11. Jonathan S. Turner. The complexity of the shortest common matching string problem. Technical report, Washington University, Department of Computer Science, 1986. Report WUCS–86–9.

12. Jan Tijmen Udding and Tom Verhoeff. Using a parital order and a metric to analyze a recursive trace set equation. Technical report, Washington University, Department of Computer Science, 1988. Report WUCS–88–17.

Polynomial-time Algorithm for Computing Translocation Distance between Genomes

Sridhar Hannenhalli[1]

Department of Computer Science and Engineering
The Pennsylvania State University
University Park, PA 16802

Abstract

With the advent of large-scale DNA physical mapping and sequencing, studies of genome rearrangements are becoming increasingly important in evolutionary molecular biology. From a computational perspective, study of evolution based on rearrangements leads to *rearrangement distance problem*, *i.e.*, computing the minimum number of rearrangement events required to transform one genome into another. Different types of rearrangement events give rise to a spectrum of interesting combinatorial problems. The complexity of most of these problems is unknown. Multichromosomal genomes frequently evolve by a rearrangement event called *translocation* which exchanges genetic material between different chromosomes. In this paper we study the *translocation distance problem*, modeling evolution of genomes evolving by translocations. Translocation distance problem was recently studied for the first time by Kececioglu and Ravi, who gave a 2-approximation algorithm for computing translocation distance. In this paper we prove a duality theorem leading to a polynomial algorithm for computing translocation distance for the case when the orientation of the genes are known. This leads to an algorithm generating a most parsimonious (shortest) scenario, transforming one genome into another by translocations.

---

[1]This work is supported by NSF Young Investigator award CCR-9457784 and NIH grant 1R01 HG00987. Author's E-mail addresses: *hannenha@cse.psu.edu*

1 Introduction

First computational attempt to analyze genome rearrangements in mamallian genomes was undertaken by Nadeau and Taylor in 1984 who estimated that just 178 ± 39 rearrangement events happened since the separation of lineages leading to human and mice 80 million years ago. This estimate was recently validated by Copeland *et al.* in 1993, based on man-mouse genetic linkage map of much higher resolution compared to the one available 10 years ago. The most common rearrangement events in mamallian evolution are *translocations*, which exchange genetic material between different chromosomes, and *reversals*, which rearrange genetic material within a chromosome.

A computational approach to evolutionary studies based on rearrangements was pioneered by Sankoff (see Sankoff *et al.*, 1990,1992 and Sankoff, 1992). Study of genomes evolving by rearrangements involves a combinatorial problem of computing the minimum number of rearrangement events transforming one genome into other and finding a shortest (most parsimonious) sequence of rearrangement events transforming one genome into other.

Although, *translocation* is a complicated biological process (see Therman and Susman, 1993 and Lewin, 1994, for the underlying biology), the following abstraction is adequate for our purpose. A chromosome can be represented as a sequence of genes, where each gene is represented by an integer. A translocation is said to act on chromosomes X and Y when the chromosomes are cleavaged as (X_1, X_2) and (Y_1, Y_2) respectively and the *segments* of the chromosomes are swapped, thus transforming chromosomes X and Y into two new chromosomes. We study the most common type of translocation, *viz.*, *reciprocal* translocation where each of the four segments, X_1, X_2, Y_1 and Y_2, is non-empty. A translocation is a *prefix-prefix* translocation if the prefix of one chromosome is swapped with the prefix of the other chromosome and a translocation is a *prefix-suffix* translocation, if the prefix of one chromosome is swapped with the suffix of the other chromosome (Fig. 1).

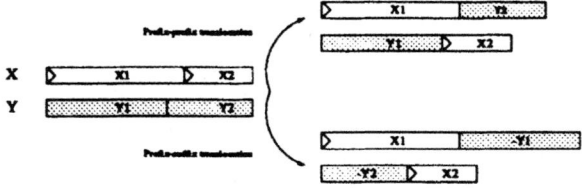

Figure 1: Examples of translocations. Notice the change in the directions of chromosomal segments, Y_1 and Y_2, after prefix-suffix translocation.

For our purposes, a genome is a set of chromosomes. A translocation on a pair of chromosomes of genome A transforms genome A into another genome. Given two genomes, A and B, *translocation distance* between A and B, $d(A, B)$, is the minimum number of translocations required to transform A into B. We refer to any sequence of translocations transforming A into B as *evolution* of A into B.

Under most of the rearrangement events, the complexity of the rearrangement distance problem is still unknown. The importance of these problems have motivated researchers to develop approximation algorithms for rearrangement distance problems for various types of rearrangements. The first steps towards a combinatorial theory of genome rearrangements have been taken very recently. Kececioglu and Sankoff, 1993, 1994, and Bafna and Pevzner, 1993, gave approximation algorithms for computing rearrangement distance for genomes evolving by reversals. (The problem is known as "sorting signed permutation by reversals".) Recently, Hannehalli and Pevzner, 1994, showed that the problem of sorting signed permutations by reversals is in **P** by proving a duality theorem that gives an efficiently computable characterization of reversal distance. Recently Kececioglu and Ravi, 1995, gave a 2-approximation algorithm for rearrangement distance problem for genomes evolving by translocations and a 1.5-approximation algorithm for rearrangement distance problem for genomes evolving by both translocations and reversals. See Bafna and Pevzner, 1995a and Hannenehalli *et al.*, 1994 for applications of genome rearrangement algorithms to analyze evolution of plant organelles, mamallian X chromosomes and herpes viruses. Also, See Bafna and Pevzner, 1995b, for computational study of genomes evolving by another type of rearrangement event called transposition.

In this paper we prove a duality theorem characterizing translocation distance for signed data. This leads to polynomial algorithm which computes a shortest sequence of translocations transforming one genome into another. We restrict our discussion to the case when both *prefix-prefix* and *prefix-suffix* reciprocal translocations are allowed. The case when only prefix-prefix translocations are allowed is amenable to similar analysis and will not be discussed in this paper.

All chromosomes contain a *centromere* which is important for cell division. A translocation is *viable* if both of the resulting chromosomes contain a centromere. This restricts the the translocations in the course of evolution. Including centromeres in our model does not present additional difficulty. For simplicity we omit centromeres from our model to be discussed elsewhere.

In the following section we present the combinatorial formulation of the problem. In section 3 we prove a lower bound on the translocation distance. In section 4 we prove a duality theorem leading to a polynomial algorithm for computing translocation distance. In section 5 we present an algorithm generating a most parsimonious (shortest) scenario of evolution, transforming one genome into other. And finally in section 6 briefly discuss the case of unsigned data.

2 Combinatorial Formulation

For the purpose of following discussion, a *gene* will be represented by a signed integer, where the sign models the direction of the gene, a *chromosome* is a sequence of genes and a *genome* is a set of chromosomes. We assume that the given genomes

$A = ((a_{11}, a_{12}, \ldots, a_{1m_1}), (a_{21}, a_{22}, \ldots, a_{2m_2}), \ldots, (a_{N1}, a_{N2}, \ldots, a_{Nm_N}))$ and
$B = ((b_{11}, b_{12}, \ldots, b_{1n_1}), (b_{21}, b_{22}, \ldots, b_{2n_2}), \ldots, (b_{N1}, b_{N2}, \ldots, b_{Nn_N}))$, contain the same set of genes and that every gene appears in each genome exactly once.

For an arbitrary sequence $S = s_1, s_2, \ldots, s_k$ of genes, we will denote the reverse ordering of S by $-S$. *i.e.*, $-S = -s_k, -s_{k-1}, \ldots, -s_1$. A chromosome Y is said to be *identical* to a chromosome $X = (x_1, x_2, \ldots, x_k)$ iff either $Y = X$ or $Y = -X$. Genomes A and B are said to be identical $(A = B)$ iff the sets of chromosomes corresponding to A and B are the same.

As a convention we illustrate a chromosome horizontally and read it from left to right. Since we do not distinguish a chromosome from its reverse ordering, any prefix-suffix translocation acting on X and Y can be visualized as a prefix-prefix translocation acting on X and $-Y$. For a pair of chromosomes $X = (x_1, x_2, \ldots, x_m)$ and $Y = (y_1, y_2, \ldots, y_n)$ denote translocation acting on X and Y as $\rho(X, Y, i, j)$, $1 < i \leq m, 1 < j \leq n$, where the cleavage occurs in X between x_{i-1} and x_i and in Y between y_{j-1} and y_j. A prefix-prefix translocation $\rho_{pp}(X, Y, i, j)$ results into chromosomes:
$(x_1, \ldots, x_{i-1}, y_j, \ldots, y_n)$ and $(y_1, \ldots, y_{j-1}, x_i, \ldots, x_m)$. A prefix-suffix translocation $\rho_{ps}(X, Y, i, j)$ results into chromosomes:
$(x_1, \ldots, x_{i-1}, -y_{j-1}, \ldots, -y_1)$ and $(-x_m, \ldots, -x_i, y_j, \ldots, y_n)$. For a genome A and a translocation ρ acting on a pair of chromosomes of A, we denote the resulting genome as $A \cdot \rho$. If ρ is a reciprocal translocation then the number of chromosomes in A and $A \cdot \rho$ is the same. Moreover, the set of *nodal* genes (first and the last gene of all the chromosomes) is the same for A and $A \cdot \rho$. Figure 2a shows an example of evolution of A into *target* genome B. In the following discussion we assume, w.l.o.g., that a target genome is fixed and refer to the translocation distance between A and the target genome as *translocation distance of A*, thus, $d(A) \equiv d(A, B)$. Also, we refer to the problem of finding a shortest sequence of translocations transforming A into the target genome as the problem of *sorting A by translocations*.

In the following, we introduce *cycle graph* of a genome, which is the basis of our analysis of translocation distance. In a chromosome $X = (x_1, x_2, \ldots, x_k)$, replace every positive integer $+x_i$ by ordered pair (x_i^t, x_i^h) of vertices (t stands for tail and h stands for head) and replace every negative integer $-x_i$ by ordered pair (x_i^h, x_i^t) of vertices (Fig 2b). We say that vertices u and v are *neighbors* in X if they are adjacent in the ordered list constructed in afore mentioned manner. Notice that u and v are neighbors in X iff u and v are neighbors in $-X$. We say that vertices u and v are *neighbors* in a genome if they are neighbors in some chromosome in this genome. For gene x, vertices x^t and x^h are always neighbors and for simplicity, we exclude them from the definition of "neighbors" in the following discussion. We construct the bicolored *cycle graph* $G_A \equiv G_{AB}(V, E)$ of a genome A (with respect to a fixed target genome B) as follows. The vertex set V contains the pair of vertices x^t and x^h for every gene x in A, *i.e.*, $V = \{u : u$ is either x^t or x^h, x is a gene in $A\}$. Edges of G_A are colored either gray or black. Vertices u and v are connected by a black (solid) edge iff they are neighbors in A. Vertices u and v are connected by a gray (dotted) edge iff they are neighbors in the target genome (Fig. 2b). Notice that x^t and x^h are not connected for any x and a pair of vertices which are neighbors in both the genomes are connected by both, a black and a gray edge. See Bafna

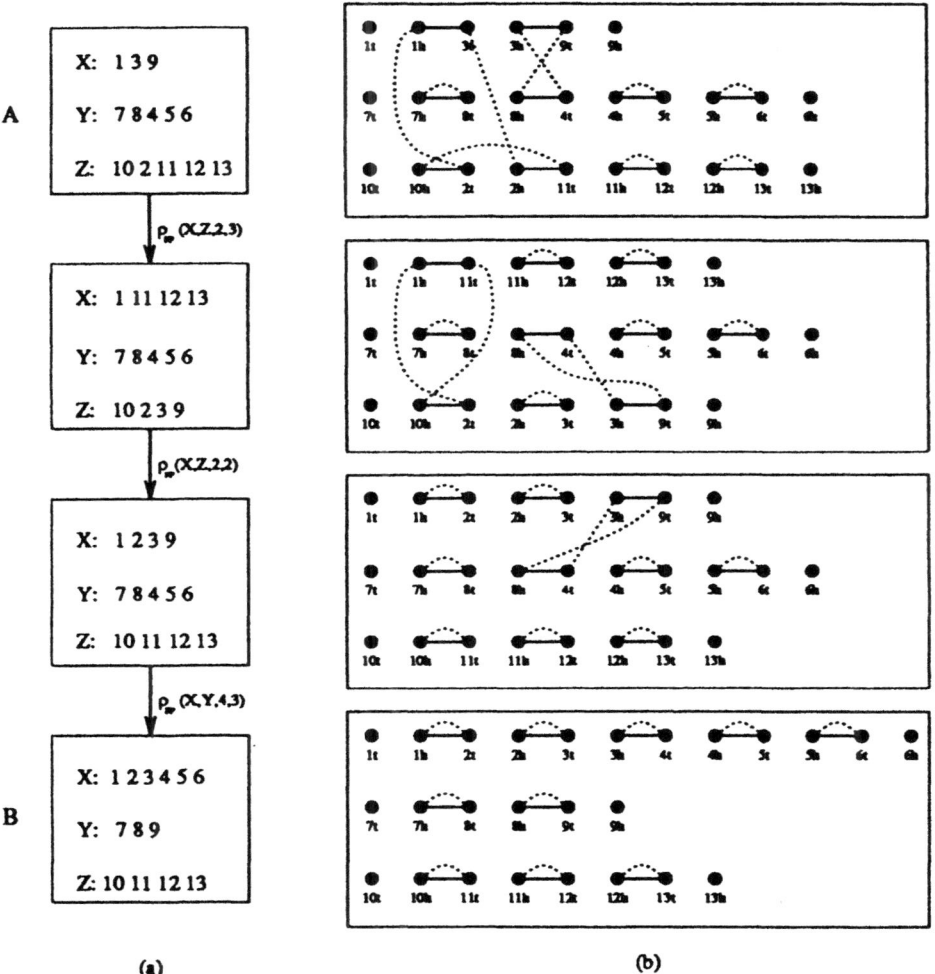

Figure 2: (a) An example of evolution by translocations (b) cycle graph corresponding to genomes at every stage of evolution with respect to the fixed target genome B.

and Pevzner, 1993, Hannenhalli and Pevzner, 1995a and Kececioglu and Ravi, 1995, for similar constructions.

The number of black (equivalently, gray) edges in G_A is $n - N$ where n is the number of genes in A. Clearly, each vertex is adjacent to exactly one black edge and one gray edge. Hence the graph can be uniquely decomposed into a number of disjoint cycles. We denote the number of cycles in G_A as c_A. Clearly, the number of cycles is maximized when A is identical to the target genome.

Lemma 1 $c_A = n - N$ iff A is identical to the target genome.

For the sake of simplicity, we will refer to all the intermediate genomes in the course of evolution of A as A. Any such sequence of translocations will be reflected as changes in the associated cycle graph until the graph is left with all cycles of length 2, i.e., $c_A = n - N$ (Fig. 2b).

Let $\rho \equiv \rho(X, Y, i, j)$ be a translocation acting on chromosomes $X = (x_1, x_2, \ldots, x_m)$ and $Y = (y_1, y_2, \ldots, y_n)$. Let $f \in \{x_{i-1}^t, x_{i-1}^h\}$ and $g \in \{x_i^t, x_i^h\}$ such that f and g are neighbors in X. Let $u \in \{y_{j-1}^t, y_{j-1}^h\}$ and $v \in \{y_j^t, y_j^h\}$ such that u and v are neighbors in Y. We say that ρ cuts black edges $(f\ g)$ and $(u\ v)$.

A prefix-prefix translocation cutting black edges $(u\ v)$ and $(f\ g)$ (read from left to right) is *proper*, if there is a cycle $(u\ v \ldots f\ g \ldots u)$ in G_A. In Fig. 2, the translocation $\rho_{pp}(X, Z, 2, 3)$ is proper since the black edges $(2^h\ 11^t)$ and $(1^h\ 3^t)$ cut by this translocation belong to the cycle $(2^h\ 11^t\ 10^h\ 2^t\ 1^h\ 3^t\ 2^h\ )$. A prefix-suffix translocation cutting black edges $(u\ v)$ and $(f\ g)$ (read from left to right) is *proper*, if there is a cycle $(u\ v \ldots g\ f \ldots u)$ in G_A. Notice that for every pair of black edges on different chromosomes but belonging to the same cycle in the cycle graph, there is proper translocation (prefix-prefix or prefix-suffix) cutting the two black edges. We call a translocation *improper*, if it cuts black edges belonging to the same cycle but is not proper. We call a translocation *bad*, if it cuts black edges belonging to different cycles (Fig. 3).

In the following we study the effect of a translocation on the structure of the cycle graph and describe the parameters that play a key role in determining the translocation distance.

3 Lower bound on translocation distance

Let ψ_A be a parameter ψ associated with genome A (or G_A). For a translocation ρ on A we denote the increase in ψ, as $\Delta(\psi)$ i.e., $\Delta(\psi) \equiv \psi_{A \cdot \rho} - \psi_A$.

Lemma 2 For a translocation ρ, $\Delta(c_A) = 1$ iff ρ is proper, $\Delta(c_A) = 0$ iff ρ is improper and $\Delta(c_A) = -1$ iff ρ is bad (Fig. 3).

Lemmas 1 and 2 imply

Theorem 1 For an arbitrary genome A, $d(A) \geq n - N - c_A$.

As it turns out, there are additional parameters associated with a genome which are important in computing the translocation distance. In particular if a set of genes occur close together within a chromosome in both the genomes but not in same order then reordering them necessitates a translocation that decreases the number of cycles. This leads to the notion of a subpermutation described in the following. Define *segment* as an interval $I = x_i, x_{i+1}, \ldots, x_j$ within a chromosome $X = x_1, x_2, \ldots, x_m$ in A. Let V_I be the set of vertices induced by the genes in I, i.e., $V_I = \{u : u \text{ is either } x_k^t \text{ or } x_k^h, i \le k \le j\}$. We refer to left vertex corresponding to x_i and right vertex corresponding to x_j as $LEFT(I)$ and $RIGHT(I)$ respectively. In Fig. 4, for the interval $I = 2, 4, 3, 5$, $LEFT(I) = 2^t$ and $RIGHT(I) = 5^h$. Define $IN(I) = V_I \backslash (LEFT(I) \cup RIGHT(I))$. An edge $(u\ v) \in G_A$ is said to be *inside* the interval I if $u, v \in IN(I)$. A *subpermutation* (SP) is an interval of genes $x_i, x_{i+1}, \ldots, x_j$ within a chromosome X in genome A such that there exists a segment $x_i, permutation(x_{i+1}, \ldots, x_{j-1}), x_j$ within some chromosome Y of target genome B and $permutation(x_{i+1}, \ldots, x_{j-1}) \ne x_{i+1}, \ldots, x_{j-1}$. Equivalently, SP is an interval I within some chromosome of A such that (i) there exists no edge $(u\ v)$ such that $u \in IN(I)$ and $v \notin IN(I)$ and (ii) there is at least one *long cycle* (of size > 2) involving edges inside I. A *minimal subpermutation* $(minSP)$ is a SP not containing any other SP. Size of a SP is the number of genes in the SP. In Fig. 4 the interval (2 4 3 5) is a $minSP$ of size 4 contained inside the SP (1 2 4 3 5 6) of size 6.

Notice that for an arbitrary partition of a SP into a non-empty prefix segment L and a non-empty suffix segment R, there must be a gray edge $(u\ v)$ such that, $u \in V_L, v \in V_R$ (the cycle containing the black edge $(RIGHT(L)\ LEFT(R))$ must contain such a gray edge). We refer to such an edge as a *connecting* gray edge from L to R.

A translocation *cuts* a segment S iff it cuts a black edge inside S. A translocation ρ *destroys* a SP S in A if S is not a SP in $A \cdot \rho$. What makes SPs interesting is that in order to destroy a SP we must do a bad translocation since there is atleast one long cycle in a SP and any translocation cutting a black edge inside a SP must cut a black edge belonging to a different cycle. $minSP$s are specially interesting in this respect since destroying a $minSP$ S destroys all the SPs containing S. We can destroy at most 2 $minSP$s on different chromosomes in a single bad translocation by choosing a translocation cutting both the $minSP$s. In the process any SP containing either of the $minSP$s is also destroyed. However a pair of $minSP$s on the same chromosome can not be destroyed simultaneously in a single bad translocation. Whenever we destroy any $minSP$ (1 or 2), the translocation must cut black edges from two different cycles and hence $\Delta(c_A) = -1$. If s_A is the number of $minSP$s in A then
Lemma 3 *For any translocation* $\Delta(c_A - s_A) \le 1$.
This leads to a slightly improved lower bound, $d(A) \ge n - N - c_A + s_A$, since $s_A = 0$ iff A is identical to the target genome. In the following we show that this bound is very tight by proving that $d(A) \le n - N - c_A + s_A + 2$. The bound, $d(A) \ge n - N - c_A + s_A$, assumes destroying two $minSP$s in a single translocation since if we destroy exactly one $minSP$ in a translocation then $\Delta(c_A - s_A) = 0$. Sometimes it may be impossible to destroy two $minSP$s in a single bad translocation and thus, unavoidable to make a translocation with $\Delta(c_A - s_A) \le 0$. If the number of $minSP$s is odd then we can not avoid such a translocation. Let $o_A = 1$ if the number of $minSP$s is odd and $o_A = 0$ otherwise. Clearly, $\Delta(c_A - s_A) = 1$ implies that $\Delta(o_A) = 0$. One could verify that

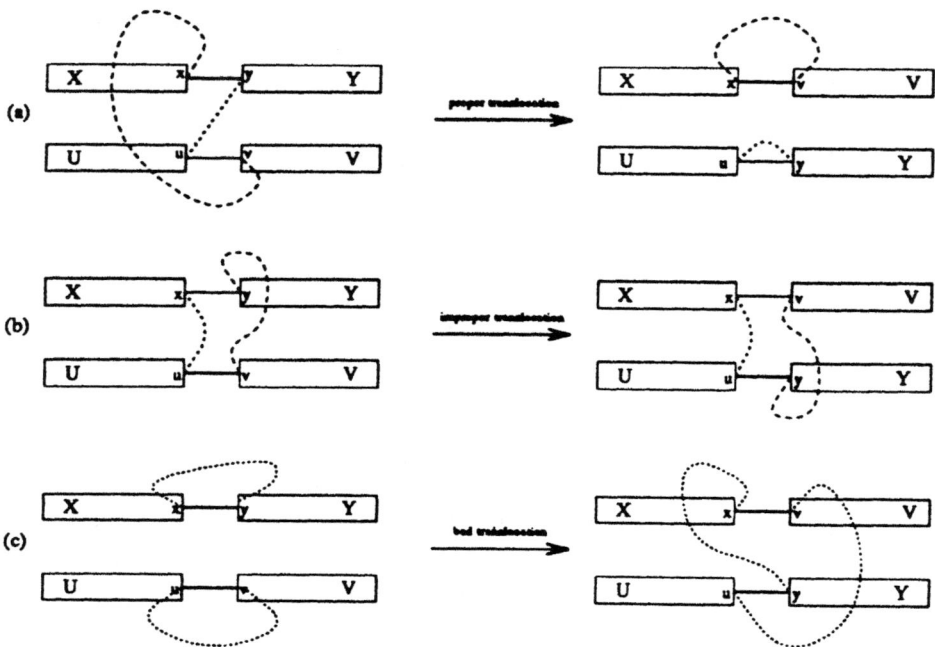

Figure 3: Prefix-prefix translocations cutting black edges $(u\ v)$ and $(x\ y)$ affects the cycle graph G_A. (a) For a proper translocation $\Delta(c_A) = 1$. (b) For an improper translocation $\Delta(c_A) = 0$. (c) For a bad translocation $\Delta(c_A) = -1$.

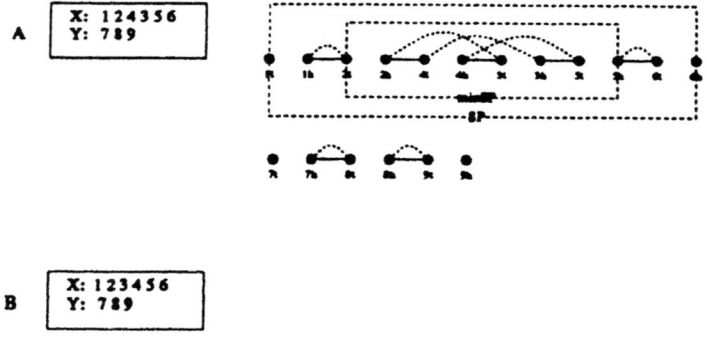

Figure 4: Examples of subpermutations.

Lemma 4 *For any translocation* $\Delta(c_A - s_A - o_A) \le 1$.

Genome A has an *even-isolation* if

1. all the $minSP$s of A reside on a single chromosome,
2. s_A is even and
3. all the $minSP$s are contained within a single SP.

Notice that if A has an *even-isolation*, we must perform a translocation with $\Delta(c_A - s_A - o_A) \le -1$. Consider the first translocation ρ destroying the even-isolation. If $\Delta(s_A) > 0$ (ρ creates one or two new $minSP$s) then $\Delta(s_A + o_A) = 2$, hence $\Delta(c_A - s_A - o_A) \le -1$. One could verify that if a translocation ρ destroys a $minSP$ then there can not be a new $minSP$ in $A \cdot \rho$. Moreover, due to property 3 in the definition of an even-isolation there can not be a proper translocation distributing the $minSP$s over two chromosomes. Hence, if $\Delta(s_A) = 0$ then ρ must be a bad translocation distributing the existing $minSP$s over two chromosomes, thus destroying even-isolation. In this case $\Delta(s_A) = \Delta(o_A) = 0$, hence $\Delta(c_A - s_A - o_A) = -1$. If $\Delta(s_A) = -1$ (ρ destroys one $minSP$s) then ρ must be a bad translocation. In this case $\Delta(s_A) = -1$, $\Delta(o_A) = 1$, hence $\Delta(c_A - s_A - o_A) = -1$.

Let $i_A = 1$ if A has an even-isolation and $i_A = 0$ otherwise. One could verify that

Lemma 5 *For any translocation* $\Delta(c_A - s_A - o_A - 2 \cdot i_A) \le 1$.

Notice that $o_A = i_A = 0$ iff A is identical to the target genome. This gives us an improved lower bound.

Theorem 2 *For an arbitrary genome A,* $d(A) \ge n - N - c_A + s_A + o_A + 2 \cdot i_A$.

4 Duality theorem for translocation distance

We call a translocation *valid* if $\Delta(c_A - s_A - o_A - 2 \cdot i_A) = 1$. In this section we prove the existence of a valid translocation for arbitrary genome A, implying that $d(A) = n - N - c_A + s_A + o_A + 2 \cdot i_A$. We say that a proper translocation ρ *acts* on the gray edge $(u\ v)$ if it cuts the black edges incident on u and v. In the following, we consider only the proper translocations acting on gray edges. Denote an arbitrary non-empty prefix (suffix) of a segment S by $pref(S)$ ($suff(S)$). Denote a segment formed by concatenating segments L and R (in that order) as $[L\ R]$.

Lemma 6 *For an arbitrary partition of a $minSP$ S into prefix segment L and suffix segment R, there exists a gray edge connecting L to R different from* $(RIGHT(L)\ LEFT(R))$.

Proof: If $(RIGHT(L)\ LEFT(R))$ is the only gray edge connecting L to R, then L (and/or R) will qualify for a SP or S did not have any long cycle inside it, implying that S was not a $minSP$, a contradiction. ∎

Lemma 7 *For an arbitrary partition of a $minSP$ S into prefix segment L, middle segment M and suffix R where all three of the segments are non empty, there must be a gray edge g such that g either connects L to M or g connects M to R.*

Proof: Since M does not contain nodal elements, there must be a vertex $v \in V_M$ which is a neighbor of some vertex $u \notin V_M$ in genome B. Clearly, $u \in V_L \cup V_R$. ∎

Theorem 3 *If there exists a proper translocation in A then there exists a proper translocation σ in A such that $A \cdot \sigma$ does not have any new minSP.*

Proof: Assume that every proper translocation leads to creation of a new $minSP$. Let ρ be the prefix-prefix (w.l.o.g.) translocation creating a *smallest* such $minSP$. Notice that ρ could create at most two new $minSPs$. Let $[L\ R]$ be the (smallest) new $minSP$ in the genome $A \cdot \rho$ where segments L and R belonged to different chromosomes in genome A (Fig. 5a). Our goal is to find an alternative proper translocation in A cutting L and R which either does not create a new $minSP$ or creates a smaller one. Since $(RIGHT(L)\ LEFT(R))$ is a gray edge, by lemma 6, there must be a gray edge $(l\ r)$ in $A \cdot \rho$ (hence in A) such that $l \in IN(L), r \in IN(R)$. A proper translocation σ (acting on $(l\ r)$) cuts L and R. Assume that the translocation σ *breaks up L* into non-empty segments L_1 and L_2 and breaks up R into non-empty segments R_1 and R_2 (Fig. 5b,c).

Case 1: σ is a prefix-prefix translocation (Fig. 5b).

We will now argue that the resulting genome $A \cdot \sigma$ either does not have any new $minSP$ or has a smaller one. We need to concentrate only on the subgraph induced by the pair of chromosomes involved in the translocation since the rest of the graph remains unchanged. Clearly, any new $minSP$ must involve parts of either L or R.

A new $minSP$ in $A \cdot \sigma$ can not be

1. $[suff(P)\ [L1\ R2]\ pref(V)]$ since there must be a gray edge $(u\ v)$ such that $u \in V_{[L1\ R2]}, v \in V_{[R1\ L2]}$ (lemma 7).
2. $[suff(U)\ [R1\ L2]\ pref(Q)]$ (similar to 1).
3. $[suff(P)\ pref([L1\ R2])]$, since, by lemma 6, there must be a connecting gray edge between $suff(P)$ and $pref([L1\ R2])$ different from $(RIGHT(P)\ LEFT([L1\ R2]))$ in $A \cdot \sigma$ (and hence in $A \cdot \rho$), implying that $[L\ R]$ is not a $minSP$ in $A \cdot \rho$, a contradiction.
4. $[suff([L1\ R2])\ pref(V)]$ (similar to 3).
5. $[suff(U)\ pref([R1\ L2])]$ (similar to 3).
6. $[suff([R1\ L2])\ pref(Q)]$ (similar to 3).

So any $minSP$ involving parts of either L or R must be within $[L1\ R2]$ or $[R1\ L2]$, hence smaller than $[L\ R]$, a contradiction.

Case 2: σ is a prefix-suffix translocation (Fig. 5c).

Arguments for this case are very similar to the previous case. Notice that there is a gray edge between $LEFT(-L2)\ (RIGHT(L2))$ and $RIGHT(-R1)$ $(LEFT(R1))$ (ρ acts on this edge in A). This prevents the creation of a new $minSP$ $[suff(P)\ [L1\ -\ R1]\ pref(-U)]$ or $[suff(-Q)\ [-L2\ R2]\ pref(V)]$ in $A \cdot \sigma$. There can not be any new $minSP$ in $A \cdot \sigma$ involving parts of either L or R by the same arguments as for the previous case. Hence we conclude that any $minSP$ involving parts of either L or R is within $[L1\ -\ R1]$ or $[-L2\ R2]$, hence smaller that $[L\ R]$, a contradiction to the assumption that ρ created the smallest $minSP$. ∎

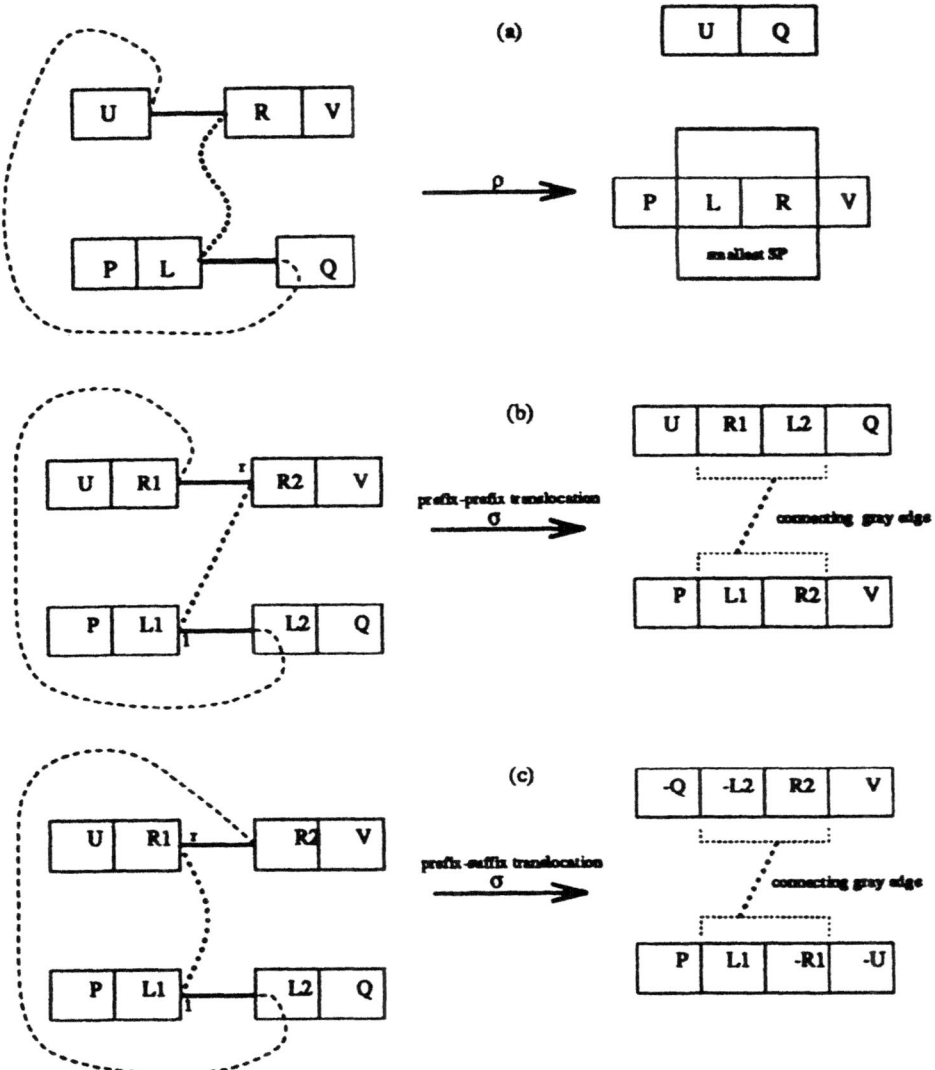

Figure 5: (a) Proper prefix-prefix translocation ρ creates a new $minSP$. (b),(c) Finding an alternate proper translocation not creating any new $minSP$.

Let S_1 and S_2 be two $minSPs$ within chromosome X of A such that S_1 is to the left of S_2. A gray edge $(u\ v)$ *separates* S_1 and S_2 if the vertex v belongs to chromosome different from X and the vertex u is in between the vertices $RIGHT(S_1)$ and $LEFT(S_2)$ on X, i.e., u is to the right of $RIGHT(S_1)$ and to the left of $LEFT(S_2)$ in the ordered list of vertices induced by genes of X. A translocation, acting on $(u\ v)$, *separates* S_1 and S_2, if $(u\ v)$ separates S_1 and S_2.

Theorem 4 *If there is a gray edge separating minSPs S_1 and S_2 in genome A then there exists a valid translocation σ separates S_1 and S_2.*

Proof By theorem 3, if a translocation ρ acting on gray edge $(u\ v)$ creates a $minSP\ [L\ R]$ (proof of theorem 3), there exists an alternative proper translocation σ connecting L to R which does not create a $minSP$ (following the arguments in the proof of theorem 3). Notice that if ρ separates $minSPs$ S_1 and S_2 then σ separates S_1 and S_2, implying that $A \cdot \sigma$ does not have an even-isolation. Hence σ is a proper valid translocation separating S_1 and S_2. ∎

Theorem 5 *If there exists a proper translocation in A then there exists a valid translocation in A.*

Proof: Notice that any proper translocation can not destroy any of the existing $minSPs$. If there exists a proper translocation separating two $minSPs$, by theorem 4, there exists a proper valid translocation. On the other hand if there is no gray edge separating two $minSPs$ then, by theorem 3, there is a proper translocation not creating any new $minSP$. Any such proper translocation ρ can not create an even-isolation, since an even-isolation in $A \cdot \rho$ implies an even-isolation in A. Hence ρ is valid. ∎

Theorem 6 *For every genome there exists a valid translocation.*

Proof: If there is a proper translocation in A then, by theorem 5, there exists a valid translocation in A. Assume that there is no proper translocation in A. Clearly, in this situation, there is no gray edge going across two different chromosomes and hence G_A is a collection of SPs distributed over the chromosomes. Let ν be the number of chromosomes containing at least one $minSP$.

Case 1: $\nu = 1$

If $s_A = 1$ then we can destroy the only $minSP$ by choosing a translocation cutting the $minSP$ and an arbitrary black edge in some other chromosome. In this case $\Delta(c_A - s_A - o_A - 2 \cdot i_A) = (-1 - (-1) - (-1) - 0) = 1$.

If $s_A > 1$ then choose a translocation ρ that destroys the $minSP$ second from the left using the same technique. Notice that, either two chromosomes in $A \cdot \rho$ contain atleast one $minSP$ or $s_{A \cdot \rho} = 1$. In either case $A \cdot \rho$ does not have even-isolation.

If s_A is odd, then $\Delta(c_A - s_A - o_A - 2 \cdot i_A) = (-1 - (-1) - (-1) - 0) = 1$. If s_A is even (A has even-isolation), then $\Delta(c_A - s_A - o_A - 2 \cdot i_A) = (-1 - (-1) - 1 - 2 \cdot (-1)) = 1$.

Case 2: $\nu = 2$

If $s_A = 2$ then choose a translocation ρ that destroys both of the $minSPs$. If

$s_A > 2$ then choose a translocation ρ that destroys the leftmost $minSP$ within one chromosome and the second $minSP$ within the other chromosome. Since $A \cdot \rho$ can not have even-isolation, $\Delta(c_A - s_A - o_A - 2 \cdot i_A) = (-1 - (-2) - 0 - 0) = 1$.

Case 3: $\nu = 3$

If $s_A = 3$ then choose a translocation ρ that destroys any two $minSPs$. If $s_A > 3$ then choose a translocation ρ acting on chromosomes X and Y where X has atleast two $minSPs$ within it. Since $A \cdot \rho$ can not have even-isolation, $\Delta(c_A - s_A - o_A - 2 \cdot i_A) = (-1 - (-2) - 0 - 0) = 1$.

Case 4: $\nu \geq 4$

In this case every translocation ρ, destroying two $minSPs$, is valid. ∎

Theorems 3, 5 and 6 imply the following duality theorem providing a characterization of translocation distance.

Theorem 7 *For an arbitrary genome A, $d(A) = n - N - c_A + s_A + o_A + 2 \cdot i_A$,* i.e.,

$$d(A) = \begin{cases} n - N - c_A + s_A + 2, & \text{if } A \text{ has an even-isolation} \\ n - N - c_A + s_A + 1, & \text{if } A \text{ has odd number of } minSPs \\ n - N - c_A + s_A, & \text{otherwise} \end{cases}$$

5 Algorithm for Sorting by Translocations

Theorems 5 and 6 motivate the algorithm *Translocation_Sort* generating a shortest sequence of translocations transforming genome A into the target genome.

Algorithm *Translocation_Sort(A)*
1.   **while** A is not identical to the target genome
2.   **if** there is a proper translocation in A
3.   select a valid proper translocation ρ (theorem 5)
4.   **else** select a valid bad translocation ρ (theorem 6)
5. $A \leftarrow A \cdot \rho$
6.   **endwhile**

The cycle graph G_A can be constructed in $O(n)$ time where n is the number of genes in A. A data structure to maintain the list of gray edges leading to proper translocations and the list of $minSPs$ can be initialized in $O(n^3)$. Clearly, there are atmost $O(n)$ iterations. Step 3 may require searching among atmost $O(n)$ proper translocations since every alternative choice of a proper translocation reduces the size of new $minSP$ created (proof of theorem 3). Checking the validity of any such translocation takes $O(n)$ time. In the absence of any proper translocation theorem 6 suggests a way to find a valid bad translocation in constant time. Performing the valid translocation in step 5 involves updating the data structures which can be done in atmost $O(n^2)$ time. Therefore the overall running time of *Translocation_Sort* is $O(n^3)$.

6 The case of unsigned data

Physical maps usually do not provide information about directions of genes, thus leading to the problem of computing rearrangement distance for unsigned data. We can construct the cycle graph for unsigned data by assuming arbitrary direction for each gene, constructing the cycle graph as described earlier and then collapsing the vertices x^t and x^h for every gene x. Notice that the resulting graph has equal number of black and gray edges incident on every vertex hence the graph can be decomposed into alternating cycles (cycles whose edges alternate colors). Any such decomposition can be viewed as assigning a direction to every gene. W.l.o.g., all genes in the target genome B have positive orientation. An assignment of directions to the genes in the source genome A dictated by the decomposition of the cycle graph is defined as a *spin* of A. Let $\hat{A}$ be the set of all spins of A. It's not hard to show that

$$d(A) = min_{\vec{A} \in \hat{A}} d(\vec{A})$$

Refer to Hannenhalli and Pevzner, 1995c for similar arguments. Hence the problem of computing the translocation distance for unsigned data is equivalent to the problem of computing an optimal spin of A, *i.e.* a spin that minimizes the translocation distance. This equivalent characterization could be used to approximate translocation distance for unsigned data. It can be shown that any decomposition of the cycle graph that attempts to maximize the number of cycles leads to an approximation of the translocation distance (discussed elsewhere). At this point, existence of a polynomial algorithm to compute translocation distance for unsigned data remains an open problem when both prefix-prefix and prefix-suffix reciprocal translocations are allowed.

7 Acknowledgments

Author is very thankful to Pavel Pevzner for many helpful suggestions, and pointing out a few mistakes in the earlier versions of this paper. Author also wishes to thank the referees for their comments.

References

[1] V. Bafna and P. Pevzner. Genome rearrangements and sorting by reversals. In *34th IEEE Symp. on Foundations of Computer Science*, pages 148–157, 1993. (to appear in SIAM J. Computing).

[2] V. Bafna and P. Pevzner. Sorting by reversals: Genome rearrangements in plant organelles and evolutionary history of X chromosome. *Mol. Biol. and Evol.*, 12:239–246, 1995.

[3] V. Bafna and P. Pevzner. Sorting by transpositions. In *Proc. of 6th Annual ACM-SIAM Symposium on Discrete Algorithms*, pages 614–623, 1995b.

[4] N. G. Copeland, N. A. Jenkins, D. J. Gilbert, J. T. Eppig, L. J. Maltals, J. C. Miller, W. F. Dietrich, A. Weaver, S. E. Lincoln, R. G. Steen, L. D. Steen, J. H. Nadeau, and E. S. Lander. A genetic linkage map of the mouse: Current applications and future prospects. *Science*, 262:57–65, 1993.

[5] S. Hannenhalli, C. Chappey, E. Koonin, and P. Pevzner. Scenarios for genome rearrangements: Herpesvirus evolution as a test case. In *Proc. of 3rd Intl. Conference on Bioinformatics and Complex Genome Analysis*, pages 91–106, 1995.

[6] S. Hannenhalli and P. Pevzner. Transforming cabbage into turnip (polynomial algorithm for sorting signed permutations by reversals). In *Proc. of 27th Annual ACM Symposium on the Theory of Computing*, 1995a. (to appear).

[7] S. Hannenhalli and P. Pevzner. Reversals do not cut long strips. Technical Report: CSE-94-074, Department of Computer Science and Engineering, The Pennsylvania State University, 1995c.

[8] J. Kececioglu and R. Ravi. Of mice and men: Evolutionary distances between genomes under translocation. In *Proc. of 6th Annual ACM-SIAM Symposium on Discrete Algorithms*, pages 604–613, 1995.

[9] J. Kececioglu and D. Sankoff. Exact and approximation algorithms for the inversion distance between two permutations. In *Proc. of 4th Ann. Symp. on Combinatorial Pattern Matching*, Lecture Notes in Computer Science 684, pages 87–105. Springer Verlag, 1993. (Extended version has appeared in Algorithmica, 13: 180-210, 1995.).

[10] J. Kececioglu and D. Sankoff. Efficient bounds for oriented chromosome inversion distance. In *Proc. of 5th Ann. Symp. on Combinatorial Pattern Matching*, Lecture Notes in Computer Science 807, pages 307–325. Springer Verlag, 1994.

[11] B. Lewin. *Genes V.* Oxford University Press, 1994.

[12] D. Sankoff. Edit distance for genome comparison based on non-local operations. In *Proc. of 3rd Ann. Symp. on Combinatorial Pattern Matching*, Lecture Notes in Computer Science 644, pages 121–135. Springer Verlag, 1992.

[13] D. Sankoff, R. Cedergren, and Y. Abel. Genomic divergence through gene rearrangement. In *Molecular Evolution: Computer Analysis of Protein and Nucleic Acid Sequences*, chapter 26, pages 428–438. Academic Press, 1990.

[14] D. Sankoff, G. Leduc, N. Antoine, B. Paquin, B. F. Lang, and R. Cedergren. Gene order comparisons for phylogenetic inference: Evolution of the mitochondrial genome. *Proc. Natl. Acad. Sci. USA*, 89:6575–6579, 1992.

[15] E. Therman and M. Susman. *Human Chromosomes, structure, behavior, and effects.* Springer-Verlag, 1993.

On the Complexity of Comparing Evolutionary Trees

(Extended Abstract)

Jotun Hein[1], Tao Jiang[2*], Lusheng Wang[3*], Kaizhong Zhang[4**]

[1] Institute for Genetics and Ecology, Aarhus University 8000 C, Denmark. Email:
jotun@hardy.pop.bio.aau.dk
[2] Department of Computer Science, McMaster University, Hamilton, Ont. L8S 4K1,
Canada. Email: jiang@maccs.mcmaster.ca
[3] Department of Electrical and Computer Engineering, McMaster University,
Hamilton, Ont. L8S 4K1, Canada. Email: lwang@maccs.mcmaster.ca
[4] Department of Computer Science, University of Western Ontario, London, Ont.
N6A 5B7, Canada. Email: kzhang@csd.uwo.ca

Abstract. We study the computational complexity and approximation
of several problems arising in the comparison of evolutionary trees. It is
shown that the maximum agreement subtree (MAST) problem for three
trees with unbounded degree cannot be approximated within ratio $2^{\log^\delta n}$
in polynomial time for any $\delta < 1$, unless NP $\subseteq$ DTIME$[2^{\text{polylog } n}]$, and
MAST with edge contractions for two binary trees is NP-hard. This an-
swers two open questions posed in [1]. For the maximum refinement subtree (MRST) problem involving two trees, we show that it is polynomial-
time solvable when both trees have bounded degree and is NP-hard when
one of the trees can have an arbitrary degree. Finally, we consider the
problem of optimally transforming a tree into another by transferring
subtrees around. It is shown that computing the subtree-transfer distance
is NP-hard and an approximation algorithm with performance ratio 3 is
given.

1 Introduction

In the analysis of molecular evolution, the evolutionary history of a set of species
is described by an *evolutionary tree* (or *phylogeny*). Let S be a set of species. An
evolutionary tree T on S is a *rooted unordered* tree such that the leaves of T are
uniquely labeled with the elements in S. The internal nodes are unlabeled and the
order among siblings is *insignificant*. Usually we require that each internal node
has at least two children. (Note that, evolutionary trees are also often viewed as
unrooted trees in the literature. All of our results hold for the unrooted version
as well.) Reconstructing the correct evolutionary tree for a set of species is one of

* Supported in part by NSERC Research Grant OGP0046613 and NSERC/MRC C-
GAT Grant GO-12278.
** Supported in part by NSERC Research Grant OGP0046373.

the fundamental yet difficult problems in evolutionary genetics. Many methods have been proposed based on various criteria. However, these methods do not always produce the same answer. Therefore, it is interesting to design metrics and automatic methods for the comparison of different evolutionary trees on the same set of species. A fruitful approach is to compute a tree that can somehow express the "intersection" of these evolutionary trees.

The notion of a *maximum agreement subtree* (MAST) was first proposed by Finden and Gordon [6]. Given an evolutionary tree T on set S and a subset $A \subseteq S$, the *restriction* of T on A, denoted $T|A$, is an evolutionary tree on set A obtained from T by eliminating the species outside A and the internal nodes with only single child. The latter operation, called *forced contraction*, is illustrated in Figure 1(a). For any two evolutionary trees T_1 and T_2 on set S, an *agreement subtree* (AST) of T_1 and T_2 is a tree T such that for some $A \subseteq S$ $T = T_1|A = T_2|A$. We call $A \subseteq S$ the set of the *agreed* species and $S - A$ the set of the *disagreed* species. A maximum agreement subtree (MAST) of T_1 and T_2 is an AST with the largest number of leaves (*i.e.* the largest number of species have been agreed upon). The notion of AST and MAST can be easily extended to more than two evolutionary trees on the same set of species [1].

The first polynomial-time algorithm for MAST on two trees was given by Steel and Warnow [16]. Their algorithm runs in $O(n^2)$ time for bounded-degree trees and $O(n^{4.5} \log n)$ for unbounded-degree trees, where n is number of species. Farach and Thorup recently improved the running time to $O(n^{1.5} \log n)$ for unbounded-degree trees and to $O(nc^{\sqrt{\log n}})$ for bounded-degree trees [4, 5]. Amir and Keselman [1] considered MAST for several trees. They showed that MAST is polynomial-time solvable for multiple bounded-degree trees and is NP-hard for three trees with unbounded degrees. An algorithm to approximate the *complement* of MAST on multiple unbounded-degree trees with ratio 4 was given. (The complement is to minimize the number of disagreed species instead of the agreed species). They also raised two questions: the approximability of MAST on multiple unbounded-degree trees and, given two trees T_1 and T_2 on set S, how to compute a tree with the largest number of *edges* which is obtainable through a sequence of *edge contractions* from both restrictions $T_1|A$ and $T_2|A$ for some subset $A \subseteq S$. An edge contraction is shown in Figure 1(b) and is also referred to as *deletion of an internal node* in tree edit [19]. Let's call the second problem maximum agreement subtree with edge contractions (MAST-EC). Here we settle these two problems by showing that MAST for three unbounded-degree trees cannot be approximated within ratio $2^{\log^\delta n}$ in polynomial time for any $\delta < 1$, unless NP $\subseteq$ DTIME[$2^{\text{polylog } n}$], and MAST-EC is NP-hard.

The *refinement* of trees is another approach towards the "intersection" of trees, and was originally introduced in the study of the *compatibility* of evolutionary trees [3, 8, 17]. Tree T is said to be a refinement of trees T_1 and T_2 if both T_1 and T_2 can be derived from T through a sequence of edge contractions. Two trees are *compatible* if they have a refinement. Polynomial-time algorithms for the tree compatibility problem have been known for a long time (*e.g.* [8, 17]). It is natural to consider the optimization version of this problem for trees which are not compatible with each other, namely, given trees T_1 and T_2 on set S, find the largest subset $A \subseteq S$ such that $T_1|A$ and $T_2|A$ are compatible [18]. Let's call a refinement of $T_1|A$ and $T_2|A$ a *maximum refinement subtree* (MRST) of T_1 and T_2 and this problem the MRST problem. One can view MRST as a natural counterpart of MAST. We show that MRST can be solved in polynomial time if T_1 and T_2 have degrees bounded by some constant and it becomes NP-hard if one of the trees is allowed to have an arbitrary degree.

When recombination of DNA sequences occurs in an evolution, the history of the evolution cannot be adequately described by a single tree. A recent proposal in attempt to solve this problem is to use a list of evolutionary trees [9, 10]. Each tree corresponds to a *region* of the DNA sequences, and each tree can be obtained from the preceding tree on the list by transferring some subtrees from one place to another. Figure 1(c) shows a subtree-transfer operation, where T_2 is moved to the branch immediately above T_4. Each such operation corresponds to a recombination event. A model for reconstructing the list of trees based on parsimony has been proposed in [9, 10]. The model requires the calculation of the subtree-transfer distance between two trees (*i.e.* the minimum number of subtrees we need to transfer). It was left open how to compute this distance. Unfortunately, we can show that computing the distance is NP-hard. We will also give a simple approximation algorithm achieving ratio 3. It turns out that this distance is also connected to the notion of agreement between trees.

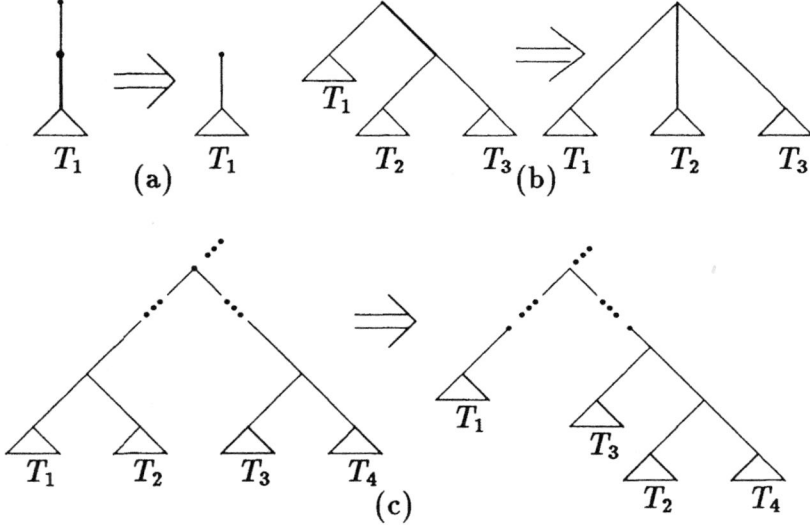

Fig. 1. The operations.

The non-approximability of MAST on multiple unbounded-degree trees is given in Section 2. Sections 3 and 4 discuss the complexity of MAST-EC and MRST. Finally, the subtree-transfer distance is considered in Section 5. Some proofs are omitted in this extended abstract and will be given in the full paper.

2 Non-approximability of MAST on 3 Unbounded-Degree Trees

In this section, we show that the following problem cannot be approximated within ratio $2^{\log^\delta n}$ in polynomial time for any $\delta < 1$, unless NP $\subseteq$ DTIME$[2^{\text{polylog } n}]$. Here, DTIME$[2^{\text{polylog } n}]$ denotes the class of problems solvable in $O(2^{\log^c n})$ time for some constant c.

Problem: MAST for Three Trees of Unbounded Degrees.
Instance: Three trees T_1, T_2 and T_3 of arbitrary degrees on set $S = \{s_1, s_2, \ldots s_n\}$.
Goal: Find a largest subset $A \subseteq S$ such that $T_1|A = T_2|A = T_3|A$.

The idea of the proof is the self-improvement technique as used in [11]. We first prove that the problem is MAX SNP-hard. Thus it cannot be approximated within ratio $1 + \epsilon$ for some $\epsilon > 0$ unless P = NP [2]. Then we define a product of trees and show that any approximation ratio r for the problem can be improved to $r^{1/k}$. Taking an appropriate k should give the desired bound.

Lemma 1. *The above problem is MAX SNP-hard.*

Proof. It can be easily verified that Amir and Keselman's construction for the NP-hardness [1] is in fact an L-reduction [15]. For the completeness of the paper, we include the construction here. The reduction is from the following variant of 3-Dimensional Matching which is MAX SNP-hard [13].
Problem: MAX 3DM-B (Maximum Bounded 3-Dimensional Matching).
Instance: A set $M \subseteq W \times X \times Y$ of ordered triples where W, X and Y are disjoint. Each element of $W \cup X \cup Y$ appears in at most B triples of M.
Goal:: Find the largest subset $M' \subseteq M$ such that no two elements of M' agree in any coordinate.

Given an instance of 3DM-B, $M \subseteq W \times X \times Y$, we construct three evolutionary trees T_1, T_2 and T_3. Let $|W| = |X| = |Y| = q$. Each T_i has $2q + 1$ children. The first $q + 1$ children are leaves labeled with new symbols $a_1, a_2, \ldots, a_{q+1}$. Each of the last q children of T_1 corresponds to an element $w \in W$, and has as its children leaves labeled with the triples of the form (w, x, y), where $x \in X$ and $y \in Y$. T_2 and T_3 are constructed in a similar way. It is easy to see that there is a 3-dimensional matching of size k if and only if there is an agreement subtree of size $k + q + 1$, and that this reduction is actually an L-reduction. $\square$

Now, we need define the *product* of two evolutionary trees. Let T_1 and T_2 be two evolutionary trees on sets S_1 and S_2 respectively. where S_1 and S_2 are the sets of labels for these two trees. For each label $s \in S_1$, let $T_{2,s}$ denote the tree obtained from T_2 by replacing each label $s' \in S_2$ with a new label (s, s').

The product of T_1 and T_2, denoted $T_1 \times T_2$, is obtained from T_1 by replacing each leaf labeled s with the tree $T_{2,s}$. For any tree T, we define $T^2 = T \times T$ and $T^{k+1} = T \times T^k$. The following lemma allows us to improve an approximation ratio for MAST by taking the product.

Lemma 2. *Let T_1, T_2 and T_3 be the three evolutionary trees and $c(T_1, T_2, T_3)$ denote the size of a MAST for T_1, T_2, T_3. Then $c(T_1^{k+1}, T_2^{k+1}, T_3^{k+1}) = c(T_1, T_2, T_3) \cdot c(T_1^k, T_2^k, T_3^k)$. Moreover, given an AST of size c for $T_1^{k+1}, T_2^{k+1}, T_3^{k+1}$, we can find in polynomial time an AST of size c_1 for T_1, T_2, T_3 and an AST of size c_2 for T_1^k, T_2^k, T_3^k such that $c_1 \cdot c_2 \geq c$.*

Proof. We give a proof for $k = 2$. The general case follows from the same argument. Let S be the set of labels in T_1, T_2, T_3. Obviously, $c(T_1^2, T_2^2, T_3^2) \geq c(T_1, T_2, T_3) \cdot c(T_1, T_2, T_3)$. Suppose that we are given an AST T of size c for T_1^2, T_2^2, T_3^2. For each label $s \in S$ such that (s', s) appears in T for some $s' \in S$, we can identify an agreement subtree of $T_{1,s}$, $T_{2,s}$, and $T_{3,s}$ in T. Let c_1 be the number of such subtrees. Without loss of generality, assume that all such subtrees have the same size c_2 (otherwise we can improve c). Then, we have $c_1 \cdot c_2 = c$. Moreover, every such subtrees gives an AST of T_1, T_2, T_3 of size c_2, and replacing each subtree in T with a single leaf labeled with an appropriate element of S also results in an AST of T_1, T_2, T_3 of size c_1. This completes the proof. $\square$

By the same argument as in [11], we have the following theorem.

Theorem 3. *For any constant $\delta < 1$, MAST for three unbounded-degree trees cannot be approximated within ratio $2^{\log^\delta n}$ in polynomial time, unless NP $\subseteq$ DTIME$[2^{\mathrm{polylog}\, n}]$.*

3 Agreement Subtrees with Edge Contractions

A natural extension of the agreement subtree approach is to allow the application of edge contractions in the formation of an "agreement" of the given trees. Intuitively, an edge contraction "loosens" the structure of a tree and thus increases the chance of having an agreement. For example, in the extreme case, we can contract all the edges in all trees to end up with a star. However, the obtained star contains little information about the evolutionary history. Therefore, it is desirable to contract a small number of edges yet to have a large number of the agreed species. This is the intuition behind the MAST-EC problem first proposed in [1]. Here, we show that MAST-EC on bounded-degree trees is NP-hard by a reduction from Exact Cover by 3-Sets [7].

Problem: Exact Cover by 3-Sets.

Instance: A collection C of subsets of a finite set S where every $c \in C$ contains three elements and every element $s \in S$ is contained in at most 3 subsets in C.

Goal: Find an exact covering $C' \subseteq C$ of S, i.e. a collection of mutually disjoint sets whose union equals S.

Theorem 4. *MAST-EC is NP-hard even if the given tree have bounded degree.*

Proof. Given an instance of Exact Cover by 3-Sets, let the set $S = \{s_1, s_2, \ldots, s_m\}$, where $m = 3q$ and $C = \{C_1, C_2, \ldots, C_n\}$, where each $C_i = \{t_{i,1}, t_{i,2}, t_{i,3}\}$, $t_{i,j} \in S$. Without loss of generality, we assume that $n > q$.

Two trees T and $\hat{T}$ are constructed as in Figure 2 and Figure 3. The top part is a binary tree whose actual structure is insignificant. (We need this part to get around the degree bound.) In order to make this part insignificant in the following calculation, introduce a large enough factor $f = n + m$. Each element $c_{i,j}$, $j = 1, 2, 3$, of C_i corresponds to a subtree $T_{i,j}$ as shown in Figure 4. Each element $s_i \in S$ corresponds to a subtree T_i as shown in Figure 5. Note that, each subtree $T_{i,j}$ is connected to the top part by a path of length $5f$ and has $f - 1$ internal nodes (including the root of the subtree) and $2f - 2$ edges. Similarly, each subtree T_i has $3f - 1$ internal nodes and $6f - 2$ edges.

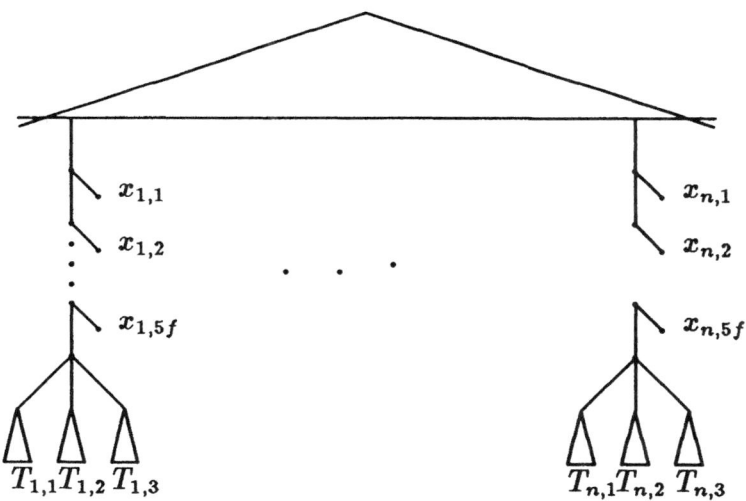

Fig. 2. The tree T constructed from $C = \{C_1, C_2, \ldots, C_n\}$, where each subtree $T_{i,j}$ corresponds to a $c_{i,j} \in C_i$, $j = 1, 2, 3$.

The following lemma, stated without proof, completes the reduction.

Lemma 5. *C has an exact cover of S if and only if there is a tree T' with at least $3(2f - 2)q + 5fq + 10f(n - q)$ edges such that, for some subset A of labels, T' can be obtained using edge contractions from both restrictions $T|A$ and $\hat{T}|A$.*
□.

A variant of MAST-EC is to construct a tree maximizing the the number of internal nodes instead of edges. Unfortunately, a simple modification of the above proves that this variant is also NP-hard.

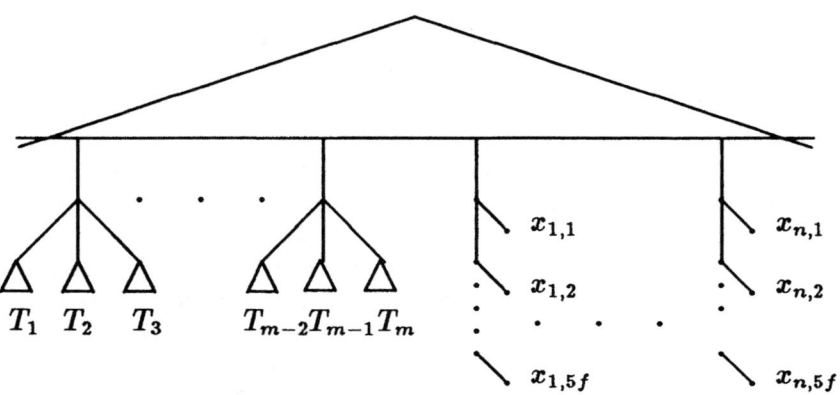

Fig. 3. The tree $\hat{T}$. Each subtree T_i corresponds to an element $s_i \in S$ and each of the n chains of length $5f$ corresponds to a subset $C_i \in C$.

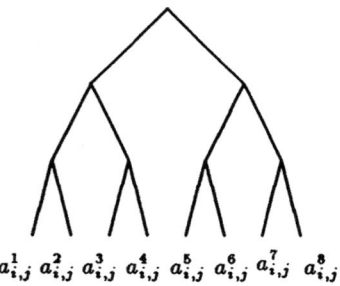

Fig. 4. The subtree $T_{i,j}$ corresponding to $c_{i,j} \in C_i$ ($f = 8$ in this case).

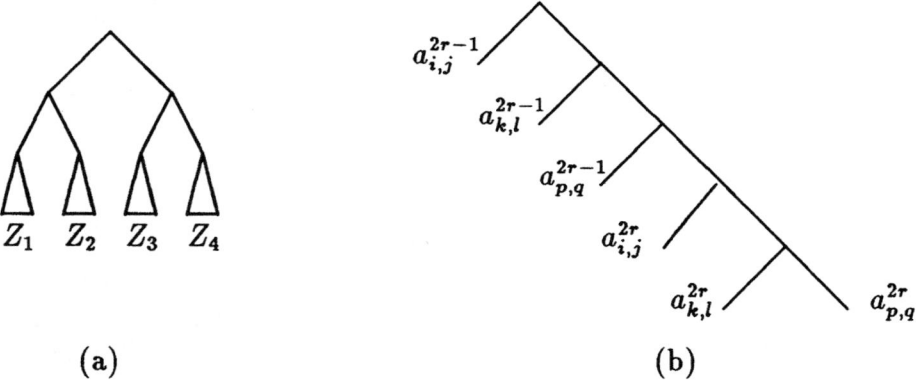

Fig. 5. Suppose that $c_{i,j}$, $c_{k,l}$ and $c_{p,q}$ are the three occurrences of s_i. (a) The subtree T_i corresponding to s_i. (b) The subtree Z_r. (Again, $f = 8$.)

4 Maximum Refinement Subtrees

The MRST problem can be reformulated as another problem, called the *alignment of trees*, recently introduced in [12]. Informally, an alignment $\mathcal{A}$ of trees T_1 and T_2 is obtained by first inserting new unlabeled nodes into T_1 and T_2 such that the two resulting trees T_1' and T_2' have the same structure, *i.e.* they are identical if the labels are ignored, and then *overlaying* T_1' on T_2'. Here inserting a new node u under an existing internal node v means we make u the parent of some children of v and then u a child of v [19]. Thus, each node in the alignment $\mathcal{A}$ is labeled with a pair of (possibly null) symbols, one from T_1' and the other from T_2'. A score scheme is defined for each pair of labels. The *value* of the alignment $\mathcal{A}$ is the sum of the scores of all pairs of opposing labels. An optimal alignment is one that maximizes the value over all possible alignments. The notion of alignment of trees was originally proposed as an alternative way of measuring the similarity of trees representing RNA secondary structures [12].

To formulate MRST as an alignment of trees problem, we need define the scores as follows: 1 for an identical pair of (non-null) labels and 0 otherwise. It is easy to see that in any alignment $\mathcal{A}$ of T_1 and T_2, the set of nodes with identical pairs of opposing labels induces a refinement subtree T of T_1 and T_2. The value of $\mathcal{A}$ is exactly the subset of labels appearing in T. Therefore, an optimal alignment of T_1 and T_2 gives an MRST of T_1 and T_2.

In [12], a polynomial-time algorithm to align trees with bounded degree is given. We hence have the following theorem.

Theorem 6. *MRST can be computed in polynomial time if both trees are of bounded degree.*

On the other hand, alignment of trees is NP-hard when one of the trees is allowed to have an arbitrary degree [12]. A slight modification of the proof of this result gives the next theorem.

Theorem 7. *MRST is NP-hard if one of the given trees can have an arbitrary degree.*

5 The Subtree-Transfer Distance

When *recombination* of DNA sequences occurs in an evolution, two sequences meet and generate a new sequence, consisting of genetic material taken left of the recombination point from the first sequence and right of the point from the second sequence [14, 9, 10]. Figure 6 (a) demonstrates a recombination. From a phylogenetic viewpoint, before the recombination, the ancestral material on the present sequence was located on two sequences, one having all the material to the left of the recombination point and another having all the material to the right of the breaking point.

In Figure 6 (b), it is assumed that a recombination has occurred in the generation of some ancestor r of s_2. Thus, the evolutionary history can no longer be

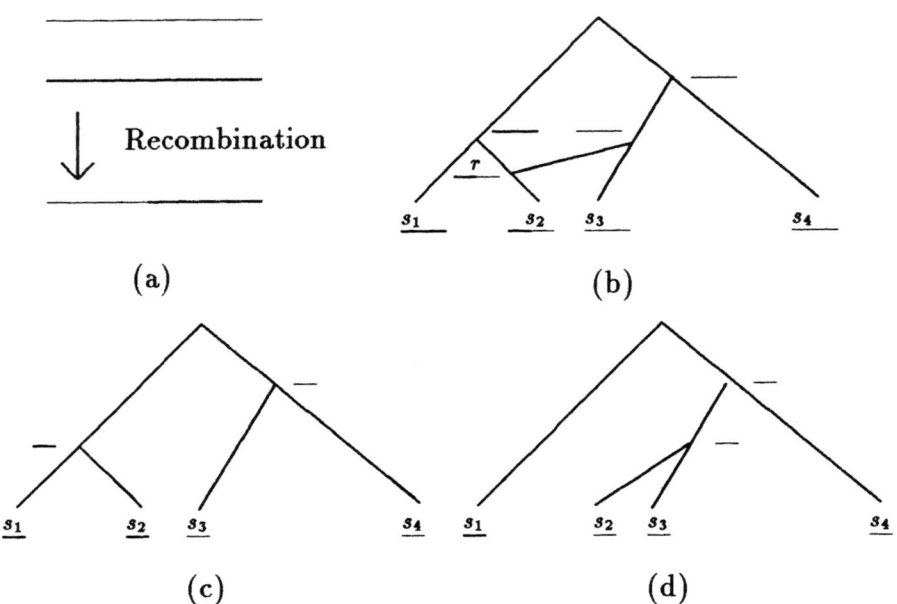

Fig. 6. (a) A recombination. (b) A recombination occurs at an ancestor r of s_2. (c) The evolutionary tree corresponding to the left region. (d) The evolutionary tree corresponding to the right region.

described by a single tree. The recombination event partitions the sequences into two *neighboring* regions. The history for the left region could be described by the evolutionary tree in Figure 6 (c), which is obtained by going back in time above the recombination following the left-going edge, while the history for the right region could be described by the tree in Figure 6 (d), which is obtained by following the right-going edge. The recombination makes the two evolutionary trees describing neighboring regions differ. However, two neighbor trees cannot be arbitrarily different, one must be obtainable from the other by a subtree-transfer operation. When more than one recombination occurs, one can describe an evolutionary history using a list of evolutionary trees, each corresponds to some region of the sequences and each can be obtained by several subtree-transfer operations from its predecessor [10]. The computation of subtree-transfer distance is useful in reconstructing such a list of trees based on parsimony [9, 10].

In this section, we show that computing the subtree-transfer distance between two evolutionary trees is NP-hard and give an approximation algorithm with performance ratio 3. Before we prove the results, it is again convenient to reformulate the problem. Let T_1 and T_2 be two evolutionary trees on set S. An *agreement forest* of T_1 and T_2 is any forest which can be obtained from both T_1 and T_2 by cutting k edges (in each tree) for some k and applying forced contractions in each resulting component trees. Define the size of a forest as the number of components it contains. Then the *maximum agreement forest* (MAF) problem is to find an agreement forest with the *smallest* size. The following lemma shows that MAF is really equivalent to computing the subtree-transfer distance.

Lemma 8. *The size of a MAF of T_1 and T_2 is one more than their subtree-transfer distance.*

The lemma can be proven by a simple induction on the number of leaves. Intuitively, the lemma says that the transfer operations can be broken down into two stages: first we cut off the subtrees to be transferred from the rest in T_1 (not worrying where to put them), then we assemble them appropriately to obtain T_2. This separation will simplify the proofs.

5.1   The NP-hardness

Theorem 9. *It is NP-hard to compute the subtree-transfer distance between two binary trees.*

Proof. The reduction is again from Exact Cover by 3-Sets. Let $S = \{s_1, s_2, \ldots s_m\}$ be a set and $C_1, \ldots, C_n$ be an instance of this problem. Assume $m = 3q$.

The tree T_1 is formed by inserting n subtrees $A_1, \ldots, A_n$ into a chain containing $2n + 2m$ leaves $x_1, \ldots, x_{2n}, y_1, \ldots, y_{2m}$ uniformly. (See Figure 7(a).) Each A_i corresponds to $C_i = \{c_{i,1}, c_{i,2}, c_{i,3}\}$, and has 9 leaves as shown in Figure 7(b). Suppose that $c_{j,j'}$, $c_{k,k'}$ and $c_{l,l'}$ are the three occurrences of an $s_i \in S$ in C. Then in T_2, we have a subtree B_i as shown in Figure 8(a). For each C_i, we also have a subtree D_i in T_2 as shown in Figure 8(b). The subtrees are arranged as a linear chain as shown in Figure 8(c).

Note that, each adjacent pair of subtrees A_i and A_{i+1} in T_1 is separated by a chain of length 2 which also appears in T_2. Thus, to form a MAF of T_1 and T_2, our best strategy is clearly to cut off $A_1, A_2, \ldots, A_n$ in T_1 and similarly cut off $B_1, B_2, \ldots, B_m$ in T_2. This then forces us to cut off $D_1, D_2, \ldots, D_n$ in T_2. Now in each A_i, we can either cut off the leaves $u_{i,1}, v_{i,1}, u_{i,2}, v_{i,2}, u_{i,3}, v_{i,3}$ to form a subtree containing three leaves $a_{i,1}, a_{i,2}, a_{i,3}$ (yielding $6 + 1 = 7$ components totally), or we can cut off $a_{i,1}$, $a_{i,2}$, and $a_{i,3}$. In the second case, we will be forced to also cut links between the three subtrees containing leaves $\{u_{i,1}, v_{i,1}\}$, $\{u_{i,2}, v_{i,2}\}$ and $\{u_{i,3}, v_{i,3}\}$ respectively, as the B_i's are already separated. Hence in this case the best we can hope for is $3 + 3 = 6$ components (if we can keep all three 2-leaf subtrees in the agreement forest).

The following lemma, which follows from the above discussion, completes the reduction.

Lemma 10. *C has an exact cover of S if and only if T_1 and T_2 have a MAF of size at most $1 + 6q + 7(n - q) = 7n - q + 1$.* □

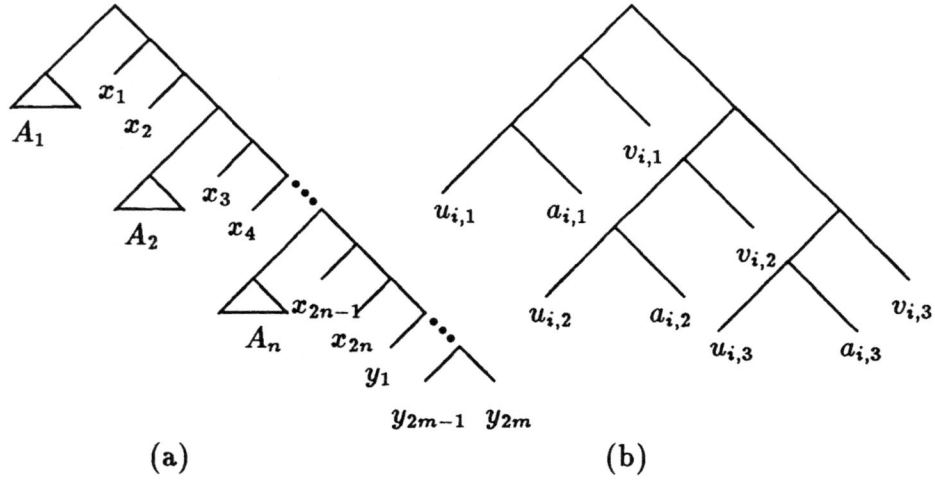

Fig. 7. (a) The tree T_1. (b) The subtree A_i.

5.2 An Approximation Algorithm of Ratio 3

Our basic idea is to deal with a pair of sibling leaves a, b in the first tree T_1 at a time. If the pair a and b are siblings in the second tree T_2, we replace this pair with a new leaf labeled by (a, b) in both trees. Otherwise, we will cut T_2 until a and b become siblings or separated. Eventually both trees will be cut into the same forest. Five cases need be considered. Figure 9 illustrate the first four cases. The last case (Case (v)) is that a and b are also siblings in T_2.

The approximation algorithm is given in Figure 10. The variable N records the number of components (or the number of cuts plus 1).

Theorem 11. *The approximation ratio of the above algorithm is 3, i.e., it always produces an agreement forest of size at most three times the size of a MAF for T_1 and T_2.*

Proof. (Sketch) We consider the number of edges cut in an agreement forest and show that the above algorithm cuts at most three times as many the edges cut by a MAF. To establish the approximation bound, the basic idea is to consider a MAF and charge the edges cut by the algorithm to the edges cut by the MAF, and make sure that each MAF edge is charged at most three edges. For convenience, we will refer to an edge according to its lower end.

We need a lemma which establishes that the algorithm is always optimal in Case (i).

Lemma 12. *There exists a MAF F which cuts all the edges cut by the algorithm in Case (i).*

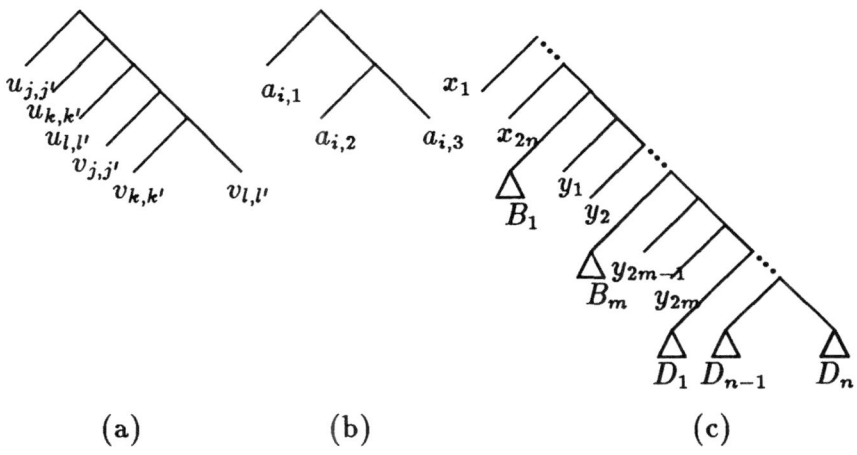

Fig. 8. (a) The subtree B_i. (b) The subtree D_i. (c) The tree T_2.

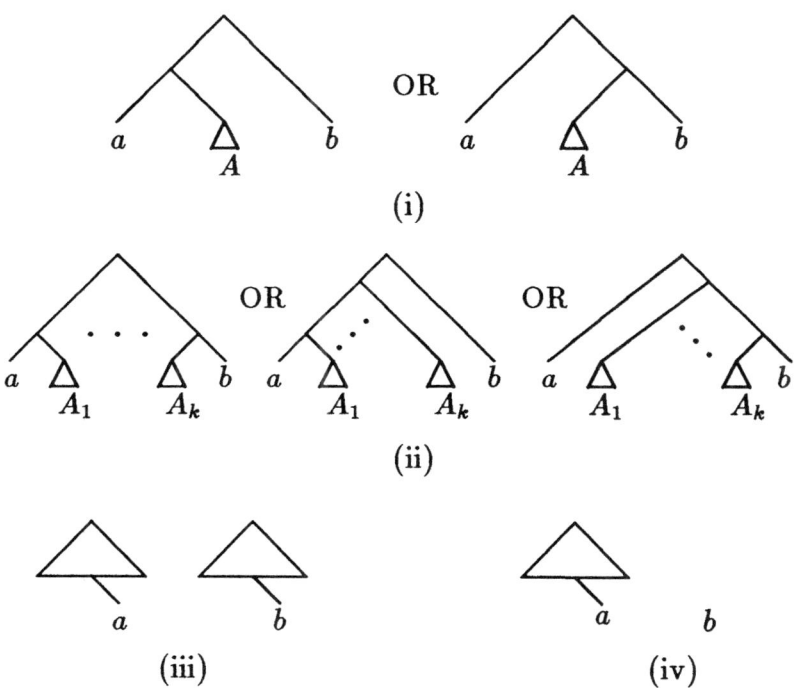

Fig. 9. The first four cases of a and b in T_2.

```
Input: T₁ and T₂.
0. N := 1;
1. For a pair of sibling leaves a, b in T₁;
 Consider how they appear in T₂ and cut the trees:
 Case (i): Cut off the middle subtree A in T₂; N := N + 1;
 Case (ii): Cut off a and b in both T₁ and T₂; N := N + 2;
 Case (iii): Cut off a and b in both T₁ and T₂; N := N + 2;
 Case (iv): Cut off b in T₁;
 Case (v): Replace this pair with a new leaf labeled (a, b) in both T₁ and T₂;
2. If some component in the forest for T₁ has size larger than 1, repeat Step 1.
Output: The forest and N.
```

Fig. 10. The approximation algorithm of ratio 3.

We only have to consider Cases (i)-(iii) because Case (iv) repeats an old cut. Let's fix the MAF F. We charge the edges as follows. In Case (i), each edge cut is simply charged to itself. Case (ii) is the most interesting. If at least one of the edges above a and b are cut by F, then we simply charge these two edges cut by the algorithms to the "correct" edge(s) cut by F. So each edge is charged with a cost of at most 2. Otherwise all the edges above the subtrees $A_1, \ldots, A_k$ $(k \geq 2)$ must be cut by F and we charge the two edges above a and b to these edges (each charged a cost of $\frac{2}{k} \leq 1$). Note that, only edges cut in Case (i) create components of size bigger than 1. Thus, if Case (iii) occurs, then a and b must belong to different components in F and hence F must cut at least one of the edges above a and b in T_1 to disconnect them. So we can charge the two edges above a and b to the "correct" one(s) cut by F (each charged a cost of at most 2).

It is easy to see that each edge cut by F is charged at most twice. Moreover, if an edge is charged twice, the first time it is charged must be in the second subcase of Case (ii). The second time can be in either Case (i), Case (iii), or the first subcase of Case (ii). Hence, each such edge is charged a cost of at most $1 + 2 = 3$ edges cut by the algorithm. That is, the algorithm cuts at most three times as many edges cut by F. □

It will be interesting to improve the approximation ratio. On the other hand, the NP-hardness proof can be easily strengthened to work for MAX SNP-hardness. Thus there is no hope for a polynomial-time approximation scheme for this problem.

References

1. A. Amir and D. Keselman, Maximum agreement subtree in a set of evolutionary trees - metrics and efficient algorithms, *IEEE FOCS'94*, 1994.

2. S. Arora, C. Lund, R. Motwani, M. Sudan, and M. Szegedy, Proof verification and hardness of approximation problems, *Proc. 33rd IEEE Symp. Found. Comp. Sci.*, 14-23, 1992.

3. G. Estabrook, C. Johnson and F. McMorris, A mathematical foundation for the analysis of cladistic character compatibility, *Math. Biosci.* 29, 181-187, 1976.

4. M. Farach and M. Thorup, Fast comparison of evolutionary trees, in *Proc. 5ith Annual ACM-SIAM Symposium on Discrete Algorithms*, 1994.

5. M. Farach and M. Thorup, Optimal evolutionary tree comparison by sparse dynamic programming, *IEEE FOCS'94*, 1994.

6. C. Finden and A. Gordon, Obtaining common pruned trees, *Journal of Classification* 2, 255-276, 1985.

7. M. R. Garey and D. S. Johnson, *Computers and Intractability: A Guide to the Theory of NP-Completeness*, W. H. Freeman, 1979.

8. D. Gusfield, Efficient algorithms for inferring evolutionary trees, *Networks* 21, 19-28, 1991.

9. J. Hein, Reconstructing evolution of sequences subject to recombination using parsimony, *Math. Biosci.* 98, 185-200, 1990

10. J. Hein, A heuristic method to reconstruct the history of sequences subject to recombination, *Journal Molecular Evolution* 36, 396-405, 1993.

11. T. Jiang and M. Li, On the approximation of shortest common supersequences and longest common subsequences, to appear in *SIAM J. Comput.*; also presented at *ICALP'94*.

12. T. Jiang, L. Wang and K. Zhang, Alignment of trees - an alternative to tree edit, *Theoretical Computer Science* 143-1, 137-148, 1995.

13. V. Kann, Maximum bounded 3-dimensional matching is MAX SNP-complete, *Information Processing Letters* 37, 27-35, 1991.

14. J. Kececioglu and D. Gusfield, Reconstructing a history of recombinations from a set of sequences, *Proc. 5th ACM-SIAM SODA*, 1994.

15. C.H. Papadimitriou and M. Yannakakis, Optimization, Approximation, and complexity classes, *Journal of Computer and System Sciences* 43, 425-440, 1991.

16. M. Steel and T. Warnow, Kaikoura tree theorems: computing the maximum agreement subtree, *Information Processing Letters* 48, 77-82, 1993.

17. T. Warnow, Tree compatibility and inferring evolutionary history, *J. of Algorithms* 16, 388-407, 1994.

18. T. Warnow, Private communication, 1994.

19. K. Zhang and D. Shasha, Simple fast algorithms for the editing distance between trees and related problems, *SIAM J. Comput.* 18, 1245-1262, 1989.

Suffix Cactus:
A Cross between Suffix Tree and Suffix Array*

Juha Kärkkäinen

Department of Computer Science, P. O. Box 26 (Teollisuuskatu 23)
FIN-00014 University of Helsinki, Finland.
E-mail: Juha.Karkkainen@cs.Helsinki.FI
URL: http://www.cs.helsinki.fi/~tpkarkka

Abstract. The suffix cactus is a new alternative to the suffix tree and
the suffix array as an index of large static texts. Its size and its performance in searches lies between those of the suffix tree and the suffix
array. Structurally, the suffix cactus can be seen either as a compact
variation of the suffix tree or as an augmented suffix array.

1 Introduction

The *suffix tree* is one of the most important data structures in stringology. The
suffix tree is an index-like structure formed from a string that allows many kinds
of fast queries about the string. What makes the suffix tree attractive is that its
size and its construction time are linear in the length of the text [19, 14, 17].
Suffix trees have a wide variety of applications. Apostolico [4] cites over forty
references on suffix trees, and Manber and Myers [13] mention several newer
ones.

The application, that we are mostly interested in in this paper, is the use of a
suffix tree as an index of a large static text to allow fast searches. The basic search
type is string matching, i.e. searching for the occurrences of a pattern string
in the text. Other useful forms of queries include regular expression matching
and approximate string matching. Examples of very large texts requiring fast
searching are electronic dictionaries [8], and biological sequence databases [16].

To work efficiently, the whole suffix tree must fit in the main memory. Thus
the space requirement of the suffix tree is an important issue. Gonnet, Baeza-
Yates and Snider [8] have studied the use of suffix trees with only a small part
at a time in the main memory, but many applications slow down unacceptably.
The exact size of the suffix tree depends on the implementation and the type of
the text. A typical size for a tight implementation on english text is about 15
bytes per text symbol.

The *suffix array* [13, 8] is a data structure which, like the suffix tree, allows
fast searches on a text. The size of an efficient implementation of a suffix array,
including the text itself, is only 6 bytes per text symbol. In string matching the

---

* Work supported by the Academy of Finland.

performance of suffix arrays is comparable to suffix trees, but other types of searches, such as regular expression matching, are slower on suffix arrays.

In this paper we present a new suffix-tree-like data structure called the *suffix cactus*. The size of a suffix cactus, 10 bytes per text symbol, lies between the sizes of suffix trees and suffix arrays. The same holds for the performance in many applications, such as regular expression matching.

The suffix cactus offers an interesting new point of view to the family of suffix structures. The structure of the suffix cactus has similarities with both the suffix tree and the suffix array. The suffix cactus could be described either as a compact version of the suffix tree or as a suffix array augmented with some extra information. The suffix cactus can therefore be called a cross between the suffix tree and the suffix array.

Recently, Anderson and Nilsson [2, 3], and Irving [9] have introduced new alternative data structures. The level compressed trie of Andersson and Nilsson takes about 12 bytes per text symbol and has matching properties comparable to the suffix cactus. The suffix binary search tree of Irving takes 14 bytes per text symbol and is similar to the suffix array in matching problems.

1.1 Basic Definitions

Let $T = t_1 t_2 \ldots t_n$ be a string over alphabet Σ. A *substring* of T is a string $T_i^j = t_i t_{i+1} \ldots t_j$ for some $1 \leq i \leq j \leq n$. The string $T_i = T_i^n = t_i \ldots t_n$ is a *suffix* of string T and the string $T^j = T_1^j = t_1 \ldots t_j$ is a *prefix* of string T. Let S and T be two strings and let j be the largest number for which $S^j = T^j$. Then the string $S^j = T^j$ is called the *longest common prefix* of S and T and its length j is denoted $\text{LCP}(S, T)$.

A *trie* (see e.g. Knuth [11]) is a rooted tree with the following properties.

1. Each node, except the root, *contains* a symbol of the alphabet.
2. No two children of the same node contain the same symbol.

A node v *represents* the string which is formed by catenating the symbols contained by the nodes on the path from the root to v, inclusive. Due to the second property, no two nodes may represent the same string. Note that, if a node v represents string S, then the ancestors of v represent the prefixes of S. The *depth* of a node v, denoted by $\text{DEPTH}(v)$, is the length of the path from the root to v, i.e., the length of the string that v represents.

The *suffix trie* $STr(T)$ of text T is a trie whose leaves represent the suffixes of T. The nodes of suffix trie $STr(T)$ represent exactly the set of substrings of T, because every substring of the text is a prefix of some suffix, i.e. $T_i^j = (T_i)^j$. An example suffix trie, for the string cabacca\$, is shown in Fig. 1.

The size of the suffix trie for a text of length n is $O(n^2)$ which makes it impractical for large texts. However, the suffix tree and the suffix cactus are basicly more compact (linear size) versions of the suffix trie. In Section 2 we will define the suffix cactus using the above description of the suffix trie.

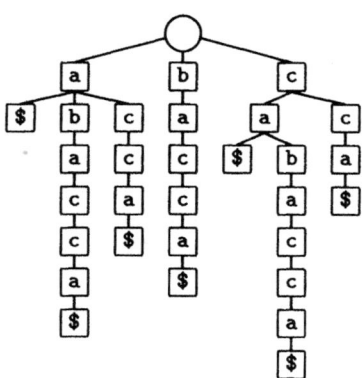

Fig. 1. The suffix trie of the string **cabacca\$**. The symbol **\$** is an extra symbol used for making all suffixes end in a leaf. The suffix **\$** is omitted from the trie.

1.2 Matching

Let a string T of length n be the *text* and a string P of length m the *pattern*. The problem of *string matching* is to find the occurrences of string P as a subtring of T. It can be solved in linear time by scanning text T using, e.g., the Knuth-Morris-Pratt algorithm [12]. For a large static text, a faster solution can be achieved by preprocessing the text. Suffix trees, suffix arrays and suffix cactuses are suitable preprocessing structures.

In *regular expression matching* the goal is to find all substrings of text T that match a given regular expression. A similar problem is *approximate string matching* where, given a string P and an integer k, one wants to find the sub-trings T_i of text T such that the edit distance between P and T_i is at most k. Both of these problems can be solved by scanning the text. Regular expression matching takes $O(n)$ time (excluding the preprocessing of the regular expression) [1] and approximate string matching $O(kn)$ time [7, 18].

Baeza-Yates and Gonnet have described methods to use the suffix tree to do both regular expression matching [5] and approximate string matching [6]. The latter idea was also independently mentioned in [10, Remark 2]. Both of these methods are based on scanning one suffix of T at a time to find whether it has a matching prefix. The methods take advantage of the fact that, if a set of suffixes has a common prefix of length d, then the state of the scan after the first d characters is the same for all of the suffixes. Therefore that part of the scan needs to be done only once. The suffix tree provides the information about common prefixes. It can be replaced by another suffix structure.

The above method for approximate string matching is more efficient than the basic text scan method only with short patterns and small values of k. However, Myers [15] has developed a method to do efficient approximate string matching even with long patterns and large k. The method divides the pattern into smaller parts whose approximate occurrences with small edit distance limit are searched separately. The results are then combined and used to restrict the area of the text that needs to be scanned. The matching of the parts can be done with the method of Baeza-Yates and Gonnet; Myers uses a slightly different method.

1.3 Suffix Tree and Suffix Array

The *suffix tree* discovered by Weiner [19] is a compact version of the suffix trie. It is formed by catenating each unary node (a node with exactly one child) with its child. An example is shown in Fig. 2(a). The strings in the catenations are substrings of the text and can thus be represented by two pointers into the text. The suffix tree has one leaf for each suffix and the number of other nodes is less than the number of leafs, because all the other nodes have at least two children. Thus the size of the suffix tree is linear in the length of the text.

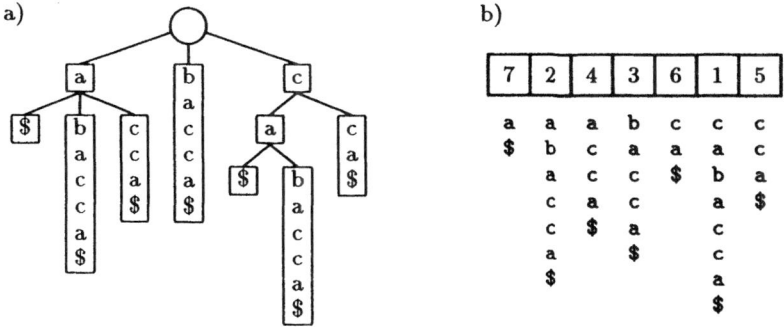

Fig. 2. a) Suffix tree and b) suffix array for string cabacca$.

If the alphabet size $|\Sigma|$ is considered constant, the suffix tree can be constructed in time $O(n)$ [19, 14, 17] and string matching takes time $O(m)$. The dependency on $|\Sigma|$ may be linear, logarithmic or constant depending on the implementation of branching. The most compact alternative uses linked lists and has linear dependency on $|\Sigma|$. In regular expression matching and approximate string matching the linked list implementation is as good as or better than other implementations.

In its basic form, the *suffix array* is just a lexicographically ordered array of the suffixes of the text. The suffixes are represented by their starting positions as illustraded in Fig. 2(b). The suffix array was discovered by Manber and Myers [13], and independently by Gonnet, Baeza-Yates and Snider [8].

String matching in suffix arrays can be done in $O(m \log n)$ time by a binary search. Manber and Myers [13] improved the string matching time to $O(m+\log n)$ by providing additional information about the lengths of the longest common prefixes (LCPs) between the suffixes. The LCPs are provided for each parent-child pair in an implicit tree structure called the interval tree. The interval tree is defined by the binary search order. The root of the interval tree is the middle suffix of the array, i.e. the first suffix processed in the binary search. The left child of the root is the middle suffix of the first half of the array and the right child is the middle suffix of the second half of the array. The next level of nodes is formed by the middle suffixes of the quarters of the array, and so on.

The above described LCP information is essential for efficient regular expression matching and approximate string matching in suffix arrays. The suffix array is still slower than the suffix tree in these tasks, in the worst case by a factor $O(\log n)$. In practice the difference is smaller, though.

The advantage of the suffix array over the suffix tree is its smaller size. Even with the LCP information the suffix array can be implemented using only 6 bytes per text symbol including the text itself.

The suffix array can be constructed in linear time by constructing first the suffix tree and then listing the suffixes in lexicographic order from the tree. Manber and Myers [13] have also described a construction algorithm that works by sorting the suffixes. It takes $O(n \log n)$ time in the worst case and $O(n \log \log n)$ time on average for random texts with even and independent distribution of characters. The advantage of this construction over the construction via the suffix tree is its smaller space requirement, 10 bytes per text symbol.

2 Suffix Cactus

The new data structure, *suffix cactus*, can, like the suffix tree, be viewed as a compact suffix trie. The suffix tree was formed by catenating the unary nodes with their children. To get a suffix cactus, every internal node is catenated with one of its children. The catenations are called the *branches* of suffix cactus.

Definition 1. Let v be a node of suffix trie $STr(T)$ of text T such that either v is the root or v is not the first child of its parent w. Then suffix cactus $SC(T)$ of T has a branch s that contains exactly the nodes on the path from v to the first leaf u under v.

Clearly, each node of $STr(T)$ is contained by exactly one branch of $SC(T)$. The branch containing the root of $STr(T)$ is called the *root branch*. The node v is called the root of branch s, u is called the leaf of s, and the parent w is called the parent node of s. The *branching depth* of s, denoted by DEPTH(s), is the depth of the parent node w. The branching depth of the root branch is 0.

Branch s *contains* the string formed by catenating the characters in the nodes contained by s. Branch s *represents* the same string as the leaf u. The leafs of $STr(T)$ represent the suffixes of T and there is thus a one-to-one correspondence between the suffixes of T and the branches of $SC(T)$. The starting point of the suffix represented by branch s will be denoted by SUFFIX(s). The string contained by s is now $T_{\text{SUFFIX}(s)+\text{DEPTH}(s)}$.

The term 'first' in Definition 1 implies the existence of an ordering among the children of a node. Any ordering can be used, which allows many alternative forms for the cactus. Two variations for string **cabacca\$** are shown in Fig. 3. The left-hand side variation uses alphabetical ordering and is the one used by the implementation described in this paper.

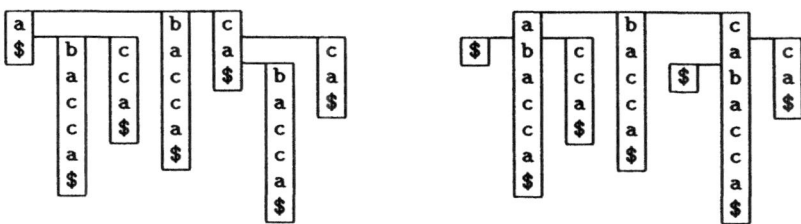

Fig. 3. Two variations of suffix cactus for the string **cabacca\$**. Turn the figure upside down to see an explanation for the name 'cactus'.

The most obvious way to define the tree structure of a suffix cactus is the following.

Definition of parent (alternative 1). Let s be a branch of $SC(T)$ and let v be its root. The parent (branch) of s is the branch containing the parent node of v.

However, for the implementation that is described in the next section, the following is a more natural definition.

Definition of parent (alternative 2). Let s be a branch of $SC(T)$ and let v be its root. The parent (branch) of s is the branch containing the preceding sibling of v. The preceding sibling is defined by the same ordering as the one used in Definition 1.

With both of the alternative definitions all branches, except the root branch, have a parent.

As an example, let us consider the third branch from left in the cactus on the left in Fig. 3. By the first definition its parent is the first branch, but by the second definition the parent is the second branch.

3 An Implementation

The name 'cactus' comes from the way the branches start in the middle of other branches. Whichever of the alternative definitions of the tree structure is used, this kind of branching needs to be implemented differently from the traditional tree branching. The implementation affects the exact space requirement of the suffix cactus and the time complexity of the different matching problems. In this paper we describe in detail an implementation that is space efficient and has, in all of the above described matching problems, the same time complexity as the linked list implementation of the suffix tree.

This implementation is based on alphabetical ordering of the children of a node and the second alternative definition of the parent branch. The children of each branch are in a linked list from the highest branching one to the lowest branching one. A key property of the second alternative definition is that a branch can have at most one child at each branching depth. Therefore, following a child list to find a specific child takes no more time than following the string contained by the branch to the point of branching of that child. The child list structure can be formalized by the operations FIRSTCHILD and NEXTSIBLING in the obvious way. Their implementation is described a little later.

The SUFFIX and DEPTH values are kept in two tables. The tables are in the lexicographic order of the suffixes. The SUFFIX table is, in fact, the basic suffix array. To simplify notation, we use the rank of a branch in the above order as the name of the branch. That is, the suffix $T_{\text{SUFFIX}(s)}$ represented by branch s is the sth suffix of T in the lexicographic order. Branch 1 is the root branch.

The following three lemmas show how the branching structure of the suffix cactus of text T can be derived straight from the text.

Lemma 2. *The branching depth* DEPTH(s) *of a branch* $s > 1$ *is* LCP$(T_{\text{SUFFIX}(s-1)}, T_{\text{SUFFIX}(s)})$.

Proof. Let v be the root, u the leaf, and w the parent node of branch s. Let v' be the alphabetically preceding sibling of v and let u' be the leaf of branch $s - 1$. Then v' must be an ancestor of u'. The paths from root to u and u' go together until node w where they get separated. Thus LCP$(T_{\text{SUFFIX}(s-1)}, T_{\text{SUFFIX}(s)}) =$ DEPTH$(w) =$ DEPTH(s). □

Lemma 3. *The parent branch of branch* $r > 1$ *is the latest branch* $s < r$ *such that* DEPTH$(s) \leq$ DEPTH(r).

Proof. Let v be the root and w the parent node of r. Let v' be the alphabetically preceding sibling of v. If s is the parent of r, then s contains v'. The parent node of s is w or an ancestor of w. Therefore the depth of s is at most DEPTH$(w) =$ DEPTH(r). Suffix $T_{\text{SUFFIX}(s)}$ precedes $T_{\text{SUFFIX}(r)}$ lexicographically and thus $s < r$. It remains to show that s is the latest branch satisfying these conditions.

Let t be a branch such that $s < t < r$. Let u'' be the leaf of t. Node v' must be an ancestor of u''. Because v' is contained by s, the root of t must be below v' on the path from v' to u''. Thus it holds DEPTH$(t) \geq$ DEPTH$(v') >$ DEPTH$(w) =$ DEPTH(r). □

Lemma 4. *A branch* s *has child branches only if branch* $s + 1$ *is a child of* s. *Let* s *be such a branch and let* $r_1, r_2, \ldots, r_k$ *be the children of* s *from the highest branching to the lowest branching. Then* $s + 1 = r_k < \cdots < r_1$.

Proof. By Lemma 3 r is a child of s if and only if

1. $s < r$,
2. $\text{DEPTH}(s) \leq \text{DEPTH}(r)$ and
3. there is no branch $t > s$ such that the first two conditions would hold if s was replaced with t.

For $r = s + 1$ the first and last condition always hold. Therefore, if $s + 1$ is not a child of s, then $\text{DEPTH}(s) > \text{DEPTH}(s + 1)$. In such a case, if any node r satisfies the first two conditions, then $t = s + 1$ violates the third condition. Thus s can have no children, if $s + 1$ is not a child of s.

The second claim of the lemma is clearly true if $k = 1$. Otherwise, let r_i and r_{i+1}, $1 \leq i < k$, be two of the children of s. Then it holds that $\text{DEPTH}(r_i) < \text{DEPTH}(r_{i+1})$. If now $r_i < r_{i+1}$, then $t = r_i$ would violate the third childhood condition of r_{i+1}. Therefore we must have $r_{i+1} < r_i$. $\qquad\square$

The last lemma enables us to describe the implementation of the branching operations FIRSTCHILD and NEXTSIBLING. The implementation consists of a single table called SIBLING. Using the notations of Lemma 4 this table can be defined by

$$\text{SIBLING}(r_i) = \begin{cases} r_1, & \text{if } i = k \\ r_{i+1}, \text{if } i < k \end{cases}$$

or alternatively by

$$\text{SIBLING}(s) = \begin{cases} \text{FIRSTCHILD}(s - 1), \text{if } s - 1 \text{ has children} \\ \text{NEXTSIBLING}(s), \quad \text{if } s \text{ has a next sibling} \end{cases}$$

In other words, the children of each branch form a cyclical list. In addition we define $\text{SIBLING}(1) = 1$. The FIRSTCHILD and NEXTSIBLING can now be defined as follows.

$$\text{FIRSTCHILD}(s) = \begin{cases} \text{SIBLING}(s + 1), \text{if } \text{SIBLING}(s + 1) \geq s + 1 \\ \text{none}, \quad \text{if } \text{SIBLING}(s + 1) < s + 1 \end{cases}$$

$$\text{NEXTSIBLING}(s) = \begin{cases} \text{SIBLING}(s), \text{if } \text{SIBLING}(s) < s \\ \text{none}, \quad \text{if } \text{SIBLING}(s) \geq s \end{cases}$$

Fig. 4 shows an example of this implementation.

s	1 2 3 4 5 6 7
SUFFIX(s)	7 2 4 3 6 1 5
DEPTH(s)	0 1 1 0 0 2 1
SIBLING(s)	1 4 3 2 5 7 6

Fig. 4. The implementation of the left-hand side suffix cactus in Fig. 3.

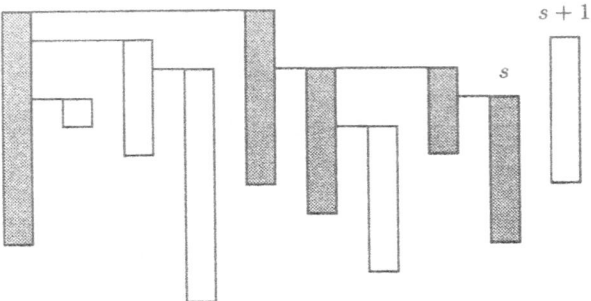

Fig. 5. The state of a suffix cactus before the processing of branch $s + 1$. The active branches are grayed.

4 Construction

In this section we will describe two construction algorithms for the above implementation of the suffix cactus. The algorithms work in two phases, the second of which is common to both. The first phases of the algorithms construct the SUFFIX and DEPTH tables. One algorithm uses the suffix tree to do this and the other uses the suffix array. The common second phase then constructs the SIBLING table from the DEPTH table. We start by describing the second phase.

At the start of the second phase the DEPTH table tells the branching depths of each branch. By Lemma 3 the parent branch of branch r is the latest branch s preceding r such that $\text{DEPTH}(s) \leq \text{DEPTH}(r)$. Therefore the DEPTH table fully defines the branching structure of the cactus and the SIBLING table can be calculated from it.

The SIBLING table is constructed in one first branch to last branch pass. Let us look at the situation when a branch s has just been processed and the processing of branch $s+1$ is about to start (Fig. 5). Let $s_1, s_2, \ldots, s_k$ be the path from branch 1 (the root) to branch s with $s_1 = 1$ and $s_k = s$. The branches on the path are called the *active branches*. The first (highest branching) children of each active branch may still be among the unprocessed branches. The first children of the other processed branches and the next siblings of all processed branches have, on the other hand, all been processed. Therefore, we can assume that the SIBLING table is finished up to the entry s, excluding the entries $s_0+1, s_1+1, \ldots, s_{k-1}+1$.

The parent of branch $s + 1$ must be one of the active branches. To be able to find the parent quickly, the active branches are on a list from the last to the first. The parent of $s+1$ is the first branch s_i on the list such that $\text{DEPTH}(s_i) \leq \text{DEPTH}(s + 1)$. The list is implemented using the so far unfinished SIBLING table entries, i.e. $\text{SIBLING}(s_i + 1) = s_{i-1}$ for $i = 1, \ldots, k - 1$.

Let us now see what happens when branch $s + 1$ is processed. If the parent of $s + 1$ is s, we make $s + 1$ active by adding it to the beginning of the list of active branches and we are done. Assume then that active branch s_i, $i < k$, is the parent of $s + 1$. Now we do the following.

1. Find s_i by following the list of active branches.
2. Remove the branches $s_{i+1}, \ldots, s_k$, that are passed during the search, from the list of active branches and finalize their first children by setting $\text{SIBLING}(s_j + 1) = s_{j+1}$ for $j = i+1, \ldots, k-1$.
3. Make s_{i+1} the next sibling of $s + 1$ by setting $\text{SIBLING}(s+1) = s_{i+1}$.
4. Add $s + 1$ to the beginning of the list of active branches.

When all branches have been processed, we travel the list of active branches once more to set the first children of the remaining active branches. The algorithm is presented in detail in Fig. 6.

```
SIBLING(1) = 1
s_{k-1} = 0
for s = 1 to n - 1 do
    if DEPTH(s) ≤ DEPTH(s + 1) then         % Is s parent of s + 1?
        SIBLING(s + 1) = s_{k-1}
        s_{k-1} = s
    else
        s_{i+1} = s
        s_i = s_{k-1}
        while DEPTH(s_i) > DEPTH(s + 1) do    % Travel the list of active branches
                                              % until the parent of s + 1 is found.
            r = SIBLING(s_i + 1)
            SIBLING(s_i + 1) = s_{i+1}         % Remove passed branches from the list
                                              % and finalize their first children.
            s_{i+1} = s_i
            s_i = r
        end
        SIBLING(s + 1) = s_{i+1}
        s_{k-1} = s_i
    end
end
s_{i+1} = n
s_i = s_{k-1}
while s_i > 0 do                              % Finalize the first children
    r = SIBLING(s_i + 1)                       % of the last active branches.
    SIBLING(s_i + 1) = s_{i+1}
    s_{i+1} = s_i
    s_i = r
end
```

Fig. 6. The construction of SIBLING table from the DEPTH table. The variables s_{k-1}, s_i and s_{i+1} are so named to help the comparison between the algorithm and the description in the text.

Excluding the **while** loops, the algorithm clearly works in linear time. Each round of the **while** loops walks one step in the list of active branches and removes one branch from the list. Once removed, a branch cannot return to the list. Thus, at most one round of the **while** loops is executed for each branch. This gives us the following theorem.

Theorem 5. *The* SIBLING *table can be constructed from the* DEPTH *table in linear time and constant additional space.*

The remaining problem with the construction of the suffix cactus is to get the SUFFIX and DEPTH tables somehow. One way is to use the suffix tree. A lexicographically ordered depth-first traversal of the tree can be used to recover the necessary information from the tree in linear time. As mentioned in Section 1.3, the suffix tree itself can be build in linear time, so the whole construction works in linear time. The construction takes at least as much space as the suffix tree construction and may take a little more depending on the details of implementation.

The SUFFIX and DEPTH tables can also be constructed from the suffix array with LCP information. The basic suffix array forms the SUFFIX table as such. As mentioned in Lemma 2, the values in DEPTH table are LCPs of lexicographically adjacent suffixes. These values can be recovered from the LCP information of the suffix array by a traversal of the interval tree in linear time. If the suffix array is build using the $O(n \log n)$ sorting method, it dominates the time complexity of the whole cactus construction. The advantage of this construction is that all stages work in the space of the final suffix cactus.

5 Experimentation

To see how the suffix cactus behaves in practice, we implemented the described variation of the suffix cactus together with the linked list version of the suffix tree and the version of the suffix array with LCP information. The tests were run on a 90 MHz Pentium PC with 16 Mbytes of memory running Linux operating system.

We implemented the standard suffix tree construction [14, 17], the suffix array construction by sorting [13], and both of the suffix cactus construction algorithms described in the previous section. Table 1 gives the execution times and the space requirements of these construction algorithms. The space requirements include the text.

The space requirement of a finished structure is 6 bytes per text symbol for the suffix array and 10 bytes per text symbol for the suffix cactus, regardless of the construction method. In principle, the space requirement of a finished suffix tree could be reduced a little from the construction time space requirement by releasing the suffix links. In our implementation this is not done because of the complications in memory management caused by not knowing the number of nodes in the suffix tree in advance.

In the implementations most numbers and pointers take 4 bytes. The exceptions are the LCPs of the suffix array and the DEPTHs of the suffix cactus, both of

Table 1. Space requirements and execution times of the construction.

text			space (bytes/n)				time (s)			
type	$|\Sigma|$	n	tree	array	cactus via tree	cactus via array	tree	array	cactus via tree	cactus via array
english	71	3000	13.48	10	14.48	10	0.08	0.21	0.09	0.23
english	74	30000	14.77	10	14.77	10	0.67	2.85	0.84	2.99
english	77	300000	15.17	10	16.17	10	6.60	36.4	8.63	37.7
random	77	300000	9.72	10	10.72	10	21.2	27.0	22.7	28.4
DNA	4	300000	17.70	10	18.70	10	5.62	41.4	7.78	42.6
random	4	300000	17.43	10	18.43	10	5.66	33.8	7.84	35.1
random	16	300000	11.80	10	12.80	10	8.10	31.2	9.91	32.5
random	64	300000	10.95	10	11.95	10	19.4	26.8	21.0	28.1

Table 2. String matching and regular expression matching times. The string matching times are total times of matching 10000 patterns.

text				string matching				regular expression matching			
type	$|\Sigma|$	n	m	matches /pattern	time (s) tree	array	cactus	matches	time (ms) tree	array	cactus
english	71	3000	8	3.87	0.82	0.38	0.79	1	1.13	2.50	1.43
english	74	30000	8	1.67	0.97	0.46	1.13	2	5.20	9.53	5.86
english	77	300000	8	4.86	1.63	0.67	1.86	33	19.5	33.1	20.8
random	77	300000	8	1.00	1.35	0.62	2.19	0	9.61	19.2	12.0
DNA	4	300000	8	8.17	0.96	0.71	0.61	19206	201	123	91.9
random	4	300000	8	5.58	0.79	0.69	0.58	18708	195	119	88.8
random	4	300000	12	1.02	0.57	0.64	0.58	"	"	"	"
random	16	300000	6	1.02	0.66	0.63	0.90	4670	740	800	730
random	64	300000	4	1.02	1.26	0.62	1.94	4	13.6	24.2	16.3

which take only one byte. The rare case that a longest common prefix between two suffixes is more than 255 is recognized and handled separately when necessary. This might affect the pattern matching time, but only when the length of the pattern exceeds 255.

To test matching performance we implemented string matching and regular expression matching algorithms for all three data structures. The results of our tests are given in Table 2. The execution times include going through the set of matches.

The string matching tests used 10000 patterns selected randomly from the text. The regular expression aS^*cS^*c, where $S = \{a, b, \ldots, z\}\setminus\{d, t\}$, was used in the regular expression tests. All the test texts contain letters a, c, and at least one of d and t. The matching times do not include the conversion of the regular expression into an automaton.

6   Concluding Remarks

We have described one variation of the suffix cactus in this paper. There are
other interesting variations, notably one which implements the branching using
hashing and another that uses a kind of binary tree structure. The main advantage of these variations would be better performance in string matching for
large alphabets. Due to the nature of the suffix cactus these other variations
need implementation stuctures and construction algorithms that are totally different from the ones described in this paper. There remains work to be done in
developing these versions.

Acknowledgements

I would like to thank Esko Ukkonen who suggested the name 'cactus'.

References

1. A. V. Aho, J. E. Hopcroft, and J. D. Ullman. *The Design and Analysis of Computer Algorithms*, chapter 9, pages 318–361. Addison-Wesley, 1974.
2. A. Andersson and S. Nilsson. Improved behaviour of tries by adaptive branching.
   Inf. Process. Lett., 46(6):295–300, July 1993.
3. A. Andersson and S. Nilsson. Efficient implementation of suffix trees. *Software—
   Practice and Experience*, 25(2):129–141, Feb. 1995.
4. A. Apostolico. The myriad virtues of subword trees. In A. Apostolico and Z. Galil,
   editors, *Combinatorial Algorithms on Words*, pages 85–95. Springer-Verlag, 1985.
5. R. A. Baeza-Yates and G. H. Gonnet. Efficient text searching of regular expressions. In *Proc. 16th International Colloquium on Automata, Languages and Programming (ICALP)*, pages 46–62, 1989.
6. R. A. Baeza-Yates and G. H. Gonnet. All-against-all sequence matching. Technical report, Department of Computer Science, University of Chile, 1990.
7. Z. Galil and K. Park. An improved algorithm for approximate string matching.
   SIAM J. Comput., 19(6):989–999, Dec. 1990.
8. G. H. Gonnet, R. A. Baeza-Yates, and T. Snider. Lexicographical indices for text:
   Inverted files vs. PAT trees. Technical Report OED-91-01, Centre for the New
   OED, University of Waterloo, 1991.
9. R. W. Irving. Suffix binary search trees. Technical report TR-1995-7, Computing
   Science Department, University of Glasgow, Apr. 1995.
10. P. Jokinen and E. Ukkonen. Two algorithms for approximate string matching in
    static texts. In *Proc. 16th International Symposium on Mathematical Foundations
    of Computer Science (MFCS)*, pages 240–248, Sept. 1991.
11. D. E. Knuth. *Sorting and Searching*, volume 3 of *The Art of Computer Program-
    ming*, chapter 6.3, pages 481–505. Addison-Wesley, 1973.
12. D. E. Knuth, J. H. Morris, and V. R. Pratt. Fast pattern matching in strings.
    SIAM J. Comput., 6(2):323–350, June 1977.

13. U. Manber and G. Myers. Suffix arrays: A new method for on-line string searches. *SIAM J. Comput.*, 22(5):935–948, Oct. 1993.
14. E. M. McCreight. A space-economical suffix tree construction algorithm. *J. ACM*, 23(2):262–272, Apr. 1976.
15. E. W. Myers. A sublinear algorithm for approximate keyword searching. *Algorithmica*, 12(4/5):345–374, Oct./Nov. 1994.
16. *Nucleic Acids Research*, 20(Sequences Supplement):2009–2210, May 1992.
17. E. Ukkonen. Constructing suffix trees on-line in linear time. In J. van Leeuwen, editor, *Algorithms, Software, Architecture. Information Processing 92*, volume 1, pages 484–492, 1992. Full version is to appear in *Algorithmica*.
18. E. Ukkonen and D. Wood. Approximate string matching with suffix automata. *Algorithmica*, 10(5):353–364, Nov. 1993.
19. P. Weiner. Linear pattern matching algorithms. In *Proc. IEEE 14th Annual Symposium on Switching and Automata Theory*, pages 1–11, 1973.

Pattern-Matching for Strings
with Short Descriptions

Marek Karpinski[1] *   Wojciech Rytter[1,2] **   Ayumi Shinohara[1,3]

[1] Department of Computer Science, University of Bonn
53117 Bonn, Germany
marek@cs.uni-bonn.de

[2] Institute of Informatics, Warsaw University
ul. Banacha 2, 02-097 Warszawa, Poland
rytter@mimuw.edu.pl

[3] Research Institute of Fundamental Information Science,
Kyushu University 33, Fukuoka 812, Japan
ayumi@rifis.kyushu-u.ac.jp

Abstract. We consider strings which are succinctly described. The description is in terms of straight-line programs in which the constants are symbols and the only operation is the concatenation. Such descriptions correspond to the systems of recurrences or to context-free grammars generating single words. The descriptive size of a string is the length n of a straight-line program (or size of a grammar) which defines this string. Usually the strings of descriptive size n are of exponential length. *Fibonacci* and *Thue-Morse words* are examples of such strings. We show that for a pattern P and text T of descriptive sizes m, n, an occurrence of P in T can be found (if there is any) in time polynomial with respect to n. This is nontrivial, since the actual lengths of P and T could be exponential, and none of the known string-matching algorithms is directly applicable. Our first tool is the periodicity lemma, which allows to represent some sets of exponentially many positions in terms of feasibly many arithmetic progressions. The second tool is arithmetics: a simple application of Euclid algorithm. Hence a textual problem for exponentially long strings is reduced here to simple arithmetics on integers with (only) linearly many bits. We present also an NP-complete version of the pattern-matching for shortly described strings.

1 Introduction

We describe algorithms for *implicit* pattern-matching problems for some well structured and exponentially long strings, given in the form of a succinct description. The *descriptive size* n of such strings is the size of their description,

---

* Research partially supported by the DFG Grant KA 673/4-1, and by the ESPRIT BR Grants 7097 and ECUS 030.
** Supported by the DFG grant.

while their *real size* N is the actual length of the string, assuming it is *explicitly written*. The size of the whole problem is n. Usually $N = \Omega(2^{c \cdot n})$.

In our algorithms we cannot write such long strings *explicitly*. Fortunately, each position in such strings can be written with only linear number of bits. Hence the size of the output is small: the output of our algorithm is an occurrence of one long string in another very long string. The input consists of short descriptions of the strings in terms of *straight-line* programs. We strengthen (in a nontrivial way) a result of [7], where quite sophisticated polynomial-time algorithm for equality of two strings generated by grammars was given. The *equality-test* algorithm from [7] is not directly applicable here since there are exponentially many positions, where the equality between the pattern and the text can happen. However we use this algorithm as one of the basic subroutines.

A *straight-line program* $\mathcal{R}$ is a sequence of assignment statements:

$$X_1 := expr_1; \; X_2 := expr_2; \ldots; \; X_n := expr_n$$

where X_i are variables and $expr_i$ are expressions of the form

- $expr_i$ is a symbol of a given alphabet Σ, or
- $expr_i = X_j \cdot X_k$, for some $j, k < i$, where $\cdot$ denotes the concatenation of X_j and X_k.

For each variable X_i, denote by $\nu(X_i)$ the value of X_i after the execution of the program. $\nu(X_i)$ is the string described by X_i. Denote by R the string described by (the value of) the program $\mathcal{R}$: $R = \nu(\mathcal{R}) = \nu(X_n)$. The size $|\mathcal{R}|$ of the program $\mathcal{R}$ is the number n, it is also called the *descriptive size* of the generated string $R = \nu(\mathcal{R})$.

R is called a *string with short description*, since usually $|R|$ is very long (exponentially) with respect to its descriptive size $n = |\mathcal{R}|$.

We say that $\mathcal{R}$ describes a *long string*. If we consider a string in a usual sense (the description is by giving the string *explicitly*) then we call such string an *explicit string*.

For a string w denote by $w[i..j]$ the subword of w starting at i and ending at j. Similarly for a long string $\mathcal{W}$ denote $\mathcal{W}[i..j] = \nu(\mathcal{W})[i..j]$.

Denote by $\mathcal{P}$ and $\mathcal{T}$ the descriptions of the pattern P and a text T. P occurs in T at position i iff $T[i..i + |P| - 1] = P$.

The *string matching problem for strings with short description* is:

given $\mathcal{P}$ and $\mathcal{T}$, check if P occurs in T, if "yes" then find any occurrence i.

The size n of the problem is the size $|\mathcal{T}|$ of the description of the text T. Assume $|\mathcal{P}| = m \le n$.

Our main result is the following theorem.

Theorem 1. *The pattern-matching problem for strings with short descriptions can be solved in polynomial time with respect to the descriptive size.*

A similar problem was considered recently in [2], where strings are described in terms of Lempel-Ziv encoding. The main difference between our results and the results in [2] is that in [2] patterns are assumed to have *explicit* representations, while we consider *implicitly* given patterns with short description.

Example 1. We refer the reader to [6] for definitions of the *Fibonacci* and *Thue-Morse* words. Let $P = F_5$ be the 5th Fibonacci word *abaab*, and $T = T_3$ be the 3rd Thue-Morse word *abbabaab*. We show below short descriptions $\mathcal{F}_5$ and $\mathcal{T}_3$ for these words. An instance of the pattern-matching problem for strings with short description is:

find any occurrence of $\mathcal{F}_5$ in $\mathcal{T}_3$.

An occurrence $i = 4$ of $\mathcal{F}_5$ in $\mathcal{T}_3$ is a solution to this instance.

The 5th Fibonacci word is described by the following program $\mathcal{P}$:

$$X_1 := b;\ X_2 := a;\ X_3 := X_2 \cdot X_1;\ X_4 := X_3 \cdot X_2;\ X_5 := X_4 \cdot X_3$$

The computation of $\mathcal{F}_5$ works as follows.

$$\nu(X_1) = b,\ \ \nu(X_2) = a,\ \ \nu(X_3) = ab,\ \ \nu(X_4) = aba,\ \ \nu(X_5) = abaab.$$

The 3rd Thue-Morse word is described by the following program $\mathcal{T}_3$.

$$X_0 := a;\qquad Y_0 := b;\qquad X_1 := X_0 \cdot Y_0;\ Y_1 := Y_0 \cdot X_0;$$
$$X_2 := X_1 \cdot Y_1;\ Y_2 := Y_1 \cdot X_1;\ X_3 = X_2 \cdot Y_2$$

The 3rd Thue-Morse word is generated as follows:

$$\nu(X_0) = a,\qquad \nu(Y_0) = b,\qquad \nu(X_1) = ab,\qquad \nu(Y_1) = ba,$$
$$\nu(X_2) = abba,\ \nu(Y_2) = baab,\ \nu(X_3) = abbabaab.$$

Using our algorithm it can be effectively found, for example, an occurrence (if there is any) of the Fibonacci word F_{220} in the Thue-Morse word T_{200}, despite the fact that *actual* lengths of these strings are astronomic: $|T_{200}| = 2^{200}$ and $|F_{220}| \geq 2^{120}$.

2 Arithmetic Progressions and Euclid's Algorithm

A crucial role in our algorithm is played by periodicities in strings. A nonnegative integer p is a *period* of a nonempty string w iff $w[i] = w[i - p]$, whenever both sides are defined. Hence $p = |w|$ and $p = 0$ are considered to be periods.

Lemma 2 (periodicity lemma, see [1]).
If w has two periods p, q such that $p + q \leq |w|$ then $\gcd(p, q)$ is a period of w, where gcd means "greatest common divisor".

Denote $Periods(w) = \{p : p \text{ is a period of } w\}$. A set of integers forming an arithmetic progression is called here *linear*. We say that a set of positive integers from $[1 \ldots N]$ is *succinct* w.r.t. N iff it can be decomposed in at most $\lfloor \log_2(N) \rfloor + 1$ linear sets. For example the set $Periods(aba) = \{0, 2, 3\}$ consists of $\lfloor \log_2(3) \rfloor + 1 = 2$ such sets.

For sets U and W define $U \oplus W = \{i + j : i \in U, j \in W\}$.

Lemma 3 (succinct sets lemma).
The set $Periods(w)$ is succinct w.r.t. $|w|$.

Proof. The proof is by induction with respect to $j = \lfloor \log_2(|w|) \rfloor$. The case $j = 0$ is trivial, one-letter string $(|w| = 1)$ has periods 0 and 1 (forming a single progression), hence we have precisely $\lfloor \log_2(|w|) \rfloor + 1$ progressions.

Let $k = \lceil \frac{|w|}{2} \rceil$. It follows directly from Lemma 2 that all periods in $A = Periods(w) \cap [1 \ldots k]$ form a single arithmetic progression, whose step is the greatest common divisor of all of them. Let q be the smallest period larger than k. Then it is easy to see that

$$Periods(w) = A \cup \{q\} \oplus Periods(w[q + 1..|w|]).$$

Now the claim follows from the inductive assumption, since $\lfloor \log_2(|w| - q) \rfloor < j$ and A is a single progression.

Observe that the structure of $Periods(w)$ corresponds to a *greedy* construction: find the first period p and take the longest progression containing consecutive periods which starts with p, then go to the next period and continue. There are at most $\lfloor \log_2(|w|) \rfloor + 1$ resulting progressions. Assume that we use such type of the representation for sets of periods. Let S_1 be a set of periods of w from $[1..k]$, and S_2 be a set of periods from an interval $[k+1..|w|]$. Then, when adding these sets, it can happen that the last linear set in S_1 continues, with the same step, in S_2, as the first linear set in S_2. We join these two progressions in S and have less linear sets in S.

Generally define the operation $compress(S)$, which for a given set of disjoint linear sets joins any two linear sets (if one is a continuation of the other) wherever it is possible. This operation is important, since we will be often adding succinct sets, and we need also a succinct representation in terms of at most logarithmically many progressions.

Denote $ArithProg(i, p, k) = \{i, i+p, i+2p, \ldots, i+kp\}$, so it is an arithmetic progression of length $k + 1$. Its description is given by numbers i, p, k written in binary. The size of the description, is the total number of bits in i, p, k.

Denote by $Solution(p, U, W)$ any position $i \in U$ such that $i + j = p$ for some $j \in W$. If there is no such position i then $Solution(p, U, W) = 0$.

Lemma 4 (application of Euclid algorithm).
Assume that two linear sets $U, W \subseteq [1 \ldots N]$ are given by their descriptions. Then for a given number $c \in [1 \ldots N]$ we can compute $Solution(c, U, W)$ in polynomial time with respect to $\log(N)$.

Proof. The problem can be easily reduced to the problem:

for given nonnegative integers a, b, c, A, B, find any integer solution (x, y) to the following equation with constraints

$$ax + by = c, \quad (1 \leq x \leq A, 1 \leq y \leq B). \tag{1}$$

It is enough to compute a solution in polynomial time with respect to the number of bits of the input constants.

We can assume that a, b are relatively prime, otherwise we can divide the equation by their greatest common divisor.

As a side effect of Euclid algorithm applied to a, b, we obtain integers (not necessarily positive, but with not too many bits) x_0', y_0' such that $ax_0' + by_0' = 1$. Let $x_0 = cx_0'$, $y_0 = cy_0'$. Then all solutions to the equation (1) are of the form

$$(x, y) = (x_0 + kb, y_0 - ka), \text{ where } k \text{ is an integer parameter.}$$

This defines a line, and we have to find any integer point in the rectangle $\{(i, j) : 1 \leq i \leq A, 1 \leq j \leq B\}$ which is *hitten* by this line. This can be done in polynomial time using operations *div* and *mod* on integers with polynomial number of bits. We refer for details to [5] (see page 325 and Exercise 14 on page 327).

3 The Pattern-Matching Algorithm

Let us fix the pattern $P = \nu(\mathcal{P})$, the length of P is M and the length of the text $T = \nu(\mathcal{T})$ is N. Observe that $N = O(2^n)$, hence all positions in T can be written using $O(n)$ bits.

Let X be a string (long or short) of the length K. Then define:

$$Prefs(X) = \{1 \leq i \leq K : X[K - i + 1..K] \text{ is a prefix of } P\}.$$
$$Suffs(X) = \{1 \leq i \leq K : X[1..i] \text{ is a suffix of } P\}.$$

Observation 1
Let $\mathcal{P}$, $\mathcal{A}$, $\mathcal{B}$ be long strings, then $\mathcal{P}$ occurs in $\mathcal{A} \cdot \mathcal{B}$ iff:
(1) $\mathcal{P}$ occurs in $\mathcal{A}$ or $\mathcal{P}$ occurs in $\mathcal{B}$;
(2) or $|P| \in Prefs(\mathcal{A}) \oplus Suffs(\mathcal{B})$.

Define the operations of the *prefix-extension* and *suffix-extension*. For a long or a short word X define

$$PrefExt(S, X) = \{i + |X| : i \in S \text{ and } P[1..i] \cdot X \text{ is a prefix of } P\}.$$
$$SuffExt(S, X) = \{i + |X| : i \in S \text{ and } X \cdot P[M - i + 1..M] \text{ is a suffix of } P\}.$$

Assume the straight line program $\mathcal{T}$ defining the text T is using variables X_1, X_2, ..., X_n. Each of variables X_i corresponds to a program $\mathcal{X}_i$ which computes X_i. We denote

$$SUFF[i] = Suffs(\mathcal{X}_i), \quad PREF[i] = Prefs(\mathcal{X}_i).$$

Observe that these tables depend on the pattern P, however it is convenient to assume further that P is fixed. We are now able to give a sketch of the whole structure of the algorithm. Assume that the lengths of all strings described by $\mathcal{X}_k$'s are computed (it can be easily done in polynomial time).

ALGORITHM *PATTERN_MATCHING* ;
for $k = 1$ **to** n **do**
    if $|\nu(\mathcal{X}_k)| \leq n$ **then** $\{\mathcal{X}_k$ can be treated as an explicit string$\}$
        test by classical methods an occurrence of P in $\mathcal{X}_k$;
        if there is an occurrence of P in $\mathcal{X}_k$ **then** report it and STOP
                else compute $PREF[k]$, $SUFF[k]$ by classical methods;
    else $\{$assume $X_k = X_i \cdot X_j$ for $i, j < k$ $\}$
        $pos := Solution(|P|, PREF[i], SUFF[j])$;
        if $pos \neq 0$ **then** report an occurrence and STOP
            else begin
                $U := PrefExt(PREF[i], \mathcal{X}_j) \cup PREF[j]$;
                $V := SuffExt(SUFF[j], \mathcal{X}_i) \cup SUFF[i]$;
                $PREF[k] := compress(U)$; $SUFF[k] := compress(V)$;
        end

Let k be the first position in $PREF[i]$, then all the other positions in $PREF[i]$ are of the form $k + p'$, where p' is a period of $P[1..k]$. Hence Lemma 3 implies directly the following fact.

Lemma 5. *The sets* $SUFF[i]$ *and* $PREF[j]$ *are succinct, for any* $1 \leq i, j \leq n$.

For a sequence of long strings $\gamma = \mathcal{X}_1, \ldots, \mathcal{X}_p$ define $\nu(\gamma) = \nu(\mathcal{X}_1)\nu(\mathcal{X}_2)\ldots\nu(\mathcal{X}_p)$. We omit the proof of the following fact. The proof employs the algorithm from [7] as a subroutine, and a kind of binary search in $[1 \ldots N]$.

Lemma 6 (subword-equality).
(a) *For two sequences of long strings* $\gamma_1 = \mathcal{X}_1, \ldots, \mathcal{X}_p$ *and* $\gamma_2 = \mathcal{Y}_1, \ldots, \mathcal{Y}_q$ *we can test equality* $\nu(\gamma_1) = \nu(\gamma_2)$ *in polynomial time with respect to the total size of all* $\mathcal{X}_i$'s *and* $\mathcal{Y}_j$'s.
(b) *For two long strings* $\mathcal{X}, \mathcal{Y}$ *and integers* i, j, k, l *we can test the equality* $\mathcal{X}[i..j] = \mathcal{Y}[k..l]$, *and find the first mismatch (if there is any) in polynomial time with respect to the size of description.*

Let us call the algorithms implied by the lemma the *equality-test algorithms*.

Our key lemma says that the operations *PrefExt* and *SuffExt* are *feasible*. Consider only the first of them, the second one is symmetric. We consider a set S which consists of one linear set. If there are polynomially many linear set-components of S, we deal with each of them separately.

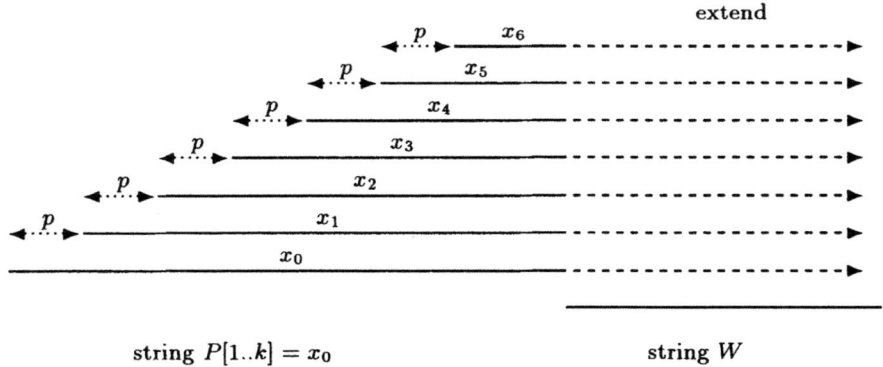

string $P[1..k] = x_0$ string W

Fig. 1. The operation $PrefExt(S, W)$, where $S = \{|x_0|, |x_1|, \ldots, |x_6|\}$.

Lemma 7 (key lemma).

Assume W is a long word, and $S = \{t_0, t_1, \ldots, t_s\} \subseteq [1 \ldots k]$ is a linear set given by its succinct representation, where $t_0 = k$ and strings $x_i = P[1..t_i]$, $0 \leq i \leq s$, are suffixes of $P[1..k]$. Then the representation of $PrefExt(S, W)$ can be computed in polynomial time.

Proof. Assume the sequence $t_0, t_1, \ldots, t_s$ is decreasing. We need to compute all possible continuation of x_i's in P which match W, see Figure 1. Denote $y_i = P[1..|x_i| + |W|]$ and $Z = P[1..k] \cdot W$. Hence our aim is to find all i's such that y_i is a suffix of Z, $(0 \leq i \leq s)$. We call such i's *good* indices. The first mismatch to the period p in a string x is the first position (if there is any) such that $x[mismatch] \neq x[mismatch - p]$. We can compute the first mismatch using an equality-test algorithm from Lemma 6. There are four basic cases:

Case A: there is no mismatch in Z but there is a mismatch in y_0.
Then good indices are all $i \geq r$, where r is the first index such that y_r contains no mismatch at all. (We have $r = 4$ in Figure 2 (case A)).

Case B: there is a mismatch in Z and y_0.
Then the only possible good index i is such that the first mismatch in y_i is exactly over the first mismatch in Z. See Figure 2 (case B), where the only good index is $i = 2$. We can easily calculate such i, it is also possible that there is no good i in this case.

Case C: there is no mismatch in Z or y_0.
Then all indices i are good.

Case D: there is a mismatch in Z but not in y_0.
Then none of indices i is good.

In this way we compute the set of good indices. Observe that it consists of a subset of consecutive indices from the set S. So the corresponding set (the required output) of integers $\{|y_i| : i$ is a good index $\}$ is linear. This completes the proof.

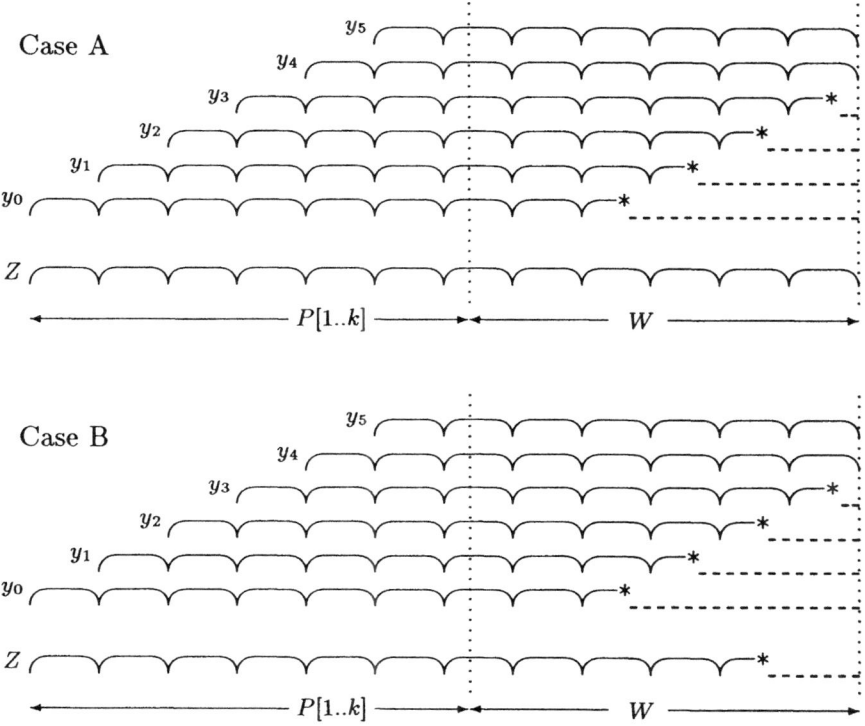

Fig. 2. Two cases: $Z = P[1..k]\cdot W$ has no mismatch in Case A, but Z has mismatch in Case B. $y_i = P[1..|x_i| + |W|]$. "$*$" denotes the mismatch to the period.

If the sets $SUFF[j], PREF[j]$ has been already computed by the algorithm, then each of them consists of a polynomial number of linear sets, for $j < i$. Hence we can compute the sets $PREF[i]$ and $SUFF[i]$ in polynomial time using polynomially many time the algorithm from Lemma 7 to each of these linear sets. In this way we have shown that the algorithm *PATTERN_MATCHING* works in polynomial time. This completes the proof of our main result (Theorem 1).

As a side effect of our pattern-matching algorithm we can compute the set of all periods for strings with short description.

Theorem 8. *Assume $\mathcal{X}$ is a string given by its description of size n. Then we can compute in polynomial time a polynomial size representation of set $Periods(\nu(\mathcal{X}))$. The representation consists of a linear number of linear sets.*

Proof. Use the algorithm *PATTERN_MATCHING* with the long pattern $\mathcal{P} = \mathcal{X}$ and the long text $\mathcal{T} = \mathcal{X}$. As a side effect we compute all suffixes of $\mathcal{T}$ which are prefixes of $\mathcal{P}$. This determines easily all periods.

4 An NP-Complete Version of Pattern-Matching

We start with a problem which has a particularly simple polynomial time algorithm, next we show that an extension of this problem is NP-complete.

The pattern–matching algorithm is much simpler if the pattern P is an *explicit word* of length $m = O(n)$, where n is the descriptive size of the *long* text T. Let X_i be the variables of a program describing T.

Denote $ShortVar = \{X_i : |\nu(X_i)| \leq m\}$. For each variable $X_i \in ShortVar$ we can compute its value by simply simulating the given straight-line program. We need $O(n \cdot m)$ time for all X_i's together.

The algorithm which looks for the explicit pattern P by searching *only* inside all words X_i (incorrectly assuming the pattern is contained totally in some of X_i's) is incorrect.

For a language L over the alphabet $ShortVar$ define $\nu(L) = \{\nu(\gamma) : \gamma \in L\}$.

Lemma 9. *Assume an (explicit) word P is of the real size $m = O(n)$. Then there is a nondeterministic finite automaton A accepting a language L over $ShortVar$ such that:*

the pattern P occurs in T iff $P \in \nu(L)$.

The constructed nondeterministic automaton A has $O(n)$ states.

Proof. We omit the proof.

We can replace each edge labeled X_i of the automaton A from the lemma above by $|\nu(X_i)|$ edges "spelling" the word $\nu(X_i)$. Then the automaton grows by a factor $O(m)$. The new automaton A' has the size $O(n^2)$. It can be applied to test if P occurs in T by simulating A' on P. A standard method can be used to test if a nondeterministic automaton accepts a text. This proves the following theorem.

Theorem 10. *Assume we have an (explicit) pattern P given explicitly of size $m = O(n)$ and a string T given by its description of size n. Then we can test if P occurs in T in $O(n^3)$ time.*

We show that the automata theoretic approach which was used above (and which corresponds to regular expressions) does not work if the pattern is a long string.

Let Var be a set of variables in some straight-line program of length n, and $\mathcal{P}$ be a long pattern (given by a straight-line program of length $m \leq n$). We consider the *regular-expression-matching* problem for shortly described strings defined as follows:

given a regular expression W over Var of size $O(n)$
test if $\nu(\mathcal{P}) \in \nu(W)$,

where $\nu(W) = \nu(L)$, and L is the language described by expression W.

Theorem 11. *The regular-expression-matching problem for shortly described strings is NP-complete, even if expressions do not contain operation *, nor empty strings and the alphabet Σ (for strings which are values of variables) is unary.*

Proof. The proof is a reduction from the SUBSET SUM problem defined as follows:

Input instance: Finite set $A = \{a_1, a_2, \ldots, a_n\}$ of integers and an integer K. The size of the input is the number of bits needed for the description.

Question: Is there a subset $A' \subseteq A$ such that the sum of the elements in A' is exactly K?

The problem SUBSET SUM is NP-complete, see [4], and [3], pp. 223. We can construct easily a straight-line program such that $\nu(X_i) = 1^{a_i}$ and $P = 1^K$. Then the SUBSET SUM problem is reduced to the membership:

$$\nu(\mathcal{P}) \in \nu((X_1 \cup \varepsilon) \cdot (X_2 \cup \varepsilon) \cdots (X_n \cup \varepsilon)).$$

The empty string ε can be easily eliminated by rescaling numbers and replacing ε by a single letter 1. This completes the proof.

References

1. M. Crochemore and W. Rytter, Text Algorithms, Oxford University Press, New York (1994).
2. M. Farach and M. Thorup, "String-matching in Lempel-Ziv compressed strings", to appear in Proc. 27th ACM STOC (1995).
3. M.R. Garey and D.S. Johnson, *Computers and Intractability: A Guide to the Theory of NP-Completeness.* W.H. Freeman (1979).
4. R.M. Karp, "Reducibility among combinatorial problems", in R.E. Miller and J.W. Thatcher (eds.), Complexity of Computer Computations, Plemum Press, New York, pp.85–103 (1972).
5. D. Knuth, *The Art of Computing, Vol. II: Seminumerical Algorithms. Second edition.* Addison-Wesley (1981).
6. M. Lothaire, *Combinatorics on Words.* Addison-Wesley (1993).
7. W. Plandowski, "Testing equivalence of morphisms on context-free languages", ESA'94, Lecture Notes in Computer Science 855, Springer-Verlag, 460–470 (1994).

Pairwise Alignment with Scoring on Tuples

Lukas Knecht

Institute for Scientific Computing
ETH Zurich, 8092 Zurich, Switzerland

Abstract. Pairwise alignment of two sequences a, b usually assumes a and b being sequences over the same alphabet A and a scoring function $s : A \times A \to \mathbb{R}$ operating on symbol pairs. The framework presented here extends this to a scoring function $s : A^p \times B^q \to \mathbb{R}$ operating on p-tuples and q-tuples of symbols, where the scoring tuples can have an arbitrary (but not complete) overlap.
We show that, if the alphabets A and B are finite and p and q are constant, the resulting algorithms have the same asymptotic time and space complexity as their single symbol counterparts.
This framework has been applied successfully to the codon-wise alignment of prokaryotic and eukaryotic genes.

1 Introduction

The pairwise alignment of two sequences a and b is usually seen as a list of matched pairs (a_x, b_y), thus always matching one symbol of sequence a with one symbol of sequence b.

However, not all alignments of two sequences can be covered by this approach:

- The direct alignment of a gene a with a protein b homologous to the protein encoded by gene a requires the matching of three bases with one amino acid to reflect the genetic code. The alignment can be seen as a list of base triples (codons) matched by single amino acids: $(a_{x_1}, a_{x_2}, a_{x_3}, b_y)$. Unlike the solutions to this problem discussed in [6], we want to allow gaps of arbitrary length between a_{x_k} and $a_{x_{k+1}}$ in order to be able to cope with introns, which often occur inside a codon.
- The alignment of a gene a with another gene b presents a similar problem where two base triples are matched. An alignment is defined by a list of pairs of matching base triples $(a_{x_1}, a_{x_2}, a_{x_3}, b_{y_1}, b_{y_2}, b_{y_3})$.
- A way to produce a context sensitive alignment of two proteins a, b is to use a dipeptide substitution matrix as proposed in [1] instead of a single amino acid substitution matrix. The alignment consists of matches of two pairs of amino acids: $(a_{x_1}, a_{x_2}, b_{y_1}, b_{y_2})$. Two subsequent matches overlap by one pair of amino acids: x_2, y_2 of one match is identical to x_1, y_1 of the next match.

A natural extension of the one to one approach is to match p symbols of a with q symbols of b. Such an alignment is equivalent to a list of matched p-tuples and q-tuples $(a_{x_1}, a_{x_2}, \ldots, a_{x_p}, b_{y_1}, b_{y_2}, \ldots, b_{y_q})$. Subsequent tuples may overlap by an arbitrary number of symbols less than p and q or not overlap.

In order to give a formal definition of such alignments, we introduce a simple and straightforward notation for sequence algebra:

- Given an alphabet A, A^* denotes all sequences over A including the empty sequence ε.
  Example: with $A = \{x, y\}, A^* = \{\varepsilon, x, y, xx, xy, yx, yy, xxx, xxy, \ldots\}$.
- A^k denotes all sequences over A of length k.
  Example: $A^2 = \{xx, xy, yx, yy\}$.
- $|a|$ denotes the length of any sequence $a \in A^*$.
  Examples: $|xyxyx| = 5, |\varepsilon| = 0$.
- a_i denotes the ith symbol of a.
  Example: with $a = xyxyx, a_2 = y, a_{|a|} = a_5 = x$.
- $a_{i\ldots j}$ denotes the contiguous subsequence of a from ith to jth symbol if $i \leq j$, otherwise the empty sequence.
  Example: with $a = xyxyx, a_{2\ldots4} = yxy, a_{1\ldots0} = \varepsilon$.

Let A and B be alphabets. Let $s : A^p \times B^q \to \mathbb{R}$ be a scoring function for matching a tuple of p symbols from A with a tuple of q symbols from B, and let $d_A : A^* \times \mathbb{N}^2 \to \mathbb{R}$ and $d_B : B^* \times \mathbb{N}^2 \to \mathbb{R}$ be cost functions for deleting (or inserting) a contiguous subsequence: $d_A(a, i, j)$ is the cost for deleting $a_{i\ldots j}$ from a.

d_A (and d_B) have the following properties:

- $d_A(a, i, j) = 0$ if $i > j$,
- $d_A(a, i, j) \leq 0$ if $i \leq j$,
- $d_A(a, i, j) \geq d_A(a, i, k) + d_A(a, k+1, j)$ if $i \leq k \leq j$.

To score an alignment of a and b, s is repeatedly applied on aligned subsequences of a and b shifted by u and v, i.e. the scoring tuples overlap by $p - u$ and $q - v$ symbols. Formally:

Definition 1. A complete alignment of two sequences a and b is given by indices $X = (x_1, \ldots, x_{(L-1)u+p})$ and $Y = (y_1, \ldots, y_{(L-1)v+q})$. Its local score (according to [7]) is

$$S_{\text{local}}(X, Y) = \sum_{k=0}^{L-1} s(a_{x_{ku+1}}, a_{x_{ku+2}}, \ldots, a_{x_{ku+p}}, b_{y_{kv+1}}, b_{y_{kv+2}}, \ldots, b_{y_{kv+q}})$$

$$+ \sum_{k=1}^{(L-1)u+p-1} d_A(a, x_k + 1, x_{k+1} - 1)$$

$$+ \sum_{k=1}^{(L-1)v+q-1} d_B(b, y_k + 1, y_{k+1} - 1) ,$$

the score of the alignment of $a_{x_1 \ldots x_{(L-1)u+p}}$ with $b_{y_1 \ldots y_{(L-1)v+q}}$.
Its global score is

$$S_{\text{global}}(X, Y) = S_{\text{local}}(X, Y)$$

$$+ d_A(a, 1, x_1 - 1) + d_A(a, x_{(L-1)u+p} + 1, |a|)$$
$$+ d_B(b, 1, y_1 - 1) + d_B(b, y_{(L-1)v+q} + 1, |b|) \ ,$$

the score of the alignment of a with b.

Example 1. Let $A = \{\mathtt{a}, \mathtt{c}, \ldots, \mathtt{y}\}$ be the alphabet of amino acids and $B = \{\mathtt{a}, \mathtt{c}, \mathtt{g}, \mathtt{t}\}$ be the alphabet of bases. We want to align a sequence a ("protein") over A with a sequence b ("gene") over B. To introduce context sensitivity, we assume a scoring function $s : A^2 \times B^3 \to \mathbb{R}$ matching an amino acid pair with a base triple. Matched pairs overlap by one amino acid. We thus have $p = 2, u = 1, q = v = 3$.

Let $a = \mathtt{knln}$ and $b = \mathtt{cagcattacggacc}$. The alignment defined by $X = (1, 2, 3, 4)$ and $Y = (2, 3, 4, 5, 6, 7, 8, 12, 13)$ can be visualized as [1]

Its scores are

$$S_{\mathrm{local}}(X, Y) = s(\mathtt{kn}, \mathtt{agc}) + s(\mathtt{nl}, \mathtt{att}) + s(\mathtt{ln}, \mathtt{aac}) + d_B(b, 9, 11) \ ,$$
$$S_{\mathrm{global}}(X, Y) = S_{\mathrm{local}}(X, Y) + d_B(b, 1, 1) + d_B(b, 14, 14) \ .$$

The two standard alignment problems solved in this paper are:

1. *Best score:* Given two sequences $a \in A^*$ and $b \in B^*$, find

$$\bar{S}_{\mathrm{local}} := \max_{X,Y} S_{\mathrm{local}}(X, Y) \text{ or } \bar{S}_{\mathrm{global}} := \max_{X,Y} S_{\mathrm{global}}(X, Y) \ .$$

2. *Best alignment:* Find the indices $\bar{X}$ and $\bar{Y}$ such that

$$S_{\mathrm{local}}(\bar{X}, \bar{Y}) = \bar{S}_{\mathrm{local}} \text{ or } S_{\mathrm{global}}(\bar{X}, \bar{Y}) = \bar{S}_{\mathrm{global}} \ .$$

2 Dynamic Programming Solution

2.1 Position Dependent Maxima

The first attempt to apply the dynamic programming paradigm to the problem stated above starts by introducing

---

[1] Visualizing tuple-alignments is generally not easy, especially if the output device is limited to the standard character set. In this case, we propose to distribute overlapping tuples over separate lines. In our example:

Definition 2.

$$M(i_1,\ldots,i_{p-u},i,\ j_1,\ldots,j_{q-v},j) \quad (i_1 < \cdots < i_{p-u} \le i,\ \ j_1 < \cdots < j_{q-v} \le j)$$

is the maximum score of a (normally complete) alignment

- ending with a_i and b_j,
- and matching exactly $a_{i_1}, a_{i_2}, \ldots, a_{i_{p-u}}$ of all a_x, $x \ge i_1$ and deleting all other such a_x,
- and matching exactly $b_{j_1}, b_{j_2}, \ldots, b_{j_{q-v}}$ of all b_y, $y \ge j_1$ and deleting all other such b_y.

The matching symbols correspond to the overlap. The special case $p = u$ and $q = v$ has no overlap, and M has just two dimensions: $M(i,j)$.

To be precise, there are M_{local} for local alignments and M_{global} for global alignments. Since the computation for both is mostly identical, we will explicitly state the difference where necessary.

Example 2. Given Example 1, we obtain for some sample values of $M(i_1, i, j)$ (global alignment):

$$M(1,1,1) = 0$$

$$M(1,2,3) = \max \left\{ \begin{array}{c} s(\text{kn}, \text{cag}) \\ d_A(a,1,2) + d_B(b,1,3) \end{array} \right\}$$

$$M(1,2,4) = \max \left\{ \begin{array}{c} s(\text{kn}, \text{agc}) + d_B(b,1,1) \\ s(\text{kn}, \text{cgc}) + d_B(b,2,2) \\ s(\text{kn}, \text{cac}) + d_B(b,3,3) \\ s(\text{kn}, \text{cag}) + d_B(b,4,4) \\ d_A(a,1,2) + d_B(b,1,4) \end{array} \right\} .$$

Given M, the score $S_{x,y}$ of the best alignment up to a_x, b_y becomes

$$S_{x,y} = \max_{\substack{i_1 < \cdots < i_{p-u} \le x \\ j_1 < \cdots < j_{q-v} \le y}} M(i_1,\ldots,i_{p-u},x,\ j_1,\ldots,j_{q-v},y) . \tag{1}$$

In the special case of $p = u$ and $q = v$: $S_{x,y} = M(x,y)$.

The dynamic programming recursion to calculate M must consider all possibilities of building an alignment corresponding to the definition of M:

$$M(i_u,\ldots,i_p,\ j_v,\ldots,j_q) = \max \Big\{$$

- starting a new local alignment if there are no gaps (omitted for global alignments):

$$\left\{ \begin{array}{l} 0 \quad \text{if } i_k + 1 = i_{k+1}\ (u \le k < p-1) \wedge i_{p-1} = i_p \\ \quad \wedge\ j_k + 1 = j_{k+1}\ (v \le k < q-1) \wedge j_{q-1} = j_q \\ -\infty \ \text{otherwise} \end{array} \right\} ,$$

– ending the alignment with a match of a_{i_p} and b_{j_q}:

$$\max_{\substack{i_1<\cdots<i_{u-1}<i_u \\ j_1<\cdots<j_{v-1}<j_v}} M(i_1,\ldots,i_{p-u},i_{p-u+1}-1,\ j_1,\ldots,j_{q-v},j_{q-v+1}-1)+$$

$$s(a_{i_1},\ldots,a_{i_p},\ b_{j_1},\ldots,b_{j_q})+\sum_{k=p-u+1}^{p-1} d_A(a,i_k+1,i_{k+1}-1)+$$

$$\sum_{k=q-v+1}^{q-1} d_B(b,j_k+1,j_{k+1}-1)\ ,$$

$$(2)$$

– ending the alignment with a deletion of a:

$$\max_{i_{p-1}\leq i<i_p} M(i_u,\ldots,i_{p-1},i,\ j_v,\ldots,j_q)$$

$$+ d_A(a,i+1,i_p)\ ,$$

$$(3)$$

– ending the alignment with a deletion of b:

$$\max_{j_{q-1}\leq j<j_q} M(i_u,\ldots,i_p,\ j_v,\ldots,j_{q-1},j)$$

$$+ d_B(b,j+1,j_q)\Big\}\ .$$

$$(4)$$

The recursion is initialized by

$$M(1,2,\ldots,p-u,p-u,\ 1,2,\ldots,q-v,q-v)=0\ .$$

$$(5)$$

In the special case of $p=u$ and $q=v$, this reduces to $M(0,0)=0$.

Assuming that s, d_A and d_B are all calculated in $O(1)$ time, each calculation of (2) requires $O(|a|^{u-1}|b|^{v-1})$ time, calculating (3) and (4) requires $O(|a|)$ and $O(|b|)$ time. Since there are $O(|a|^{p-u+1}|b|^{q-v+1})$ values of M, we find the following

Theorem 3. *The dynamic programming algorithm calculating the best score based on position dependent scores requires*

$$O\big(|a|^{p-u+1}|b|^{q-v+1}\max(|a|^{u-1}|b|^{v-1},|a|+|b|)\big)\ time$$

and

$$O\big(|a|^{p-u+1}|b|^{q-v+1}\big)\ space.$$

Some special cases might be of interest:

– If $p=q=u=v=1$, the recursion reduces to the standard dynamic programming recursion for sequence alignment, and its time complexity is $O(|a||b|(|a|+|b|))$, a well known result for arbitrary deletion cost functions.
– If $u>1$ and $v>1$, the complexity is $O(|a|^p|b|^q)$.

However, in the case of $p,q>1$, the algorithm to calculate M requires far too much time and space for most practical purposes. A more efficient solution is required.

2.2 Sequence Dependent Maxima

We start with the key observations that

- the value of $s(a_{i_1}, \ldots, a_{i_p}, b_{j_1}, \ldots, b_{j_q})$ does not depend on the sequence positions i_k and j_k, but only on the actual symbols $a_{i_1}, \ldots, a_{i_p}$ and $b_{j_1}, \ldots, b_{j_q}$,
- A and B are usually finite.

This leads to the following definitions:

Definition 4. A partial alignment of two sequences a and b is given by indices $X = (x_1, \ldots, x_m)$ and $Y = (y_1, \ldots, y_n)$, with arbitrary $m \leq |a|$ and $n \leq |b|$. It can be split into a complete alignment of length $L = \lfloor \min(\frac{m-p}{u}, \frac{n-q}{v}) \rfloor + 1$ with indices $X' = (x_1, \ldots, x_{(L-1)u+p})$, $Y' = (y_1, \ldots, y_{(L-1)v+q})$ and unaligned ends. Its score is

$$S(X,Y) = S(X',Y') + \sum_{k=(L-1)u+p}^{m-1} d_A(a, x_k + 1, x_{k+1} - 1) +$$

$$\sum_{k=(L-1)v+q}^{n-1} d_B(b, y_k + 1, y_{k+1} - 1) \ .$$

Example 3. In Example 1, $X = (1,2,3)$ and $Y = (2,3,4,5,6,7,8,10)$ define a partial alignment

The unaligned ends are ε and $b_8.b_{10} = $ **ag**. Its (local) score is

$$S(X,Y) = s(\mathbf{kn}, \mathbf{agc}) + s(\mathbf{nl}, \mathbf{att}) + d_B(b, 9, 9) \ .$$

Definition 5. $N_{i,j}(\alpha, \beta)$ is the maximum score of all partial alignments ending with a_i and b_j restricted to those which match exactly α (a subsequence of a) and β (a subsequence of b) at the end. If no such partial alignment exists, $N_{i,j}(\alpha, \beta) = -\infty$.

Example 4. Assuming that all the alignments and subalignments in Example 3 are optimal and we are calculating global alignments, we have

$$N_{1,1}(\varepsilon, \varepsilon) = d_A(a, 1, 1) + d_B(b, 1, 1)$$
$$N_{1,1}(\mathbf{k}, \mathbf{c}) = 0$$
$$N_{1,1}(\mathbf{w}, \mathbf{c}) = -\infty$$
$$N_{2,4}(\mathbf{n}, \varepsilon) = N_{2,5}(\mathbf{n}, \mathbf{a}) = N_{2,6}(\mathbf{n}, \mathbf{at}) = d_B(b, 1, 1) + s(\mathbf{kn}, \mathbf{agc}) \ .$$

Given N, the score $S_{x,y}$ of the best alignment up to a_x, b_y becomes

$$S_{x,y} = \max_{\substack{\alpha \in A^{p-u} \\ \beta \in B^{q-v}}} N_{x,y}(\alpha, \beta) \ .$$

The dynamic programming recursion must consider all possibilities of building a partial alignment corresponding to the definition of N:

$$N_{i,j}(\alpha, \beta) = \max \Big\{$$

- ending the alignment with a complete match of a_i and b_j if:
  - α and β have the length of an overlap
  - and α is the empty sequence or ends with a_i,
  - and β is the empty sequence or ends with b_j.

$$\begin{cases} \max_{\substack{\alpha_0 \in A^u, \beta_0 \in B^v \\ \bar{\alpha} := \alpha_0.\alpha, \bar{\beta} := \beta_0.\beta \\ \bar{\alpha}_{|\bar{\alpha}|} = a_i, \bar{\beta}_{|\bar{\beta}|} = b_j}} N_{i-1,j-1}(\bar{\alpha}_{1...|\bar{\alpha}|-1}, \bar{\beta}_{1...|\bar{\beta}|-1}) \text{ if } |\alpha| = p - u \\ \qquad\qquad\qquad\qquad + s(\bar{\alpha}, \bar{\beta}) \qquad\qquad \wedge |\beta| = q - v \\ \\ \qquad -\infty \qquad\qquad\qquad\qquad\qquad\qquad \text{otherwise} \end{cases} ,$$

$$(6)$$

- ending the alignment with a partial match of a_i or b_j or starting a new local alignment:

$$\begin{cases} N_{i-1,j-1}(\alpha_{1...|\alpha|-1}, \beta_{1...|\beta|-1}) & \text{if } \alpha \neq \varepsilon \wedge \alpha_{|\alpha|} = a_i \wedge \beta \neq \varepsilon \wedge \beta_{|\beta|} = b_j \\ N_{i-1,j}(\alpha_{1...|\alpha|-1}, \varepsilon) & \text{if } \alpha \neq \varepsilon \wedge \alpha_{|\alpha|} = a_i \wedge \beta = \varepsilon \\ N_{i,j-1}(\varepsilon, \beta_{1...|\beta|-1}) & \text{if } \alpha = \varepsilon \wedge \beta \neq \varepsilon \wedge \beta_{|\beta|} = b_j \\ 0 & \text{if } \alpha = \varepsilon \wedge \beta = \varepsilon \wedge \text{local alignment} \\ -\infty & \text{otherwise} \end{cases} ,$$

$$(7)$$

- ending the alignment with a deletion of a:

$$\max_{x<i} N_{x,j}(\alpha, \beta) + d_A(a, x+1, i) \ , \tag{8}$$

- ending the alignment with a deletion of b:

$$\max_{y<j} N_{i,y}(\alpha, \beta) + d_B(b, y+1, j) \Big\} \ . \tag{9}$$

The recursion is initialized by

$$N_{0,0}(\varepsilon, \varepsilon) = 0 \ . \tag{10}$$

This recursion finds the *best score* $S_{x,y}$ for all x, y. To find the *best alignment*, we have to store for each $N_{i,j}(\alpha, \beta)$ information on how it was obtained in order to reproduce the path leading to it.

For which α and β is N required? Evaluating (6) requires $N_{i,j}(\alpha, \beta)$ where $|\alpha| = p - 1$ and $|\beta| = q - 1$, (7) uses α and β of decreasing length, (8) and (9) use α and β of the same length. The $N_{i,j}(\alpha, \beta)$ required are therefore:

- $N_{i,j}(\alpha, \beta)$ for $\alpha \in A^{p-k}$, $\beta \in B^{q-k}$, $k = 1, 2, \ldots, \min(p, q)$,
- $N_{i,j}(\alpha, \varepsilon)$ for $\alpha \in A^{p-k}$, $k = \min(p, q) + 1 \ldots, p$,
- $N_{i,j}(\varepsilon, \beta)$ for $\beta \in B^{q-k}$, $k = \min(p, q) + 1 \ldots, q$,

resulting in a total number of

$$n_{\text{tot}} := \sum_{k=1}^{\min(p,q)} |A|^{p-k} |B|^{q-k} + \sum_{k=\min(p,q)+1}^{p} |A|^{p-k} + \sum_{k=\min(p,q)+1}^{q} |B|^{q-k}$$

values of $N_{i,j}(\alpha, \beta)$ for each i, j.

To calculate the time needed, we notice that some of the calculations for (8) or (9) are not required:

Lemma 6. *Some of the expressions (8) and (9) in the above recursion have upper bounds:*

$$\alpha = \varepsilon \wedge (|\beta| \neq q - v \vee \beta \neq \varepsilon \wedge \beta_{|\beta|} \neq b_j) \Rightarrow (8) \leq \max((7), (9))$$
$$\beta = \varepsilon \wedge (|\alpha| \neq p - u \vee \alpha \neq \varepsilon \wedge \alpha_{|\alpha|} \neq a_j) \Rightarrow (9) \leq \max((7), (8)) .$$

Proof. Since the two cases are symmetric, we proof the case $\alpha = \varepsilon$ only:
If (8) evaluates to $-\infty$, the lemma is trivially true. Otherwise, there exists a smallest i_0 such that

$$(8) = \max_{x < i} N_{x,j}(\varepsilon, \beta) + d_A(a, x + 1, i)$$
$$= N_{i_0,j}(\varepsilon, \beta) + d_A(a, i_0 + 1, i) .$$

Since $|\beta| \neq q - v$ or $\beta_{|\beta|} \neq b_j$, $N_{i_0,j}(\varepsilon, \beta)$ cannot be obtained by (6), and since $d_A(a, i, j) \leq d_A(a, i, k) + d_A(a, k + 1, j)$, it cannot be obtained by (8). Three cases must be distinguished:

1. $\beta = \varepsilon$: $N_{i_0,j}(\varepsilon, \beta)$ is calculated via (7) (local alignment only) or via (9), and a smallest j_0 exists such that:

$$(8) = \max(0, N_{i_0,j_0}(\varepsilon, \varepsilon) + d_B(b, j_0 + 1, j)) + d_A(a, i_0 + 1, i)$$
$$\leq \max(d_A(a, i_0 + 1, i), N_{i,j_0}(\varepsilon, \varepsilon) + d_B(b, j_0 + 1, j))$$
$$\leq \max((7), (9)) .$$

2. $\beta \neq \varepsilon \wedge \beta_{|\beta|} \neq b_j$: $N_{i_0,j}(\varepsilon, \beta)$ is calculated via (9), and a smallest j_0 exists such that:

$$(8) = N_{i_0,j_0}(\varepsilon, \beta) + d_B(b, j_0 + 1, j) + d_A(a, i_0 + 1, i)$$
$$\leq N_{i,j_0}(\varepsilon, \beta) + d_B(b, j_0 + 1, j)$$
$$\leq (9) .$$

3. $\beta \neq \varepsilon \wedge \beta_{|\beta|} = b_j$: $N_{i_0,j}(\varepsilon,\beta)$ is calculated via (7) using b_j or via (9), and a smallest j_0 exists such that:

$$(8) = \max(N_{i_0,j-1}(\varepsilon,\beta_{1\ldots|\beta|-1}), N_{i_0,j_0}(\varepsilon,\beta) + d_B(b,j_0+1,j)) + d_A(a,i_0+1,i)$$
$$\leq \max(N_{i,j-1}(\varepsilon,\beta_{1\ldots|\beta|-1}), N_{i,j_0}(\varepsilon,\beta) + d_B(b,j_0+1,j))$$
$$\leq \max((7),(9)) \ .$$

In any case, $(8) \leq \max((7),(9))$. $\qquad\qquad\qquad\qquad\qquad\qquad\qquad\qquad\square$

Assuming that s, d_A and d_B are all calculated in $O(1)$ time, the time we need for each i,j is:

- for (6): $O(|A|^{p-1}|B|^{q-1})$,
- for (7): $O(n_{\text{tot}} - |A|^{p-1}|B|^{q-1})$,
- for (8): $O(n_A|a|)$, where (Lemma 6) $n_A = n_{\text{tot}} - \sum_{k=\min(p,q)+1}^{q} |B|^{q-k} + 1$,
- for (9): $O(n_B|b|)$, where (Lemma 6) $n_B = n_{\text{tot}} - \sum_{k=\min(p,q)+1}^{p} |A|^{p-k} + 1$,

resulting in a total of $O(n_A|a| + n_B|b|)$ time for each i,j. Since there are $|a||b|$ pairs i,j, and $O(n_{\text{tot}}) = O(n_A) = O(n_B) = O(|A|^{p-1}|B|^{q-1})$, we obtain:

Theorem 7. *If the alphabets A and B are finite, there exists a dynamic programming algorithm calculating the best alignment which requires*

$$O\big(|a||b|(|a| + |b|)|A|^{p-1}|B|^{q-1}\big) \ time \qquad\qquad (11)$$

and

$$O\big(|a||b||A|^{p-1}|B|^{q-1}\big) \ space. \qquad\qquad (12)$$

If $|A|$, $|B|$, p and q are considered constant, the complexities are $O(|a||b|(|a|+|b|))$ and $O(|a||b|)$. Finding the best alignment with scoring on tuples of arbitrary size then has the same time and space complexity as scoring on tuples of size 1. Of course, the constant factor n_{tot} grows rapidly with growing alphabet sizes and tuple sizes. Therefore, extensive space requirements can make the algorithm useless for practical purposes.

The shift parameters u and v have been assumed fixed. The only place we made use of this so far is (6). Allowing u and v to vary for each single match requires the evaluation of (6) for all α, β where $|\alpha| = 0, \ldots, p-1$ and $|\beta| = 0, \ldots, q-1$. Lemma 6 does not hold any longer. The time requirements in (11) thus are multiplied by pq. If p and q are fixed, the asymptotic complexity remains the same.

3 Special Cases

3.1 Additive Deletion Costs

For practical purposes, the cubic complexity of the algorithm is undesirable. It is caused by the general form of the deletion cost functions d_A and d_B. With a simple restriction based on an idea published in [3], the alignment can be calculated in quadratic time and linear space:

Definition 8. A function $f : A^* \times \mathbb{N}^2 \to \mathbb{R}$ is additive if there exists a function $f^+ : A^* \times \mathbb{N} \to \mathbb{R}$ such that

$$f(a, i, j) = f(a, i, j - 1) + f^+(a, j) \quad \forall a, \forall i < j .$$

The most widely used example for an additive deletion function is $d(a, i, j) = \mu + (j - i)\lambda$ for a deletion of length $j - i + 1$. If d_A is additive, (8) becomes

$$
\begin{aligned}
D_{i,j}^{(A)}(\alpha, \beta) &:= \max_{x < i} \ N_{x,j}(\alpha, \beta) + d_A(a, x + 1, i) \\
&= \max(N_{i-1,j}(\alpha, \beta) + d_A(a, i, i), \\
&\qquad \max_{x < i-1} N_{x,j}(\alpha, \beta) + d_A(a, x + 1, i - 1) + d_A^+(a, i)) \\
&= \max(N_{i-1,j}(\alpha, \beta) + d_A(a, i, i), \ D_{i-1,j}^{(A)}(\alpha, \beta) + d_A^+(a, i)) .
\end{aligned}
$$

The same is true for d_B and (9) by introducing $D^{(B)}$.

Lemma 6 and the following lemma determine those α, β for which we need to store $D_{i,j}^{(A)}(\alpha, \beta)$ and $D_{i,j}^{(B)}(\alpha, \beta)$.

Lemma 9.

$$
\begin{aligned}
\alpha \neq \varepsilon \wedge \alpha_{|\alpha|} \neq a_i &\Rightarrow D_{i,j}^{(A)}(\alpha, \varepsilon) = N_{i,j}(\alpha, \varepsilon) \\
\beta \neq \varepsilon \wedge \beta_{|\beta|} \neq b_j &\Rightarrow D_{i,j}^{(B)}(\varepsilon, \beta) = N_{i,j}(\varepsilon, \beta) .
\end{aligned}
$$

Proof. The two cases are symmetric, we only proof $\beta = \varepsilon$. Since $\alpha_{|\alpha|} \neq a_i$, $N_{i,j}(\alpha, \varepsilon) = \max((8),(9))$. From Lemma 6 and $\alpha_{|\alpha|} \neq a_i$ follows $(9) \leq (8)$. Thus, $N_{i,j}(\alpha, \varepsilon) = (8) = D_{i,j}^{(A)}(\alpha, \varepsilon)$. $\qquad\square$

Calculating (8) and (9) now requires $O(n_A)$ and $O(n_B)$ time, leading to

Corollary 10. *If the alphabets A and B are finite and the deletion cost functions d_A and d_B are additive, a dynamic programming algorithm exists calculating the best alignment in*

$$O(|a||b||A|^{p-1}|B|^{q-1}) \ time$$

and

$$O((|a| + |b|)|A|^{p-1}|B|^{q-1}) \ space.$$

Since computation of $N_{i,j}$ and $D_{i,j}$ is based on $N_{i-1,j-1}$, $N_{i-1,j}$, $N_{i,j-1}$, $D_{i-1,j}$, and $D_{i,j-1}$ only, it is obvious that computing the optimal *score* requires linear space. Computing the optimal *alignment* in linear space can be done by a relatively straightforward adaptation of the recursive algorithm in [5]. We omit the details.

With additive deletion costs, an efficient parallel version of the algorithm calculating N, $D^{(A)}$ and $D^{(B)}$ can be obtained by assigning symbol a_i (or b_i) to processor P_i. The only communication taking place is sending b_{j-i+1}, $N_{i,j-i+1}$, $N_{i-1,j-i+1}$, and $D^{(A)}_{i,j-i+1}$ from P_i to P_{i+1}. If there are $\min(|a|,|b|)$ processors, the running time decreases to $O((|a|+|b|)|A|^{p-1}|B|^{q-1})$.

3.2 $p = 1$ and $q = v$

If the size of one scoring tuple is 1 (w.l.o.g. $p = 1$ and therefore $u = 1$), and if there is no overlap ($q = v$), from Lemma 6 follows that for given i, j we only need to consider

- $N_{i-1,j-1}(\varepsilon, \beta)$ ($|\beta| = q - 1$) for (6),
- $N_{i,j-1}(\varepsilon, \beta)$ ($|\beta| < q - 1$) for (7),
- $N_{x,j}(\varepsilon, \varepsilon)$ ($x < i$) for (8),
- $N_{i,y}(\varepsilon, \beta)$ ($y < j$) for (9).

If the inner loop follows j, we only have to store $N_{i,j}(\varepsilon, \varepsilon)$ for all i, and the space requirements are reduced from (12) to

$$O(|a||b| + |b||B|^{q-1}) \ .$$

Of course, the time complexity remains the same.

Furthermore, if the gap costs are additive, the space requirements are $O(|b| + |B|^{q-1})$ or $O(|b|)$ if q is fixed, and thus identical to the requirements of the classical sequence alignment algorithm with scoring on single symbols. This is even true for the recursive algorithm finding the best alignment [5] if a is split. In this case, a parallel version should distribute a over the processors, minimizing the communication from P_i to P_{i+1} to four scalars: b_{j-i+1}, $N_{i,j-i+1}(\varepsilon, \varepsilon)$, $N_{i-1,j-i+1}(\varepsilon, \varepsilon)$ and $D^{(A)}_{i,j-i+1}(\varepsilon, \varepsilon)$.

3.3 Special Structure of s

Sometimes the scoring function $s : A^p \times B^q \to \mathbb{R}$ can be replaced by some simpler function: Given

- a mapping function

$$f : A^{p-1} \times B^{q-1} \to C \text{ where } C \subset A^{p-1} \times B^{q-1} \ ,$$

– a simpler scoring function

$$\bar{s} : C \times A \times B \to \mathbb{R}$$

with

$$\bar{s}(f(a_1, \ldots, a_{p-1}, b_1, \ldots, b_{q-1}), a_p, b_q) = s(a_1, \ldots, a_p, b_1, \ldots, b_q) \ ,$$

it is obvious that instead of $N_{i,j}(\alpha, \beta)$ where $\alpha \in A^{p-1}, \beta \in B^{q-1}$, we can use $\bar{N}_{i,j}(\gamma)$ where $\gamma \in C$. Hence, the complexity of the algorithm can be reduced to

$$O(|a||b|(|a| + |b|)|C|) \text{ time}$$

and

$$O((|a||b|)|C|) \text{ space.}$$

Example 5. Nucleotide sequences are not always uniquely determined. This is reflected by the additional symbol x for any unknown base. The base alphabet then grows from 4 to 5 symbols, and the constant factor n_{tot} for the alignment of a nucleotide sequence and a protein increases from $4^2 + 4 + 1 = 21$ to $5^2 + 5 + 1 = 31$. All codons having an x in the first or second position are ambiguous and mapped to the unknown amino acid. Thus we can set $C = \{a, c, g, t\}^2 \cup \{xa, ax\}$, and

$$f(b_1, b_2) = \begin{cases} b_1.b_2 & \text{if } b_1 \neq x \wedge b_2 \neq x \\ xa & \text{if } b_1 = x \wedge b_2 \neq x \\ ax & \text{if } b_2 = x \end{cases} .$$

The constant factor increases by 2 for the two additional symbols xa and ax [2].

Example 6. Let $A = B = \{\text{amino acids}\}$. A simple context dependent scoring of an alignment of two proteins might be achieved by setting

$$s(a_1, a_2, b_1, b_2) = \begin{cases} s_{\text{match}}(a_2, b_2) & \text{if } a_1 = b_1 \\ s_{\text{mismatch}}(a_2, b_2) & \text{if } a_1 \neq b_1 \end{cases} .$$

With $C = \{aa, ac\}$ and

$$f(a_1, b_1) = \begin{cases} aa & \text{if } a_1 = b_1 \\ ac & \text{if } a_1 \neq b_1 \end{cases} ,$$

$s : A^4 \to \mathbb{R}$ can be replaced by $\bar{s} : \{aa, ac\} \times A^2 \to \mathbb{R}$, and only two values of $\bar{N}$ are required for each i, j: $\bar{N}_{i,j}(aa)$ (i.e. the maximum score of an alignment ending in a match) and $\bar{N}_{i,j}(ac)$ (i.e. the maximum score of an alignment ending in a mismatch).

---

[2] This is true if the dynamic programming proceeds forward, i.e. from first to third base of a codon. Proceeding backwards is slightly more complicated.

4 Applications

4.1 Gene Alignments

Since selective pressure operates on amino acids, but not on base pairs, genes coding for proteins should be aligned codon-wise. Formally: given

- $A = \{\text{amino acids}\}$ and $B = \{\text{nucleotide bases}\}$,
- the genetic code $g : B^3 \to A \cup \{\$\}$ where $\$$ stands for a stop codon,
- an amino acid and stop codon similarity score $s_A : (A \cup \{\$\})^2 \to \mathbb{R}$ and deletion costs d_A, d_B,

we can define two basic alignment problems:

1. $A = B = B$, $p = q = u = v = 3$ and

$$s(a_1, a_2, a_3, b_1, b_2, b_3) := s_A(g(a_1, a_2, a_3), g(b_1, b_2, b_3))$$

   define the codon-wise alignment of a gene a with another gene b. Since the time factor $(|B|^2)^2 = 256$ is relatively small, such alignments are feasible if d_B is additive.
2. $A = A$, $B = B$, $p = u = 1$, $q = v = 3$ and

$$s(a_1, b_1, b_2, b_3) := s_A(a_1, g(b_1, b_2, b_3))$$

   defines the alignment of amino acids of a protein a with codons of a gene b. The time factor $|B|^2 = 16$ is small, and with additive d_A and d_B, all simplifications of Sect. 3.2 apply.

   A solution to a similar problem is described in [6]. It uses a simplified variant of the position dependent algorithm looking at the most likely position combinations only for building codons. Since introns can appear anywhere within a codon, this approach fails in detecting introns.

If d_B has a sensitivity for potential introns (i.e. the cost for a deletion being smaller if the sequence deleted is a potential intron), it can be used to detect exons and introns. The intron cost function $d_I : B^* \times \mathbb{N}^2 \to \mathbb{R}$ of the form

$$d_I(a, i, j) = \begin{cases} \alpha(a, i) + \sum_{k=i+1}^{j-1} \delta(a, k) + \omega(a, j) & \text{if } j - i + 1 \geq L_0 \\ -\infty & \text{if } j - i + 1 < L_0 \end{cases}$$

scores introns having a minimum length L_0. Since it is essentially additive, it does not increase the asymptotic time and space complexity of the algorithm.

Both algorithms have been implemented as part of the computational biochemistry tool Darwin [2]. They simultaneously detect reading frame(s) and introns of a gene. Example results are shown in Figs. 1 and 2. They visualize the alignments of a gene (CAXYLG) with the protein encoded by it (according to the database) and homologous proteins. Two additional exons (bp 1188 to 1311 and 1381 to 1454) not contained in the protein and an additional intron (bp 1657 to 1716) appear. This might be due to mutations, but much more likely it is an erroneous interpretation of the gene, i.e. an error in the protein sequence of XYNA_CRYAL. Thus, the gene-protein alignment algorithm allows a systematic verification of gene interpretations by alignments with homologue proteins.

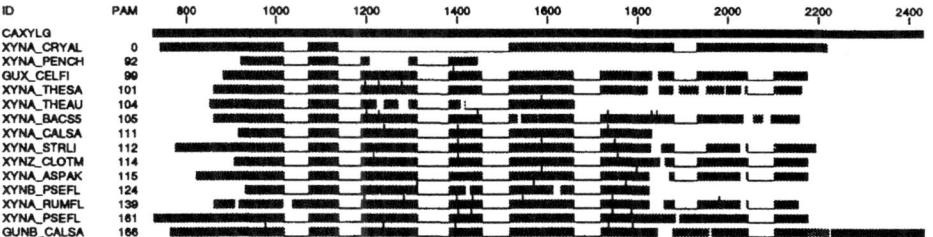

Fig. 1. Alignment of a gene with encoded and homologue proteins

```
ID=CAXYLG standard; DNA; FUN; 2984 BP.
DE=Cryptococcus albidus xylanase gene.  OS=Cryptococcus albidus
ID=GUX_CELFI    DE=EXOGLUCANASE PRECURSOR (EC 3.2.1.91)
(EXOCELLOBIOHYDROLASE) (1,4-BETA- CELLOBIOHYDROLASE).    OS=CELLULOMONAS FIMI.

0....|...70....|...80....|...90....|.1000....|...10....|...20....|...30....|...4
AAGTGGGAAGTCGTTGAGCCGACCGAGGGTAACTTTGATTTCACCGGTACCGACAAGgtgagcctggagtcataagtgac
<K><W><E><V><V><E><P><T><E><G><N><F><D><F><T><G><T><D><K>
 |  |  !  .  |  |  :  :  .  :  |  |  .  .  .  |  !
 K  W  D  A  T  E  P  S  Q  N  S  F  S  F  G  A  G  D  R <...................(5

0....|.1050....|...60....|...70....|...80....|...90....|.1100....|...10....|...
aaagcagtcagacttaccctacttgctccagATCGTCGCCGAAGCCAAGAAGACTGGTTCTCTGCTCCGAGGCCACAAC
                              <I><V><A><E><A><K><K><T><G><S><L><L><R><G><H><N>
                                 !  :  |  |  .  |  |  .  |  |  .  |  |
4/-12.1)...................> V  A  S  Y  A  A  D  T  G  K  E  L  Y  G  H  T

20....|...30....|...40....|.1150....|...60....|...70....|...80....|...90....|.
ATCTGCTGGGACTCTCAGgtaagtggtccaagaaagtaaagcgatacgacgactgatcatgggctctagACCCCTGCA
<I><C><W><D><S><Q>                                              <T><P><A>
 :  .  |  .  |  |                                               .  |  .
 L  V  W  H  S  Q <...................(51/-9.4).....................> L  P  D
```

Fig. 2. Alignment of gene CAXYLG and protein GUX_CELFI (partial)

4.2 Dipeptide Scoring in Protein Alignments

The classical models for aligning proteins assume mutation at position i to happen independently of mutation at position $i + 1$. Recent analysis of observed dipeptide mutations reveals that amino acid substitutions do not happen independently [1].

One way of expressing this dependency is by using dipeptide substitution matrices to score the alignments. With $s : \mathcal{A}^2 \times \mathcal{A}^2 \to \mathbb{R}$ and $u = v = 1$, the above framework allows to score correctly for dipeptides with gaps. Simplifications like the one in Example 6 can be used to obtain efficient algorithms.

However, experiments do not show a clear advantage of context dependent scoring over simpler approaches ([4] and own experiments). For closely related sequences, the best alignment is generally fixed and rather independent of the scoring scheme used. On the other hand, the differences when aligning distantly related sequences are very difficult to analyze.

5 Conclusions

The dynamic programming framework presented here generalizes the well known standard algorithms scoring on single symbols. Given fixed size alphabets and fixed p and q, the dynamic programming algorithm has the same asymptotic time and space complexity as the single symbol algorithm.

The algorithm presented in Sects. 2.2 and 3 has been applied successfully to gene alignment and intron detection, allowing verification of gene interpretations and improved mutation analysis.

While the framework is conceptually simple, actual implementations tend to be rather complex. A tool that given A, B, p, q, u, v and an optional mapping function f (as defined in Sect. 3.3) automatically generates implementations to find the optimum score and alignment for local and global alignments is desirable.

References

1. Mark A. Cohen, Steven A. Benner, and Gaston H. Gonnet. Analysis of mutation during divergent evolution: The 400 by 400 dipeptide mutation matrix. *Biochem. Biophys. Res. Commun.*, 199:489–496, 1994.
2. Gaston H. Gonnet. A tutorial introduction to computational biochemistry using Darwin. E.T.H. Zurich, Switzerland, 1994.
3. O. Gotoh. An improved algorithm for matching biological sequences. *Journal of Molecular Biology*, 162:705–708, 1982.
4. Xiaoqiu Huang. A context dependent method for comparing sequences. In M. Crochemore and D. Gusfield, editors, *Combinatorial Pattern Matching*, volume 807 of *Lecture Notes in Computer Science*, pages 54–63. Springer-Verlag, 1994.
5. E. W. Myers and M. Miller. Optimal alignments in linear space. *Comput. Applic. Biosci.*, 4:11–17, 1988.
6. Hannu Peltola, Hans Söderlund, and Esko Ukkonen. Algorithms for the search of amino acid patterns in nucleic acid sequences. *Nucleic Acids Research*, 14:99–107, 1986.
7. T. F. Smith and M. S. Waterman. Identification of common molecular subsequences. *Journal of Molecular Biology*, 147:195–197, 1981.

Matching a Set of Strings
with Variable Length Don't Cares

Gregory Kucherov and Michaël Rusinowitch *

1   Introduction

Given an alphabet A, a *pattern* p is a sequence $(v_1, \ldots, v_m)$ of words from A^* called
keywords. We represent p as a single word $v_1 @ \ldots @ v_m$, where $@ \notin A$ is a distinguished
symbol called *variable length don't care symbol*. Pattern p is said to *match* a text $t \in A^*$
if $t = u_0 v_1 u_1 \ldots u_{m-1} v_m u_m$ for some $u_0, \ldots, u_m \in A^*$. In this paper we address the
following problem: given a set P of patterns and a text t, test whether one of the
patterns of P matches t.

Quoting Fisher and Paterson in the concluding section of [10], "a good algorithm
for this (problem) would have obvious practical applications". For instance, as it was
reported by Manber and Baeza-Yates [13], the DNA pattern TATA often appears after
the pattern CAATCT within a variable length space. It may therefore be interesting to
look for the general pattern CAATCT@TATA. If we are given a set of such general patterns, it is desirable to have an algorithm that searches for all of them simultaneously
instead of searching consecutively for each one.

In this paper we propose an algorithm that solves the problem in time $O((|t| +
|P|) \log |P|)$, where $|t|$ is the length of the text and $|P|$ is the total length of all
keywords of P.

Several variants of the problem have been considered in the literature. Matching
set of strings with "unit length don't care symbols" that match any individual letter,
was studied in [10, 15]. Bertossi and Logi [5] have proposed an efficient parallel
algorithm for finding in a text the occurrences of a single pattern with variable length
don't-care symbols. Their algorithm has an $O(\log |t|)$ running time on $O(|t||P|/ \log |t|)$
processors.

Our problem can also be viewed as testing membership of a word in a regular lan-
guage of type $\cup_{i=1}^{n} A^* u_1^i A^* u_2^i A^* \ldots A^* u_{m_i}^i A^*$. Note that any regular expression where
the star operation only applies to the subexpression A (i.e. the union of all letters)
can be reduced to the above form by distributing concatenation over union. An
$O(|t||E|/ \log |t|)$ solution for the case of a general regular expression E has been given
by Myers [14].

The algorithm we propose here reads the text and the patterns from different
tapes in the left-to-right fashion. The text is searched on-line, which means that
the match is reported immediately after reading the shortest matched portion of the

---

*INRIA-Lorraine and CRIN/CNRS, Campus Scientifique, 615, rue du Jardin Botanique, BP 101,
54602 Villers-lès-Nancy, France, email: {kucherov,rusi}@loria.fr

text. Moreover, every pattern is read in the on-line fashion too, in the sense that the algorithm starts reading a keyword in a pattern only when all previous keywords of this pattern have been found in the text. This allows keywords to be specified dynamically, possibly depending on the search situation, for example on the keywords of other patterns that have been found by that moment. This feature of the algorithm makes it, we believe, particularly useful for some applications.

In contrast to most of the existing string matching algorithms(see [1]) our algorithm is not composed of two successive stages – preprocessing the pattern (resp. the text) and reading through the text (resp. the pattern) – but has these two stages essentially interleaved. The basic data structure used in the algorithm is the DAWG (Directed Acyclic Word Graph) [6, 7]. The DAWG is a flexible and powerful data structure related to suffix trees and similar structures (see [1, section 6.2] for references and [17] for one of the most recent works). In particular, the DAWG was used in [6, 7] as an intermediate structure for constructing in linear time the minimal *factor automaton* for a (set of) word(s). An elegant linear time on-line algorithm for constructing the DAWG was proposed in these papers. Independently, the DAWG for a single word was studied by Crochemore [8, 9] under the name of *suffix automaton*. In particular, in [9] he has extended the DAWG to a matching automaton, similar to the well-known Aho-Corasick automaton, to derive a new string matching algorithm. The algorithm we propose in this paper uses on the one hand, Crochemore's idea of using the DAWG for string matching and on the other hand, the efficient DAWG construction given in [6, 7].

The paper is organized as follows. In Section 2 we present the DAWG and define on top of it our basic data structure. Section 3 explains how to modify the DAWG, namely how to append a letter to a keyword and how to unload a keyword from the DAWG. In Section 4 the DAWG is further extended to be used as a matching automaton for solving the variable length don't care problem. The pattern matching algorithm is then detailed, its correctness is proved, and its complexity is evaluated. Finally, concluding remarks are made in the last section.

2 The DAWG

2.1 Definitions and main properties

Our terminology and definitions of this section basically follow [7].

$|v|$ denotes the length of $v \in A^*$. If $v = v_1 w v_2$, then w is said to occur in v at *position* $|v_1|$ and at *end position* $|v_1 w|$. For $D = \{v_1, \ldots, v_n\}$, a position (resp. end position) of w in D refers to a pair $< i, j >$, where j is a position (resp. end position) of w in v_i. $end\text{-}pos_D(w)$ is the set of all possible end positions of w in D. $pref(v)$ (resp. $pref(D)$) stands for the set of prefixes of v (resp. prefixes of the words from D). Similarly, $suff(v)$ $(suff(D))$ and $sub(v)$ $(sub(D))$ denote the set of suffixes and subwords respectively. ε denotes the empty word.

Our basic data structure is the *Directed Acyclic Word Graph (DAWG)* [6, 7].

Definition 1 *Let* $D = \{v_1, \ldots, v_n\} \subseteq A^*$. *For* $u, v \in sub(D)$, *define* $u \equiv_D v$ *iff* $end\text{-}pos_D(u) = end\text{-}pos_D(v)$. $[u]_D$ *denotes the equivalence class of* u *w.r.t.* $\equiv_D$. *The DAWG* $\mathcal{A}_D$ *for* D *is a directed acyclic graph with set of nodes* $\{[u]_D | u \in sub(D)\}$ *and*

set of edges $\{(([u]_D, [ua]_D)|u, ua \in sub(D), a \in A\}$. *The edges are labeled by letters in*
A *so that the edge* $([u]_D, [ua]_D)$ *is labeled by* a. *The node* $[\varepsilon]_D$ *is called the* source *of*
$\mathcal{A}_D$.

An example of a DAWG is given in Appendix A0. Viewed as a finite automaton
with every state being accepting, the DAWG is a deterministic automaton recognizing
the subwords of D. Moreover, except for accepting states, the DAWG is isomorphic
to the *minimal* deterministic automaton recognizing the suffixes of D, where syntac-
tically equal suffixes of different keywords are considered to be different. Formally,
this automaton can be obtained by appending at the end of each $v_i \in D$ a distinct
fresh symbol $\$_i$, then constructing the minimal deterministic automaton for the suf-
fixes of the modified set, and then forgetting the accepting sink state together with
all incoming $\$_i$-transitions. This construction ensures the uniqueness of the DAWG,
the property that will be tacitly used throughout the paper. If D consists of a single
keyword, this automaton called *suffix automaton* is just the minimal deterministic
automaton recognizing the suffixes of D [8, 9].

The reader is referred to [6, 7] for a more detailed analysis of the DAWG, in
particular for linear bounds on its size and the relationship between the DAWG and
the suffix tree. The following property allows us to define an important tree structure
on the nodes of $\mathcal{A}_D$.

Proposition 1 *For* $u, v \in suff(D)$, *if* $end\text{-}pos_D(u) \cap end\text{-}pos_D(v) \neq \emptyset$, *then either*
$u \in suff(v)$ *or* $v \in suff(u)$. *This implies that*

(i) *every* $\equiv_D$-*equivalence class has a longest element called* the representative *of
this class,*

(ii) *if* $end\text{-}pos_D(u) \cap end\text{-}pos_D(v) \neq \emptyset$, *then either* $end\text{-}pos_D(u) \subseteq end\text{-}pos_D(v)$ *or*
$end\text{-}pos_D(v) \subseteq end\text{-}pos_D(u)$.

Property (ii) ensures that the subset relation on equivalence classes defines a tree
structure on the nodes. If $end\text{-}pos_D(u) \subset end\text{-}pos_D(v)$, and for no $w \in sub(D)$,
$end\text{-}pos_D(u) \subset end\text{-}pos_D(w) \subset end\text{-}pos_D(v)$, then we say that there exists a *suffix
pointer* going from $[u]_D$ to $[v]_D$. $[u]_D$ is said to be a *child* of $[v]_D$ and $[v]_D$ the *parent*
of $[u]_D$. The source of $\mathcal{A}_D$ is the root of this tree. The sequence of suffix pointers going
from some node to the source is called the *suffix chain* of this node. The following
lemma clarifies the relation between two nodes linked by a suffix pointer.

Lemma 1 ([7]) *Let* u *be the representative of* $[u]_D$. *Then any child of* $[u]_D$ *can be
expressed as* $[au]_D$ *for some* $a \in A$.

If $au \in sub(D)$ (resp. $ua \in sub(D)$) for some $a \in A$, $u \in A^*$, then we say that a is
a *left context* (resp. *right context*) of u in D. The lemma above shows that if u is the
representative of $[u]_D$, then every child of $[u]_D$ corresponds to a distinct left context
of u in D. This implies that each node has at most $|A|$ children. In contrast, edges
of the DAWG refer to the *right context*: for every right context a of $u \in sub(D)$, the
DAWG contains the edge $([u]_D, [ua]_D)$ labeled by a.

The following fact related to lemma 1 is also enlightening.

Lemma 2 $u \in sub(D)$ *is the representative of* $[u]_D$ *iff either* u *is a prefix of some keyword in* D, *or* u *has two or more distinct left contexts in* D.

The edges of the DAWG are divided into two categories: Assume that u is the representative of $[u]_D$. The edge $([u]_D, [ua]_D)$ is called *primary* if ua is the representative of $[ua]_D$, otherwise it is called *secondary*. The primary edges form a spanning tree of the DAWG rooted at the source. This tree can be also obtained by taking in the DAWG only the longest path from the source to each node. With each node $[u]_D$ we associate a number $depth([u]_D)$ which is defined as the depth of $[u]_D$ in the tree of primary edges. Equivalently, $depth([u]_D)$ is the length of the representative of $[u]_D$. Note that if the edge $([u]_D, [ua]_D)$ is primary, then $depth([ua]_D) = depth([u]_D) + 1$, otherwise $depth([ua]_D) > depth([u]_D) + 1$.

If $w \in pref(v_i)$ for some $v_i \in D$, then we call $[w]_D$ a *prefix node for* v_i. Note that by lemma 2, w is the representative of $[w]_D$. Besides, if $w = v_i$ for some $v_i \in D$, then the node $[w]_D$ is also called a *terminal node* for v_i.

2.2 Data structure

We assume that each node α of the DAWG is represented by a data structure providing the following attributes:

out(α, a): a reference to the target node of the edge issuing from α and labeled by a; out$(\alpha, a) =$ undefined when there is no such edge,

type(α, a): type$(\alpha, a) =$ primary if the edge issuing from α and labeled by a is primary, otherwise type$(\alpha, a) =$ secondary,

suf-pointer(α): a reference to the node pointed by the suffix pointer of α; suf-pointer$(\alpha) =$ undefined if α is the source,

depth(α): $depth(\alpha)$,

terminal(α): terminal$(\alpha) =$ null if α is not a terminal node, otherwise terminal(α) refers to a list of keywords for which α is terminal (we do not assume that all keywords are different and therefore a node can be terminal for several keywords). The list will be defined more precisely in section 4.2.

origin(α): a reference to the node that the primary edge to α comes from,

last-letter(α): the label of incoming edges to α (equivalently, the last letter of any word in α),

number-of-children(α): has three possible values $\{0, 1, \text{more-than-one}\}$. number-of-children$(\alpha) = 0$ (respectively number-of-children$(\alpha) = 1$, number-of-children$(\alpha) =$ more-than-one) if there are no (respectively one, more than one) suffix pointers that point to α,

child(α): refers to the only child of α when number-of-children$(\alpha) = 1$,

prefix-degree(α): the number of keywords in D for which α is a prefix node. prefix-degree$(\alpha) = 0$ if α is not a prefix node.

out and type implement the DAWG itself, the other attributes are needed for different purposes that will become clear in the following sections. We will use the same denotation $\mathcal{A}_D$ for the whole data structure described above. An example of the data structure is given in Appendix A0.

We always assume the alphabet to be of a fixed size. We assume the uniform RAM model of computation and then assume that retrieving, modifying and comparing any attribute values as well as creating and deleting a DAWG node is done in constant time.

3 Modifying a DAWG

3.1 Appending a letter to a keyword

A.Blumer, J.Blumer, Haussler, McConnell and Ehrenfeucht [7] (BBHME for short) proposed an algorithm to construct $\mathcal{A}_D$ for a given set D in time $O(|D|)$. The algorithm processes consecutively the patterns of D such that if $\{v_1, \ldots, v_i\}$ have been already processed, then the constructed data structure is $\mathcal{A}_{\{v_1,\ldots,v_i\}}$. Processing a pattern v_{i+1} (equivalently, extending $\mathcal{A}_{\{v_1,\ldots,v_i\}}$ to $\mathcal{A}_{\{v_1,\ldots,v_i,v_{i+1}\}}$) is called *loading* v_{i+1} (to $\mathcal{A}_{\{v_1,\ldots,v_i\}}$). Loading v_{i+1} to $\mathcal{A}_{\{v_1,\ldots,v_i\}}$ is done by scanning v_{i+1} from left to right such that if $w \in pref(v_{i+1})$ is an already processed prefix of v_{i+1}, then the constructed data structure is $\mathcal{A}_{\{v_1,\ldots,v_i,w\}}$. Therefore, processing patterns in the set as well as letters in the pattern is done in the *on-line* fashion, and a basic step of the algorithm amounts to extending $\mathcal{A}_{\{v_1,\ldots,v_i,w\}}$ to $\mathcal{A}_{\{v_1,\ldots,v_i,wa\}}$ for some $a \in A$.

The BBHME data structure has only attributes **out**, **type** and **suf-pointer**. However, the BBHME algorithm can be easily extended to maintain the additional attributes that we need for our purposes. Since the BBHME algorithm is fundamental for this paper, we give its pseudocode in appendix A1. We shortly comment the algorithm below.

Function APPEND-LETTER implements the main step. It takes the terminal node of w in $\mathcal{A}_{\{v_1,\ldots,v_i,w\}}$ and a letter a and outputs the terminal node for wa in $\mathcal{A}_{\{v_1,\ldots,v_i,wa\}}$. APPEND-LETTER creates, if necessary, a new node for $[wa]_{\{v_1,\ldots,v_i,wa\}}$, and then traverses the suffix chain of the node $[w]_{\{v_1,\ldots,v_i,w\}}$ (installing secondary edges to the new node) up to the first node with an outcoming a-edge. If this edge is primary, no further traversals have to be done. If it is secondary, the function SPLIT is called which creates another new node, installs its outcoming edges, updates suffix pointers, and then continues the traversal unless a node with a primary outcoming a-edge is found. Thus, at most two new nodes are created and the suffix chain of $[w]_{\{v_1,\ldots,v_i,w\}}$ is traversed up to the first primary outcoming a-edge.

In the paper we will be modifying the BBHME algorithm. In particular, some instructions will be added to the SPLIT function, which is indicated at line 4 of its code in appendix A1.

Functions APPEND-LETTER and SPLIT maintain additional attributes **origin**, **last-letter**, **depth**, **number-of-children** and **child**. As for **origin**, **last-letter** and **depth**, this is explicitly shown in the algorithm. **number-of-children** and **child** are updated every time the tree of suffix pointers is modified (lines 12,14,15 in APPEND-LETTER and lines 6,7 in SPLIT). Maintaining **number-of-children** is trivial. **child** can be implemented by organizing the set of children of each node in a double-linked list and keeping a pointer to the first child in the list. Deleting a child then takes time $O(1)$.

LOAD-KEYWORD(v) loads a keyword v by scanning it and iterating the APPEND-LETTER function. Also, LOAD-KEYWORD maintains the **prefix-degree** attribute. Maintaining **terminal** will be considered later.

The remarkable property of the algorithm, shown in [6, 7], is that it builds the DAWG for a set D in time $O(|D|)$ by iterating LOAD-KEYWORD(v) for every $v \in D$, starting with a one-node DAWG. Actually, loading an individual keyword v into $\mathcal{A}_D$ takes time linear on $|v|$ *regardless of the set D.*

Lemma 3 LOAD-KEYWORD(v) *runs in time* $O(|v|)$.

3.2 Unloading a keyword

In this section we give an algorithm that unloads a keyword v_{i+1} from $\mathcal{A}_{\{v_1,\ldots,v_i,v_{i+1}\}}$. Starting from the terminal node $[v_{i+1}]_{\{v_1,\ldots,v_i,v_{i+1}\}}$, the algorithm traces back the chain of primary edges and at each step undoes the modifications caused by appending a corresponding letter. Thus, the main step is inverse to APPEND-LETTER and amounts to transforming $\mathcal{A}_{\{v_1,\ldots,v_i,wa\}}$ into $\mathcal{A}_{\{v_1,\ldots,v_i,w\}}$ for $wa \in pref(v_{i+1})$. The modifications to be applied to the nodes of $\mathcal{A}_{\{v_1,\ldots,v_i,wa\}}$ are described in the following lemma which is in a sense inverse to lemma 2.1 from [6].

Lemma 4   *(i) wa is not the representative of an equivalence class w.r.t. $\equiv_{\{v_1,\ldots,v_i,w\}}$ iff either $wa \notin sub(\{v_1,\ldots,v_i,w\})$ or wa has only one left context in $\{v_1,\ldots,v_i,w\}$ (and hence has only one child) and $wa \notin pref(\{v_1,\ldots,v_i,w\})$. In the first case the class $[wa]_{\{v_1,\ldots,v_i,wa\}}$ is deleted. In the second case this class is merged with its child.*

  (ii) Let $wa = u_1u_2a$ and u_2a is the representative of the class pointed to by the suffix pointer of $[wa]_{\{v_1,\ldots,v_i,wa\}}$. Then u_2a is not the representative of an equivalence class w.r.t. $\equiv_{\{v_1,\ldots,v_i,w\}}$ iff u_2a has only one left context in $\{v_1,\ldots,v_i,w\}$ and $u_2a \notin pref(\{v_1,\ldots,v_i,w\})$. In this case $[u_2a]_{\{v_1,\ldots,v_i,wa\}}$ is merged with its single child.

  (iii) There are no other modifications in the equivalence classes except those given in (i) and (ii).

The transformation of $\mathcal{A}_{\{v_1,\ldots,v_i,wa\}}$ into $\mathcal{A}_{\{v_1,\ldots,v_i,w\}}$ is done by modifying the DAWG according to lemma 4. The algorithm is given in appendix A2. Its short account follows.

Let *activenode* be $[wa]_{\{v_1,\ldots,v_i,wa\}}$. A main function DELETE-LETTER takes *activenode*, finds the node $[w]_{\{v_1,\ldots,v_i,wa\}}$ called *newactivenode* by retrieving *origin(activenode)*, and then proceeds by case analysis. If *activenode* is a prefix node for a keyword other than v_{i+1}, or has two or more children, then no more work has to be done. If *activenode* is not a prefix node of any other keyword and has only one child, then it should be merged with this child which is done by an auxiliary function MERGE. Finally, if *activenode* has no children and is not a prefix node of any other keyword, this means that $wa \notin \{v_1,\ldots,v_i,w\}$ and therefore *activenode* should be deleted. Before it is deleted, the suffix chain of *newactivenode* is traversed and outcoming secondary a-edges leading to *activenode* are deleted. Let *varnode* be the first node on the suffix chain with outcoming primary a-edge. It can be shown that *varnode* is actually $[u_2]_{\{v_1,\ldots,v_i,wa\}}$ (lemma 4(ii)), and the node that this edge leads to, called *suffixnode*, is $[u_2a]_{\{v_1,\ldots,v_i,wa\}}$. Once *suffixnode* is found, *activenode* is deleted together with its suffix pointer pointing to *suffixnode*. If *suffixnode* is a prefix node or has more than one child left, the algorithm terminates. Otherwise *suffixnode* has to be merged with its single child which is done by the MERGE function. MERGE acts inversely to the SPLIT function (see appendix A2 for details). Note that MERGE continues the traversal of *varnode* up to the first node with outcoming primary *a*-edge.

Similarly to the loading case, we will be extending the algorithm afterwards. In particular, instructions will be added to DELETE-LETTER and MERGE at line 9 and 2 respectively.

Maintenance of additional attributes origin, last-letter, depth, number-of-children and child is done similarly to the loading case.

UNLOAD-KEYWORD($[v]_D$) unloads pattern $v \in D$ from $\mathcal{A}_D$ by iterating DELETE-LETTER. Also, UNLOAD-KEYWORD maintains the prefix-degree attribute. Like in the case of insertion, UNLOAD-KEYWORD runs in time $O(|v|)$ *regardless of D*.

Lemma 5 UNLOAD-KEYWORD($[v]$) *runs in time $O(|v|)$.*

The proof goes along the same lines as for the case of insertion.

4 Matching a set of strings with variable length don't cares

4.1 Extending the DAWG for string matching

Crochemore noticed [9] that in the case of one keyword the DAWG can be used as a string matching automaton similar to that of Aho-Corasick, where suffix pointers play the role of failure transitions. The idea is to extend the current state *currentnode* with a counter *length* updated at each transition step. The procedure below describes a basic step of the algorithm. *currentletter* is assumed to be the current letter in the text.

```
UPDATE-CURRENT-NODE(currentnode, length, currentletter)
1    while out(currentnode, currentletter) = undefined do
2        if currentnode = source then
3            return < currentnode, 0 >
4        else currentnode := suf-pointer(currentnode)
5            length := depth(currentnode)
6    currentnode := out(currentnode, currentletter)
7    length := length + 1
8    return < currentnode, length >
```

The meaning of *currentnode* and *length* is given by the following proposition which is an extension of Proposition 2 of [9] for the multiple keyword case.

Proposition 2 *Assume that D is a set of keywords and $t = t_1 \ldots t_n$ is a text. Consider $\mathcal{A}_D$ and iterate* UPDATE-CURRENT-NODE(*currentnode, length, currentletter*) *on t with initial values currentnode = source, length = 0, and currentletter = t_1. At any step, if $t_1 \ldots t_i$ is the prefix of t scanned so far, and w is the longest word from suff($\{t_1 \ldots t_i\}$) $\cap$ sub(D), then w belongs to currentnode (regarded as an equivalence class) and length = $|w|$.*

Crochemore used proposition 2 as a basis for a linear string matching algorithm in the case of a single keyword. An occurrence of the keyword is reported iff *currentstate* is terminal and, in addition, the current value of *length* is equal to depth(*currentnode*). The current position in the text is then the end position of the keyword occurrence. The linearity of the algorithm of proposition 2 can be shown using the same arguments as for the Aho-Corasick algorithm.

However, this idea does not extend to the multiple keyword case, since one or several keywords may occur at the current end position in the text even if *currentnode* is not terminal. This is the case for the keywords that are suffixes of $t_1 \ldots t_i$ shorter than the current value of *length*. To detect these occurrences, at every call to UPDATE-CURRENT-NODE the suffix chain of *currentnode* should be traversed and a match should be reported for every terminal node on the chain. A naive implementation of this traversal would lead to a prohibitive $O(|t||D|)$ search time.

One approach to the problem would be to attach to each node a pointer to the closest terminal node on the suffix chain. When the set of keywords is fixed once for all, this approach amounts to an additional preprocessing pass which can be done in time $O(|D|)$ and therefore does not affect the overall linear complexity bound. However, when the set of keywords is changing over time, which is our case, this approach becomes unsatisfactory, since modifying a single keyword may require $O(|D|)$ operations.

String matching for a changing set of keywords has been recently studied in the literature under the name of *dynamic dictionary matching*. Several solutions have been proposed [2, 3, 11, 4]. All of them, however, had to face a difficulty similar to the one described above. In terms of data structure, the problem amounts to finding for a node of a dynamically changing tree (in our case, tree of suffix pointers) the closest marked ancestor node (in our case, terminal node), where nodes are also marked and unmarked dynamically.

In this paper we borrow the solution proposed in [3] which consists in using the dynamic trees of Sleator and Tarjan [16]. The tree of suffix pointers is split into a forest by deleting all suffix pointers of terminal nodes. Thus, every terminal node in the tree becomes the root node of a tree in the forest. The forest is implemented using the dynamic trees technique of [16]. For shortness, we will call the DAWG augmented with this data structure the *extended DAWG*. Since finding the closest terminal node on the suffix chain of a node amounts to finding the root of its tree in the forest, this operation takes $O(\log |D|)$ time. We will denote by CLOSEST-TERMINAL(α) a function which implements this operation. It returns the closest terminal node on the suffix chain of α if such a node exists, and returns **undefined** otherwise.

On the other hand, creating, deleting and redirecting suffix pointers takes no longer a constant time, but time $O(\log |D|)$. Since both APPEND-LETTER and DELETE-LETTER requires a constant number of such operations, we restate lemmas 3 and 5 as follows.

Lemma 6 *On the extended DAWG,* LOAD-KEYWORD(v) *and* UNLOAD-KEYWORD($[v]$) *run in time* $O(|v| \log |D|)$.

4.2 Pattern matching algorithm

Assume that P is a finite set of strings $\{p_1, \ldots, p_n\}$ over $A \cup \{@\}$, where each $p_i \in P$ is written as $v_1^i @ v_2^i @ \ldots @ v_{m_i}^i$, for some $v_1^i, v_2^i, \ldots, v_{m_i}^i \in A^*$. According to our terminology, p_i's are called *patterns* and v_j's *keywords*. A pattern p_i matches a text $t \in A^*$, if $t = u_1 v_1^i u_2 \ldots u_{m_i} v_{m_i}^i u_{m_i+1}$ for some $u_1, u_2, \ldots, u_{m_i}, u_{m_i+1} \in A^*$. We address the following problem: given a set of patterns P and a text $t = t_1 \ldots t_k$, test whether one of the patterns of P matches t.

We assume that the text and every pattern is read from left to right from a separate input tape.

Let us first give an intuitive idea of the algorithm. At each moment of the text scan, the algorithm searches for a group of keywords, one from each pattern, represented by the DAWG. The search is done using the DAWG as an automaton similar to proposition 2. Every time a keyword is found, it is unloaded from the DAWG and the next keyword in the corresponding pattern is loaded instead. The crucial point is that the loading process is "spread over time" so that loading one letter of the keyword alternates with processing one letter of the text. In this way the correctness of the algorithm is ensured. Thus, unlike the usual automata string matching technique, the underlying automaton evolves over time adapting to the changing set of keywords.

Let us turn to a formal description. Let $t[1:l] = t_1 \ldots t_l$ be a prefix of t scanned so far. For every $p_i = v_1^i @ v_2^i @ \ldots @ v_{m_i}^i \in P$, consider a decomposition

$$t[1:l] = u_1^i v_1^i u_2^i \ldots v_{j_i-1}^i u_{j_i}^i \tag{1}$$

for $j_i \in [1, m_i], u_1^i, u_2^i, \ldots, u_{j_i}^i \in A^*$, such that

- for every $r \in [1, j_i - 1]$, $v_r^i \notin sub(u_r^i v_r^i) \setminus suff(u_r^i v_r^i)$,

- $v_{j_i}^i \notin sub(u_{j_i}^i)$.

Clearly, under the conditions above, decomposition (1) is unique. The intuition is that the *leftmost* occurrence of each pattern is looked for, that is the leftmost occurrence of every keyword that follows the occurrence of the preceding keyword.

Consider decompositions (1) for every $i \in [1, n]$. We now define the state of the matching process after l letters of the text have been processed. We first introduce some terminology. For every $i \in [1, n]$, $v_{j_i}^i$ is called an *active* keyword. If $|u_{j_i}^i| < |v_{j_i}^i|$ for $i \in [1, n]$, then both the pattern p_i and its active keyword $v_{j_i}^i$ are said to be *under loading*. For each $i \in [1, n]$, define $\bar{v}_{j_i}^i = v_{j_i}^i[1:q]$ where $q = min\{|u_{j_i}^i|, |v_{j_i}^i|\}$. Thus, if p_i is under loading then $\bar{v}_{j_i}^i$ is the prefix of $v_{j_i}^i$ of length $|u_{j_i}^i|$, otherwise $\bar{v}_{j_i}^i = v_{j_i}^i$. The current situation of the matching process is represented by a data structure consisting of three components given below together with their invariant conditions:

1. The extended DAWG for the set $V = \{\bar{v}_{j_1}^1, \ldots, \bar{v}_{j_n}^n\}$, defined as in Sections 2, 4.1, except that if $v_{j_i}^i$ is a keyword under loading, then the node $[\bar{v}_{j_i}^i]_V$ is not considered terminal. We call these nodes *port nodes*. Intuitively, a port node refers to a node in the DAWG to which the next letter of the corresponding keyword under loading should be appended. If $v_{j_i}^i$ has been completely loaded, that is $\bar{v}_{j_i}^i = v_{j_i}^i$, then $[\bar{v}_{j_i}^i]_V$ is terminal for $v_{j_i}^i$ and the list terminal$([\bar{v}_{j_i}^i])$ contains a reference to p_i. In other words, if α is a terminal node, terminal(α) is the list of patterns p_i such that α is terminal for the active keyword of p_i.

2. A distinguished node in the DAWG called *currentnode*, together with a counter *length*. *currentnode* is the node $[w]_V$, where w is the longest word in $suff(t[1:l]) \cap sub(\{\bar{v}_{j_1}^1, \ldots, \bar{v}_{j_n}^n\})$, and *length* = $|w|$.

3. A double linked list of patterns under loading each element of which has a reference to the corresponding pattern p_i and a reference to the corresponding port node in the DAWG.

A basic step of the algorithm consists of three stages. First, for each keyword under loading, the next letter is inserted into the DAWG using the APPEND-LETTER procedure and the port node is updated. If the keyword has been loaded completely, then the corresponding port node becomes terminal and the corresponding pattern is deleted from the list of patterns under loading. Secondly, *currentnode* and *length* are updated using the UPDATE-CURRENT-NODE procedure. Finally, the suffix chain of *currentnode* is traversed looking for terminal nodes. Each such node corresponds to one or several active keywords that occur at the current end position in the text. Each detected matching keyword is unloaded from the DAWG using the UNLOAD-KEYWORD algorithm, and the following keyword in the pattern becomes under loading with the source being the port node.

To define the algorithm, functions SPLIT, UNLOAD-KEYWORD and MERGE from Section 3 should be slightly modified. The reason for modifying SPLIT is that the node which has to be split (*targetnode* in the SPLIT algorithm) may happen to be the actual value of *currentnode*. *currentnode* should then be updated so that condition 2 above be preserved. The following instruction has to be inserted into the SPLIT algorithm at line 4 (Appendix A1).

4.1   if *currentnode* = *targetnode* then
4.2   if *length* ≤ **depth**(*newtargetnode*) then *currentnode* := *newtargetnode*

Similarly, each of the functions DELETE-LETTER and MERGE may be led to delete a node which is actually *currentnode*, in which case *currentnode* must be updated. Again, the new value is computed in order to preserve condition 2. The following instructions have to be inserted into the DELETE-LETTER algorithm at line 9 (Appendix A2).

9.1   if *currentnode* = *activenode* then
9.2   *currentnode* := **suf-pointer**(*activenode*)
9.3   *length* := **depth**(*currentnode*)

The instruction below has to be inserted into the MERGE algorithm at line 2 (Appendix A2).

2   if *currentnode* = *targetnode* then *currentnode* := *newtargetnode*

Note that modified functions SPLIT, DELETE-LETTER and MERGE may now change *currentnode* and *length* as a side effect. The modifications will be further discussed in Section 4.3.

We are now ready to give the complete algorithm. t denotes a subject text and READ-LETTER(t) returns the scanned letter. For a pattern p under loading, READ-LETTER(p) returns the next letter of p, and PORT-NODE(p) refers to the corresponding port node.

MATCH$(t, P = \{p_1, \ldots, p_n\})$

```
1    create a node currentnode
2    length := 0
3    set the list of patterns under loading to be {p_1, ..., p_n}
4    for each p_i do PORT-NODE(p_i) := currentnode
5    while the end of t has not been reached do
         %STAGE 1
6        for each pattern under loading p_i do
7            portnode := PORT-NODE(p_i)
8            patternletter := READ-LETTER(p_i)
9            newportnode := APPEND-LETTER(portnode, patternletter)
10           prefix-degree(newportnode) := prefix-degree(newportnode) + 1
11           if all letters of the active keyword of p_i have been read then
12               delete p_i from the list of patterns under loading
13               mark newportnode as a terminal node unless it was already the case
14                   and add p_i to the list terminal(newportnode)
15           else PORT-NODE(p_i) := newportnode
     %STAGE 2
16       currentletter := READ-LETTER(t)
17       < currentnode, length >:=
                 UPDATE-CURRENT-NODE(currentnode, length, currentletter)
     %STAGE 3
18       if currentnode is terminal and depth(currentnode) = length then
19           closestterminal := currentnode
20       else closestterminal := CLOSEST-TERMINAL(currentnode)
21       while closestterminal ≠ undefined do
22           unmark closestterminal as a terminal node
23           currentterminal := closestterminal
24           closestterminal := CLOSEST-TERMINAL(closestterminal)
25           for each p_i from the list terminal(currentterminal) do
26               if all keywords of p_i have been read then
27                   output "p_i occurs in t" and stop
28               else UNLOAD-KEYWORD(currentterminal)
29                   add p_i to the list of patterns under loading
30                   PORT-NODE(p_i) := source
31   output "t does not have occurrences of P"
```

4.3 Correctness of the algorithm

To prove the correctness and completeness of MATCH we verify by induction that conditions 1-3 of Section 4.2 are invariant under the main **while**-loop. More precisely, we prove that the set $\{\bar{v}_{j_1}^1, \ldots, \bar{v}_{j_n}^n\}$, currentnode, length, and the list of patterns under loading satisfy conditions 1-3 at every step of the algorithm. The correctness and completeness would then follow from decomposition (1). Below we give an outline of the proof.

We first consider conditions 1 and 3. Consider decomposition (1) for some $i \in [1, n]$ and assume that $u_{j_i}^i = \bar{v}_{j_i}^i = \varepsilon$. (In other words, consider the first step when $v_{j_i}^i$ is active.) Since one letter is read from the text at every iteration (line 16), $u_{j_i}^i$ in decomposition (1) is extended by one letter (unless an occurrence of $v_{j_i}^i$ is found, see below). While $v_{j_i}^i$ is under loading, $\bar{v}_{j_i}^i$ is extended by one letter at every iteration too (line 8). Therefore, the algorithm keeps $\bar{v}_{j_i}^i$ to be a proper prefix of $v_{j_i}^i$ of length $|u_{j_i}^i|$. When all letters of $v_{j_i}^i$ are loaded, then $\bar{v}_{j_i}^i = v_{j_i}^i$, and since $|u_{j_i}^i| = |\bar{v}_{j_i}^i|$, then $v_{j_i}^i$ should cease to be under loading (condition 1). This corresponds to instruction 12 in the algorithm that deletes p_i from the list of patterns under loading as soon as $v_{j_i}^i$ has been read completely. At subsequent iterations, $\bar{v}_{j_i}^i$ does not change, unless $v_{j_i}^i$ is unloaded, which happens iff $[v_{j_i}^i]$ is on the suffix chain of *currentnode*. Since by condition 2 the member of *currentnode* (regarded as an equivalence class) of length *length* is a suffix of the text read so far, then so is $v_{j_i}^i$. With respect to decomposition (1), this means that j_i is incremented by 1 with $u_{j_i+1}^i = \varepsilon$. This completes the induction. The above arguments together with the uniqueness of the DAWG (Section 2.1) shows that at every moment the values of $\bar{v}_{j_i}^i$'s and the list of patterns under loading are correct.

We now prove that *currentnode* and *length* verify condition 2 at every step of the algorithm. Assume that condition 2 is verified at the beginning of the iteration of the while-loop. Let $V = \{\bar{v}_{j_1}^1, \ldots, \bar{v}_{j_n}^n\}$ be the current underlying set of words and w be the longest word from $\text{suff}(t[1:l]) \cap \text{pref}(V)$, where $t[1:l]$ is the prefix of the text read so far. By condition 2, *currentnode* is $[w]_V$ and $\text{length} = |w|$.

The first stage extends the DAWG to a set $\tilde{V} = \{\tilde{v}_1, \ldots, \tilde{v}_n\}$, where each $\tilde{v}_i$ is either $\bar{v}_{j_i}^i$ or $\bar{v}_{j_i}^i a$ for some $a \in A$. During this stage, *currentnode* is kept to be the equivalence class of w w.r.t. the changing set V. The only point when *currentnode* may need to be updated is when this node is split into two by the SPLIT function. In this case we should decide which of the two resulting classes contains w and then becomes a new value of *currentnode*. This is decided according to the value of *length*. The update of *currentnode* is done by instruction 4 added to SPLIT in the previous section. The correctness follows from a more detailed analysis of SPLIT that can be found in [6].

At the second stage, the next letter is read from the text, which means that index l in condition 2 is incremented, and then *currentnode* and *length* are updated by UPDATE-CURRENT-NODE. We have to show that after that, *currentnode* and *length* verify condition 2, which means that *currentnode* is $[\tilde{w}]_{\tilde{V}}$, where $\tilde{w}$ is the longest word from $\text{suff}(t[1:l+1]) \cap \text{pref}(\tilde{V})$, and *length* is $|\tilde{w}|$. The proof of this part is similar to that of proposition 2.

At the third stage the keywords are detected which occur at the current end position in the text. Clearly, all these keywords are suffixes of w. w itself is a keyword iff *currentnode* is a terminal node and w is its representative, that is $\text{depth}(\text{currentnode}) = \text{length}$. The keywords which are proper suffixes of w have their terminal nodes on the suffix chain of w. Thus, all matching keywords are detected by the algorithm. Each matching keyword, when detected, is unloaded from $\tilde{V}$. We have to show again that condition 2 is preserved under this transformation. Consider an elementary deletion step (DELETE-LETTER) which consists in transforming the DAWG from some set $\hat{V} = \{\hat{v}_1, \ldots, \hat{v}_{n-1}, \hat{v}_n a\}$ to $\hat{V}' = \{\hat{v}_1, \ldots, \hat{v}_{n-1}, \hat{v}_n\}$. If $w \in \text{suff}(\hat{v}_n a)$ and $w \notin \text{sub}(\hat{V}')$, then w should be reset to its longest suffix which belongs to $\text{sub}(\hat{V}')$. The corresponding modification of *currentnode* and *length* is done by the instructions added to DELETE-LETTER in the previous section. The instruction added to MERGE updates *currentnode* whenever this class is merged with another one. This modification is inverse to the one done by the SPLIT function.

We summarize the discussion above in the following theorem.

Theorem 1 *The algorithm* MATCH(t, P) *is correct and complete, i.e. it detects an occurrence of patterns of P in t iff there is one.*

It is important to note that the correctness of the algorithm is essentially due to the fact that the process of keyword loading is synchronized with the text scan. If a whole keyword had been loaded immediately after the previous one has been found, a correct maintenance of *currentnode* would become impossible.

4.4 Complexity of the algorithm

In this section we evaluate the time complexity of MATCH(t, P). Define $|P| = \sum_{i=1}^{n} \sum_{j=1}^{n_j} |v_j^i|$ and $d = \sum_{i=1}^{n} \max\{|v_j^i| \mid j \in [1 : n_i]\}$. Distinguishing d and $|P|$ in the complexity analysis is useful for applications in which patterns of P are long sequences of short keywords.

Let us first focus on the time taken by stage 2, that is on the total time of executing UPDATE-CURRENT-NODE (instruction 17). One can show that this time is $O(|t|)$ using a standard argument of amortizing the number of iterations of the **while**-loop in UPDATE-CURRENT-NODE over all letters of t (cf [1]).

Let us analyse stage 3 now. CLOSEST-TERMINAL(α) runs in time $O(\log d)$, and within one execution of stage 3 there is one more call to CLOSEST-TERMINAL than terminal nodes on the suffix chain of *currentnode*. Therefore, each call to CLOSEST-TERMINAL but one is followed by unloading at least one keyword. Each iteration of the **for**-loop (line 25) either unloads one keyword or stops the whole run of the algorithm. Clearly, every keyword of P can be loaded and unloaded at most once during the run of MATCH. Unloading a keyword v_j^i using UNLOAD-KEYWORD takes $O(|v_j^i| \log d)$ time by lemma 6. Since the list of patterns under loading is implemented as a double linked list, instruction 29 as well as instruction 12 of stage 1 is done in time $O(1)$. To sum up, the time spent on stage 3 during the whole run of MATCH can be evaluated as $O(|t| \log d + \sum_{i=1}^{n} \sum_{j=1}^{n_j} |v_j^i| \log d) = O((|t| + |P|) \log d)$.

Now let us turn to stage 1. Each iteration of the **for**-loop (line 6) calls to APPEND-LETTER (line 9) which is the only individual step taking non-constant time. Thus, it remains to evaluate the complexity of the loading process. Here, however, we face a difficulty. To describe it, we forget for a moment about the auxiliary dynamic tree structure defined in Section 4.1 which introduces a $\log d$ factor in appending a letter (lemma 6). The problem is that although by lemma 3, loading a keyword takes time linear on its length, this result does not generally hold for our mode of loading. The reason is that in our case, loading letters of a keyword alternates with loading letters of other keywords and unloading some keywords, while the proof of lemma 3 in [6] assumes tacitly that the DAWG does not change between loadings of two consecutive letters of a keyword. In the rest of this section we outline a solution to this problem which takes linear time *with respect to the set of all loaded keywords*. The complete description of the solution is far too long to be given here, and is left to the full version of the paper.

Recall from section 3.1 that calling APPEND-LETTER(α, a) provokes a traversal of the suffix chain of α up to the first node with an outcoming primary a-edge. During stage 1, such a traversal is made for the port node α and the current letter a of every keyword under loading. The solution consists in synchronizing the traversals and imposing an order of traversing parts of the suffix chains of the port nodes.

Assume that the same letter a is appended to two port nodes α_1 and α_2, and assume that the suffix chains of α_1 and α_2 have a common part. A careful analysis of possible situations shows that after both loadings the branching node will have an outcoming primary a-edge and all other nodes of the common part will have outcoming secondary a-edges going to the same node as the primary one does. This suggests the principle that *common parts of the suffix chains of nodes extended by the same letter can be treated once.*

It is not too difficult to see how this principle can be implemented. The simplest way is to perform the traversal in two passes. In the first pass the suffix chain of every port node is traversed and the visited nodes are marked with the letter to be appended to the port node. The traversal stops if the node is already marked with the same letter. This node is additionally marked as a *branching node*. In the second pass the loading process is performed using the marking so that the common parts (delimited by branching nodes) are traversed once. A more complicated task is to prove that the above principle preserves linearity. The idea of the proof is to amortize all suffix chain traversals over the *total length of all keywords under loading*. In this way we show that the total time taken by loading keywords during the matching algorithm is $O(|P|)$.

If the auxiliary dynamic tree structure of suffix pointers has to be maintained (Section 4.1), appending each letter requires additional $O(\log d)$ time, and the whole loading time is then $O(|P| \log d)$.

Summarizing the complexity of all stages, we state the following result :

Theorem 2 MATCH(t, P) *runs in time* $O((|t| + |P|) \log d)$.

5   Concluding Remarks

In this paper we have designed an efficient algorithm for matching a text against a set of patterns with variable length don't cares. Note that this problem can be considered as a generalization of the dynamic dictionary matching (DDM) problem [2, 3, 11, 4] in that the dictionary (underlying set of words) changes *during* the text search. In particular, the technique of using the DAWG as a matching automaton together with the algorithms of modifying the DAWG used in this paper, constitute yet another solution of the DDM problem, that matches the same complexity bounds as in [2, 3, 11]. Also, our method meets a similar obstacle as the DDM algorithms (see section 4.1), which gives rise to the $\log d$ factor in the complexity bound (theorem 2). The obstacle amounts to the detection of keywords that are prefixes (in case of [3, 4]) or suffixes (in our case) of a given subword of another keyword. In [4] a new solution to this problem was proposed, based on the reduction to the *parenthesis maintenance problem*, which improved the complexity bounds by the $\log \log d$ factor. This solution can be plugged into our algorithm, allowing a similar improvement.

Note that our algorithm detects the leftmost occurrence of the patterns in the text. This is an obvious drawback for those applications that require to find all occurrences. However, it can be easily seen that even for a single pattern the number of occurrences may be exponential, which makes impossible an efficient algorithm that outputs all of them. Note however that the number of occurrences may be computed in polynomial time using the dynamic programming technique.

References

[1] A. V. Aho. Algorithms for finding patterns in strings. In J. van Leeuwen, editor, *Handbook of Theoretical Computer Science*. Elsevier Science Publishers B. V. (North-Holland), 1990.

[2] Amihood Amir and Martin Farach. Adaptive dictionary matching. In *Proceedings of the 32nd Annual IEEE Symposium on Foundations of Computer Science, San Juan (Puerto Rico)*, pages 760–766. IEEE computer society press, October 1991.

[3] Amihood Amir, Martin Farach, Zvi Galil, Raffaele Giancarlo, and Park Kunsoo. Dynamic dictionary matching. To appear in Journal of Computer and System Sciences, June 1993.

[4] Amihood Amir, Martin Farach, Ramana M. Idury, Johannes A. La Poutré, and Alejandro A. Schäffer. Improved dynamic dictionary matching. In *Proceedings of the 4th Annual ACM-SIAM Symposium on Discrete Algorithms, Austin (TX)*, pages 392–401, January 1993. to appear in Information and Computation.

[5] Alan A. Bertossi and Filippo Logi. Parallel string matching with variable length don't cares. *Journal of parallel and distributed computing*, 22:229–234, 1994.

[6] A. Blumer, J. Blumer, D. Haussler, A. Ehrenfeucht, M. T. Chen, and J. Seiferas. The smallest automaton recognizing the subwords of a text. *Theoretical Computer Science*, 40:31–55, 1985.

[7] A. Blumer, J. Blumer, D. Haussler, R. McConnell, and A. Ehrenfeucht. Complete inverted files for efficient text retrieval and analysis. *Journal of the ACM*, 34(3):578–595, July 1987.

[8] Maxime Crochemore. Transducers and repetitions. *Theoretical Computer Science*, 45:63–86, 1986.

[9] Maxime Crochemore. String matching with constraints. In *Proceedings International Symposium on Mathematical Foundations of Computer Science*, volume 324 of *Lecture Notes in Computer Science*, pages 44–58. Springer-Verlag, 1988.

[10] Michael J. Fisher and Michael S. Paterson. String-matching and other products. In R. M. Karp, editor, *Complexity of Computation*, volume 7 of *SIAM-AMS Proceedings*, pages 113–125. American Mathematical Society, Providence, RI, 1974.

[11] Ramana M. Idury and Alejandro A. Schäffer. Dynamic dictionary matching with failure functions. *Theoretical Computer Science*, 131:295–310, 1994.

[12] G. Kucherov and M. Rusinowitch. Complexity of testing ground reducibility for linear word rewriting systems with variables. In *Proceedings 4th International Workshop on Conditional and Typed Term Rewriting Systems, Jerusalem (Israel)*, July 1994. to appear in the LNCS series.

[13] Udi Manber and Ricardo Baeza-Yates. An algorithm for string matching with a sequence of don't cares. *Information Processing Letters*, 37:133–136, 1991.

[14] Gene Myers. A four russians algorithm for regular expression pattern matching. *Journal of the ACM*, 39(4):430–448, April 1992.

[15] Ron Y. Pinter. Efficient string matching with don't-care patterns. In A. Apostolico and Z. Galil, editors, *Combinatorial Pattern Matching*, volume F12 of *ASI Series*, pages 11–29. Springer-Verlag, 1985.

[16] Daniel D. Sleator and Robert Endre Tarjan. A data structure for dynamic trees. *Journal of Computer and System Sciences*, 26:362–391, 1983.

[17] Esko Ukkonen. On-line construction of suffix-trees. Report A-1993-1, University of Helsinki, Department of Computer Science, February 1993.

A0 Example of a DAWG

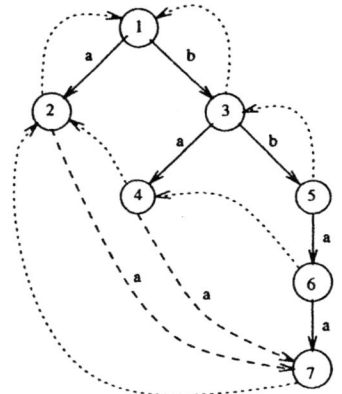

This is the DAWG for the set $\{ba, bbaa\}$. Nodes 1-7 correspond respectively to the equivalence classes $\{\varepsilon\}$, $\{a\}$, $\{b\}$, $\{ba\}$, $\{bb\}$, $\{bba\}$, $\{aa, baa, bbaa\}$. Primary edges are drawn with normal arrows, secondary edges with dashed arrows, and suffix pointers with dotted arrows. Depth of nodes 1-7 is 0, 1, 1, 2, 2, 3, 4 respectively. Nodes 3, 4, 5, 6, 7 are prefix nodes with prefix-degree(3) = 2 and prefix-degree(4) = prefix-degree(5) = prefix-degree(6) = prefix-degree(7) = 1. Nodes 4 and 7 are terminal nodes.
number-of-children(1) = number-of-children(2) = more-than-one,
number-of-children(3) = number-of-children(4) = 1,
number-of-children(5) = number-of-children(6) = number-of-children(7) = 0

A1 Extending the DAWG

APPEND-LETTER($activenode, a$)

```
1      if out(activenode, a) ≠ undefined then
2          if type(activenode, a) = primary then
3              return out(activenode, a)
4          else return SPLIT(activenode, out(activenode, a))
5      else create a new node newactivenode and set number-of-children(newactivenode) := 0
6          create a new primary a-edge (activenode, newactivenode) and set
               origin(newactivenode) := activenode, last-letter(newactivenode) := a,
               depth(newactivenode) := depth(activenode) + 1
7          varnode := suf-pointer(activenode)
8          while varnode ≠ undefined and out(varnode, a) = undefined do
9              create a new secondary a-edge (varnode, newactivenode)
10             varnode := suf-pointer(varnode)
11         if varnode = undefined then
12             create a suffix pointer from newactivenode to source
13         elseif type(varnode, a) = primary then
14             create a suffix pointer from newactivenode to out(varnode, a)
15         else create a suffix pointer from newactivenode
                                 to SPLIT(varnode, out(varnode, a))
16         return newactivenode
```

SPLIT($originnode, targetnode$)

```
1      create a new node newtargetnode
2      appendedletter := last-letter(targetnode)
3      replace the secondary edge (originnode, targetnode) by a primary edge
           (originnode, newtargetnode) with the same label and set
           origin(newtargetnode) := originnode, last-letter(newtargetnode) := appendedletter,
           depth(newtargetnode) := depth(originnode) + 1
4      instructions are added at this line in section 4.2
5      for every outcoming edge of targetnode, create a secondary outcoming
               edge of newtargetnode with the same label and going to the same node
6      create a suffix pointer of newtargetnode pointing to suf-pointer(targetnode)
7      redirect the suffix pointer of targetnode to point to newtargetnode
8      varnode := suf-pointer(originnode)
9      while varnode ≠ undefined and type(varnode, appendedletter) = secondary do
10         redirect the secondary edge of varnode (labeled by appendedletter)
                                 to point to newtargetnode
11         varnode := suf-pointer(varnode)
12     return newtargetnode
```

LOAD-KEYWORD($v = v_1 \ldots v_n$)

```
1      activenode := source
2      for i := 1 to n do
3          activenode := APPEND-LETTER(activenode, vi)
4          prefix-degree(activenode) := prefix-degree(activenode) + 1
```

A2 Reducing the DAWG

DELETE-LETTER(*activenode*)
1   *newactivenode* := **origin**(*activenode*)
2   *deletedletter* := **last-letter**(*activenode*)
3   **if** prefix-degree(*activenode*)>0 **or** number-of-children(*activenode*) = more-than-one **then**
4   **return** *newactivenode*
5   **elseif** number-of-children(*activenode*) = 1 **then**
6   MERGE(*activenode*, *newactivenode*, *deletedletter*)
7   **return** *newactivenode*
8   **else** delete the primary edge (*newactivenode*, *activenode*) labeled by *deletedletter*
9   instructions are added at this line in section 4.2
10   *varnode* := **suf-pointer**(*newactivenode*)
11   **while** *varnode* ≠ **undefined** and **type**(*varnode*, *deletedletter*) = **secondary do**
12   delete the secondary edge (*varnode*, *activenode*) labeled by *deletedletter*
13   *varnode* := **suf-pointer**(*varnode*)
14   delete *activenode*
15   **if** *varnode* ≠ **undefined then**
16   *suffixnode* := **out**(*varnode*, *deletedletter*)
17   **if** prefix-degree(*suffixnode*) = 0 **and** number-of-children(*suffixnode*) = 1 **then**
18   MERGE(*suffixnode*, *varnode*, *deletedletter*)
19   **return** *newactivenode*

MERGE(*targetnode*, *originnode*, *deletedletter*)
1   *newtargetnode* := **child**(*targetnode*)
2   instructions are added at this line in section 4.2
3   replace the primary edge (*originnode*, *targetnode*) labeled by *deletedletter*
         by a secondary edge (*originnode*, *newtargetnode*) with the same label
4   *varnode* := **suf-pointer**(*originnode*)
5   **while** *varnode* ≠ **undefined** and **type**(*varnode*, *deletedletter*) = **secondary do**
6   redirect the secondary edge (*varnode*, *targetnode*) labeled by *deletedletter*
                                   to *newtargetnode*
7   *varnode* := **suf-pointer**(*varnode*)
8   redirect the suffix pointer of *newtargetnode* to point to **suf-pointer**(*targetnode*)
9   delete the suffix pointer of *targetnode*
10   delete *targetnode*

UNLOAD-KEYWORD(*terminalnode*)
1   *activenode* := *terminalnode*
2   **while** *activenode* ≠ *source* **do**
3   prefix-degree(*activenode*) := prefix-degree(*activenode*) − 1
4   *activenode* := DELETE-LETTER(*activenode*)

Three-Dimensional Pattern Matching in Protein Structure Analysis

Arthur M. Lesk

University of Cambridge Clinical School, MRC Centre, Hills Road,
Cambridge CB2 2QH, United Kingdom

Abstract. Many pattern-matching problems that arise in "one dimension" in the analysis of genomic sequences have three-dimensional analogs in the analysis of protein structures. This report focuses on the identification and matching of common substructures, and treats two problems: the probing of a database of structures with a segment of a protein to identify regions from other proteins with conformations similar to that of the probe, and the determination of the maximal common "rigid subunit" in comparing alternative conformations of a single protein. Approaches based on the representation of structures in terms of lists of coordinates or as a distance matrices are compared.

1 Introduction

Each protein structure contains a chain folded into a space curve; the course of the chain and the interactions of separated regions of the chain define a topology or "folding pattern." The classification of the topologies of folding patterns, the integration of each newly-reported structure into the scheme of things, and the discovery of general principles of protein architecture are foci of current research. Upon the arrival via the internet of newly solved structures, we are like the comparative anatomists of 200 years ago who upon the return to Europe of expeditions of exploration would receive a set of specimens of exotic animals and plants.

The amount of data is growing quickly indeed: the Protein Data Bank currently comprises 3500 entries, each entry typically containing coordinates of 2000 atoms [4]. This has motivated the search for efficient algorithms for the analysis of the structures. In order to suggest the flavour of the activity in the field, we will focus on two related problems, treatable by complementary methods: (1) detection of a structural pattern in a large structure or set of structures, and (2) identification of the maximum common substructure of two proteins. The first problem has applications to the general statistical analysis of common conformational patterns in the database, and to particular model-building tasks in protein engineering; and the second is useful in analysing the differences between related structures, or between two different conformational states of a single molecule.

Approaches to these problems can be based on representations of the structural information by sets of coordinates, or by distance matrices. (The $(i, j)^{th}$

element of the distance matrix is the distance between points i and j.) Many algorithms for complex problems are built from certain basic operations that can be applied to coordinate sets or distance matrices, including the determination of the best "fit" of two structural fragments in the least-squares sense, or the matching of submatrices of distance matrices. These are discussed in the next section.

2 Elementary operations

Structural superposition

As in the case of sequences, a fundamental question in analyzing structures is to devise and compute a measure of similarity. If two structures are represented by point sets:

$$p_i = (x_i, y_i, z_i), i = 1, \ldots n \quad \text{and} \quad q_i = (x'_i, y'_i, z'_i), i = 1, \ldots n$$

a suitable measure of similarity is the root-mean-square deviation Δ of corresponding atoms after optimal superposition. Because the most general motion of a rigid body is a combination of a rotation $\mathbf{R}$ and translation $\mathbf{t}$:

$$\Delta^2 = \min_{\mathbf{R},\mathbf{t}} \{ \sum_{n=1}^{n} |\mathbf{R}p_i + \mathbf{t} - q_i|^2 \}$$

where $\mathbf{R}$ is a proper rotation matrix ($\det \mathbf{R} = 1$) and $\mathbf{t}$ is a translation vector.

This calculation is straightforward provided we know the correspondence between the atoms. Several solutions are known, including those based on the Polar or Singular Value decompositions of the correlation matrix $\sum p_i q_i^\dagger$, or on quaternions (see [14]; [22]). An important feature of these methods is that they produce a *global* optimum, and cannot get trapped in local minima.

Distance matrices and their subsets

Distance matrices are an alternative to lists of coordinates for representing the structure of an ordered set of points. The advantage of distance matrices is that they are independent of position and orientation, so that comparison of two structures does not require solving the superposition problem discussed in the previous paragraph. Their disadvantage is that one has lost structural information because any set of points has the identical distance matrix as its enantimorph (that is, left- and right-handed versions of the same structure have the same distance matrices). Nevertheless, given a complete and exact distance matrix, it is possible to recover the coordinate set, up to the enantimorph ambiguity [8]. Therefore, two structures with identical distance matrices must be either identical or mirror images. What is more, in practice the enantiomorph ambiguity is rarely troublesome and structures with similar distance matrices have similar conformations (see [28, 25]).

A submatrix of a distance matrix created by crossing out certain rows and the corresponding columns is the distance matrix of a subset of the original set of points. The extraction of the maximal common substructure of two structures corresponds to determining the maximal common submatrices of their distance matrices. This is equivalent to the "maximal clique" problem of graph theory. Certain calculations involving matching of submatrices of distance matrices can be expressed as linear Boolean programming problems (see section 5).

3 Classification of structural pattern-matching problems

Many pattern-matching problems in structural analysis can be built up from the elementary operations. They generally fall into three categories, of increasing difficulty:

• Determining the similarity of two sets of n atoms with known correspondences:

$$p_i \longleftrightarrow q_i, i = 1, \dots n.$$

The analog, for sequences, is the Hamming distance: mismatches only. This can be solved directly by elementary operations as described in the previous section.

• Determining the similarity of two sets of atoms (containing n and m points) with unknown correspondences, but for which the molecular structure – specifically the linear order of the residues – restricts the possible correspondences to those that retain order along the chain:

$$p_{i(k)} \longleftrightarrow q_{j(k)}, k = 1, \dots K \leq n, m$$

with: $k_1 > k_2 \Rightarrow i(k_1) > i(k_2)$ and $j(k_1) > j(k_2)$. This corresponds to the Levenshtein distance, or to sequence alignment with gaps.

• Determining the similarities between two sets of atoms with unknown and unrestricted correspondence:

$$p_{i(k)} \longleftrightarrow q_{j(k)}$$

This problem arises in the following important case: Suppose two (or more) molecules have similar biological effects, such as a common pharmacological activity. It is often the case that the structures share a common constellation of a relatively small subset of their atoms that are responsible for the biological activity, called a *pharmacophore*. The problem is to identify them: to do so it is useful to be able to find the maximal subsets of atoms from the two molecules that have a similar structure. In the case of proteins, this type of problem would arise in detection of similar substructures produced by proteins of different topology; that is, when two substructures are similar but the correspondence between atoms does not obey the constraint of the preceding paragraph. Alt, *et al.*, [1] presented an $O(n \ln n)$ algorithm to solve this problem; see also Willett [30].

Here we will consider two problems: (1) identification of probe structures in a large data base, and (2) determination of the maximal common substructure of two structures.

4   Searching for regions similar in structure to a probe

The mainchain conformation of a protein is defined by the positions of four consecutive atoms N, Cα, C and O, in each residue. The problem of substructure searching is as follows: Given the coordinates of the mainchain atoms of a probe region, residues i through j (in many cases of interest $j - i \sim 10$) and a database of protein structures each of length $\gg |j - i|$, find all regions in the database with conformations similar to that of the probe, as measured by the size of the r.m.s. deviation Δ. Two versions of this problem are deceptively similar: (1) Find *some* examples of substructures similar to the probe, and (2) Find *all* examples that match the probe to within some specified threshold on Δ. The second is considerably harder. Numerous practical "tricks" are effective for the first problem but do not address the second.

It is a straightforward generalisation to assign to each atom in the target segment a nonnegative weight. Using weights consisting of blocks of 1's and 0's, one can search for two or more regions separated by fixed numbers of residues of unspecified conformation, *i.e.*, "wild-cards". This version is useful in searching a data base for short segments of chain that span fixed endpoints.

Approaches to the problem of substructure searching fall into several categories:

1. The conformations of individual residues in a protein can be classified into discrete states. Therefore the mainchain conformation of the entire protein may be written as a character string specifying the sequence of residue conformations [19, 23]. Alternatively, the conformation may be characterized by a sequence of sets of distances between a point within each residue (*e.g.*, the α-Carbon position) to corresponding points in other residues nearby in the sequence [24]. These formulations reduce the problem to that of aligning strings.

2. In their original work, Jones and Thirup [18] used the matrix of distances between Cα positions of a fragment to characterise the structure of the fragment. They used similarity in distance matrices as a sieve to select candidates for optimal superposition calculations to identify structures with low root-mean-square deviations from the target structure. The assumption is that similarity in interatomic distances implies similarity in structure. This is not true, if only because distance matrices do not determine the enantiomorph, but it nevertheless provides a useful practical approach.

Besides the enantimorph ambiguity, there is no rigorous relationship between deviations of distance matrices and r.m.s. deviations. Maiorov and Crippen [25] have published empirical relationships which vary from structure to structure.

3. A geometric hashing technique originally developed for object recognition in computer vision has been applied to protein structure comparison [2, 9].

4. Sequential searching through the data base. This "brute force" approach can be speeded up by applying lower bounds to the r.m.s. deviation of point sets [29]. We shall discuss this approach in more detail here. Our work has dealt exclusively with the more stringent formulation of the substructure search problem: to find *all* – rather than just *some* – substructures similar to a probe structure.

There are several sources of lower bounds to the r.m.s. deviation of two point sets. First, the typical experimental accuracy in the coordinates implies that we need compute Δ to only two or three significant figures. If one is using an iterative procedure for computing the eigenvalues or singular values of a matrix, Gershgorin bounds can provide useful termination criteria based on the desired accuracy of the result. Second, in screening a data bank, it is often satisfactory to treat the following problem: Given a probe structure, determine all regions in proteins of known structure that match the probe structure within a root-mean-square deviation $\Delta \leq M^2$, where M^2 is a prespecified threshold. A cheap-to-compute lower bound to Δ, even if not a sharp one, avoids many explicit determinations of Δ. One such bound is derived in [29]:

$$n\Delta^2 \leq \sum_i (|p_i| - |q_i|)^2 = \sum_i (|p_i|^2 + |q_i|^2) \cdot \{1 - [4|p_i|^2|q_i|^2/(|p_i|^2 + |q_i|^2)^2]^{\frac{1}{2}}\}.$$

Here n is the number of points, and the probe structure p_i and the fragment of the database q_i are assumed to be translated to their respective centers of gravity. Note that calculation of $|q_i|$ in the first inequality requires a square root calculation, but that the second inequality contains only $|q_i|^2$. The final expression is convenient for numerical computation because $4|p_i|^2|q_i|^2/(|p_i|^2 + |q_i|^2)^2$ is always ≤ 1, so that a table-look-up procedure may replace the computation of the square root.

The sharper the bound imposed on Δ, the more segments it will reject. In a benchmark calculation searching a large set of protein structures for a 7-residue region from the antibody fragment McPC603, setting the bound $\Delta \leq 1.0$ Å led to the exclusion of 98.4% of potential matches, whereas $\Delta \leq 2.0$ Å rejected only 46% of the potential matches [29]. In practical cases, the threshold value 1.0 Å is the appropriate one (see [21], p. 131) and therefore the thresholding is an effective way of reducing the computational cost of the superposition calculations.

Nevertheless, this approach still requires the examination of every region of every structure. Preferable would be methods in which candidate matches could be brought together in a list. One way to accomplish this without losing any valid matches is to make use of other bounds to Δ available from the moment invariants of the pattern derived by Guo [13]. We assume without loss of generality that the mean position of of both sets of points p and q is $(0, 0, 0)$. The first of Guo's invariants is the second moment of the point set. If $\mu_2(p) = \sum |p_i|^2$ is the second moment of the set of points p_i, then

$$\Delta^2 \geq [(|\mu_2(p) - \mu_2(q)| + \mu_2(p))^{\frac{1}{2}} - \mu_2(p)^{\frac{1}{2}}]^2.$$

To show this, suppose $p_i = q_i + \Delta_i$. Then

$$p_i \cdot p_i = q_i \cdot q_i + 2\Delta_i \cdot q_i + \Delta_i^2$$

$$\sum_i (p_i \cdot p_i - q_i \cdot q_i) = \sum_i \Delta_i^2 + 2\sum_i \Delta_i \cdot q_i$$

Letting $\quad \Delta^2 = \sum_i \Delta_i^2, \mu_2 = \sum_i p_i \cdot p_i, \quad$ and $\quad \mu_2' = \sum_i q_i \cdot q_i,$

$$\mu_2' - \mu_2 = \Delta^2 + 2(\Delta^2 \mu_2)^{\frac{1}{2}} \frac{\sum_i \Delta \cdot p_i}{[(\sum_i \Delta_i)^2 \sum_i (q_i \cdot q_i)]^{\frac{1}{2}}}$$

Then

$$|\mu_2' - \mu_2| \le \Delta^2 + 2|(\Delta^2 \mu_2)|^{\frac{1}{2}} \times \left| \frac{\sum_i \Delta \cdot p_i}{[(\sum_i \Delta_i)^2 \sum_i (q_i \cdot q_i)]^{\frac{1}{2}}} \right|$$

$$\le \Delta^2 + (4\Delta^2 \mu_2)^{\frac{1}{2}} = (|\Delta^2|^{\frac{1}{2}} + \mu_2^{\frac{1}{2}})^2 - \mu_2.$$

Hence:

$$|\mu_2' - \mu_2| + \mu_2 \le (|\Delta^2|^{\frac{1}{2}} + \mu_2^{\frac{1}{2}})^2$$

and

$$[|\mu_2' - \mu_2| + \mu_2]^{\frac{1}{2}} \le |\Delta^2|^{\frac{1}{2}} + \mu_2^{\frac{1}{2}}$$

which is equivalent to:

$$\Delta^2 \ge [(|\mu_2' - \mu_2| + \mu_2)^{\frac{1}{2}} - \mu_2^{\frac{1}{2}}]^2$$

To apply this result, suppose that p is the probe structure, and q is a candidate match, and that we are interested only in matches for which $\Delta^2 \le |M|^2$. Then we can calculate $\mu_2(p)$, and prepare a sorted list of $\mu_2(q)$ for all possible q in the data base. Only that slice of the list for which

$$|\mu_2(q) - \mu_2(p)| \le [|M| + \mu_2(p)^{\frac{1}{2}}]^2 - \mu_2(p)$$

need be considered.

As an example we consider the same benchmark problem treated by Rustici and Lesk, a seven-residue loop from the antibody McPC603 [29]. Taking four mainchain atoms per residue, the probe object contained 28 points, and had a second moment around its center of gravity of 462.2 Å^2. For this value of $\mu_2(p)$, the preceding equation gives the values:

M	0.6	0.8	1.0	1.2	1.4	1.6	1.8	2.0
$|\mu_2(q) - \mu_2(p)|$	26.2	35.0	44.0	53.0	62.2	71.4	80.6	90.0

For each consecutive 7-residue region in the structure of McPC603, we calculated the second moment and $|\delta\mu_2| = |\mu_2(q) - \mu_2(p)| =$ the absolute value of the difference between the second moment of the region and the second moment of the probe structure. We also calculated $\Delta =$ the r.m.s. deviation of the region and the probe region. Sorting the results in order of increasing $\delta\mu_2$, the top values were:

$|\delta\mu_2|$	Δ	$|\delta\mu_2|$	Δ	$|\delta\mu_2|$	Δ	$|\delta\mu_2|$	Δ
0.00	0.00	22.28	2.05	50.26	3.08	75.26	2.54
1.16	2.02	24.86	2.00	50.82	2.15	78.57	2.55
1.89	2.67	30.84	2.31	52.17	1.89	82.41	2.21
10.58	1.78	35.97	2.92	59.81	2.08	98.84	2.40
13.08	2.24	37.67	3.51	66.28	1.83		
17.64	1.65	41.94	2.16	70.50	2.06		
17.75	1.79	44.89	3.98	71.14	2.14		

Comparing these tables, it is clear that if the threshold $M \leq 1.0$ there are 20 candidates not excluded on the basis of second moments, one of which is the target region itself, and no others fit within $\Delta = 1.0 \overset{\circ}{A}$. If the threshold $M \leq 2.0$, a value we consider unrealistically high, there are 24 candidates. As there are a total of 229 regions in the entire coordinate set of McPC603, we have not even considered 94% of the possibilities, for $M = 1.0$. (Of the 20 candidates, all but 6 would be excluded by the previous test developed by Rustici and Lesk [29].) Again we emphasize that the advantage of using the second moment test as a screen over alternatives such as end-to-end distance is that we have shown that there can be no exceptions.

We are currently working to apply the other moment invariants given by Guo, in order to sharpen the screen.

It is of interest to ask how this problem could be treated alternatively using distance matrices. Suppose the probe has length n and we wish to search a structure of length N. The distance matrix of the probe is of dimension $n \times n$ and we must look for matches by sliding it along the band of width n around the diagonal of a distance matrix of dimension $N \times N$. (Of course only elements above the diagonal need be considered, as the matrices are symmetric.) Candidate "hits" identified by this screen can then be fitted to the probe structure using the superposition techniques. In practice, subsets of the distance matrix can provide efficient screens. For instance, limiting the screen to a test of the difference between the $(1, n)$ element of the distance matrix of the probe and the $(i, i+n-1)$ element of the distance matrix of the structure being searched amounts to a screen on similar end-to-end distances.

5 Analysis of conformational change in proteins

Many proteins can exist in two or more conformations, and in some cases changes in structure are essential components of the mechanism of function of the molecule [11]. An important class of conformational changes is a kind of "hinge motion" in which two regions of the protein move, as rigid bodies, between different relative positions. A typical example is the enzyme citrate synthase, which can exist in "open" and "closed" forms; in this case the "open" form has a cleft between two domains that can receive ligands, and in the closed form the domains have moved together to close the cleft over the bound substrate and cofactor.

Here we consider the extraction of the largest similar substructure as an application of distance matrices. This approach was introduced and developed by Nichols, Rose, TenEyck and Zimm [26], who present algorithms and numerical examples.

Statement of the problem: $(p_i, i = 1, \cdots, n)$ and $(q_i, i = 1, \cdots, n)$ are two sets of coordinates, where p_i and q_i each represent the x, y and z coordinates of a point. The sets contain the same number of points, and we assume that they correspond to each other consecutively; that is $p_i \longleftrightarrow q_i, i = 1, \ldots n$. We wish select subsets of points, that is, a set of indices $j(i), i = 1, \cdots m$, with $m \leq n$, such that the two subsets of points $p_{j(i)}$ and $q_{j(i)}$ are similar in structure. This will be an interesting question if the full sets of points p and q are not similar in structure, but subsets of them are.

In the field of protein structure, this arises in two situations: First, in two proteins related by evolution there is a common core of the structure that retains a similar geometry, which may comprise numerous regions separated by other regions of variable structure and length. In the absence of insertions and deletions, the procedures described would extract the common core. Second, a protein may exist in two conformational states – perhaps changing structure when it binds to a small molecule – in which several large pieces of the protein move as rigid bodies. Extraction of the maximum common substructure would identify the largest of these rigid units; application of the procedure again to what is left would identify the next largest rigid unit disjoint from the first, and this could be continued as far as desired.

Definitions: The **distance matrix** $D(p)$ of a set of points p is defined by:

$$D_{ij}(p) = |p_i - p_j|.$$

The **difference distance matrix** $\Delta D(p, q)$ of two sets of points is the matrix of the absolute values of the differences between the distance matrices of the point sets p and q.

$$\Delta D_{ij}(p, q) = |D_{ij}(p) - D_{ij}(q)|.$$

The **thresholded difference distance matrix** $\Theta^T(p, q)$ is a matrix containing only 1's and 0's according to whether the corresponding distances are below or above the threshold T:

$$\Theta_{ij}^T(p, q) = \begin{cases} 1 & \text{if } \Delta D_{ij}(p, q) \leq T \\ 0 & \text{if } \Delta D_{ij}(p, q) \geq T \end{cases}$$

Note that $D(p)$, $\Delta D(p, q)$ and $\Theta^T(p, q)$ are all symmetric matrices.

An approach to the problem of determining the maximal common substructure of p and q is to ask how to cross out selected rows and the corresponding columns to leave a matrix containing only 1's. For the subsets of points remaining, the corresponding sets of pairwise distances differ by no more than T.

There are several possible approaches to this problem. Computer scientists will recognize it as equivalent to the problem of finding a maximal clique in a graph – let the atoms of the protein label the vertices of a graph, and let any two vertices be connected iff the distance between the atoms is less than the threshold; then one seeks the largest subset of nodes for which each pair is connected by edges. This is a well-studied problem in computer science, and is known to be NP-complete [27]. Willett and his colleagues have explored the application of graph theory to problems of protein structure (see [16, 15]).

It is known that the problem can also be cast in the form of a linear Boolean programming problem [3, 5, 27, 6]. We suggest that there are a number of advantages of this approach. The first is that algorithms are known that find global optima (of course the optima may not be unique). Programs based on these algorithms have acceptable running times on contemporary equipment – see section

7. Secondly, it is fairly easy to translate additional desirable features of the solution, suggested by what we know about the molecular aspects of the problem, into algebraic expressions that fit cleanly into the formalism. Here is an example: What we seek in comparing two protein structures is a set of substantial regions of similar structure. Occasionally large structural changes may bring odd residues to the same position in space by accident, and this is really "noise" in the structural comparison. It is shown in what follows that it is easy to add constraints to the Boolean programming problem that exclude regions of length one. In this way the formulation can be considered as a kind of workbench in which the effect of different constraints suggested by molecular considerations can be explored.

We have focused on the question of finding the maximal common substructure, and have noted that it may not be unique. Suppose that the maximal common substructure of two point sets contains N elements. One may wish to ask for all common substructures (in the sense of non-zero subsets of the difference distance matrix) containing at least $K \leq N$ elements. These can be found by solving a set of simultaneous linear Boolean inequalities, a problem related to the linear Boolean programming problem, for which algorithms are also known and for which software is available [17, 20].

6 Formulation of the maximal common substructure or maximal clique problem as a linear Boolean programming problem

Let $x_i, i = 1, \cdots n$ be a binary vector: $x_i = 0$ or 1. The interpretation will be that $x_i = 1$ means that point i is a member of the selected subset of points, and $x_i = 0$ means that i is not a member of the subset. To maximize the number of elements in the subset, maximize $\sum x_i$. Next, consider the product of the thresholded difference distance matrix Θ with the vector x. Each element of the product cannot exceed the number of values of x that are 1, that is, $\sum x_i$. It can equal this value only if the element in every column of Θ that corresponds to a 1 value of x is also 1. Therefore for each j such that $x_j = 1$, we must demand that the jth element of $\Theta x = \sum x_i$. If however $x_j = 0$ we need impose no constraints on the jth element of Θx.

This is equivalent to the problem:

$$\text{Maximise} \sum x_i,$$

subject to the constraints:

$$x_i = 0 \text{ or } 1$$

$$(\Theta x)_j \geq \sum x_i \text{ for each } j \text{ such that } x_j = 1$$

We can rewrite the last condition as:

$$(\Theta x)_j \geq \sum x_i - C(1 - x_j) = \sum x_i + Cx_j - C,$$

where C is a constant larger than the greatest possible value of $\sum x_i$; for instance take $C = 2n$. If $x_j = 1$ the condition reduces to $(\Theta x)_j \geq \sum x_i$. If $x_j = 0$ the condition reduces to $(\Theta x)_j \geq \sum x_i - C$, but since $(\Theta x)_j \geq 0$ and $\sum x_i - C < 0$ for proper choice of C the condition will be satisfied for any values of the x_i.

We can thereby formulate the problem as a standard linear Boolean programming problem:

$$\text{Maximise} \sum x_i,$$

subject to the constraints:

$$x_i = 0 \text{ or } 1, \quad Ax + b \geq 0,$$

where:

$$A_{ij} = \Theta^T_{ij} - C\delta_{ij} - 1, \text{ and } b_j = C \text{ for all } j.$$

The Boolean programming formulation also provides a framework for tailoring the solution by adding additional constraints – provided that these can be expressed as linear inequalities involving the x_i. For instance, we may wish to exclude from the solution any "singletons" – that is, we don't want to include a residue if at least one of its immediate neighbors is not also included. A singleton will appear in the solution as a value of j such that $x_j = 1$ but $x_{j-1} = x_{j+1} = 0$, (with obvious modifications for $j = 1$ and $j = n$.) For this to be true, $x_j - x_{j-1} = 1$ and $x_j - x_{j+1} = 1$. The constraint to add is:

$$\text{for all } j, \quad 2x_j - x_{j-1} - x_{j+1} < 2$$

or, in the standard form:

$$\text{for all } j, \quad x_{j-1} - 2x_j + x_{j+1} + 1 \geq 0$$

To include these constraints one adds additional rows to the matrix A.

7 Results

Human antithrombin is a serine protease inhibitor that plays a crucial role in the control mechanisms of blood coagulation. In a recent crystal structure determination two conformations of the molecule were observed, called the "latent" and "native" states. The conformational changes in antithrombin and related molecules are essential to their function.

Extracting one atom per residue from these structures produced two sets of 402 points. These are relatively large structures.

We wrote a program for the linear Boolean programming problem based on the general implicit enumeration algorithm of Balas as formulated by Geoffrion [10]. Despite the fact that this is not the best available algorithm for large problems and that no special care was taken in the programming, the software runs in time acceptable for casual use. All computations were carried out on the Alliant FX-2800 at the MRC Laboratory of Molecular Biology, in single-processor mode.

For the two states of the serpins, we determined the maximal common substructures, excluding isolated residues, for thresholds in the range 0.5 Å– 1.2 Å. The linear Boolean programming problem had 402 variables and 804 constraints. The following table contains the results of these calculations.

Extraction of maximum common substructure from 402 $C\alpha$ atoms from the latent and native states of antitrypsin.

Threshold	0.5	0.6	0.7	0.8	0.9	1.0	1.1	1.2
Number of points fit	43	56	80	91	108	128	142	157
r.m.s. deviation (Å)	0.43	0.45	0.49	0.48	0.49	0.57	0.58	0.59
cpu time elapsed (s)	11.59	11.39	10.53	10.11	9.27	10.45	9.27	8.39

These results show that the extraction of a maximal common substructure from two conformations of a protein of this size is possible using this approach.

8   Conclusions

The analysis of protein structures requires the development of effective algorithms for substructure matching. Many of the problems that arise in practice can be formulated as problems of known mathematical structure, and it is therefore possible to identify state-of-the-art algorithms to apply.

Acknowledgements
I thank the Kay Kendall Foundation for generous support.

References

1. Alt, H., Melhorn, K., Wagener, H. and Welzl, E.: Congruence, similarity, and symmetries of geometric objects. Discrete Comput. Geom. **3**, 237–256 (1988)
2. Bachar, O., Fischer, D., Nussinov, R. and Wolfson, H.J.: A computer vision based technique for 3-D sequence independent structural comparison of proteins. Prot. Eng. **6**, 279–288 (1993)
3. Balas, E. and Yu, C.S. Finding a maximal clique in an arbitrary graph. SIAM J. Comput. 4 1054–1068 (1986).
4. Bernstein, F.C., Koetzle, T.F., Williams, G.J.B., Meyer, E.F. Jr., Brice, M.D., Rodgers, J.R., Kennard, O., Shimanouchi, T., Tasumi, M. The protein databank: A computer-based archival file for macromolecular structure. J. Mol. Biol. **112**, 535–542 (1977)
5. Bron, C. and Kerbosch, J. Algorithm 457: Finding all cliques of an undirected graph J. Assoc. Comp. Mach. **16**, 575–577 (1973)
6. Carraghan, R. and Pardalos, P.M. An exact algorithm for the maximum clique problem. Op. Res. Lett. **9**, 375–382 (1990)

7. Carrell, R. W., Stein, P. E., Fermi, G. and Wardell, M. R. Biological implications of a 3Å structure of dimeric antithrombin. Structure **2**, 257–270 (1994)
8. Crippen, G.M. and Havel, T.F. *Distance Geometry and Molecular Conformation.* New York: John Wiley and Sons, 1988
9. Fischer, D., Bachar, O., Nussinov, R. and Wolfson H.J. An efficient automated computer vision based technique for detection of three-dimensional structural motifs in proteins. J. Biomol. Str. Dyn. **9**, 769–789 (1992).
10. Geoffrion, A.M. Integer programming by implicit enumeration and Balas' method. SIAM Review 9 (1967) 178–190
11. Gerstein, M., Lesk, A. M. and Chothia, C. Structural mechanisms for domain movements in proteins. Biochemistry **33**, 6739–6749 (1994)
12. Gusfield, D. and Pitt, L. Equivalent approximation algorithms for node cover. Inf. Proc. Lett. **22**, 291–294 (1986)
13. Guo, X. Three dimensional moment invariants under rigid transformation. In, Computer Analysis of Images and Patterns, D. Chetverikov and W. G. Kropatsch (eds.). Springer-Verlag, Berlin, 1993, pp. 518–522
14. Golub, G. and Van Loan, C.F. Matrix Computations. 2nd Ed. Baltimore, The Johns Hopkins University Press, 1989, Chap. 12
15. Grindley H., Artymiuk P.J., Rice D. and Willett P. Identification of tertiary structure resemblance in proteins using a maximal common subgraph isomorphism algorithm. J. Mol. Biol. **229** 707–721 (1993).
16. Mitchell E.M., Artymiuk P.J., Rice D.W. and Willett P. Use of techniques from graph theory to compare secondary structure motifs in proteins. J. Mol. Biol. **212** 151–166 (1989).
17. Hammer, P. and Rudeanu, S. Boolean methods in operations research and related areas. New York, Springer-Verlag, 1968.
18. Jones, T.A. and Thirup, S. Using known substructures in protein model building and crystallography. EMBO J. **5**, 819–822 (1986)
19. Karpen, M.E., de Haseth, P.L. and Neet, K.E. Comparing short protein substructures by a method based on backbone torsion angles. Proteins: Structure, Function, Genetics **6**, 155–167 (1989)
20. Lesk, A.M. A FORTRAN program for the solution of simultaneous linear boolean inequalities by the algorithm of Hammer and Rudeanu J. Comp. Phys. 12 (1973) 150–152.
21. Lesk, A.M. Protein Architecture: A Practical Approach. IRL Press, Oxford, 1991.
22. Lesk, A.M. Computational Molecular Biology. In: Encyclopedia of Computer Science and Technology A. Kent and J.G. Williams, (eds.) New York, Marcel Dekker, Inc. 1994, Volume 31, pp. 101–165.
23. Levine, M., Stuart, D. and Williams, J. A method for systematic comparison of the three-dimensional structures of proteins and some results. Acta crystallographica **A40**, 600–610 (1984)
24. Liebman, M. N., Venanzi, C.A., Weinstein, H., Structural analysis of carboxypeptidase A and its complexes with inhibitors as a basis for modelling enzyme recognition and specificity. Biopolymers **24**, 1721–1758 (1985)
25. Maiorov, V.N. and Crippen, G. M. Significance of root-mean-square deviation in comparing three-dimensional structures of globular proteins. J. Mol. Biol. **235**, 625–634 (1994).
26. Nichols, W.L, Rose, G.D., Ten Eyck, L.F. and Zimm, B.H. Rigid Domains in Proteins: An Algorithmic Approach to their Identification. Proteins, in press (1995).

27. Parker, R.G. and Rardin, R.L. Discrete Optimization. Academic Press, New York, 1988.
28. Pastore, A., Atkinson, R.A., Saudek, V. and Williams, R.J.P. Topological mirror images in protein structure computation: an underestimated problem. Proteins **10**, 22–32 (1991).
29. Rustici, M. and Lesk, A.M. Three-dimensional searching for recurrent structural motifs in databases of protein structures. J. Comp. Biol. **1**, 121–132 (1994)
30. Willett, P. Three-Dimensional Chemical Structure Handling. Research Studies Press, Taunton, Somerset, U.K. (1991)

Genome Analysis:
Pattern Search in Biological Macromolecules

H.W. Mewes & K. Heumann

Max-Planck-Inst. f. Biochemie, 82152 Martinsried, Germany

Abstract. Biological sequence data analysis has developed into an inevitable tool for macromolecular biology, key to any detailed understanding of the living cell. A brief survey on the biological macromolecules and their function is given. Sequence data analysis is introduced as a basic tool for the experimental bench biologist. So far, most queries for such analyses are issued on flat files and static indices. We discuss position tree structures and their potential in sequence data analysis. The hash position tree is introduced as a persistent, dynamic data structure for pattern searches in large sequence databases in biology.

1 Introduction

Biology followed physics and chemistry and reached atomic resolution to gain insight into living matter at the molecular level. After the introduction of the basic methodologies, the exhaustive analysis of complex biological systems becomes feasible. Sequences of biological macromolecules, their three dimensional structure and their physico-chemical properties determine the functionality of any organism. The current level of knowledge in the biosciences makes its industrial exploration practicable. Biotechnology has become a rapidly growing, important economic sector. Progress in health research, particularly in the diagnosis and therapy of genetic defects and carcinoma, raises hopes for direct treatment on the molecular level based on the understanding of the functionality of biological macromolecules. These nascent developments will have a serious impact on the social, ethical, ecological and economic developments in the future.

Biological systems are highly complex when expressed in terms of chemistry and the physico-chemical dimensions: several hundred reactants, thousands of different products and many thousands of different molecules occur at the same time in a highly structured, multi-phase space. While biochemists revealed metabolic pathways, its intermediate steps and the biological catalysts (enzymes) involved, molecular biology has opened the ability to analyze the genetic information. It makes it possible to build the dictionary of cellular components and to start the detailed analysis of the functions of all individual elements. These functions cover the complete spectrum of activities, including reproduction, energy metabolism, structural elements, active transport, signal transduction, communication, protein synthesis and degradation, biosynthesis of organic compounds, DNA repair and many others.

Classical experiments in biology were designed to answer specific questions. For example, how is the energy stored in a carbohydrate source converted into cellular fuel (the energy rich molecule ATP)? What is the three-dimensional structure of the human oxygen binding protein (hemoglobin)? While trying to identify and understand the properties of every genetic element, the systematic analysis of genomic information applies an entirely different paradigm. It tries to read the blueprint of life; the elements are known, but not the instruction language.

Genetic elements are entities that behave as objects with a variety of disjunct properties: e.g. coding regions (translatable into proteins), regulatory elements (promotors to control biosynthesis), and replication start sites. Genome analysis generates raw data first; the sequence of the four nucleotides in DNA is not informative per se. Only the application of the set of rules, valid in the particular species of its origin, allow us to draw conclusions about its properties. The knowledge of the translation start and stop signals and the applied genetic code allows the artificial translation into a hypothetical protein. This process might immediately lead to a model of its function and even of its three dimensional structure: If a protein of a closely related species is known and its structure experimentally determined, one can assume that structure and function of the homologous proteins are similar as the consequence of a common evolutionary inheritance.

As the most important biological macromolecules, nucleic acids and proteins are built from a limited set of building blocks. Genomes contain 4 nucleotides, the equivalent alphabet for proteins is composed of 20 amino acids. However, the three dimensional structure of proteins varies widely as a result of the flexibility of the peptide bond that builds the backbone of the amino acid chain. This chain might also be subject to additional posttranslational processing which might include modifications of aminoacid side chains, the formation of chemical bonds between distant amino acids or specific proteolytic processing of precursor molecules. Protein-protein interactions further modify the functional properties by forming complexes of homo- and heterologous multimers.

The very first biological sequences were determined by chemical degradation [1] and subsequent analysis of N-terminal amino acids in proteins in 1953 [2,3]. Due to incomplete chemical degradation, sensitivity of the method is limited and the length of the determined sequences surpassed 40-50 amino acid residues only under ideal conditions. After more than 20 years of intense laboratory work, several hundred independent sequences were published [4]. The basic deficiencies of chemical protein sequencing were overcome by the introduction of DNA sequencing [5] which allowed the reliable generation of sequence data supported by highly sensitive analytical equipment linked to sophisticated, computerized data processing devices. Thousands of laboratories worldwide apply DNA sequencing technology today, generating sequence data between a few thousand and several hundred thousand base pairs per year. The current volume of nucleic acid sequences in the public databases is approximately 200 million bases (Mbases). Increase in information flow is rapid; the size of the databases doubles every 2.5 years. The evolution of DNA sequencing techniques has been paralleled by the technological revolution in microelectronics and computer manufacturing. Since the first attempts to apply computers to investigate evolutionary relations in the late 70s [6], 'biocomputing' has become an inevitable tool in the daily work of the molecular biologist. Standard sequence data analysis packages have been developed [7,8] and major public databases for nucleic acid and protein sequences have been established [9,10,11,12,13].

2 Systematic Analysis of Genomes

The systematic exploration of genomic information raised public discussions when the Human Genome Initiative in the United States launched its program to analyze the 3.3 billion base pairs in man. However, the size of the task and the lack of a suitable international organization to set up its realization rendered the project impractical with respect to the given means and technology. The first approach to analyze the entire genome of a model organism was initiated by A. Goffeau, who presented a feasibility study on 'Sequencing the Yeast Genome' in 1988 to the European Commission [14]. Following this proposal, a sequencing network was established in Europe, which completed the sequence of the first eukaryotic chromosome in 1992 [15] followed by the publication of chromosomes XI and II in 1994 [16,17,18]. The 16 chromosomes of the yeast genome are expected to be completed by 1997 in a joint effort of European, Canadian, American and Japanese laboratories[1]. In the meantime, systematic sequencing of several other model organisms has been launched: Bacillus subtilis (EU), Caenorhabditis elegans (UK/US), Arabidopsis thaliana (EU), and Oryza sativa (Japan). The size of these genomes ranges from the simple, unicellular bacterium B. subtilis of 4 Mbases to a simple, differentiating, multicellular worm (C. elegans) of 100 Mbases to plants (A. thaliana and O. sativa; 100, resp. 500 Mbases).

The first complete sequence of a yeast chromosome has proven the impact of systematic sequencing, the publication [15] has already become a citation classic. A significant part of the work to follow up the sequencing task has been dedicated to sequence data analysis (e.g. [19,20]). Systematic analysis of a newly sequenced gene starts with a number of basic questions:

- Is the sequence already known in the databases? Do differences reflect genetic variation or sequencing errors? What information is available from the databases (e.g. classification, motifs, features, secondary or three dimensional structure of derived proteins)?

- Does the sequence contain an open reading frame which is probably translated

---

[1] At the time of writing this manuscript, chromosomes I, II, III, V, VIII, IX, XI were accessible in the public databases.

into a protein? Does the deduced hypothetical protein show any significant similarities to proteins existing in the databases? Is it a new member of a known protein family? Does the protein contain patterns that have been characterized as functional sites? Is it possible to deduce the function of the protein from detailed sequence analysis? Is the sequence consistent with all characteristics of the known family members (e.g. conserved residues)?

- What are the characteristics of the non-coding regions surrounding the open reading frame? Is it possible to identify promotors, transcription factor binding sites or other genetic elements?

The interpretation of the yeast chromosomes published so far has proven that the public databases are obviously biased towards sequences that have been described earlier by biochemical studies or were subject of particular research interest (e.g. transcription factors, protein kinases, etc.). In case of the yeast chromosome III sequence, Koonin et al. [20] claimed to find 50% of the detected open reading frames to be similar to sequences of known function. However, detailed analysis shows [21] that only 30% of the proteins found in a yeast chromosome can be assigned with confidence to a well defined function. Although these similarities may help as a starting point for further experimental evaluation, the assignment to a certain class of proteins (e.g. transcription factor) remains incomplete and insufficient. Experimental work conducted by computer analysis will help to elucidate sequence properties in detail and achieve a clear detailed understanding of their functionality.

The functional analysis of the yeast genome also asks for tools to identify complex relationships between distant genetic elements. These relationships are often encoded on the genomic sequence as specific sequence patterns. Some functionally relevant sequence signals (e.g. regulatory elements, ARS-elements, sites of DNA-protein interaction, promotors, etc.) have been identified. However, the set of known sequence elements alone does not yet allow us to fully understand the mechanisms at play. Many relevant sequence patterns are yet undiscovered. The analysis of model organisms provides not only a template for the understanding of more complex, differentiated genomes. It also allows a great variety of genetic experiments to verify hypotheses based on computer analysis that cannot be undertaken in organisms of long generation times.

3 Computational Approaches in Sequence Data Analysis

Excellent introductions to the subject of sequence data analysis in biology have been published [21,22,23]. The basic analytical tool provided by the informatic approach to the molecular biologist is sequence comparison. If a new sequence can be aligned to a known one, it is possible to follow the fundamental paradigm 'similarity in sequence reflects similarity in three dimensional structure and structural similarity results in similar functional properties'. The concept of *homology* describes the fact of a reliable evolutionary relationship, i.e. the inheritance of the sequences from a common anchestor is assumed. Homology is always reflected by a significant sequence similarity. Information from well characterized proteins can be transferred to sequences related by homology. The methodology of sequence comparisons has been reviewed in the literature (e.g. [24,25]). The task to characterize structure/function relationships applies first of all to protein sequences, since evolutionary constraints apply to protein function as determined by the amino acid sequence and only indirectly to the nucleic acid equivalent.

Macromolecular sequences are commonly symbolized by strings of characters. Sequence alignments are performed by the comparison of pairs of amino acids. The scoring matrix of all possible combinations (20x20, equivalent to the number of amino acids in the alphabet) is applied to weight frequently observed substitutions in homologous proteins [26]. Scoring is performed by filling the cells of an alignment matrix (i,j) where columns and rows represent the individual amino acid positions of the sequences s_1 and s_2 to be compared. The value of each individual pair of amino acids in (i,j) is equivalent to the matrix value of the pair of amino acids $(s_{1,i}, s_{2,j})$. The optimal alignment of the sequence pair is the highest score along the path through the matrix that does not violate the boundary conditions: (i) each amino acid residue must be part of the alignment (ii) each residue is allowed to occur only once in the alignment and (iii) the order of residues in both sequences must be retained. The problem to find the optimized path through the matrix has been solved by Needleman and Wunsch [27] who described a dynamic programming algorithm.

Practically, the path through the matrix is found by starting from the lower right of the matrix calculating the sum in each cell of the matrix as the result of all previous alignment steps, including insertions and deletions (gaps). After filling the matrix it is easy to find the path backwards from the top score. However, if gaps are not taken into account for the alignment score, practically any short sequence can be aligned to a long sequence at 100 % identity by the introduction of gaps. Gap penalties P set an initial value I for the initiation of a gap and add a constant C for each gap extension. Gap penalty $P = I + CS$, where S represents the integer size of the gap. As long as closely related homologous sequences are compared, the Needleman-Wunsch algorithm provides perfectly satisfactory results. However due to the fact that a global alignment over the entire length of the sequences is assumed, weak similarities are found to score close to the noise of random alignments. The dynamic programming algorithm has been modified [28] to allow the detection of short, significant homologies that may be the result of conserved domains or motifs in proteins. Also the alignment of fragments and its statistical evaluation has been applied to increase sensitivity [29]. Neither the Needleman-Wunsch nor the Smith-Waterman algorithms are suitable for homology searches through large databases of more than 10.000 sequences, except for the application of massive parallel architectures [30]. Methods for rapid database searches by the application of lookup tables to detect the highest scoring diagonals of any possible pairwise alignment are computationaly fast and became very popular (e.g. FASTA [31]). It is feasible to build dynamically updated databases of sequence similarity scores based on FASTA comparisons for more than 100.000 sequences [32].

The most critical question in sequence comparison is to decide if a found similarity can be interpreted as homology (equivalent to evolutionary relationship). No decisive algorithm is available. Guidelines have been established, but particularly the choice of the gap penalty parameters is critical for the evaluation of pairwise alignments [33,34,35]. The decision if a sequence belongs to a related group of proteins (a family or a domain) can be greatly facilitated by the application of multiple alignments. The present biochemical literature is rich in publications that draw conclusions based on the multiple alignments (e.g. [36,37]). The application of the Needleman-Wunsch algorithm to an n-dimensional transformed matrix for multiple alignments becomes computationaly highly complex and has been shown to be limited to 3 sequences [38]. Improved approaches have been presented by successive pairwise alignments [39], the profile approach [40] or CLUSTAL [41].

With the massive increase in the volume of sequence information, the organization of all sequences into alignments becomes increasingly important. As pointed out, sensitive comparison tools will be required to cluster all database sequences into families. Therefore, it is necessary to distinguish between reliably conserved regions and unrelated ones and to identify family indicating motifs and patterns to detect distinct members [42].

4 Pattern Oriented Sequence Data Analysis

An alternative concept to methods of pairwise or multiple sequence comparisons is to analyze a database for significant patterns to find all occurrences of a pattern of length m in a text of length n. The search for evolutionary conserved short regions (motifs, domains) delineates the sequence, increases signal to noise ratios in the search and allows the identification of distantly related proteins.

4.1 Position Trees as Suitable Data Structures

Many queries can be expressed as pattern matching tasks [43]. Pattern matching within strings is a classic topic in computer science [44,45]. Two general principles of data access can be distinguished. According to the first principle, an automaton is generated that accepts the query pattern to search a given text string [46,47,48,49,50]. This accepting automaton reads sequentially through the text to identify the matching patterns. The time to process a query is proportional to the length n of the text. The second principle requires the text to be indexed prior to the retrieval of the pattern [51, 52]. Provided that the index is already generated, the time to process a query depends merely on the length m of the query pattern. Query optimization using indices is most favorable for large text strings, where high query frequency and low update rates are experienced. These conditions apply either for data retrieval from complete sequence databases or for the analysis of large contiguous sequences, such as those corresponding to chromosomes or other large contiguous genomic data. Any optimized index data structure used by a given application must reflect the characteristics of the data as well as the nature of the queries to be expected. Current index systems applied to sequence information in databases do not allow versatile queries and cannot be incrementaly updated [53,54].

Index data structures including position trees have been subjected to extensive research, mostly to address searching problems in text processing [52,55,56]. The set of elements from which the strings are built is called the alphabet [57]. Text data are characterized by the alphabet, the frequency of words or subwords, the average length of words, and the range in the length of words. These attributes effect the balance of such tree structures. To gain efficiency for data access speed, the data structure must be tuned to limit access to persistent storage. Our recent efforts have been directed to develop a novel method to apply dynamic and persistent position trees in the context of sequence data analysis [58].

Pattern matching problems based on character string comparisons have been the subject of research in sequence analysis [59,60]. Queries may ask for exact matches of a pattern or reflect fuzziness by allowing mismatches and gaps. It is also possible to implement algorithms that use the specific properties of position trees to analyze data, e.g. to find sequences that are similar to a given motif. Current search methods developed to query biological sequence databases vary widely with respect to these capabilities. In general, the sensitivity to detect weak relationships is linked to performance, equivalent to the time required to satisfy the query. Known implementations follow fixed strategies to balance these factors. A preferable query tool would take advantage of a supporting index structure and provide the ability to balance sensitivity and performance within a unified approach.

From a computational point of view, pattern matching in text is linear[2]. Sequence data analysis has a number of more challenging problems, e.g. clustering of sequence data into evolutionary related protein families, structure prediction, and fragment assembly in sequencing experiments. Fragment assembly has been selected as an example, since large scale sequencing projects require a substantial amount of human interaction to assemble very small portions of DNA (around 300 nucleotides) generated by shotgun[3] sequencing techniques. Computational fragment assembly infers the contiguous sequence by searching for overlapping subsequences. Combination of all fragments is computationally intractable. In its idealized form, assuming error-free fragments and ignoring the existence of dyadic complements, the fragment-assembly-problem has

---

[2] Depending on the implementation, linear to the query or the searched text.

[3] 'Shotgun' sequencing generates sequence data from 200-600 residues out of DNA inserts into vectors of size up to 50 kBases. The resulting sequence data are randomized, i.e. unordered and display high error rates (0.3 - 2%).

been shown to belong to the class of problems that are NP-hard [44]. Therefore under practical conditions, this problem is at least NP-hard.

In this work we propose a novel approach to efficient sequence data analysis by the flexible and efficient application of pattern searches. The approach is based on the formulation of sequence analysis queries as pattern specifications. A formal decomposition procedure permits mapping of query definitions to a high performance search engine, the Hashed Position Tree (HPT), a dynamic, persistent index data structure for large data sets. The application of the HPT in functional analysis of complete genomes (i), as search engine for entire sequence databases (ii), and as a highly efficient heuristic solution to the NP-hard problem of sequence fragment assembly (iii) will be demonstrated.

4.2 Data and Query Representation

Any protein or nucleic acid sequence data collection can be represented as a continuous concatenated string with sequence boundaries encoded with specially designated symbols. Thus, any sequence database search problem is equivalent to the analysis of a single contiguous string within this context. In the following, the *text* string refers to the character string representing the data set, whereas the sequence string s refers to the substring between two boundary symbols. Queries to text strings are often expressed as pattern matching problems. Such patterns, describing biologically significant features in proteins, have been collected and published (e.g. nucleotide binding sites, loops, active centers etc.) [59]. In order to expand the approach of simple pattern retrieval, we take pattern matching as a starting point and view the text string as a set of signal vectors of discrete values, assigning each text position a characteristic signal. To search a pattern in the database, we define a pattern m as a list of substrings w_j, where $\mathbf{p}$ is the number of substrings in m.

$$m = (w_1, w_2, ..., w_p)$$

With each substring w_j a substring recognizer is associated that is capable of identifying a substring specified by w_j. Substring and recognizer are connected by the functional relation $f(w_j) = w_j$. If recognizer $f(w_j)$ fails to find w_j in the text, it evaluates to ε (the empty string) else it returns the substring w_j.

The substrings w_j are constructed from the alphabet[4] Σ. In a pattern m two distinct types of substrings w_j exist. The substrings x_i are of fixed length l_i containing specific symbols of Σ. The substrings e_i are of variable length k_i containing *don't care* symbols[5].

---

[4] For proteins Σ is the set of 20 standard amino acids. For nucleic acids Σ is the set of the 4 bases {A,G,T,C}.

[5] Symbols that match any character from alphabet Σ.

In related biological sequences common substrings x_i represent conserved, functionally relevant regions, while the spacer substrings e_i stand for variable, non-conserved regions of a pattern. Variation in distantly related sequences within the conserved regions x_i can be expressed by utilizing a different type of recognizers for x_i that allow for mismatches.

For a given pattern m we express a relation to the sequence s_t of the text by $f_t(m) = s_t$. This expression is *true* if and only if a sequence entry s_t of the text matching pattern m can be recognized. The index t is used to relate substring recognizers to a specific sequence entry s_t of the text.

For a given pattern definition a query may then be formulated as follows:

$$FIND \{ s_t \mid f_t(m) = s_t, \text{ with } m = (e_1, x_1, e_2, x_2, ..., x_p, e_{p+1}) \}$$

The query is evaluated by unifying [61] a pattern m with sequences s_t using f_t. The functions f_t represent specific substring recognizers for the conserved elements x_i: $f_t(x_i) = x_i$ and the variable elements e_i: $f_t(e_i) = e_{t,i}$. Unification means that $e_{t,1} x_1 e_{t,2} x_2 ... x_p e_{t,p+1}$ and s_t are made textualy identical, i.e. any valid configuration t must hold $f_t(m) = s_t$.

The query expression is evaluated by decomposing the expression $f_t(m) = s_t$ as:

$$f_t(m) = s_t$$
$$f_t(e_1, x_1, e_2, x_2, ..., x_p, e_{p+1}) = s_t$$
$$f_t(e_1) f_t(x_1) f_t(e_2) f_t(x_2) ... f_t(x_p) f_t(e_{p+1}) = s_t$$
$$e_{t,1} x_1 e_{t,2} x_2 ... x_p e_{t,p+1} = s_t$$

In order to achieve efficient evaluation of the query, a two step strategy (*select before join*) is applied. In step one the recognizers $f_t(x_i)$ are evaluated. These recognizers are mutually independent and therefore can be evaluated in parallel. In step two the subwords x_i are joined within position ranges specified by the recognizers $f_t(e_i)$. This evaluation is performed for each sequence of the data set. To resolve such a query it is necessary to read the complete sequence database from a mass storage device, a condition that becomes prohibitive for large sequence data bases. Therefore the recognizers f_t must be optimized by the application of an index data structure that renders the set of all substrings in the data base efficiently accessible. For those classes of recognizers $f(x_i)$, that allow only a limited number of mismatches to a given pattern, position trees can be used to perform Step 1 rapidly. Computation time is of order p, the length of the pattern. As an important consequence, this formalism allows us to expand a given pattern by variation to identify previously unrecognized examples of a conserved region (motif) in a biological sequence.

The formalism applied allows the pattern matching problem to be inverted. Queries applying frequency as a constraint may be posed. The frequency $h(m)$ of the pattern m

must be greater than the threshold n, i.e. the minimal rate of the occurrence of pattern m in the data set. Functions g_i select pattern components of length l_i that fullfil this constraint. A pattern detection query may be expressed as follows:

$$FIND \; \{ \; s_t, x_i \,|\, f_t(m) = s_t, \text{ where } m = (e_1, g_1(l_1), e_2 g_2(l_2), ..., g_p(l_p), e_{p+1}), h(m) > n \; \}$$

The result of the query is a set of $(p+1)$-tupels, each composed of a sequence s_t and p pattern components x_i that match the pattern frequency constraint. The decomposition of the query expression is as follows:

$$f_t(m) = s_t$$
$$f_t(e_1, g_1(l_1), e_2 g_2(l_2), ..., g_p(l_p), e_{p+1}) = s_t$$
$$f_t(e_1) f_t(g_1(l_1)) f_t(e_2) f_t(g_2(l_2)) ... f_t(g_p(l_p)) f_t(e_{p+1}) = s_t$$
$$f_t(e_1) f_t(x_1) f_t(e_2) f_t(x_2) ... f_t(x_p) f_t(e_{p+1}) = s_t$$
$$e_{t,1} x_1 e_{t,2} x_2 ... x_p e_{t,p+1} = s_t$$

The processing time to evaluate functions f_t and g_p is critical for the realization of this approach. To allow the evaluation of constraint parameters, an efficient search engine is required.

4.3 The Multiple Entry Position Tree

We introduce the position tree as an efficient index to allow time optimal evaluation of recognizers $f(x_i)$. We extend the data structure to realize functions $g(l_i)$. The position tree is an index data structure that generates a key to every individual position of a text. The classical position tree is characterized by two features. (i) All character positions in a text are labeled at the leaves of the tree. (ii) The path from the root to a given leaf is a unique identifier, a key, for this text position. This key is defined as the position identifier. The path encodes the minimal length substring required to uniquely identify that position in the text string. The branches are labeled by sequence characters; the terminal nodes are labeled by the position of the specified substring in the text string.

Finding a substring in the position tree is equivalent to following the path indicated by the query. The time dependence to resolve the query is always less than or equal to m, the length of the query string.

$$\underset{\substack{1\;2\;3\;4\quad\;5\;6\;7\;8\;9\quad10\;11\;12\;13}}{\textbf{a b a b \$ b a b a b \$ b a b a \$}}$$

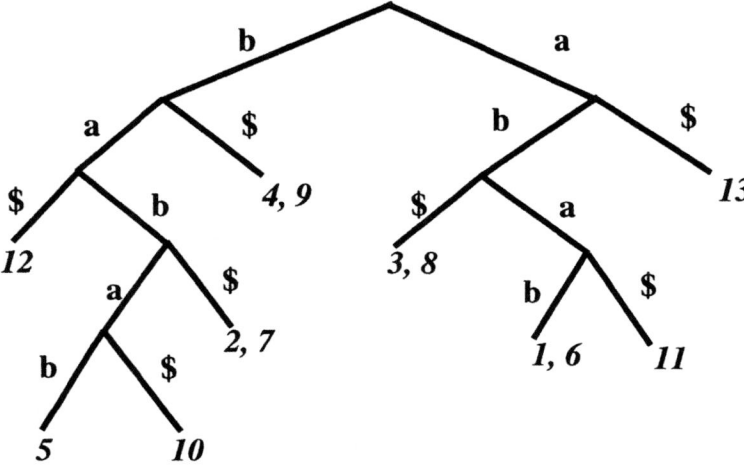

Figure 1 **An example of a multiple entry position tree for three sequence entries**

The classical position tree ignores any logical structure of the text. Sequence databases correspond to sets of distinct sequence units. We encode this structure in the text by delimiting sequence boundaries by the symbol '$'; no position identifier may contain '$' as a non-terminal symbol; thus position identifiers are not always unique in this case. Positions in the text corresponding to '$' delimiters are not numbered. Figure 1 shows an example of this encoding scheme. The set of all position identifiers is partitioned into logically disjoint sets according to the structure the sequence boundaries imposed on the data set. As an important consequence, it is possible to remove/insert entries without corrupting position identifiers of entries other than the modified ones; a feature which is imperative for on-line updating[6] of the position tree.

It is well known [62,63,64,65] that the position tree is theoretically capable of dramatically improving the performance of many queries to biological sequence data. However, most published implementations of position trees are memory resident, requiring massive computer resources to accommodate available data sets[7]. As data collections currently increase exponentially, this approach becomes even less feasible. We introduce a dynamic, persistent form of the position tree, the hashed position tree (HPT), suitable for implementation on systems equipped with slow mass-storage devices.

---

[6] In this case, text postitions must be numbered relative to the beginning of the sequence entry.

[7] An independen approach of directly mapping suffix trees to disk is presented in [65].

4.4 The Hashed Position Tree

Two practical considerations are critical to the application of position trees to large data sets stored on slow mass storage devices:

- A *find substring* operation must not require more than a single disk access[8].

- The space requirements for the position tree must scale linearly with the size of the database.

In addition, a general purpose data structure should be defined to allow the optimization of individualized search strategies through parametrization. Biological sequences differ from natural language text. Natural languages are restricted to a relatively small subset of short substrings that can be constructed from a given alphabet. Biological sequences represent the word space for short substrings more uniformly. In the case of proteins, up to a word length of five, all possible subwords are represented in the current collection of data. The corresponding position tree is complete to a depth of five. However, biological sequence data are also biased. The evolutionary conservation by inheritance resulted in a considerable redundancy of the long subsequences realized in nature. This situation leads to very long branches in position trees required to address all occurrences explicitly. Such branches can be compressed by the introduction of data segments (buckets) located on mass storage devices to decrease memory requirements and to improve the performance of tree operations.

The hashed position tree (HPT) incorporates these particular characteristics through parametrization. It is a hybrid data structure that combines a hash directory, a set of collision class position trees, and a set of data pages. The HPT has the following properties:

- The hash directory is used to represent the root of the tree down to the level where it becomes incomplete; for proteins this is level five. The entries of the hash directory either directly address a segment of a data page or refer to an entry point of a collision class position tree.

- The collision class position trees roughly separate the data into classes of redundant sequences. The leaves of such collision class position trees point to segments of a data page on the mass storage device.

- The data pages, organized by segments are the transport units. These segments store the positions associated with the partial position identifier. Segments are not allowed to span pages.

---

[8] Here, only disk accesses required to identify the location of the sequence or a given sequence component are considered.

It can be shown that any position tree can be transformed into an equivalent HPT (see Fig. 2).

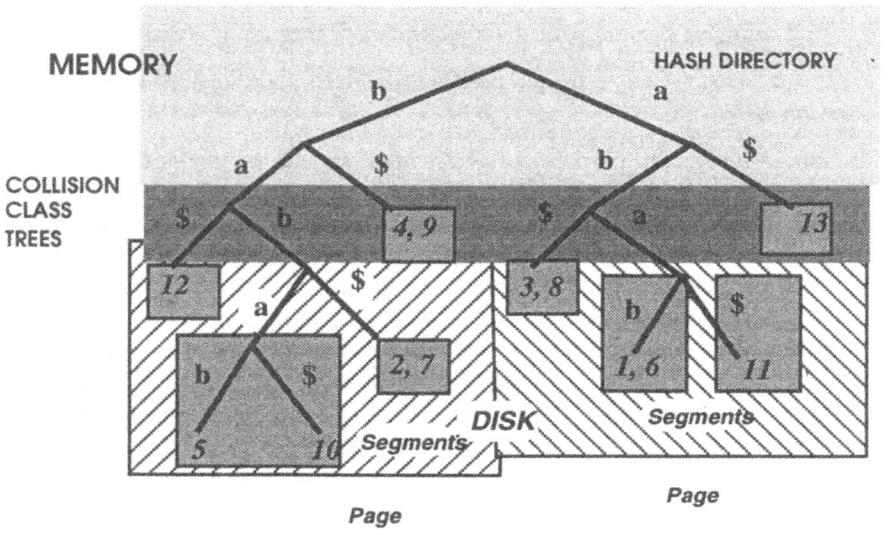

Figure 2 Transformation of a position tree into the corresponding HPT

The HPT can be balanced to assure retrieval of any position identifier in a single disk access as long as hash directory and collision classes are kept in main memory. A second disk access will be necessary to retrieve the sequence from the data set.

The HPT is implemented using the commercial object-oriented database system ObjectStore [66] to generate an persistent and updatable index data structure. Indexing the current complete protein sequence database consisting of 80,000 sequences with more than 22 million residues is performed in less than 48 hours. 400 MB of mass storage space is required (PIR-International, Rel. 43 [9]; DEC Workstation AXP 3000).

Since CPU requirements are negligible for performance considerations, disk access is the only critical parameter for performance analysis. The number of disk accesses required for the basic operations of the HPT was evaluated as follows[9]:

- find substring: The *find* operation is performed with at most one disk accesses[10]. Under practical conditions about 5,000 find operations per second are achieved.

---

[9] A detailed performance analysis will be given in [58].

[10] Disk access to fetch the original sequence object from physical storage is neglected. Also, unspecific queries addressing data that span across pages may require additional disk access operations.

- update entry: The *update* operation is based on a *delete* and *add* operation each requiring a *find* operation. Therefore an *update* of an entry of length *l* requires *2l* disk accesses for the *find* operations. If a data page must be split during an *add* operation another disk access is required resulting in at most 3 disk accesses. About 2 sequences (average size 400 residues) per second can be updated, which is satisfactory for practical purposes.

- generate HPT: The *generate* operation can be performed as successive *add* operations. To achieve *generate* times at higher rates than subsequent *add* operations a 'split and merge' strategy is employed by selecting subsets of the database. As a result *generate* times are reduced by a factor of 5-10.

In order to demonstrate the flexibility and strength of this approach, the HPT has been parametrized to support (i) sequence data processing for efficient pattern matching against complete sequence databases and chromosomes, (ii) efficient fragment assembly for large contigs in DNA sequencing projects.

4.5 Pattern Recoginition

Figure 3 shows a comparison of the left telomeric region of chromosome II of Saccharomyces cerevisiae (position 1 to 6,000) against the fungi sequences of the EMBL Data Library using an HPT index. The method was applied to identify published sequences corresponding to the sequence generated in the genomic sequencing project [17]. The HPT-index was generated from the sequence of chromosome II and the EMBL data were searched against this index. Matching entries that share regions longer that 200 nucleotides with less than 20 mismatches (i.e. at least 90% sequence similarity) were selected.

Each EMBL sequence was expressed as a pattern incorporating the length and mismatch constraints. Because the set of all possible pattern configurations consistent with these criteria can be very large, it was required to define adequate subgoals to generate a limited set of selective substring recognizers. The fixed-length substrings of the sequence entry constitute an optimal set of subgoals. In a first step these subgoals are partially unified with the chromosome sequence. In a second step the subgoals are joined with respect to the match criteria. In this respect the process of unification is equivalent to aligning a sequence entry with the chromosome. A pattern of blackened bands in the display (e.g., sequences SCYKL225W or SCYKL224C) graphically represents the configuration of pattern *m*, as inferred from the sequence entry. For these sequences the expression $f(m) = s_t$ evaluates to *true*, thus satisfying the match criteria. s_t corresponds to the chromosome sequence. Query sequences may align to more than one region of the chromosome representing gene duplications.

It was observed that a large number of sequences matched in the region of positions 4,000-5,000 of the yeast chromosome. This region was easily identified as a Ty-element, displaying significant sequence conservation throughout the genome [17].

A consensus pattern can be inferred from the multiple sequence alignment of these matching segments. Although the initial query was not formulated to infer a consensus pattern, the pattern emerged from the unification process. Similar cases can be observed frequently; other genetic elements (e.g. t-RNA's) can be identified and yet uncharacterized signals can be selected for further investigation.

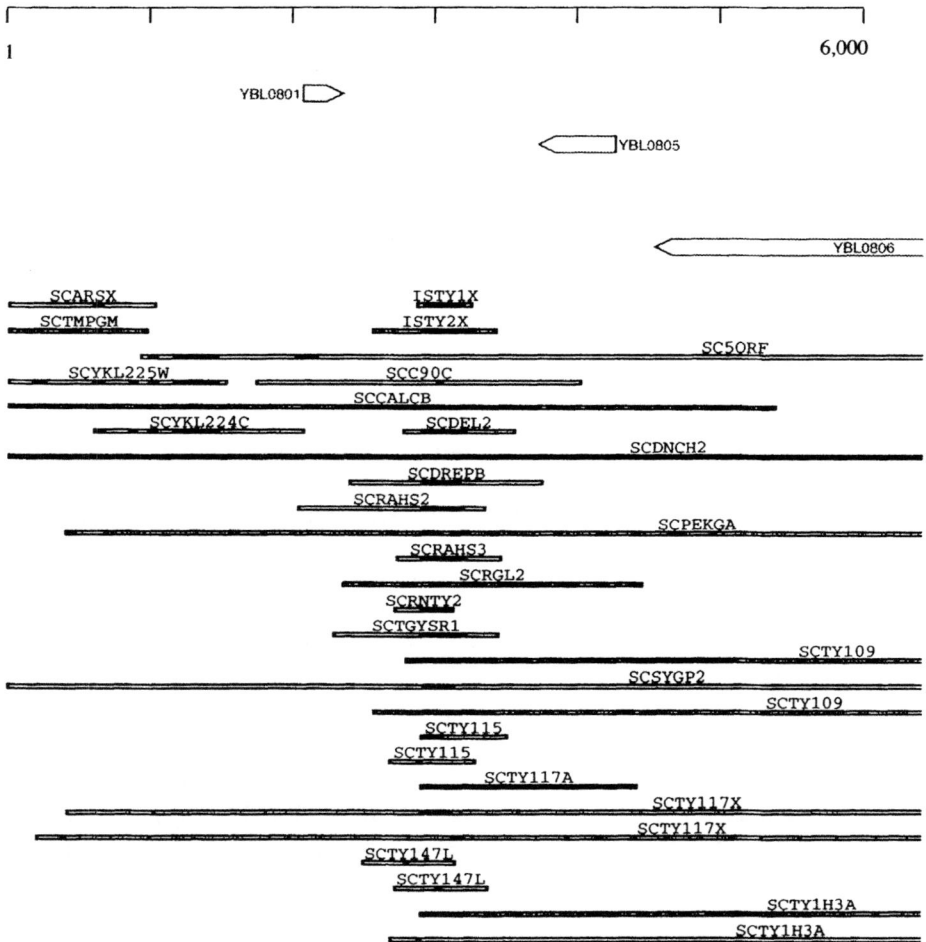

Figure 3 ***Alignment of the left arm of chromosome II of S. cerevisiae against the fungi set of the EMBL nucleic acid data library.*** The graphical display was produced with the program XChromo. Arrows represent the identified ORFs and their orientation on the chromosome. The bars represent entries of the EMBL-fungi data library aligned against the chromosome. Blackened bands within these bars identify regions of sequence identity.

XChromo-displays similar to that shown in Figure 3, revealed two ORF (open reading frame) homologies that had not been recognized previously [17] due to a sequencing error resulting in a reading frame shift. Sequence inconsistencies detected in this way can be corrected using a voting algorithm. The application of HPT can provide a valuable quality-control tool for the analysis of data generated from large scale sequencing projects in comparison to a large set of database entries.

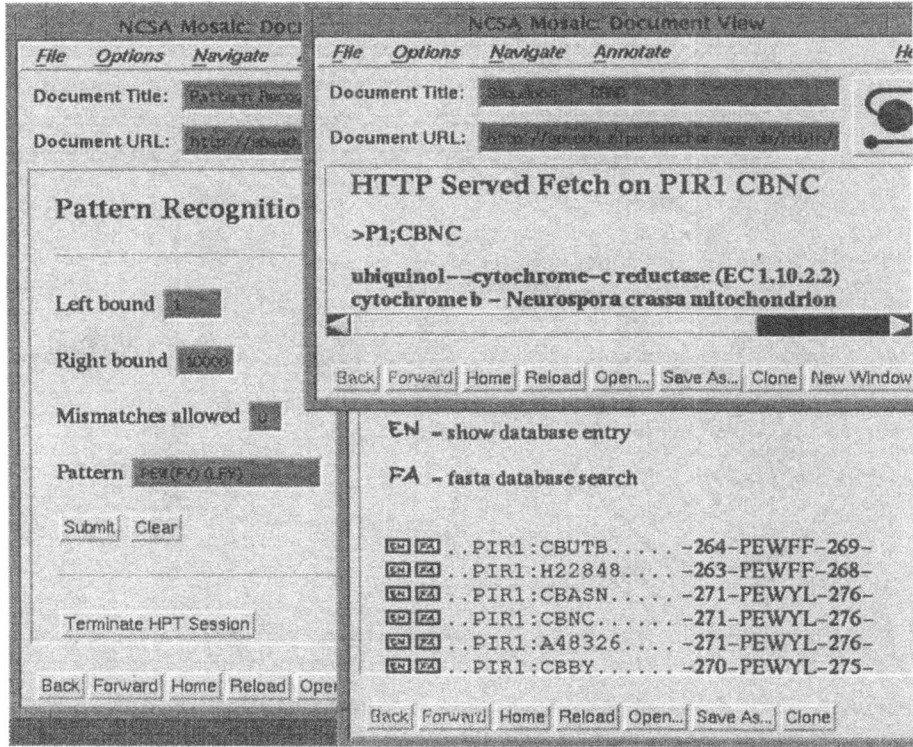

Figure 4   WWW-gateway to interactive pattern retrieval using the HPT

A World Wide Web (WWW) interface was implemented to provide user-friendly, platform independent access to the pattern retrieval facilities of the HPT via MOSAIC. Figure 4 displays a protoype version of this interface. The example shows a simple pattern composed of a single element. The pattern PEW(FY)(LFY) (a signature for cytochrome b/b6-type proteins) [59] is retrieved from the protein sequence database [9].

The output consists of a list of entry identification codes corresponding to sequences containing the specified pattern. The location of the match within the sequence is also listed. The associated database entries can be displayed by clicking on the code. The interactive nature of this workbench environment enables variations of the pattern to be explored. The present implementation is restricted to patterns of the form $m = (e_1, x_1, e_2)$. Figure 3 shows $x_1 = PEW(FY)(LFY)$; e_1 and e_2 represent the boundary conditions on the location of the pattern in the target sequence.

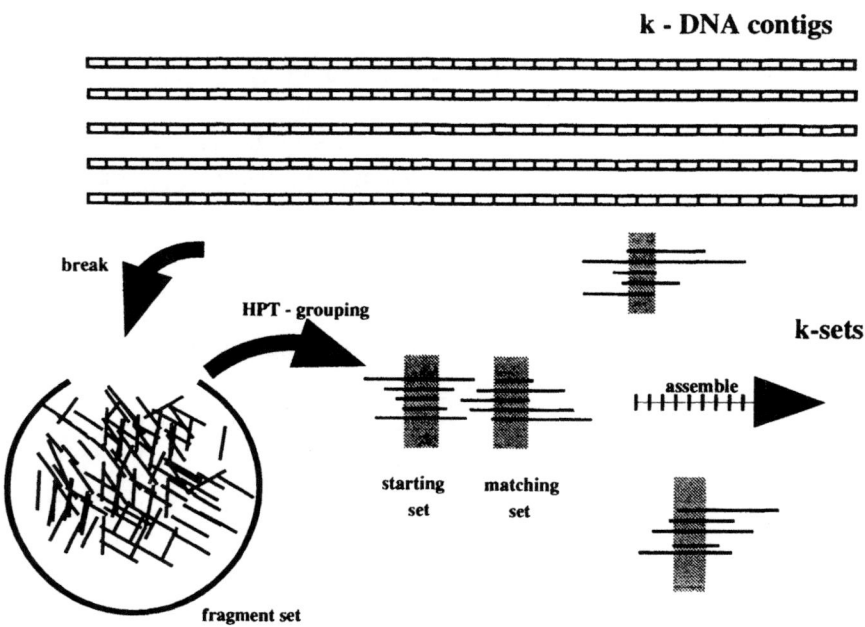

k - DNA contigs

k-sets

Figure 5   DNA fragment assembly by the application of HPT

4.6 Fragment Assembly

Figure 5 demonstrates the principle of the fragment assembly experiment. k-replicates of double stranded DNA-contig are split by the sequencing technique into pieces of about equal size that can be submitted to the automated sequencing machine. The data from all reads constitute the fragment set. When breaking up the original DNA-contigs into fragments, the orientation and position information of the sequence fragment is lost. This information must be restored by the reconstruction algorithm.

Thus during the reconstruction it must be guaranteed that only fragments of the same strand are being assembled to generate the original sequence. The rate of randomly distributed sequencing errors is approximately 1%. The reconstruction algorithm should be capable of eliminating errors by taking advantage of the redundancy provided by the random selection of clones sequenced of the original contig. The coverage of the contig by its fragments is not necessarily complete, some fragments might be missing because of structural properties of the DNA that may cause rearrangement of single strand sequences preventing correct insertion into the vector. Those missing links can not be reconstructed. As a result the reconstruction process will generate an unordered set of large contigs which can be positioned on the restriction map[11]. Gaps must be closed either by selection of additional clones or alternative sequencing techniques (PCR, primer walking).

Efficient fragment assembly during the course of the experiment may significantly reduce the overall effort. Time consuming additional approaches can be reduced to a limited number of defined regions. The combination of directed sequencing strategies and rapid fragment assembly is expected to accelerate sequencing at reduced costs and increased quality.

The HPT provides a favorable solution to the fragment assembly problem. It organizes related but unordered fragments into sets according to the principle of greatest similarity indicated by bands in figure 5. In the terminology of the query mechanism this is expressed by pattern definitions as follows $m_0=(x_1,...,x_{k-1},g(l),x_{k+1},...,x_{2k-1})$ with $g(l)=x_k$ provided the frequency $h(x_k)$ is equal to k. Note that for the fragment assembly the e_i are all empty; therefore they are omitted. The x_i with i not equal to k represent the overlapping edges of the k-fragment set. Similarity is defined with accordance to the expected experimental error rate on the raw data allowing for mismatches in the x_i. It is unlikely that a k-cluster incorporates fragments generated from complementary DNA strands. The subsequence that is covered best by a k-set is selected as a starting point of the assembly process. This is the case for l to become maximal for $g(l) = x_k$.

Provided a starting set has been determined, the HPT is used as an index to lookup the neighboring fragment set. This matching fragment set is retrieved by pattern $m_1=(x'_1,...,x'_{k-1},g'(l),x'_{k+1},...,x'_{2k-1})$ with $x'_1=x_{k+1},...,x'_{k-1}=x_{2k-1}$ and $g'(l)=x'_k$ expressing the overlap relationship between pattern m_0 and pattern m_1. This inductive process is repeated for all fragment sets expressed as subsequent pattern specifications where m_j is expressed by m_{j-1}. During the process of reassembly all fragments that have been incorporated into a contig will be deleted from the HPT. As a result, the remaining fragment sets are less likely to contain ambiguous configurations. Since the time complexity of functions f and g depends only on the length of x_i, the reconstruction algorithm operates under optimal conditions in linear time $O(n)$ with n representing the length of the DNA-contig. The overall time requirements depend on the construction of the HPT. In our current implementation, the HPT is constructed in $O(nlog(n))$ time. Thus, assembly of double stranded DNA-contigs from fragments of coverage k, this results in $O(2knlog(2kn))$ overall time complexity.

Within this approach it can not be guaranteed that the reconstruction algorithm will

---

[11] The restriction map of a DNA sequence locates markers (cutting patterns of restriction enzymes) on the DNA sequence.

always generate a solution in linear time. Ambiguities may result for the following reasons:

- No fragment set is available to complete the reassembly. Reconstruction can be continued by resuming the assembly from a new starting set. With the remaining fragments other non-overlapping contigs can be generated.

- More than one fragment set fits with equal probability to its neighbor. In this case both alternatives are investigated in a backtracking fashion.

In a simulation using randomly generated fragments, the HPT was able to reconstruct a contig of size $n=50,000$ (equivalent to shotgun sequencing practice). The average size of the fragments was set to 300, coverage of the complete sequence contig was set to $k=5$. This is equivalent to about 1700 fragments. The sequencing error rate was estimated at 2%. Under these conditions the time required for the construction of the HPT overweighed the time required for the reconstruction. The algorithm was implemented on a DEC station 3000; for the given example, the processing time was less then 1 hour. Currently the system is under investigation using experimental data provided by a sequencing laboratory. The variation of the coverage k throughout the contig appears to be the most difficult problem for contig assembly from experimental data. In reality, the parameter k varies within the boundaries of a tolerable interval. Therefore the fragment assembly may only provide a partial solution. Additional sequencing experiments are mandatory in order to yield the complete contig.

5 Summary

The systematic analysis of biological information, particularly in genome analysis, requires profound support from informatics. The interpretation of biological sequences is often difficult for the molecular biologist since the available information is either buried in the literature or cannot be deduced from the comparison of sequence data alone. Such analyses are currently subject of intense research. Computer analysis of the rapidly growing data sets will allow predictions of increasing reliability that have to be experimentally verified.

Pattern analysis of biological sequences is a well established method in biological sequence data analysis and plays an important role in the methodology outlined. The application of suitable data structures will allow more efficient usage of the available data. We have presented our initial experiences with the HPT, a special position tree data structure suitable to store, retrieve and update large sets of unstructured textual data. HPT is particularly suitable for finding substring patterns in sequence databases. We expect this new method to emerge as the versatile tool to be employed in many aspects of genome analysis.

Acknowledgments

This work was supported by BIOTECH program of the Commission of the European Union (BIO2-CT93-0003). We like to acknowledge that the WWW-programming was done by Cynthia Harris. XChromo and ORFex programs were developed by Susanne Liebl. The support of Alfred Zollner in the chromosomal analysis and the contributions of Andreas Maierl and David George while carefully reading the manuscript are gratefully acknowledged.

References

1. P. Edman: A method for the determination of the amino acid sequences in peptides Arch. Biochem. 22, 457 (1949)

2. F. Sanger: The arrangement of amino acids in proteins. Adv. ProteinChem. 7:1-67 (1952)

3. F. Sanger, E.O.P. Thompson: The amino-acid sequence in the phenylalanyl chain of insulin. Biochem. J. 53, 366-374 (1953)

4. M. Dayhoff (edt.): Atlas of Protein Sequence and Structure" National Biomedical Research Foundation. Silver Spring, Maryland (1978)

5. A.M. Maxam, W. Gilbert W.: A new method for sequencing DNA. Proc. Natl. Acad. Sci. USA 74, 560-564 (1977)

6. R.M. Schwartz, M.O. Dayhoff: Origins of Prokaryotes, Eukaryotes, Mitochondria, and Chloroplasts. Science 199, 355 (1978)

7. J. Devereux, P. Haeberli; O. Smithies: A comprehensive set of sequence analysis programs for the VAX. Nucleic Acids Res. 12, 387-395 (1984)

8. C. Rawlings: "Software Directory for Molecular Biologists." MacMillan, London (1986)

9. W.C. Barker, D.G. George, H.W. Mewes, F. Pfeiffer, A. Tsugita: The PIR-International databases: Nucl. Acids Res. 22, 3089-3092 (1994)

10. D.A. Benson, M. Boguski, D.J. Lipman, J. Ostell: GenBank. Nucl. Acids Res. 22, 3441-3444 (1994)

11. D.B. Emmert, P.J. Stoehr, G. Stoesser, G. Cameron: The European Bioinformatics Institute (EBI) databases. Nucl. Acids Res. 22, 3445-3449 (1994)

12. K.H. Fasman, A.J. Cuticchia, D.T. Kingsbury: The GDB (TM) human genome data base anno 1994. Nucl. Acids Res. 22, 3462-3469 (1994)

13. A. Bairoch, B. Boeckmann: The SWISS-PROT protein sequence data bank: current status. Nucl. Acids. Research 22, 1994: 22, 3578-3580

14. Goffeau A. (edt.): Sequencing the Yeast Genome, A detailed assessment. Commission of the European Communities (1988)

15. S.G. Oliver, Q.J.M. van der Aart, M.L. Agostoni-Carbone, M. Aigle, L. Alberghina, D. Alexandraki, G. Antoine, R. Anwar, J.P.G. Ballesta, P. Benit, G. Berben, E. Bergantino, N. Biteau, P.A. Bolle, M. Bolotin-Fukuhara, A. Brown, A.J.P. Brown, J.M. Buhler, C. Carcano, G. Carignani, H. Cederberg, R. Chanet, R. Contreras, M. Crouzet, B. Daignan-Fornier, E. Defoor, M. Delgado, C. Doira, J. Demolder, E. Dubois, B. Dujon, A. Dusterhoft, D. Erdmann, M. Esteban, F. Fabre, C. Fairhead, G. Faye, H. Feldmann, W. Fiers, M.C. Francingues-Gaillard, L. Franco, L. Frontali, H. Fukuhara, L.J. Fuller, P. Galland, M.E. Gent, D. Gigot, V. Gilliquet, N. Glansdorff, A. Goffeau, M. Grenson, P. Grisanti, L.A. Grivell, M. de Haan, M. Haasemann, D. Hatat, J. Hoenicka, J. Hegemann, C.J. Herbert, F. Hilger, S. Hohmann, C.P. Hollenberg, K. Huse, F. Iborra, K.J. Indge, K. Isono, C. Jacq, M. Jacquet, C.M. James, J.C. Jauniaux, Y. Jia, A. Jimenez, A. Kelly, Kleinhans U., Kreisl P., G. Lanfranchi, C. Lewis, C.G. van der Linden, G. Lucchini, K. Lutzenkirchen, M.J. Maat, G. Mannhaupt, E. Martegani, A. Mathieu, C.T.C. Maurer, D. McConnell, R.A. McKee, H.W. Mewes, F. Messenguy, F. Molemans, M.A. Montague, M. Falconi, F. Muzi, L. Navas, C.S. Newlon, D. Noone, C. Pallier, L. Panzeri, B.M. Pearson, Perea J., P. Philippsen, A. Pierard, R.J. Planta, P. Plevani, B. Poetsch, F. Pohl, B. Purnelle, M. Ramezani-Rad, S.W. Rasmussen, A. Raynal, M. Remacha, P. Richterich, A.B. Roberts, F. Rodriguez, E. Sanz, I. Schaaff-Gerstenschlager, B. Scherens, B. Schweitzer, Y. Shu, J. Skala, P.P. Slonimski, F. Sor, C. Soustelle, R. Spiegelberg, L.I. Stateva, H.Y. Steensma, S. Steiner, A. Thierry, G. Thireos, M. Tzermia, L.A. Urrestarazu, G. Valle, I. Vetter, J.C. van Vliet-Reedijk, M. Voet, G. Volckaert, P. Vreken, H. Wang, J.R. Warmington, D. von Wettstein, B.L. Wicksteed, C. Wilson, H. Wurst, G. Xu, F.K. Zimmermann, J.G. Sgouros: The complete DNA sequence of yeast chromosome III. Nature 357, 38-46 (1992)

16. B. Dujon, D. Alexandraki, B. Andre, W. Ansorge, V. Baladron, J.P.G. Ballesta, A. Banrevi, P.A. A. Bolle, M. Bolotin-Fukuhara, P. Bossier, G. Bou, J. Boyer, M.J. Bultrago, G. Cheret, L. Colleaux, B. Dalgnan-Fornier, F. del Rey, C. Dion, H. Domdey, A. Duesterhoeft, S. Duesterhus, K.D. Entian, H. Erfle, P.F. Esteban, H. Feldmann, L. Fernandes, G.M. Fobo, C. Fritz, H. Fukuhara, C. Gabel, L. Gaillon, J.M. Carcia-Cantalejo, J.J. Garcia-Ramirez, M.E. Gent, M. Ghazvini, A. Goffeau, A. Gonzalez, D. Grothues, P. Guerreiro, J. Hegemann, N. Hewitt, F. Hilger, C.P. Hollenberg, O. Horaitis, K.J. Indge, A. Jacquier, C.M. James, J.C. Jauniaux, A. Jimenez, H. Keuchel, L. Kirchrath, K. Kleine, P. Koetter, P. Legrain, S. Liebl, E.J. Louis, A. Maia e Silva, C. Marck, A.L. Monnier, D. Moestl, S. Mueller, B. Obermaier, S.G. Oliver, C. Pallier, S. Pascolo, F. Pfeiffer, P. Philippsen, R.J. Planta, F.M. Pohl, T.M. Pohl, R. Poehlmann, D. Portetelle, B. Purnelle, V. Puzos, M.R. Rad, S.W. Rasmussen, M. Remacha, J.L. Revuelta, G.F. Richard, M. Rieger, C. Rodrigues-Pousada, M. Rose, T. Rupp, M.A. Santos, C. Schwager, C. Sensen, J. Skala, H. Soares, F. Sor, J. Stegemann, H. Tettelin, A. Thierry, M. Tzermia, L.A. Urrestarazu, L. van Dyck, J.C. van Vliet-Reedijk, M. Valens, M. Vandenbol, C. Vilela, S. Vissers, D. von Wettstein, H. Voss, S. Wiemann, G. Xu,

J. Zimmermann, M. Haasemann, I. Becker, H.W. Mewes H.W; "The complete sequence of chromosome XI of Saccharomyces Cerevisiae", Nature (1994) 396, 371-378

17. H. Feldmann, M. Aigle, G. Aljinovic, B. Andre, M.C Baclet, A. Barthe, C. Baur, A.M. Becam, N. Biteau, E. Boles, T. Brandt, M. Brendel, M. Bruckner, F. Busereau, C. Christiansen, R. Contreras, M. Crouzet, C. Cziepluch, N. Demolis, T. Delaveau, F. Doignon, H. Domdey, S. Dusterhus, E. Dubois, B. Dujon, M. Elbakkoury, K.D. Entian, M. Feuermann, W. Fiers, G.M. Fobo, C. Fritz, H. Gassenhuber, N. Glansdorff, A. Goffeau, L.A. Grivell, M. Dehaan, C. Hein, C.J. Herbert, C.P. Hollenberg, K. Holmstrom, C. Jacq, M. Jacquet, J.C. Jauniaux, J.L. Jonniaux, T. Kallesoe, P. Kiesau, L. Kirchrath, P. Kotter, S. Koroll, S. Liebl, M. Logghe, A.J.E. Lohan, EJ. Louis, ZY. Li, M.J. Maat, L. Mallet, G. Mannhaupt, F. Messenguy, T. Miosga, F. Molemans, W. Muller, S. Nasr, B. Obermaier, J. Perea, A. Pierard, E. Piravandi, F.M. Pohl, T.M. Pohl, S. Potier, M. Proft, B. Purnelle, M.R. Rad, M. Rieger, M. Rose, I. Schaaff-Gerstenschlager, C. Scherens, B. Schwarzlose, J. Skala, P.P. Slonimski, P.H.M. Smits, J.L. Souciet, H.Y. Steensma, R. Stucka, A. Urrestarazu, Q.J.M. Vanderaart, L. Vandyck, A. Vassarotti, I. Vetter, S. Vierendeels, F. Vissers, G. Wagner, P. Dewergifosse, K.H. Wolfe, M. Zagulski, F.K. Zimmermann, H.W. Mewes, K. Kleine: 'Complete DNA-Sequence of Yeast Chromosome-II', EMBO JOURNAL (1994) 13, 5795-5809

18. M. Johnston, S. Andrews, R. Brinkman, J. Cooper, H. Ding, J. Dover, Z. Du, A. Favello, L. Fulton, S. Gattung, C. Geisel, J. Kirsten, T. Kucaba, L. Hillier, M. Jier, L. Johnston, Y. Langston, P. Latreille, E.J. Louis, C. Macri, E. Mardis, S.Menezes, L. Mouser, M. Nhan, L. Rifkin, L. Riles, H. St. Peter, E. Trevaskis, K. Vaughan, D. Vignati, L. Wilcox, P. Wohldman, R. Waterston, R. Wilson, M. Vaudin: Compltete Nucleiotide Sequence of Saccharomyces cerevisiae Chromosome VIII. Science 256, 2077-2082 (1994)

19. P. Bork, C. Ouzounis,, C. Sander, M. Scharf, R. Schneider, E. Sonnhammer: Comprehensive sequence analysis of the 182 predicted open reading frames of yeast chromosome III. Protein Science 1:1677-1690 (1992)

20. E.V. Koonin, P. Bork, C. Sander: Yeast chromosome III: new gene functions. EMBO Journal 13, 493-503 (1994)

21. Dujon B. et al.,: Detailed evalutation of the complete sequence of chromosome XI of S. cerevisiae'. Manuscript in preparation.

22. R.F. Doolitle: Of URFs and ORFs: A Primer on How to Analyze Derived Amino Acid Sequences. University Science Books, Mill Valley, CA (1987)

23. A.M. Lesk: Computational Molecular Biology. In: Encyclopedia of Computer Science and Technology Vol. 31, Marcel Dekker, New York (1994)

24. R.F. Doolittle: Searching through sequence databases, in: Methods in Enzymology (R.F. Doolittle edt.) 183, 99-110 (1990)

25. P. Argos, M. Vingron, G. Vogt: Protein sequence comparison: methods and significance. Protein Engineering 4, 375-383 (1991)

26. D.G. George, W.C. Barker, L.T. Hunt: Mutation Data Matrix and Its Uses. In: Methods in Enzymology (R.F. Doolittle edt.) 183, 333-351 (1990)

27. S.B. Needleman, C.D. Wunsch: A general method applicable to the search for similarities in the amino acid sequence of two proteins. J. Mol. Biol. 48, 443-453 (1970)

28. T.F. Smith, M.S. Waterman, W.M. Fitch: Comparative biosequence metrics. J. Mol. Evol 18, 38-46 (1981)

29. P. Argos: A sensitive procedure to compare amino acid sequences. J. Mol. Biol. 193, 385-396 (1987)

30. J.F. Colllins, S.F. Reddaway: High-Efficiency Sequence Database Searching: Use of the Distributed Array Processor. In: G.I. Bell, T.G. Marr (eds): Computers and DNA, Addison-Wesley (1990)

31. W.J. Wilbur, D.J. Lipman: Rapid similarity searches of nucleic acid and protein data banks. Proc. Natl. Acad. Sci. USA 80, 726-730 (1983)

32. S. Liebl, H.W. Mewes: A dynamic database of sequence similarities. Manuscript in preparation

33. M.S. Waterman, M. Vingron: Rapid and accurate estimates of statistical siginificance for sequence data base searches. Proc. Natl. Acad. Sci. USA 91, 4625-4628 (1994)

34. C. Sander, R. Schneider: Database of homology-derived protein structures and the structural meaning of sequence alignment. Protens 9, 56-68 (1991)

35. M. Vingron, M.S. Waterman: Sequence alignment and penalty choice. J. Mol. Biol. 235, 1-12 (1994)

36. P. Bork, R.F. Doolittle R.F.: Proposed acquisition of an animal protein domain by bacteria. Proc. Natl. Acad. Sci. USA 89, 8990-8994 (1992)

37. P. Bork, C. Sander, A. Valencia: An ATPase domain common to prokaryotic cell cycle proteins, sugar kinases, actin, and hsp70 heat shock proteins. Proc. Natl. Acad. Sci. USA 89, 7290-7294 (1992)

38. M. Murata, S.S. Richardson, J.L. Sussman: Simultanous comparison of three protein sequences. Proc. Natl. Acad. Sci. USA 82, 2444-2448 (1985)

39. G.J. Barton, M.J.E. Sternberg: Flexible Protein Sequence Patterns, A Sensitive Method to Detect Weak Structural Similarities. J. Mol. Biol. 212, 389-402 (1990)

40. M. Gribskov, R. Luthy, D. Eisenberg: Profile analysis: Detection of distantly related proteins. Proc. Natl. Acad. Sci. USA 84, 4355-4359 (1987)

41. J.D. Thompson, D.G. Higgins, T.J. Gibbson: "Multiple sequence alignment", Nucleic Acids Res. 22 , 4673-4680 (1994)

42. P. Argos, M. Vingron, G. Vogt. Protein sequence comparison: methods and significance. Protein Engineering 4, 375-383 (1991)

43. Bishop J.: Nucleic Acid and Protein Sequence Analysis. A practical approach. IRL Press (1987)

44. Meier, D., "The compelxity of some problems on subsequences and supersequences", Jour. Assoc. Comput. Mach. 25 (2) (1978), 322-336.

45. Knuth D.E.: The Art of Computer Programming, Vol.3, Sorting and Searching, Addison-Wessley, Reading Mass. (1973)

46. S.F. Altschul, W. Gish, W. Miller, E.W. Myers, D.J. Lipman: Basic Local Alignment Search Tool. J. Mol. Biol. 215, 403-410 (1990)

47. R. Baeza-Yates, G.H. Gonnet: A new Approach to Text Searching. Com. ACM 35, 10, 74-82 (1992)

48. U. Manber, R. Baeza-Yates: An algorithm for string matching with a sequence of don't cares. Information Processing Letters 37, 133-136 (1991)

49. R. Pearson: Rapid and Sensitive Sequence Comparision with FASTP and FASTA. In: Methods in Enzymology (R.F. Doolittle edt.) 183, 63-98 (1990)

50. S. Wu, U. Manber Fast Text Searching Allowing Errors. Com. AC 35, 83-91 (1992)

51. A. Califano, I. Rigoutsos: FLASH: A Fast Look-UP Algorithm for String Homology. In: Proceedings, First International Conference on Intelligen Sysem for Molecular Biology (Hunter L., Searls D., Shavlik J. eds.) AAAI Press, Menlo Park, CA, 56-64 (1993)

52. U. Manber, E.W. Meyers: Suffix Arrays: A New Method for On-Line String Searches . Proceedings: First Annual ACM-SIAM Symposium on Diskrete Algorithms. 319-327 (1990)

53. GCG, Genetic Computer Group. GCG-Manual Release 8. Madison, Wisconsin (1994)

54. ATLAS--User's Guide. Document Version 10.0. NBRF Washington D.C. (1994)

55. E.M. McCreight: A space-economical suffix tree construction algorithm; J. As soc. Comp. Mach. 23, 262-272 (1976)

56. M. Kempf, R. Bayer, U. Güntzer: Time Optimal Left to Right Construction of Position Trees. Acta Informatica 24, 461-474 (1987)

57. T.A. Sudkamp: Languages and Machines. Addison-Wesley (1988)

58. K. Heumann: 'The hashed position tree: a dynamic, persistant variant of position trees. Mansucript in preparation.

59. A. Bairoch: PROSITE: a dictionary of sites and patterns in proteins. Nucleic Acids Research 20, 2013-2018 (1992)

60. J.T.L. Wang, T.G. Marr, D. Shasha, B.A. Shipiro, G.-W. Chirn: Discovering active motifs in sets of related protein sequences and using them for classification; Nucl. Acids Res. 22, 2769-2775 (1994)

61. J.D. Ullman: Principles of Dtabase and Knowledge-Base Systems, Vol. I. Computer scinece Press, Rockville. (1988)

62. G. Gonnet, A. Mark, S. Benner: Exhaustive Matching of the Entire Protein Sequence Database. Science 256, 1443-1445 (1992)

63. C. Lefevre, J. Ikeda: Pattern recognition in DNA sequences and its application to consensus foot-printing. Comp. Appl. Biosc. 9, 349-354 (1993)

64. C. Lefevere, J. Ikeda: The position end-set tree: A small automaton for ward recognition in biological sequences. Comp. Appl. Biosc. 9, 343-348 (1993)

65. P. Bieganski, J. Riedl, J.V. Cartis: Generalized suffix trees for biological sequence data: applications and implementation. In: Proceedings of the Twenty-Seventh Hawaii International Conference on System Sciences. Vol.V: Biotechnology Computing; IEEE Comput. Soc. Press, 35-44. (1994)

66. Object Design, Inc. (1993) Reference Manual. ObjectStore Release 3.0 Beta. For VAX/VMS Systems. Burlington.

Matching Patterns of An Automaton

Mehryar Mohri[*]

AT&T Bell Laboratories
600 Mountain Avenue
Murray Hill, NJ 07974, U.S.A.
E-mail: mohri@research.att.com

Abstract. We present an algorithm permitting to search in a text for the patterns of a regular set. Unlike many classical algorithms, we shall assume that the input of the algorithm is a deterministic automaton and not a regular expression. Our algorithm is based on the notion of failure function and mainly consists in efficiently constructing a new deterministic automaton. This construction is shown to be more efficient both in time and space than naive algorithms such as the powerset algorithm used for determinization. In particular, its space complexity is linear in the size of the obtained automaton.

1 Introduction

Pattern-matching consists in finding the occurrences of a set of strings in a text. Two general approaches have been used to perform this task given a regular expression r describing the patterns. Both require a preprocessing stage which consists in constructing an automaton representing the language described by the regular expression A^*r, where A is the alphabet of the text. This automaton is then used to recognize occurrences of the patterns in a text t (see [5], [1], and [10] for a general survey of string-matching). The first approach aims at the construction of a non-deterministic automaton while the second yields a deterministic one. The choice between these methods depends on time-space tradeoff considerations. In general, the construction of a non-deterministic automaton corresponding to A^*r is linear in time and space $(O(|r|))$, but its use in recognition is quadratic $(O(|r| \cdot |t|))$. The preprocessing in the deterministic case, namely the construction of the automaton, is exponential in time and space $(O(2^{|r|}))$, but the recognition of the patterns in a text t is then linear $(O(|t|))$.

In the particular case where the set of patterns is finite, more efficient algorithms have been designed. Aho and Corasick ([2]) gave an algorithm (AC for short) permitting to find occurrences of n patterns P_i $(0 \leq i \leq n)$ in a text in linear time $(O(|t|))$. Their algorithm is based on an efficient construction of an automaton recognizing $(A^*(P_1 + ... + P_n))$ represented with a failure function. It can be considered as a generalization of the well-known algorithm of Knuth, Morris and Pratt used in string-matching ([12]) to multi-pattern matching. The complexity of the construction of the automaton required in AC is linear

---

[*] This work was done while the author was teaching at Institut Gaspard Monge-LADL in France.

in time and space in the sum of the lengths of all patterns, more precisely in $O(\log |A| \cdot \sum_{i=1}^{n} |P_i|)$, where A is the alphabet of the patterns. Commentz-Walter ([7]) gave an algorithm which is the extension of the Boyer-Moore [6] algorithm to the case of a finite set of strings. The complexity of her algorithm is quadratic though more efficient in practice than AC for shorts strings ([1]). Crochemore *et al.* gave a linear time version of this algorithm combining the use of the AC automaton with that of a suffix automaton ([9]).

The input of all these algorithms is a regular expression describing the set of patterns to search for. Here, we are concerned with the problem of searching for a set of patterns in the case where this set is directly given in the shape of a deterministic automaton. Although in many practical operations such as those used under the UNIX operating system regular expressions constitute a very convenient way of representing the patterns, there are many applications especially in Natural Language Processing in which such a representation does not seem appropriate. Indeed, in those applications the deterministic automaton representing the patterns may have been obtained as a result of several complex operations. Besides, in most applications to this field the size of automata exceeds a hundred states and can reach hundreds of thousand. Thus, users cannot conveniently provide the corresponding regular expressions as the input of a pattern-matching algorithm.

We shall describe, in the following, algorithms which help to search in a text for the patterns described by a deterministic automaton G. In order to do so, we shall indicate how to construct a deterministic automaton representing the language $A^*L(G)$, where $L(G)$ represents the language recognized by G. This automaton can be used to determine whether t contains a pattern of $L(G)$. Our representation of this automaton is such that it also helps to find easily the occurrences of the patterns found in t.

Constructing a deterministic automaton representing $A^*L(G)$ from G can be done by adding loops labeled with elements of the alphabet A at the initial state and then use the classical powerset construction permitting to obtain a deterministic automaton [5]. However, the size of the alphabet A can be very large in some applications and this may lead to deterministic automata with a huge number of transitions. In some Natural Language Processing problems for instance, A represents the size of the whole dictionary of inflected forms of the considered language. This size reaches several hundreds of thousand for languages such as English or French. Moreover, in other applications such as in error correction the alphabet itself may not be known because it could depend on the considered text. Thus, one would like to provide a representation of the deterministic automaton such that it be independent of the alphabet of the texts and such that its number of transitions be as small as possible.

The problem of an efficient determinization of automata has been pointed out by D. Perrin ([14]) and M. Crochemore ([10]). In case G represents a single string, one can provide a linear time linear space deterministic automaton representing $A^*L(G)$ using failure functions ([12]). In the following we shall extend the use of failure functions to the general case. In some cases the number of states of the minimal automaton representing $A^* L(G)$ is exponential in the size of G. Hence, in what follows we shall be mainly concerned with complexities depending on the size of the resulting automaton. Notice that typical examples of such blow-up cases are those corresponding to languages such as $(a+b)^* a(a+b)^n$, so they are of the shape $A^* L(G)$.

We shall first present an algorithm permitting to obtain the desired deterministic automaton from an acyclic automaton G, hence one representing a finite set of patterns. This algorithm can be considered as an extension of the AC algorithm to the case of automata. We then generalize the algorithm to deal with any deterministic automaton G.

2 Case of acyclic automata

In the following, we shall consider a deterministic automaton $G = (V, i, F, A^*, \delta)$ with V the set of its states, $i \in V$ its initial state, $F \subseteq V$ the set of its final states, A a finite alphabet and δ the state transition function which maps $V \times A$ to V. For any $u \in V$, we denote by $Trans[u]$ the set of transitions leaving u, and for any t in $Trans[u]$ by $t.v$ the vertex reached by t, and by $t.l$ its label. E stands for the set of edges of G. In this section, we shall assume that the automaton G is acyclic.

2.1 Algorithm

Our algorithm is close to the AC algorithm. Indeed, as in this algorithm we define a failure function s associating with each state q of G the longest proper suffix of the strings leading to q when read from the initial state which are also prefixes of those recognized by G. However, since we deal with automata several distinct strings may reach the same state q, and these strings might have longest suffixes prefixes of strings of $L(G)$ which lead to different states when read from the initial state. In order to deal with such cases and make it possible to define s, we gradually transform the initial automaton G. Each state q is duplicated as many times as necessary such that we have a single possible default state associated with each state. Figure 1 and 2 illustrate this construction in a particular case[2]. The state 2 of G has been duplicated so as to take into account the different possible default states (0 and 1).

The use of the resulting automaton in recognition is the same as usual except that at a given state q $(q \neq i)$, if the input symbol a corresponds to none of the transitions leaving q, a failure transition is made, that is the current state of the automaton becomes the default one $s(q)$.

In order to compute the failure function and duplicate states whenever necessary we use a dynamic traversal over the automaton G. We use a breadth-first traversal of G so that states of level l be visited before those of level $l+1$.

---

[2] At each state of the automaton of Figure 2, the first number refers to q and the second one following the slash to the default state $s(q)$.

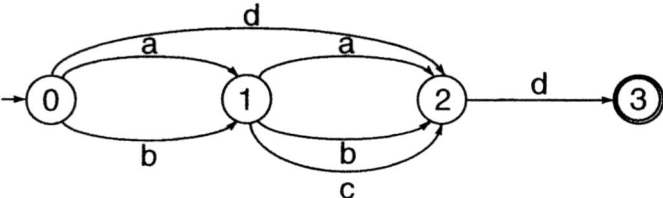

Fig. 1. Automaton G.

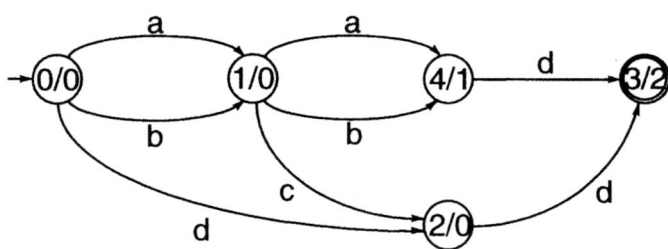

Fig. 2. Deterministic automaton representing $A^*L(G)$.

The choice of the type of traversal is motivated by the following property of s which can be easily proved from its definition:

1- $s[i] = i$ and $s[q] = i$ for all states q of level 1;

2- for a given state q with a level ≥ 2, $s[\delta(q, a)] = \delta(s^k[u], a)$, k being the minimal integer such that $\delta(s^k[u], a)$ is defined, or $s[\delta(q, a)] = i$ if none of these is defined.

Since the level of $s(q)$ is by definition always less than that of q $(q \neq i)$, a breadth-first traversal guarantees that $s(q)$ can be computed following this process. Figure 3 gives a pseudocode for the algorithm MATCHER computing a deterministic automaton recognizing $A^*L(G)$ from G.

We here denote by UNDEFINED a constant different from all states of G. A first-in first-out queue Q is used for managing the set of states to visit at each step according to a breath-first search. The level of each state is computed and stored in the array d. The computation of the levels is only useful in the case of cyclic automata examined in the next section.

The algorithm gradually modifies G by computing the default states at each step and by duplicating states whenever necessary. The duplication is performed by the function COPY-STATE which creates a new state q' copy of q with the same transitions as those q originally had. Since transitions leaving q may have been changed because of other previous duplications, for each state q we need to keep track of the original state of G which q is a copy of. This is done[3] through the function f. Initially, $f[q] = q$ for all q since no copy has been done. In order to limit duplications to what is actually necessary, the list of duplicated states

---

[3] So, when copying q function COPY-STATE creates a new state q' with leaving transitions $\delta(q', t.l) = f[t.v]$ for each $t \in Trans[q]$.

```
MATCHER(G)
1   for each u ∈ V
2   do s[u] ← UNDEFINED
3       f[u] ← u
4   s[i] ← i
5   d[i] ← 0
6   Q ← {i}
7   while Q ≠ ∅
8   do u ← head[Q]
9       for each t ∈ Trans[u]
10      do v ← s[u]
11          while (v ≠ i) and (δ(v, t.l) not defined)
12          do   v ← s[v]
13          if (u ≠ i) and (δ(v, t.l) defined)
14          then v ← δ(v, t.l)
15          if (s[t.v] = UNDEFINED)
16          then s[t.v] ← v
17              d[t.v] ← d[u] + 1
18              if (v ∈ F)
19              then F ← F ∪ {t.v}
20              LIST-INSERT(list[t.v], t.v)
21              ENQUEUE(Q, t.v)
22          else if (there exists w ∈ list[f[t.v]] such that s[w] = v)
23          then t.v ← w
24          else if (v ≠ t.v)
25          then w ← COPY-STATE(t.v) ▷ copy of t.v with same transitions using f
26              s[w] ← v
27              f[w] ← f[t.v]
28              d[w] ← d[u] + 1
29              if (v ∈ F)
30              then F ← F ∪ {w}
31              LIST-INSERT(list[f[t.v]], w)
32              t.v ← w
33              ENQUEUE(Q, w)
34      DEQUEUE(Q)
```

Fig. 3. Algorithm for the construction from G_2 of a deterministic automaton for $A^*L(G_2)$, the acyclic case.

of each state q is stored in $list[q]$. Only if none of the elements of $list[q]$ has the desired default state is the state q duplicated. Notice also that in case the default state is q itself no duplication is necessary (see condition at line 24).

Theorem 1 *Let G be an acyclic deterministic automaton. Algorithm MATCHER permits to correctly compute a representation of a deterministic automaton recognizing $A^*L(G)$.*

Proof. The loop of lines 7-34 corresponds to a dynamic breadth-first traversal of G. Each state q is enqueued exactly once in Q and corresponds to one or more prefixes of strings of $L(G)$. Since the automaton is acyclic, the number of these prefixes is finite. Thus, the loop of lines 7-34 terminates. The loop of lines 9-33 is executed exactly $out - degree(u)$ times for each state u.

Lines 10-14 of the algorithm correspond to the computation of the default state v for each state $t.v$ reached by a transition from state u as previously described. The termination of the loop of lines 11-12 is ensured since for $q \neq i$ the level of $s(q)$ is strictly inferior to that of q.

The algorithm terminates since all loops do. Final states of the resulting automaton are or final states of the initial automaton, or those whose default states are final (see lines 18-19 and 29-30). Hence, they correspond exactly to those paths from the initial state to these states which are prefixes of $L(G)$ and which have a suffix in $L(G)$.

The resulting automaton permits to recognize exactly $A^* L(G)$. Indeed, the definition of s ensures that the state q reached after reading an input string w corresponds to a path labeled with the longest suffix w' of w which is a prefix of $L(G)$. w is in $A^* L(G)$ iff w' has a suffix in $L(G)$, that is iff q is a final state. This ends the proof of the theorem.

Notice that the recognition is independant of the alphabet of the string to recognize. Although the construction of an automaton following MATCHER only involves the alphabet of the strings represented by G the resulting automaton permits to recognize $A^* L(G)$, for any A including the alphabet of G.

2.2 Complexity and optimization

Thanks to the use of a failure function, the space complexity of the algorithm MATCHER is linear in the size of the obtained automaton. Indeed, the size of the queue Q does not exceed the number of states V' of the resulting automaton since each state is exactly enqueued once. The total size of the lists $list[q]$ involved in the algorithm is also bounded by V' as two distinct lists have no element in common. The required size of the arrays s, f and d is equal to V'. Thus, the space complexity of MATCHER is in $O(|V'| + |E'|)$, where E' is the set of the transitions of the resulting automaton.

This is an interesting advantage of this algorithm especially in the case of automata containing more than several hundred thousands of states or transitions such as those involved in some applications. The naive determinization algorithm applying the classical powerset construction to the automaton G provided with a loop labeled with all elements of A at the initial state is quadractic. Indeed, in this algorithm each of the states of the final automaton is represented by a subset of V'. The size[4] of each subset is less than $|V'|$, hence the algorithm is in $O(|V'|^2)$. The sum of the sizes of all subsets is indeed equivalent to $|V'|^2$ even in simple cases such as $A^* a^n$. Another advantage of the algorithm MATCHER is

---

[4] This size is in fact bounded by $|V|$ since elements of the subset all belong to the initial automaton. There exists k such that: $k \cdot \log |V'| \leq |V| \leq |V'|$. However, to be consistent with other expressions, we shall give all complexities in terms of the sizes of the resulting automaton.

that it permits to save the space required for the representation of many transitions which need to be explicitly indicated in the case of the naive algorithm, thanks to the use of the failure function.

In general, even if the initial automaton G is minimal the resulting automaton of the algorithm MATCHER is not the minimal deterministic one representing $A^* L(G)$. For example, the application of this algorithm to the minimal automaton of the set $X = a(a + b + c) + bc(a + b) + cc$ does not provide the minimal one of $A^* X$. Notice that in any case the resulting automaton of MATCHER still permits to recognize the language $L(G)$ when used in the usual way without having recourse to the failure function. But this is not necessarily the case of the minimal automaton representing $A^* L(G)$.

Since it also represents $L(G)$, the resulting automaton of the algorithm constitutes a single device permitting not only to know whether a given text contains a pattern of $L(G)$ but also to determine the occurrences of these patterns. Indeed, if a final state is reached while reading a prefix t' of a text t using the representation corresponding to $A^* L(G)$, then $|t'|$ corresponds to the ending position of some patterns of $L(G)$. Then reading t'^R the reverse string of t' from the reached state to the initial state in the reverse way using this automaton and the final states permits to obtain the list of all occurrences of the patterns ending at that position.

As in the case of the AC algorithm, this automaton offers linear time recognition. Indeed, each time a failure transition is made the level of the control state of the automaton decreases. At most n forward transitions are made when processing a string of length n. Since the level of the state reached after processing this string is positive, the total number of (failure or forward) transitions is bounded by $2n$.

Using the same argument, it can be easily shown that the total number of failure transitions made in the algorithm MATCHER to compute s along each path from the initial state to a state with no leaving transition is bounded by the length of this path. So, if we assume that the test at line 22 is done in constant time $O(1)$ and that the insertions of lines 20 and 31 as well are in $O(1)$ on the average using perfect hashing methods (see [4] and [11]), then on the average the time complexity[5] of the algorithm is in $O(\log |A| \cdot L)$ where L is the sum of the lengths of all the strings recognized by G. Namely, the algorithm has the same average complexity as the algorithm AC. However, one would like here to describe the complexity in terms of the size of the automaton since L might be exponentially larger than V'. The complexity of the algorithm AC is in $O(log|A| \cdot |V'|^2)$, if $|V'|$ denotes the number of nodes or edges of the trie representing the finite set of patterns. Indeed, at most $|V'|$ failure transitions are made for the computation of each default state. The complexity of the algorithm MATCHER is similar. It is in $O(\log |A| \cdot |E'| \cdot |V'|)$ where E' denotes the final number of edges. Indeed, the loop of lines 9-33 is performed exactly $|E'|$ times,

---

[5] The $\log |A|$ factor corresponds to the cost of searching for a transition. In all the following complexities $|A|$ can be replaced with $min\{|A|, e_{max}\}$, where e_{max} denotes the maximum out-degree of all states.

the number of failure transitions at each step is bounded by $|V'|$, and the size of the lists considered at line 22 is bounded by $|V'|$. Notice that this complexity is clearly better than that of the classical powerset construction ($O(\log|A| \cdot |A| \cdot |V'|^2 \cdot |V|^2)$).

In practice, a simple modification of the algorithm permits to speed up the construction. Rather than computing s for each adjacent state $t.v$ of u independently, one can compute them during a single series of failure transitions. Default states are successively examined from the state $s[u]$ until each state $t.v$ is attributed the value of the initial state or that of a state admitting the transition $t.1$, intermediate results being stored in an array of size $|A|$.

Also, once the automaton is constructed, the failure function s of the obtained automaton can be replaced with an optimized one r which avoids unnecessary failure transitions in a way similar to the one described for the AC algorithm ([1]).

3 The general case

We here consider the general case where G is a deterministic automaton not necessarily acyclic. The algorithm presented in the previous section cannot be applied in the general case. Indeed, if G contains cycles, there are states which can be reached by an infinite number of paths starting at the initial state. There might be then an infinite number of paths with distinct longest suffixes prefixes of $L(G)$. Using the algorithm of the previous section can lead to the creation of an infinite new copies of some states. Figures 4 and 5 illustrate this case.

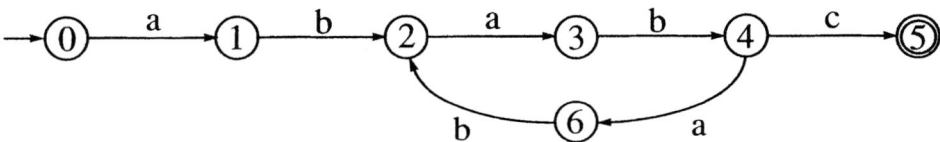

Fig. 4. Cyclic automaton G.

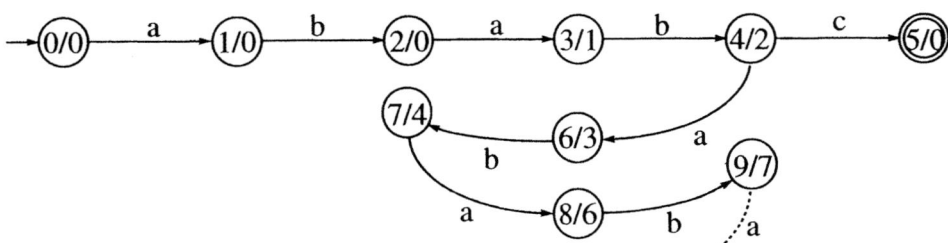

Fig. 5. Blow-up in the determinization.

The first states of the automaton obtained by applying MATCHER to the one of figure 4 are shown in figure 5. They indicate the endless creation of new states corresponding for instance to new suffixes of $(ab)^2((ab)^2)^*$ prefixes of $L(G)$. In the following we shall indicate how to modify this algorithm in order to avoid this explosion and to obtain a deterministic automaton such as the automaton of figure 6 which represents $A^*L(G)$ using the same failure function.

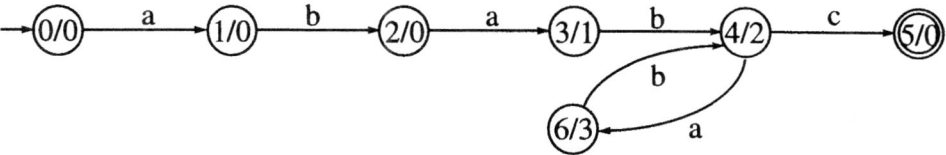

Fig. 6. Deterministic automaton representing $A^*L(G)$.

3.1 Algorithm

The determinization we are interested in can be performed using the powerset construction. This determinization has a particular property here since we deal with a specific type of automata. The powerset associated with each state q contains exactly the states corresponding to strings w which, in addition to be suffixes of the strings leading to q from the initial state, are prefixes of those recognized by the automaton. Thus, each state q, or equivalently each string w leading to q from the initial state can be identified with its corresponding set $set(q)$. Members of $set(q)$ can be ordered following their levels. They correspond exactly to the states obtained by cascaded composition of the failure function s at that state: $set(q) = \{q, s(q), ..., s^k(q) = i\}$.

Let w be a string such that $\delta(i, w) = q$, and assume that $\delta(q, a) = q'$. We shall determine in what follows in which case it is necessary during the application of the powerset construction to create a new state copy of q', or to modify the transition (q, q').

The set S_{wa} corresponding to the string wa is of the form :

$$\{\delta(q, a) = q'\} \cup \{\delta(s^{i_1}(q), a), ..., \delta(s^{i_{k'}}(q), a) = i\}$$

where i_1 is the minimal integer such that $\delta(s^{i_1}(q), a)$ be defined, i_2 the next such integer, etc. We can write:

$$S_{wa} = \{q_1, ..., q_m\} \cup \{q'\}$$

with $level(q_1) < ... < level(q_m)$. States $q_1, ..., q_m$ all correspond to suffixes of wa which are also prefixes of strings recognized by G. The following lemma indicates in which case no new state or change of the (q, q') transition is needed.

Lemma 1. $S_{wa} = set(q')$ iff $q' = q_m$ or $s(q') = q_m$.

Proof. The equality $S_{wa} = set(q')$ holds iff q' is the state of S_{wa} with the highest level. The state of S_{wa} with the highest level is either q_m or q'. The former is possible if and only if the next state of S_{wa} with the highest level is $s(q')$. This ends the proof of the lemma.

Notice that the conditions indicated in the lemma are equivalent to those considered in the case of acyclic automata. However, here, sometimes when these conditions do not hold it is still not necessary to create a new state. Only the reached state of the transition needs then to be modified. For this to occur during the determinization there must exist a state q'' such that $set(q'') = \{q_1, ..., q_m\} \cup \{q'\}$. The following lemma gives the precise condition for this case to appear, when the conditions of lemma 1 do not hold.

Lemma 2. *If $q' \neq q_m$ and $s(q') \neq q_m$, then there exists q'' such that $set(q'') = S_{wa}$ iff $q'' = q_m$ and $q' \in \{q_1, ..., q_{m-1}\}$.*

Proof. By definition of $set(q'')$, q'' is the state of $set(q'')$ with the highest level. Hence, if such a state q'' exists, considering the longest levels of $set(q'')$ and S_{wa} leads to: $q'' = q'$ or $q'' = q_m$. But $q'' = q'$ cannot be since it would require $q_m = s(q')$ (q_m would be the state with next to the highest level in S_{wa}) or $q' = q_m$. Hence, $q'' = q_m$. This implies that $q' \in \{q_1, ..., q_{m-1}\}$.
Conversely, if $q'' = q_m$ and $q' \in \{q_1, ..., q_{m-1}\}$ then we can choose $q'' = q_m$ since $set(q_m) = \{q_1, ..., q_m\} = S_{wa}$. This proves the lemma.
Notice that if the conditions of lemma 2 hold then $q' \in \{q_1, ..., q_{m-1}\}$ implies that the level of q' is strictly less than that of q. Thus, this can only occur if (q, q') is a *back edge* in the sense of a breadth-first traversal of the automaton. These two lemmas give a precise description of the conditions corresponding to each of the three following cases:
  - no new state is created, the transition remains unchanged (lemma 1);
  - no new state is created, the destination state becomes one of the states already defined (lemma 2), the considered tranition corresponds to a back edge;
  - a new state is created which becomes the destination state of the transition (all other cases).
Since the condition of the lemma 1 is identical to the one used in the algorithm MATCHER, slight modifications of this algorithm permit to obtain one which deals with the general case. To that end, line 24 should be transformed into:

24 **else if** $(v \neq t.v)$ **and** $((d[t.v] \geq d[u])$ **or** $(t.v \notin \{s(v), s^2(v), ..., s^k(v) = i\}))$

in order to take into account the condition of lemma 2, and a new line 33' should be inserted between 33 and 34:

33   ENQUEUE(Q, w)
33' **else** $t.v \leftarrow v$
34 DEQUEUE(Q)

Theorem 2 *Let MATCHER2 be the algorithm obtained after making these modifications. The algorithm MATCHER2 permits to correctly compute a representation of a deterministic automaton recognizing the language $A^* L(G)$ in the general case.*

Proof. The use of the array d storing the levels of the states permits to determine whether the considered transition is a back edge $(d[t.v] < d[u])$. Hence, the new line 24 corresponds to the complementary of the conditions of lemma 1 and lemma 2, namely the third case recalled above (creation of a new state). In case the second condition above applies, as mentioned in lemma 2, the destination state is simply the state q_m of lemma 2, that is v in the algorithm. The new line 33' permits to treat this case. That the algorithm terminates is simply a consequence of the definition of the failure function s and of the fact that the same states as in the powerset construction are constructed.

3.2 Complexity

Here again the space complexity of the algorithm is clearly linear in the size of the resulting automaton $O(|V'| + |E'|)$, thanks to the use of a failure function. The algorithm is also more efficient in time than that of the naive determinization. Its time complexity is in $O(\log |A|.|E'|.|V'|)$ since the number of default transitions made at line 12 or for the condition of line 24 is bounded by the total number of final states V', and since the size of the lists of line 22 is also bounded by $|V'|$. While in the powerset construction, at a given state $\{q_1, ..., q_m\}$ and for a given letter, one needs to consult each state q_i to find possible transitions leaving this state labeled with that letter, this occurs only in the worst case using MATCHER2. Indeed, only if the default state q_m admits no transition by the considered letter is the following state considered, etc. Also, thanks to the use of a failure function, not all the transitions created in the naive algorithm need to be considered here either.

4 Conclusion

The construction of an automaton representing $A^*L(G)$ is needed in many algorithms dealing with lexical or syntactic analysis, or in speech processing. In particular, they are very useful when dealing with local grammars represented by automata or transducers ([13]).

We have fully implemented the algorithms described in the previous sections. Results in practical cases in Natural Language Processing show them to be very efficient. We used an implementation of our algorithm to improve a version of a program close to the Unix command *sed*. On large texts, it turned out to be 4 times faster than those obtained with the Unix command without having recourse to any optimization. Also, this algorithm permitted us to successfully achieve in practice the determinization of automata involved in speech processing where the naive powerset construction showed to be computationally intractable. The resulting sizes of the determinized automata exceed 1.5 million states in these cases.

Although we did not show it here, these algorithms can be adapted so as to create according to the *lazy transition evaluation technique* described by A. Aho [1] only the necessary part of the deterministic automaton.

5 Acknowledgements

The author wishes to thank Maxime Crochemore for several discussions on this work, and Ron Kaplan and Lauri Karttunen for their encouraging interest in the general case of this algorithm.

References

1. A. V. Aho. Algorithms for finding patterns in strings. In J. Van Leuwen, editor, *Handbook of Theoretical Computer Science, Volume A: Algorithms and Complexity*, pages 255–300. Elsevier, Amsterdam, 1990.
2. A. V. Aho and M. J. Corasick. Efficient string matching: An aid to bibliographic search. *Communication of the Association for Computing Machinery*, 18 (6), 333-340, 1975.
3. A. V. Aho, J. Hopcroft, and J. D. Ullman. *The design and analysis of computer algorithms*. Addison Wesley, 1974.
4. A. V. Aho and D. Lee. Storing a dynamic sparse table. In *Proceedings of the 27th Annual IEEE Symposium on Foundations of Computer Science*, 55-60, 1986.
5. A. V. Aho, R. Sethi, and J. D. Ullman. *Compilers, Principles, Techniques and Tools*. Addison Wesley, 1986.
6. R.S. Boyer and J.S. Moore. A fast string-searching algorithm. *Communication of ACM*, 20: 762-772, 1977.
7. B. Commentz-Walter. A string matching algorithm fast on the average. *Automata, Languages and Programming*, Lecture Notes in Computer Science, Springer-Verlag, Berlin: 118-132, 1979.
8. T. Cormen, C. Leiserson, and R. Rivest. *Algorithms*. Mc Graw Hill, 1992.
9. M. Crochemore, A. Czumaj, L. Gasieniec, S. Jarominek, T. Lecroq, W. Plandowski, and W. Rytter. Fast multi-pattern matching. Technical Report IGM 93-3, Institut Gaspard Monge, 1993.
10. M. Crochemore and Wojciech Rytter. *Text Algorithms*. Oxford University Press, 1994.
11. M. Dietzfelbinger, A. Karlin, K. Mehlhorn, F. Meyer Auf Der Heide, H. Rohnert and R. E. Tarjan. Dynamic perfect hashing: upper and lower bounds. In *Proceedings of the 29th Annual IEEE Symposium on Foundations of Computer Science*, 524-531, 1988.
12. D.E. Knuth, J.H. Morris Jr, and V.R. Pratt. Fast pattern matching in strings. *SIAM Journal of Comput. Syst. Sci.*, 6: 323-350, 1977.
13. M. Mohri. Syntactic analysis by local grammars automata: an efficient algorithm. In *Proceedings of the International Conference on Computational Lexicography (COMPLEX 94)*. Linguistic Institute, Hungarian Academy of Science, Budapest, Hungary, 1994.
14. D. Perrin. Finite automata. In J. Van Leuwen, editor, *Handbook of Theoretical Computer Science, Volume B: Formal Models and Semantics*, pages 1–57. Elsevier, Amsterdam, 1990.

New Results and Open Problems Related to Non-Standard Stringology*

S. Muthukrishnan

DIMACS, Rutgers University
Internet: muthu@dimacs.rutgers.edu

Abstract

There are a number of string matching problems for which the best known algorithms rely on algebraic convolutions (an approach pioneered by Fischer and Paterson [FP74]). These include for instance the classical string matching with wild cards and the k-mismatches problem. In [MP94], the authors studied generalizations of these problems which they called the *non-standard stringology*. There they derived upper and lower bounds for non-standard string matching problems.

In this paper, we pose several *novel* problems in the area of non-standard stringology. Some we have been able to resolve here; others we leave open. Among the technical results in this paper are:

1. improved bounds for string matching when a symbol in the string matches at most d others (motivated by noisy string matching),

2. first-known bounds for approximately counting mismatches in noisy string matching as above, and

3. improved bounds for the k-witnesses problem and its applications.

Our results are obtained by using the probabilistic proof technique and randomized algorithmic methods; these techniques, although standard, have seldom been used in combinatorial pattern matching.

1 Introduction

Fischer and Paterson [FP74] proposed a general algorithmic approach for solving string matching problems. Their approach was to reduce the string matching problems to a (possibly) small number of boolean convolution operations[1]. They proposed their approach in the context of standard string matching and string

---

*Supported by DIMACS (Center for Discrete Mathematics and Theoretical Computer Science), a National Science Foundation Science and Technology Center under NSF contract STC-8809648.

[1]Boolean convolution of two vectors is formally defined in Section 2. Informally, it is the product of two boolean vectors over the boolean operators AND and OR.

matching with wild cards; it has been extended to several other problems since [Ab87,AF91,AL88,GG88,Ko89,Pi85,MR92,MP94]. These, for example, include string matching with wild cards, ranges, complements and subsets. The standard string matching problem itself can be solved faster using alternate approaches (for instance using [KMP77]). However, for its close variants such as string matching with wild cards, the Fischer-Paterson approach gives the best known complexity in the worst case. In [MP94], the authors extended the Fischer-Paterson approach to a general graph-theoretic setting which they called *non-standard stringology*. In this paper, we pose several novel problems in the area of non-standard stringology and provide efficient solutions for some of them. Interestingly, our results employ probabilistic proof techniques and randomized sampling methods which are seldom seen in combinatorial pattern matching.

In the problems in non-standard stringology, the input is a text t of length n, a pattern p of length m and a *match relation* M which specifies the pairs of text and pattern positions that match. The match relation can be thought of as a bipartite graph called the *match graph* or its bipartite complement called the *conflict graph*. The problem may be the *general string matching* (**gsm**) problem or the *general count matching* (**gcm**) problem. In the former, the output is the set of positions in t where p occurs under M, and in the latter, the output for each text position i is the number of mismatches (again, under M) between the text and the pattern placed on the text at i. By appropriately defining M, the **gsm** and **gcm** problems can model a number of other well studied problems. For example, the **gsm** problem can model standard string matching, string matching with wild cards, with ranges, and so on. In [MP94], the authors studied the complexity of **gsm** and **gcm** problems in terms of the structure of M. In order to understand the problems we solve here and our results, it would be helpful to know the following about the results in [MP94]: they showed that the complexity of **gsm** and **gcm** were proportional to ccn and cpn of the conflict graph respectively. Here, $\text{ccn}(G)$ is the minimum number of (not necessarily edge-disjoint) bipartite cliques whose union yields G and $\text{cpn}(G)$ is the minimum number of edge-disjoint bipartite cliques whose union yields G.

1.1 Our Contribution and Results

Our contribution is two-fold: we pose novel problems related to non-standard stringology and we provide efficient solutions for some of them. Our technical results are as follows.

1. Motivated by noisy string matching we consider **gsm** problem in which each symbol matches say at most d others. We give a randomized algorithm for this problem that takes time $O(nd \text{ polylog } m)$. In contrast, the previously best known algorithm for this case takes $O(nd^3 \text{polylog} m)$ time [MR92].

   We derive this result by proving an upper bound of $O(d \log m)$ on the ccn of a certain graph. We prove that by the probabilistic proof technique [AS93]; this bound is nearly tight.

2. We pose an approximate version of the **gcm** problem. Again for the case when each symbol matches at most d others, we provide a Las vegas type randomized algorithm that provides a $\log m$ factor approximation and takes time $O(nd \text{ polylog } m)$. In contrast, counting the mismatches exactly takes time $\Omega(n\sqrt{m} \text{ polylog} m)$ time for this case.

We obtain this result by proving a bound of $O(d \log m)$ on the size of a clique cover in which each edge appears in at most $\log m$ cliques. Here too, we provide a probabilistic proof. This bound is nearly tight since there exists graphs for which in any clique cover of size $O(d \log m)$, some edge appears in roughly $\Omega(\frac{\log m}{\log \log m})$ of them.

3. We consider the k-witnesses problem of returning for each text location i, a list of at most k positions in the pattern where it mismatches the text when placed on i. For this problem, we provide an $O(nk^2 \text{ polylog } m)$ time Las Vegas type algorithm for a fixed size alphabet.

We use this algorithm to improve the main result in [AF91] for approximate matching of nonrectangular two dimensional matrices at the expense of randomization. Our algorithm for the k-witnesses problem is based on random sampling; it uses the recent intuition in finding witnesses for boolean matrix multiplication in the context of the all-pairs-shortest paths problem [Se92,AGMN92].

4. We provide a purely *combinatorial* algorithm (that is, one which does not use algebraic convolutions) for a *relaxed approximation* for **gsm** problems in $o(nm)$ time.

Designing $o(nm)$ time combinatorial algorithms for **gsm** has been open for long. Here, we provide a randomized algorithm which is approximate, that is, it finds all positions where the pattern occurs with sufficiently large number witnesses (say, $\geq \sqrt{m}$). In addition, the problem is relaxed since the algorithm may return some positions where the pattern occurs with fewer witnesses. Our algorithm for this relaxed approximation takes time $O(n\sqrt{m} \text{ polylog } m)$. It appears that both relaxed and approximate notions are necessary for our algorithm to achieve $o(nm)$ time.

5. We consider an instance of the **gsm** problem which subsumes all instances studied thus far (string matching with complements, subsets, ranges etc.). This instance contains patterns in which a position is specified as a set of intervals of the alphabet set. For this more general instance, we have designed a $o(nm)$ time algorithm.

We prove other technical results as well which we have not summarized here. In addition, a significant contribution of ours is the set of open problems we have posed. Note that these problem are not the obvious ones that arise out of our study. We have carefully devised simple versions of problems so they clearly identify certain fundamental difficulties in non-standard stringology (for example, see Sections 9.1, 9.3 and 9.6).

Our techniques. We use probabilistic techniques to a great extent. These techniques, although standard, have not been extensively used before in combinatorial pattern matching. There is a caveat in designing algorithms for string matching based on random sampling. If we require that the error probability to be polynomially small in n (rather than m), then we are sometimes required to restrict the size of the text to be polynomial in m; this is relevant especially in Sections 5 and 6.

Related Work. A number of practical algorithms are known for problems in non-standard stringology [WM92,BYG89]. For some non-standard string matching problems, algorithms (that do not rely on convolutions) are known that are fast in general, but they take $O(nm)$ time in the worst case [GG88,Uk85,Pi85]. Also, $O(nm/\log n)$ time algorithms are known for some of the problems using the Four Russians' trick [MaP80]. Finally, if you generalize the non-standard string matching problems allowing gaps etc., convolutions do not help; dynamic programming does (see for example [LV89]).

Map. For definitions, see Section 2. The five technical results above are described in Sections 3 through 7. Our contribution of open problems can be found in the Appendix, Section 9.

2   Definitions and Preliminaries

Given two vectors $a_1 a_2 \ldots a_n$ and $b_1 b_2 \ldots b_m$, $n \geq m$, their convolution $(\bigotimes, \bigoplus)$ is the vector $c_1 c_2 \ldots c_{n-m+1}$, where

$$c_i = \bigoplus_{1 \leq j \leq m} [\bigotimes(a_{i+j-1}, b_j)]$$

for $1 \leq i \leq n-m+1$. In *boolean convolution*, vectors are boolean, $\bigotimes$ is $\bigwedge$ (logical AND) and $\bigoplus$ is $\bigvee$ (logical OR). In *integer convolution*, vectors are binary (or composed of numbers represented in $O(\log n)$ bits), $\bigotimes$ is $\times$ (multiply operation) and $\bigoplus$ is $+$ (add operation). Both these convolutions can be done in $O(n \log m)$ time [AHU74]. (Our definition of convolution is different from the standard one in [AHU74], although it is equivalent to the one there.)

The general string matching and counting matching problems, **ccn**, and **cpn** have already been defined in Section 1. We use the following result.

Proposition 1 *[MP94] The **gsm** problem on text t of length n, pattern p of length m, and conflict graph G_C, can be solved using **ccn**(G_C) boolean convolutions of vectors of length n and m. The **gcm** problem on the same input can be solved using **cpn**(G_C) integer convolutions.*

We denote a complete bipartite graph with n vertices on the left and n on the right by $K_{n,n}$. We denote a perfect matching in this graph by $M_{n,n}$. The graph $K_{n,n} - M_{n,n}$ denotes the bipartite clique on n vertices in which a set of perfect matching edges has been removed.

3 On Clique Covers and NS String Matching

In this section, we explore problems on cliques covers in non-standard string matching. We identify an interesting class of conflict graphs (motivated by noisy string matching) and establish a nontrivial bound on their ccn.

In practice (such as in DNA sequencing), noise is unavoidable in sensing underlying strings. In such scenarios, it is reasonable to assume that a symbol in the observed string in fact represents a small set of (say, d) possible symbols in the underlying string that get mapped on to this symbol owing to noise. Motivated by string matching in such settings, we consider match graphs as in the theorem below.

Theorem 1 *Consider a bipartite graph* $G = (L, R, E)$, $|L| = n_1$, $|R| = n_2$, *in which each node in L has degree* $\geq n_2 - d_1$ *and each node in R has degree* $\geq n_1 - d_2$. *Then,*

$$\mathbf{ccn}(G) = O((\min\{d_1, d_2\})\log(n_1 n_2))$$

In particular, when $n_1 = n_2$ and $d_1 = d_2$, $\mathbf{ccn}(G) = O(d \log n)$.

Proof. Our proof uses the probabilistic method [AS93] to prove the existence of a clique cover of appropriate size for G. Assume without loss of generality that $d_1 \geq d_2$.

Pick cliques $C_1, \ldots, C_l$ independently and randomly. Each clique C_i is picked uniformly as follows. Each vertex in L is picked with probability p independently to be in the left set of C_i. The right set of C_i is defined to be the intersection of the neighbor sets of each of the vertices in the left set. Clearly that defines a bipartite clique. Since an edge (i, j) is in a clique C_k if and only if i is chosen and none of the neighbors of j are chosen,

$$\Pr(\text{an } (i, j) \text{ in } G \text{ is in a clique } C_k) \geq p(1 - p)^{d_2}$$

Therefore,

$$\Pr(\text{an } (i, j) \text{ in } G \text{ is not in any clique } C_k) \leq \left(1 - p(1 - p)^{d_2}\right)^l$$

Then,

$$\Pr(\text{some edge } (i, j) \text{ in } G \text{ is not in any clique } C_k)$$
$$\leq n_1 n_2 (1 - p(1 - p)^{d_2})^l \leq n_1 n_2 e^{-ple^{-pd_2}}$$

For this probability to be strictly less than 1, we choose $p = 1/2d_2$ and $l = O(d_2 \log(n_1 n_2))$. Then by the standard probabilistic argument [AS93], it follows that there exists a clique cover of size $O(d_2 \log(n_1 n_2))$ for G; that completes the proof. □

It is easy to see that our proof gives a Las Vegas type randomized algorithm for picking $O(d \log n)$ cliques in a cover. As a corollary, general string matching over the graphs in the Theorem 1 can be done in $O(nd\text{polylog}m)$ time by a Las Vegas type randomized algorithm. In contrast, previously best known algorithms take time $\Omega(nd^3\text{polylog}m)$ time [MP94].

Problem 1 *Design a deterministic algorithm to obtain the clique cover in Theorem 1.*

Remark. There exist graphs satisfying the conditions in the Theorem 1 for which $\Omega(d\log(n/d))$ cliques are needed in any clique cover (Mike Paterson, personal communication).

4 Approximate Clique Partitions

In general, it seems harder to count the mismatches between the pattern and text substrings than to detect occurrences of the pattern in the text. For example, the standard string matching problem can be solved in $O(n + m)$ time while counting the mismatches takes $O(n\sqrt{m}\text{polylog}m)$ time. In what follows, we study approximations to the problem of counting the mismatches and provide first nontrivial general (probabilistic) upper bounds.

Definition 1 *Given a bipartite graph G, define k-clique cover of G to be a clique cover of G of minimum size in which each edge appears in at most k of the cliques. Let k-ccn be the size of such a cover.*

Theorem 2 *The general counting matching problem on text t, pattern p, and conflict graph G_c can be approximated to a multiplicative factor k using k-ccn(G_c) integer convolutions.*

Proof. Simple (therefore omitted). □

That provides the motivation for obtaining tight bounds for k-cpn(G). Trivially, cpn$(G) \geq k$-ccn$(G) \geq$ ccn(G) for a graph G. Consider the graph $H = K_{n,n} - M_{n,n}$. Then for any k, $\Omega(n^{1/k}) \leq k$-ccn$(H) \leq kn^{1/k}$ (The lower bound is in [KNSW92] and the upper bound is implicit in [AL88] and explicit in [KNSW92].)[2] We consider the match graph for noisy string matching from the previous section.

Theorem 3 *Consider a bipartite graph $G = (L, R, E)$, $|L| = |R| = n$, in which each node in L and R has degree $\geq n - d$, $d \geq 1$. Then,*

$$O(\log n)\text{-ccn}(G) = O(d\log n)$$

Moreover, there exists a graph such that no collection of $O(d\log n)$ bipartite cliques can be a $O\left(\frac{\log n}{\log\log n + \log d}\right)$-clique cover.

Proof. The proof of the former claim is probabilistic and similar to that of Theorem 1. In fact, consider picking the bipartite cliques in the manner described there. Let X_{ij} be the random variable denoting the number of these cliques that contain the edge (i, j). Set $l = O(d\log n)$ and $p = 1/(2d)$. Then,

---

[2] For the special case of $H = K_{n,n} - M_{n,n}$, Karloff [K93] has an elegant approach for approximately counting mismatches that does not directly rely on k-clique covers.

$$\mathbf{E}(X_{ij}) = lp(1-p)^d = \Theta(\log n).$$

By Chernoff bounds [AS93], there exists a constant c such that

$$\Pr(\mathbf{E}(X_{ij}) \geq c \log n) < \tfrac{1}{n^2}$$

Therefore,

$$\Pr(\text{ for some } (i,j) \in E, \mathbf{E}(X_{ij}) \geq c \log n)$$
$$\leq n^2 \Pr(\text{ for a particular edge } (i,j), \mathbf{E}(X_{ij}) \geq c \log n) < 1$$

Thus there exists a collection of $O(d \log n)$ cliques that cover the edges of G in such a way each edge lies in $O(\log n)$ cliques.

The lower bound follows easily. Consider the graph $K_{n,n} - M_{n,n}$ for which k-ccn is $\Omega(n^{1/k})$ as shown by [KNSW92]; here, $d = 1$. For that graph, our upper bound shows the existence of a k-clique cover of size $O(d \log n)$. Therefore, $d \log n = \Omega(n^{1/k})$. Hence, $k = \Omega\left(\frac{\log n}{\log \log n + \log d}\right)$ □

Problem 2 *Is there a small k-clique cover (say of size $n^{O(1/k)}$) for the graph in the Theorem above when $k = o(\log n)$ (in particular, when $k = O(1)$).*

Our probabilistic construction above does not yield nontrivial bounds for this case. We give a partial result based on an alternate construction.

Theorem 4 *Consider a bipartite graph $G = (L, R, E)$, $|L| = |R| = n$, in which each node in L and R has degree $\geq n - d$, $d \geq 1$. Then for some constant C, $k = O(\log n)$ and $\sqrt{d} \leq 2^{k/C}$,*

$$O(k)\text{-ccn}(G) = O\left(\frac{nd^{3/2}k}{2^{k/C}}\right)$$

Proof. Fix C to be specified later. Divide the problem into $L = \frac{n\sqrt{d}}{2^{k/C}}$ subproblems each of which has $l = \frac{2^{k/C}}{\sqrt{d}}$ disjoint vertices on the left together with all the n vertices on the right. Consider one such subproblem. Divide the vertices on the right into two sets S and S'; the former comprises vertices not connected to at least one of the vertices on the left, and the latter comprises the rest. Clearly $|S| \leq ld$ (since each vertex is not connected to at most d vertices). Let all the vertices on the left together with those in S' on the right be one subgraph (this subgraph is a clique and each edge in this clique appears only once in the clique cover); the problem is then reduced to finding the k-clique cover of a graph of $l \times ld$ vertices (on the vertices in S on the right). Now we follow the argument in Theorem 3. That is, there is a k-clique cover of this graph of size s if

$$dl^2 e^{-sp(1-p)^d} < 1$$

and $sp(1 - p)^d = k/c$, for some constant c (that appears in applying the Chernoff's bound). We choose $p = 1/2d$, $s = O(d \log l) = O(dk)$ and set $C = c/3$ to satisfy both these constraints.

We omit the argument that the (probabilistic) construction above gives a k-clique cover for the entire graph. The total size of this cover is

$$L(1 + dk) = O\left(\frac{n\sqrt{d}dk}{2^{k/C}}\right)$$

which gives the theorem. $\qquad\square$

To understand the bound in the theorem above, let $k = c \log n$, for a sufficiently small constant $c < 1$ and $d = O(1)$. Then k-ccn of such a graph is $O(|\Sigma|^\epsilon \text{polylog} m)$, where $\epsilon < 1$ is a positive constant. Thus, $c \log n$–approximation for such a graph can be computed in $O(n|\Sigma|^\epsilon \text{polylog} m) = O(nm^\epsilon \text{polylog} m)$ time. In contrast, the best known algorithms for *exactly* computing the number of mismatches takes $O(nm)$ in the worst case for many graphs even with $d = O(1)$.

5   Finding k-witnesses for convolution

In this section we consider the k-witnesses problem.

Definition 2 *Given a text t, a pattern p and a match relation M, determine for each text location, k locations in the pattern where the pattern mismatches the aligned text location (under M) when placed on the text beginning at that location. (If there are fewer than k mismatches for some text location, all the mismatching locations in the pattern must be returned). We call this the k-witnesses problem; each of the pattern locations in the output for a text location is called a* witness.

The k-witnesses problem, besides being a natural generalization of the k-mismatches problems, is interesting in its own right. For instance, in applications in computational biology, one might look at the witnesses and use specialized knowledge supplied by the Biologist to resolve the significance of the mismatch. In what follows, we provide two algorithms for this problem: the first is a randomized one that is efficient when k is small, and the other is a deterministic one that is efficient in the worst case (that is, for general value of k). We can use of randomized algorithm to improve the best-known bounds for two dimensional approximate matching of non-rectangular figures [AF91].

If the match graph were $M_{n,n}$, then the k-witnesses problem is easy. The algorithm of Landau and Vishkin [LV89] that relies on suffix trees can pull out the k-witnesses for each text location with no more expense than it takes to solve the k-mismatches problem, that is, in $O(|t|k)$ time. (In fact, this algorithm determines the *leftmost* k witnesses for each text location).

The procedure above does not work when the match graph is nontrivial. In what follows, we provide alternate algorithms; our algorithms do not necessarily determine the k leftmost witnesses.

Theorem 5 *Given a threshold matching problem on text t, pattern p and a conflict graph G_C, known algorithms solve it using $\mathbf{cpn}(G_C)$ integer convolutions of two 0/1 vectors of length $|t|$ and $|p|$ respectively. The k-witnesses problem on t, p, and G_C can be solved using $\mathbf{cpn}(G_C)$ invocations of the k-witnesses problem for the boolean convolution of two 0/1 vectors of length $|t|$ and $|p|$ respectively.*

Proof. We claim that the desired witnesses for a text location is the union of the witnesses for that location derived from each of the $\mathbf{cpn}(G)$ boolean convolutions. The (simple) proof is omitted. □

Therefore, it suffices to focus on the k-witnesses problem for the standard boolean convolution. Let $a[1 \cdots n]$ and $b[1 \cdots m]$ be the two vectors where the length of the text and the pattern are n and m respectively.

We first describe our randomized algorithm. It is derived using the intuition in [Se92] where Seidel developed a 1-witness algorithm for boolean matrix multiplication in the context of the all-pairs shortest path problem. The following simple proposition is the key.

Proposition 2 *[Se92,AGMN92] Consider a set $s = \{s_1, \cdots, s_l\}$ of positions in b, that is, $1 \le s_i \le m$, for each i. Let b' be defined as follows: $b'[i] = 0$ if $i \notin s$ and $b'[i] = i$ otherwise. Let c be the integer convolution of a and b'. If for some text location l there exists only one witness s_i among the positions in s, then $c[l] = s_i$.*

Proof. Follows from the definition of integer convolution easily. □

Theorem 6 *There exists a Las Vegas type algorithm to the k-witnesses problem that takes $O(nk^2 \log^3 m \log n)$ expected time with high probability.*

Proof. The overall strategy in the algorithm is to randomly sample the positions in b (several times) so that the witnesses are discovered one after another, each witness being discovered using the lemma above, that is, by making sure that witness is the *only* witness in some sample. The key is to determine the size of the random sample (S) as well as the number of times (R) samples of a particular size have to be drawn in order to insure the condition above.

In what follows, we consider a *fixed* text location f. Say it has w witnesses in all and *for now assume* that w is known. We are required to determine k of these witnesses ($k \le w$). We choose a S such that $m/2 \le wS \le m$; clearly this choice of S is possible. Now we are required to determine R; for this, we perform two calculations. Consider a sample of S positions in b picked independently, uniformly and randomly from $1 \cdots m$. Consider a particular witness s for f.

$$\Pr(s \text{ is picked exactly once in a } S \text{ sample}) = S\frac{1}{m}(1 - \frac{w}{m})^{S-1}$$

$$\geq \frac{Se^{-w(S-1)/m}}{m}$$

$$\geq \frac{1}{2we} \quad (\text{using } m/2 \leq wS \leq m)$$

We now consider the failure probability for f, that is, the probability that k witnesses are not chosen.

$$\Pr(k \text{ witnesses are not chosen})) \leq \sum_{w-k+1 \leq i \leq w} \binom{w}{i}(1 - \frac{1}{2we})^{iR}$$

$$\leq \sum_{w-k+1 \leq i \leq w} \binom{w}{i}(1 - \frac{1}{2we})^{(w-k+1)R}$$

$$\leq \sum_{w-k+1 \leq i \leq w} \binom{w}{i}e^{-\frac{(w-k+1)R}{2we}}$$

Now consider the failure probability for any text location that has w witnesses. Let $P = \Pr(\text{for some location, } k \text{ witnesses are not found})$.

$$P \leq n \sum_{w-k+1 \leq i \leq w} \binom{w}{i}e^{-\frac{(w-k+1)R}{2we}}$$

$$\leq n \sum_{0 \leq j \leq k-1} \binom{w}{j}e^{-\frac{(w-k+1)R}{2we}}$$

$$\leq nw^k e^{-\frac{(w-k+1)R}{2we}} = O(n^{-c})$$

for some constant c, when $R = O\left(\frac{w}{w-k+1}k \log w \log n\right)$. Since $\frac{w}{w-k+1} \leq 2k$ and $w \leq m$, $R = O\left(k^2 \log m \log n\right)$ suffices.

In general different text locations need not have the same number of witnesses. Say the text location i has w_i witnesses. For each such w_i, we require a sample of size S_i in the calculations above in order to determine the k witnesses, where $m/2 \leq w_i S_i \leq m$. A simple observation is that it suffices to choose S to be each of the values 2^i, $i = 0, \cdots, (\log m - 1)$. Then for each w_i there always exists some S_i from among these values for which $m/2 \leq w_i S_i \leq m$.

That completes the description of the major details. Now we can summarize the algorithm. Consider each $S_i = 2^i$, $i = 0, \cdots, (\log m - 1)$ in succession. For a given S_i, pick $R = O\left(k^2 \log m \log n\right)$ random samples of size S_i. Each sample of size S_i is picked by uniformly, randomly, choosing locations from $1 \cdots m$ with replacement. For each such choice, an integer convolution is performed as in Proposition 2 to determine a witness, if any, for each text location. Finally, all the witnesses found in this manner are returned.

In all, there are $O(R \log m) = O(k^2 \log^2 m \log n)$ random samples and for each sample, we need to perform an integer convolution in using the Proposition 2 which takes $O(nk^2 \log^3 m \log n)$ time in all. There are a number of algorithmic details which we omit here; these have to do with, for instance, maintaining the distinct set of witnesses (for each text location) and ensuring the algorithm is of Las Vegas type. We claim that the algorithm can be made to work in $O(nk^2 \log^3 m \log n)$ time in all. That gives the theorem. □

Problem 3 *Design a deterministic algorithm for the k-witnesses problem with comparable performance.*

We believe our randomized algorithm can be derandomized in the same manner as that in [AGMN92]. Here we present a different deterministic algorithm which is efficient for large k.

Theorem 7 *There exists a deterministic algorithm for the k-witnesses problem that takes $O(n\sqrt{km \log m})$ time.*

Proof. (Sketch) Divide the pattern into disjoint blocks of size z each. There are m/z such blocks. For each text location, determine the number of witnesses in each block by m/z integer convolutions in all. Following that, we can determine for each text location, a set of blocks in which the number of witnesses is k (or possibly, greater). There are at most k such blocks per text location. For each text location, check each such block brute force for the (up to) k witnesses. The total time taken is $O(nkz + n\frac{m}{z} \log m)$. Set $z = \sqrt{\frac{m \log m}{k}}$ to balance the two terms; that yields the time complexity in the theorem. □

We claim that our randomized algorithm can be used to improve the result in [AF91] on approximate non-rectangular two dimensional matching under the edit distance metric. We merely note that the $O(\sqrt{k \log k})$ factor there can be replaced by a factor $O(\text{polylog}\, m)$, which yields an improved algorithm for their problem for a significant range of values for k (at the expense of randomization). We defer details to the final version.

6 Combinatorial algorithms for non-standard matching

All $o(nm)$ algorithms known for non-standard string matching problems use convolutions (except those which rely on the Four Russian's trick). These algorithms are complex since $o(nm)$ time algorithms for convolutions rely on complex operations such as the FFT. So, the following problem is open and it has been mentioned by several researchers:

Problem 4 *Design a $o(nm)$ time combinatorial algorithm (that is, one which does not use algebraic convolutions) for string matching with wildcards or boolean convolution or one of the non-standard string matching problems with nontrivial match graph.*

Here we are not able to settle the open problem above; but we provide a combinatorial algorithm to an approximation that we define below.

Definition 3 *For a non-standard string matching problem with text t, pattern p, match relation M and a parameter δ, determine a subset (possibly all) of those positions in the text where the pattern mismatches under M; however, this subset is guaranteed to contain all those positions where the pattern mismatches the text in at least δ positions under M. We call this the δ-approximation problem.*

Comments on the δ-approximation problem. Note that solving the 1-approximation is equivalent to solving the exact version. The δ-approximation is an approximation of the exact matching in the following sense: instead of finding all positions where the pattern occurs in the text (under M) without any distortions (i.e., no mismatches), we are guaranteed to find at least all those positions where the pattern occurs in the text sufficiently distorted (that is, with more than δ mismatches).

There is a difference between the δ-approximation and standard approximation for exact matching, namely, the k-mismatches problem[3]. If we insist that *only* those locations where the pattern matches the text with δ or more witnesses, then the δ-approximation is precisely the $m - δ$-mismatches problem. However, in the δ-approximation we are given the extra flexibility to return some text locations where the pattern occurs with fewer than δ mismatches. This flexibility is crucial in our $o(nm)$ combinatorial upper bound for this problem.

Theorem 8 *There is a Monte Carlo type randomized algorithm (using only text to pattern comparisons) that takes $O(n(m/δ)\text{polylog}n)$ time to solve the δ-approximation of a non-standard string matching problem on a text of length n and a pattern of length m.*

Proof. (Sketch) For a position i in the text, if $p[j]$ does not match $t[i+j-1]$, for some $1 \leq j \leq m$, then j is called a *witness* for i. Let W_i be the set of witnesses of i. Note that each text position has possibly several witnesses.

The algorithm relies on random sampling to search for witnesses for each text position. Pick κ locations $l_1, \cdots, l_\kappa$, each independently, uniformly and randomly from the range $1 \cdots m$. For each of these positions l_j and for each text position i, determine if l_j is a witness for i (that is, if $l_j \in W_i$). Output all text positions for which a witness was found in the process.

Now we determine the probability that our algorithm makes a mistake, that is, some position in the text with at least δ witnesses is not in the output. Clearly,

$$\Pr(\text{for a fixed } i, j \in W_i \text{ and } j \neq l_k \text{ for some } k) \quad \leq \quad \left(1 - \tfrac{δ}{m}\right)^\kappa$$

---

[3] In the k-mismatches problem, we are required to return the positions in the text where the pattern occurs with at most k mismatches.

$$\Pr(\exists i,\ j \in W_i \text{ and } j \neq l_k \text{ for some } k) \leq n\left(1 - \frac{\delta}{m}\right)^{\kappa}$$
$$\leq ne^{\frac{-\delta\kappa}{m}}$$
$$= O(n^{-c}),\ \text{for some constant } c,$$

when $\kappa = O(\frac{m}{\delta}\text{polylog}n)$.

The running time of the algorithm is $O(n\kappa) = O(\frac{nm}{\delta}\text{polylog}n)$, since each test of the form $j \in W_i$ can be done in $O(1)$ time. □

Setting the value of δ to be $\sqrt{m}$ for instance, we get a $\sqrt{m}$-approximation algorithm that takes time $O(nm^{1/2}\text{polylog}n)$; importantly, our algorithm is purely combinatorial.

7 String matching with sets of ranges

There is no $o(nm)$ time algorithm for the general string matching problem. The best known upper bound is $o(nm)$ only when the match graph is not very dense (See [MR92] for details). Therefore, it is of interest to identify instances of general string matching that have $o(nm)$ time algorithms even if the match graph is dense. Instances that have been studied so far include string matching in which the positions of the pattern are (not necessarily singleton) subsets of the alphabet set [Ab87] and string matching in which the positions of the pattern are each a range (i.e., a consecutive subset) of the ordered alphabet set [AF91,MP94]. Let the former be Problem **A** and the latter, Problem **B**. Problems **A** and **B** have also been studied with extensions: wild cards, complements of symbols etc.

Here we study a *gsm* problem that includes all the above mentioned variants, that is, we consider string matching in which each pattern position is a (possibly singleton) subset of ranges of the ordered alphabet set. Let this be Problem **C**.

Example. The string $aba\langle c + a\rangle d$ is an example for the pattern in Problem **A**; here, the fourth location of the pattern can match either c or a in the aligned text. The string $ab[c\cdots f]d$ is an example for the pattern in Problem **B**; here, the third location of the pattern can match c, d, e or f in the aligned text. The string $ab\langle a + [c\cdots f]\rangle d$ is an example for the pattern in Problem **C** with the natural interpretation.

It is easy to verify that instances of Problems **A** and **B** (with extensions mentioned earlier) are instances of Problem **C**. In what follows, we provide an efficient algorithm for Problem **C**. Let M denote the total number of elements in the pattern, that is, counting separate symbols and ranges as one element each. For example, for the pattern $ab\langle a + [c\cdots f]\rangle d$ above, $M = 1 + 1 + (1 + 1) + 1 = 5$.

Theorem 9 *Problem* **C** *can be solved in* $O(n + M + M^{1/3}(nm)^{2/3}\text{polylog}m)$ *time deterministically.*

Proof.(Sketch) Let $n \leq 2m$ (as is standard, the text can always be broken into pieces to satisfy this constraint); we show that Problem **C** can be solved in $O(M + M^{1/3}m^{4/3}\text{polylog}m)$ time.

First we claim that the text and the pattern symbols can be renumbered so the alphabet size is $O(m)$ irrespective of M. We omit the details of this claim except to note that since there are only $O(m)$ symbols in the text, there are only $O(m)$ "interesting" endpoints of ranges. Next we claim that Problem C can be solved in $O(n|\Sigma|\text{polylog}m)$ time using the Fischer-Paterson technique where $\Sigma = O(m)$ is the number of distinct symbols in the text even when the pattern and/or the text have wild cards. In what follows, we will use this claim.

Consider all those symbols in the text which appear at least z times (z will be fixed later); there are $O(m/z)$ such symbols. The mismatches of the pattern in the text, if any, due to these symbols can be determined in $O(n(m/z)\text{polylog}m)$ time using the claim above. The rest of the symbols appear at most z times each. Divide the alphabet set into disjoint $O(m/y)$ segments of length y. Replace each range (or each separate symbol) by a (possibly singleton) subset of the minimum number of disjoint segments that subsume the range (or the symbol, respectively). Determine all the mismatches with this new alphabet set of size m/y; that takes $O(n(m/y)\text{polylog}m)$ time. Some mismatches have been missed in this process for the following reason. Some singleton symbols have been replaced by the segment they fall in. Therefore, they have matched positions wherever this segment matches and not necessarily where the symbol matches. We can easily correct for this in $O(Mzy)$ time since each of the at most M ranges has to consider at most $O(y)$ symbols, each of which appears at most z times. Thus the total time taken is $O(n(m/y)\text{polylog}m+n(m/z)\text{polylog}m+Mzy)$. Set $y = z$ and the running time is $O(n(m/z)\text{polylog}m + Mz^2)$. Set $z = O\left(\left(\frac{nm}{M}\right)^{1/3}\text{polylog}m\right)$ and since $n \leq 2m$, the running time is $O(M^{1/3}m^{4/3}\text{polylog}m)$; that gives the theorem. We have omitted a number of details of our algorithm and its formal description. $\qquad\Box$

Note that the running time in the theorem above is $o(nm)$ as long as $M = o(nm/\text{polylog}m)$. (This condition is the same as the case in [Ab87] for Problem A). If $M = O(m)$, the running time above is $O(nm^{2/3}\text{polylog}m)$ for Problem C.

8 Acknowledgement

Sincere thanks to Babu Narayanan for discussions, insights and help with Theorem 3. Thanks to Peter Winkler and Mike Paterson as well for discussions.

References

[A89]   A. Aho. Algorithms for finding patterns in strings. *Handbook of theoretical computer science*, Vol 1, Van Leeuwen Ed., 1989.

[Ab87]   K. Abrahamson. Generalized string matching. *SIAM J. Comp.*, 1987, 1039-1051.

[AC75]   A. Aho and M. Corasick. Efficient string searching: An aid to biblio-graphic search. *Comm. of the ACM*, 18(6), 1975, 333-340.

[AF91]   A. Amir and M. Farach. Efficient 2-dimensional Approximate Match-ing of Non-rectangular Figures. *Proc of 2nd Ann ACM Symp on Dis-crete Algorithms*, 1991, 212-222.

[AGMN92] N. Alon, Z. Galil, O. Margalit and M. Naor. Witnesses for boolean matrix multiplication and for shortest paths. *Proc. 33rd Ann. IEEE Symp. Foundations of CS*, 1992, 417–426.

[AHU74]   A. Aho, J. Hopcroft, and J. Ullman. The design and analysis of computer algorithms. *Addison-Wesley Publishers*, 1974.

[AL88]   A. Amir and G. Landau. Fast serial and parallel multidimensional approximate array matching. *Theoretical Computer Science*, 81, 1991, 97-115.

[AS93]   N. Alon and J. Spencer. The probabilistic method. Wiley, 1993.

[BYG89]   R. Baeza-Yates and G. Gonnet. A new approach to text searching. *Proc. ACM SIGIR*, Cambridge, Mass., 12:168-175, 1989.

[C71]   S. Cook. Linear time simulation of deterministic two-way pushdown automata. *Proc IFIP Congress*, 1971.

[DSO79]   M. Dayhoff, R. Schwartz and B. Orcutt. A model for evolutionary change in proteins, in Dayhoff, ed., *Atlas of Protein Sequence and Structure*, 5, 1979, 345–352.

[FP74]   M. Fischer and M. Paterson. String Matching and other Products. *SIAM-AMS Proceedings*, Vol. 7, 113-125, 1974.

[Ga79]   Z. Galil. Some open problems in the theory of computation as ques-tions about two-way deterministic pushdown automaton languages. *Mathematical Systems Theory*, 1979, 211–228.

[Ga85]   Z. Galil. Open problems in stringology. *Combinatorial Algorithms on Words*, A. Apostolico and Z. Galil Eds, Springer-Verlag Lecture Notes, 1985. 1-8.

[GG88]   Z. Galil and R. Giancarlo. Data structures and algorithms for approx-imate string matching. *Journal of Complexity*, 4(1988), 33-72.

[HU79]   J. Hopcroft and J. Ullman. *Introduction to Automata Theory, Lan-guages and Computation*, Addison-Wesley, Reading, Mass., 1979.

[K93]   H. Karloff. Fast algorithms for approximately counting mismatches. *Manuscript*, 1993.

[Ko87]   S.R. Kosaraju. Efficient string searching. Manuscript, 1987.

[Ko89]   S.R. Kosaraju. Efficient tree pattern matching. *Proc IEEE Ann. Symp. on FOCS*, 1989, 178-183.

[KMP77] D. E. Knuth, J. H. Morris, and V. R. Pratt. Fast pattern matching in strings. *SIAM J. Computing*, 6:323–350, 1977.

[KNSW92] M. Karchmer, I. Newman, M. Saks and A. Wigderson. Non-deterministic communication complexity with few witnesses. *Manuscript*, 1992.

[KR87]   R. Karp and M.O. Rabin. Efficient randomized pattern matching algorithms. *IBM Journal of Research and Development*, 31(2), 249-260.

[Lov]   L. Lovasz. Communication complexity - a survey. *Paths, Flows and VLSI Layout*, Korte, Lovasz, Promel, Schrijver Eds., Springer-Verlag (1990), 235-266.

[LV89]   G.M. Landau and U. Vishkin. Fast parallel and serial approximate string matching. *Journal of Algorithms*, Vol.10 2(1989), 262-272.

[LW75]   R. Lowrance and R. Wagner. An extension of the string-to-string correction problem. *Journal of Association of Computing Machinery*, 22, 1975, 177–183.

[MP94]   S. Muthukrishnan and K. Palem. Non-standard stringology: algorithms and complexity. *Proc. 26th Annual ACM Symp. on the Theory of Computing*, 1994, 770-779.

[MR92]   S. Muthukrishnan and H. Ramesh. String matching under general match relation. *Proc 12th FST & TCS*, India, LNCS, Springer-Verlag, Vol. 652, 1992, 356-367.

[MaP80] W. Masek and M. Paterson. A faster algorithm for computing string-edit distances. *Journal of Computer and System Sciences*, 20(1), 1980, 18–31.

[P94]   V. Pan. *Personal Communication*, 1994.

[Pi85]   R.Y. Pinter. Efficient string matching with don't-care in patterns. *Combinatorial Algorithms on Words*, NATO-ASI series, pp. 11-29, 1985. Editors: A. Apostolico and Z. Galil.

[Se92]   R. Seidel. On the all-pairs-shortest-path problems. *Proc. 24th Ann. ACM Symp. Theory of Computing*, 1992, 745–749.

[Uk85]   E. Ukkonen. Finding approximate patterns in strings. *Journal of Algorithms*, Vol.6, 1985, 132-137.

[WM92]   S. Wu and U. Manber. Fast text searching allowing errors. *Communications of ACM*, 35, 1992, 83-91.

9 Appendix A: Some open problems

9.1 Weighted k-mismatches problem

Consider the following weighted k-mismatches problem.

Problem 5 *Given a text $t = t[1 \cdots n]$ and a pattern $p = p[1 \cdots m]$ (both over the alphabet set Σ), a real parameter Δ, and a weight function $f : \Sigma \times \Sigma \Rightarrow \Re$, determine all positions i in the text where $\sum_{j=1}^{j=m} f(t[i+j-1], p[j]) \leq \Delta$. For what functions f can this problem be solved in $o(nm)$ time?*

Consider the simple function $f(\sigma, \tau) = 1$ when $\sigma \neq \tau$ and $f(\sigma, \tau) = 0$ otherwise. In this case, this problem is the standard k-mismatches problem and it can be solved in $O(n\sqrt{m}\text{polylog}m)$ time [Ab87,Ko89]. We do not know of any other functions for which the trick in [Ab87,Ko89] can be applied to get an $o(nm)$ time algorithm. Consider another simple function f over the alphabet set of natural numbers where $f(\sigma, \tau) = |\sigma - \tau|$. Can this be solved in $o(nm)$ time?

The notion of weighting pairwise matches is common in Computational Biology, for example, the PAM matrices in [DSO79]. Note that using convolutions, this problem can be solved (in general) trivially in $O(n|\Sigma|\text{polylog}m)$ time when the range of f can be represented using $O(\log m)$ bits.

9.2 String Matching with Wildcards on 2DPDA

In this section, we pose a string matching problem on one two-way head deterministic (non-) pushdown automata, denoted 2DPDA (2NPDA respectively). See [HU79] for a description of these machines.

Problem 6 *Can we prove that string matching with wildcards cannot be solved on a 2DPDA? More specifically, let*

$$L = \{t\#p \mid p \text{ is a substring of } t \text{ and } p \text{ contains wildcards}\}$$

Can we prove that L cannot be recognized by a 2DPDA?

Remark. Cook [C71] proved (the amazing result!) that any language recognized by a 2PDPA can be recognized in *linear* time on a RAM irrespective of the time taken on the 2DPDA.[4] We suspect string matching with wildcards cannot be solved in linear time on a RAM (Some evidence for this can be found in [MR92,MP94]). This would imply that string matching with wildcards cannot be recognized by a 2DPDA. Since proving lower bounds for problems on the RAM model is difficult, as a weak evidence that string matching with wildcards cannot be solved in linear time on a RAM, we can attempt to prove that string matching with wildcards cannot be recognized on a 2DPDA (even with unbounded time!).

---

[4]Historically, this was the first linear time algorithm for the standard string matching problem since string matching could be performed on a 2DPDA in quadratic time naively.

Solving this problem would mean separating the 2DPDA from the 2NPDA, which is a longstanding open problem. See [Ga79] in this connection. The difficulty in proving the lower bounds for 2DPDAs lies in the ability of the heads to move in both directions. Can meaningful lower bounds for this problem be proved by restricting the head to making at most k reversals?

9.3 String Matching with Swaps

In this section, we define an approximate string matching problem based on only neighbor-swap operations. Formally,

Problem 7 *Given a text string of length n and a pattern string of length m, determine all positions i in the text where $t[i] \cdots t[i + m - 1] = f(p[1] \cdots p[m])$, where the function f swaps some* disjoint *set of neighbors in p. Design an $o(nm)$ time algorithm for this problem.*

Example. If $t = ababcd$, patterns $baacb$, $aabcb$ and $babac$ occur at the leftmost position. The pattern $abbca$ does not occur at the leftmost position.

Remark. Note that this problem is practically motivated since neighbor-swaps are prevalent in typing errors. For this reason, they were considered together with other edit mistakes in [LW75] where an $O(nm)$ time algorithm was developed. To our knowledge, no $o(nm)$ time algorithm is known for this problem even when the alphabet set is fixed. Since no insertions and deletions are allowed, the pattern occurs in the text with its length preserved; therefore, one would expect the Fischer-Paterson [FP74] technique to yield an $O(n\mathrm{polylog}m)$ time algorithm for a fixed alphabet. Surprisingly, the Fischer-Paterson [FP74] approach does not apply; also, it is not apparent that convolutions are *necessary* to solve this problem. It is intriguing that such a simply stated problem appears hard.

We claim that this problem can be solved in time $O(nS)$ where S is the maximum number of disjoint neighbor-swaps in any occurrence of the pattern in the text. Details are similar to [LV89] and they are omitted.

9.4 Algebraic Clique Covers

Motivated by the problem of solving gsm problems using polynomial convolutions (rather than boolean convolutions as is standard), we define a generalization of clique covers; these clique covers have not been explored in literature thus far.

Consider a bipartite graph G and its adjacency matrix M (each row corresponds to a vertex on the left and each column to a vertex on the right). Let $C_1, \ldots, C_l$ be a collection of bipartite cliques and $M_1, \ldots M_l$ be their respective adjacency matrices. We want a decomposition of M such that, there is an assignment sgn(i) of $+$ or $-$, to each C_i, so the following holds: $M \equiv \sum_i \mathrm{sgn}(i)M_i$, where for two $n \times n$ matrices A and B,

$$A \equiv B \quad \Rightarrow \quad \begin{array}{l} B(i,j) \geq 1 \text{ if } A(i,j) = 1 \\ B(i,j) = 0 \text{ if } A(i,j) = 0. \end{array}$$

Any collection of such cliques of minimum cardinality for G is called the *algebraic clique cover* of G, denoted $\text{acc}(G)$, and its cardinality is the *algebraic clique cover number* of G, denoted $\text{accn}(G)$.

Remark. If $\text{sgn}(i) = 1$ for all i, then $\text{acc}(G)$ ($\text{accn}(G)$, respectively) is the same as $\text{cc}(G)$ ($\text{ccn}(G)$, respectively). Nothing is known about the accn of graphs. Normally problems that have both algebraic and graph-theoretic aspects are hard to handle; see literature in Communication Complexity [Lov].

Theorem 10 *Consider a bipartite graph $G = (L, R, E)$ and let $N(v)$ denote the set of neighbors of a vertex v in G. Let L' (R') be the number of distinct $N(v)$'s where $v \in L$ ($v \in R$, respectively). Then,*

$$\text{accn}(G) \geq \log_2\left(\max\{L', R'\}\right)$$

Proof. The proof is information-theoretic; omitted. □

Remark. As a corollary, note that for the special bipartite graph $H = K_{n,n} - M_{n,n}$, $\text{accn}(H) \geq \log_2 n$, since each of the vertices on the left have distinct neighbor sets.[5] Since $\text{accn}(H) \leq \text{ccn}(H)$ trivially and it has been proved that $\text{ccn}(H) \leq \log_2 n + 0.5 \log\log n + O(1)$ [MP94],

$$\log_2 n \leq \quad \text{accn}(H) \quad \leq \log_2 n + 0.5 \log\log n + O(1).$$

The exact value of $\text{accn}(H)$ is unknown.

Problem 8 *Is there a graph G for which $\text{accn}(G) \ll \text{ccn}(G)$? The following is easy to see: $\text{accn}(G) \leq 1 + \text{cpn}(\overline{G})$; does this help? Also, what is the best upper bound on $\text{accn}(G)$ where G is as in Theorem 1?*

9.5 Multiple pattern matching with wild cards

Problem 9 *Given a text t of length n and l patterns of length m each possibly containing wild cards, determine for each text location, all the patterns which occur there. Design an efficient algorithm for this problem.*

This is a generalization of string matching with wild cards to include multiple patterns. The best known complexity for this problem is the same as the naive one, namely, $O(nl\text{polylog}m)$, that corresponds to matching each pattern separately against the text using the result in [FP74]. Can this bound be improved? This problem is one of the simplest cases where we face fundamental issues in extending convolution-based approach to multiple pattern strings.

Consider the (simpler?) goal of obtaining the bound of $O(nW + n + ml)$ where W is the maximum number of wild cards in any pattern. If $W = 0$, this

---

[5] This was independently observed by Noga Alon in a personal communication.

problem corresponds to the standard multiple pattern matching problem which can indeed be solved in $O(n + ml)$ time [AC75]. If $l = 1$, then the algorithm in [Pi85] gives this bound although the bound in [FP74] is better for this case. Perhaps the approach in [Pi85] can be generalized to $l > 1$ strings to derive what we want. It is not hard to see that the natural approach of searching a trie of the wild card-free substrings of the patterns starting from each text location does not work.

9.6 Sparse Convolutions

Problem 10 *Given vectors* **a** *and* **b** *of length* $n \leq 2m$ *and* m *respectively, each containing at most* r *1's, determine their boolean convolution. Consider, in particular, the case when* r *is polynomially small in* m. *Can an* $o(n \log m)$ *time algorithm be developed for this problem?*

Trivially $O(nr)$ time suffices to solve this problem in general. Using the ideas in Section 6, we can get a randomized algorithm for $O(\sqrt{r})$–approximation that takes $O(n\sqrt{r}\text{polylog}r)$ time. Recently, Victor Pan [P94] has found hardness results for sparse convolutions.

Another interesting problem for standard boolean convolutions is the sparse output case.

Problem 11 *Given two vectors* **a** *and* **b** *of length* n *and* m *respectively and a set* $S = \{l_1, \ldots, l_s\}$ *of marked positions in* **a**, *determine the boolean convolution of* **a** *and* **b** *only at the* s *positions, that is, determine only* $c[l_1], \cdots, c[l_s]$ *where* **c** *is the boolean convolution vector of* **a** *and* **b**. *Accomplish this in* $o(n \log m)$ *time for appropriate size* s, *say* s *polynomially small in* n.

String Matching in Hypertext

Kunsoo Park* Dong Kyue Kim

Department of Computer Engineering, Seoul National University
Seoul 151-742, Korea

Abstract. In this paper we consider the string matching problem in hypertext, which is a nonlinear structure of text. We model the hypertext as a directed graph $G = (V, E)$, where each node $v \in V$ has text T_v associated with it and each link $(v, w) \in E$ connects the end of text T_v to the start of text T_w. We define the string matching problem in hypertext as follows: Given a graph G modeling a hypertext and a pattern P, find all occurrences of the pattern in graph G. The pattern length is m and the sum of the lengths of all texts T_v in G is N. The main difficulty in the hypertext string matching problem is that the pattern may occur across links.

There is a linear time algorithm for the case when graph G is a tree. In this paper we present a linear $O(N + |E|)$ time algorithm when $n_v = \text{length}(T_v)$ is larger than or equal to m for all v, and a more involved algorithm that takes $O(N + |E|m)$ time when there exist some nodes v with $n_v < m$. To obtain the results, we combine the notion of witnesses and duels with the suffix tree, which enables us to eliminate possible occurrences of any substring of the pattern.

1 Introduction

Hypertext has recently become very popular and is being used in a number of fields related to computer science and many applications [8]. Hypertext can be defined as text-nodes with connection links between them, and the search problem in hypertext is an important problem for efficient utilization of hypertext systems. Therefore, the development of efficient algorithms for finding data in hypertext is one of the central problems in hypertext research.

In the string matching problem, we want to find all occurrences of a pattern string in a text string. The Knuth-Morris-Pratt algorithm [15] and the Boyer-Moore algorithm [5, 11] are well-known classical algorithms for the problem. Algorithms for this problem and related problems are being used in many applications such as word processing, storage and transmission of data, compiler construction, computer vision, and molecular biology [1].

In this paper we consider the string matching problem in hypertext. We first give the computation model for the hypertext. We model the hypertext as a directed

---
* Work supported by NON-DIRECTED RESEARCH FUND, Korea Research Fund.
Email: kpark@theory.snu.ac.kr

graph $G = (V, E)$, where each node $v \in V$ has text T_v associated with it and each link $(v, w) \in E$ connects the end of text T_v to the start of text T_w [17]. We define the string matching problem in hypertext as follows: Given a graph G modeling a hypertext and a pattern P, find all occurrences of the pattern in graph G. The pattern length is m and the sum of the lengths of all texts T_v in G is N. Compared with the normal string matching problem, there can be pattern occurrences across links in the hypertext string matching problem.

There is a linear time algorithm for the case when graph G is a tree [10]. In this paper we present a linear $O(N + |E|)$ time algorithm for the case when $n_v = \text{length}(T_v)$ is larger than or equal to m for all v, and a more involved algorithm that takes $O(N + |E|m)$ time for the general case, i.e., it may be possible that there exist some nodes v with $n_v < m$. To obtain the results, we combine the notion of witnesses and duels with prefix matching and the suffix tree, especially combining duels and the suffix tree enables us to eliminate possible occurrences of *any substring* of the pattern.

The paper is organized as follows. In Section 2 we define the computation model of hypertext, and in Section 3 we describe the basic concepts used in developing the algorithms. In Section 4 we describe the algorithms for the case when $n_v \geq m$ for all node v and the case when there exist some nodes v with $n_v < m$. In Section 5 we conclude with some remarks.

2 Hypertext modeling

In general, hypertext is composed of text-nodes and links which connect the text-nodes. If we model one text-node in the hypertext as a node in a graph then there is no way to represent the location of links. There should be a secondary structure to represent the multiple links existing inside one text-node. To go around the problem of unbounded number of links associated with one node, we use the following model for hypertexts proposed by Manber and Wu [17].

In the directed graph $G = (V, E)$ that represents a hypertext, each node v has text T_v associated with it. Here, the text T_v for node v is not a text in the original hypertext but defined as the largest substring that does not contain a link. Every link in $G = (V, E)$ starts at the end of a text T_v and ends at the start of another (possibly the same) text T_w. Fig. 1 shows that one text-node in the original hypertext that contains multiple links is represented as a subgraph in this model.

In a hypertext there must be a text-node TN such that all other text-nodes can be reached from this text-node. In a modeled directed graph $G = (V, E)$, we define the node $v \in V$ corresponding to the first part of the text-node TN as *the source node*. See Fig. 1. In the matching algorithms we will start the search from the source node.

If there is no pattern occurrence across links, the string matching problem in hypertext is not much different from the normal string matching problem. By

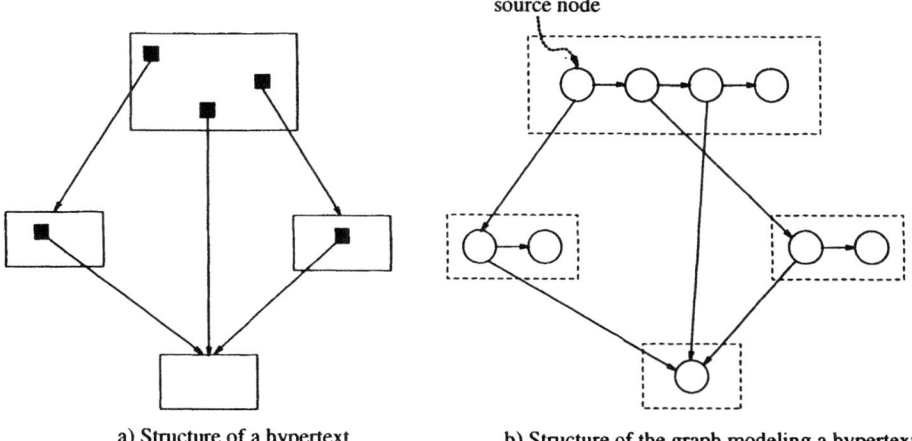

a) Structure of a hypertext   b) Structure of the graph modeling a hypertext

Fig. 1. A hypertext and its model.

the model proposed in this paper, a text-node in the original hypertext may be divided into multiple nodes in the modeled directed graph. In other words, it is possible that a text can be divided by links at any location. Thus what we want to do in this paper is to find the occurrences of the pattern across multiple nodes. For example, if the end of a text T_v is "data", there exists a link from v to w, and the start of T_w is "base", then we should have an occurrence of pattern "database" in the hypertext.

3 Basics

In this section we describe three basic concepts which will be used for string matching in a hypertext, i.e., KMP matching, witness & duel, and suffix tree.

The KMP matching algorithm was proposed by Knuth, Morris and Pratt [15], and was the first algorithm to solve the string matching problem in linear time. For strings x, w, and u, if $x = wu$ we say that w is a *prefix* of x and u is a *suffix* of x. In the preprocessing stage, the KMP algorithm computes the *failure function* $F(i)$ for each location i in the pattern. $F(i)$ is defined as the length of the longest proper substring that is a prefix and a suffix of $P[1..i]$. In the matching stage, the algorithm computes the *prefix value* q_j for each location j in the text. The prefix value q_j is defined as the length of the longest substring of the text ending at location j which is a prefix of the pattern. The KMP algorithm proceeds by comparing the pattern with the text character by character. In case of a match at text location j, q_j is the prefix value at the previous location $j - 1$ (q_{j-1}) plus one. In case of a mismatch at location j, $F(q_{j-1})$ is the length of the longest proper suffix and prefix at location $j - 1$, so the algorithm sets q_{j-1} as $F(q_{j-1})$ and repeats a comparison at location j until there is a match or q_{j-1} becomes

0. In doing that, if $q_j = m$ then there is an occurrence of the pattern ending at location j.

The notion of witness and duel was first proposed by Vishkin [20] and it has been used in many sequential and parallel algorithms, e.g., [3, 6, 7, 12, 13]. Consider two copies of string $P[1..m]$, upper one shifted by i positions to the right. If there is a position w in the lower copy where the two copies of P differ (i.e., $P[w] \neq P[w-i]$), then position w is a *witness* against i. We call the length $k = m - w$ a *witness distance* (from the right) against i. We define *the witness distance* against i to be the smallest witness distance against i.

Let $j_1 < j_2$ be two locations of the text such that $j_2 - j_1 < m$. A *duel* between locations j_1 and j_2 of text T is as follows. If there is witness distance k against $j_2 - j_1$ (i.e., $P[m-k] \neq P[m-k-(j_2-j_1)]$) then it is impossible that the pattern ends at both locations j_1 and j_2 of the text. If $T[j_1 - k] \neq P[m-k]$ then there is no occurrence of the pattern ending at j_1, and if $T[j_1 - k] \neq P[m - k - (j_2 - j_1)]$ then there is no occurrence of the pattern ending at j_2. Hence by two comparisons we can eliminate at least one of end positions j_1 and j_2 as possible occurrences of the pattern.

The suffix tree was proposed by McCreight [18] as a space-efficient alternative to Weiner's position tree [21]. McCreight also gave a linear-time algorithm for the construction of the suffix tree. Let $\#$ be a special symbol that is not in an alphabet Σ. The *suffix tree* of string $x\#$ is a rooted tree satisfying the following conditions [4]:

(a) Each edge is labeled with a nonempty substring of $x\#$, which is represented by its start position and length.

(b) Each internal node has at least two children.

(c) No two sibling edges have the same first symbol in their labels.

(d) For each node v, let $L(v)$ be the string obtained by concatenating the labels on the path from the root to v. Every leaf v is associated with i such that i is a distinct position of $x\#$ and $L(v)$ is the suffix of $x\#$ starting at position i.

4   The Algorithms

It is known that if the graph $G = (V, E)$ modeling a hypertext is a tree, one can find all the occurrences of the pattern in linear time by way of an application of the KMP algorithm [10]. In this section we describe matching algorithms when the modeled graph $G = (V, E)$ is a general graph. We consider the two cases when the length of text T_v of every node v is larger than or equal to m and when there exist some nodes v such that the length of text T_v is less than m, since witness and duel will be applied differently in each case.

4.1 The Case of Long Texts

In this subsection we present an algorithm for the case when the length n_v of text T_v is larger than or equal to the pattern length m for every node v.

Consider a node v in modeled graph $G = (V, E)$. There can be many directed paths from the source node u_1 to v. Let's suppose that a directed path from u_1 to a parent node u_r of v is a sequence of nodes $u_1, u_2, ..., u_r$. We can obtain a string x_i associated with the path from u_1 to u_r by concatenating texts $T_{u_1}, T_{u_2}, ..., T_{u_r}$ of nodes $u_1, u_2, ..., u_r$. Let X_v be the set of strings such that each string $x_i \in X_v$ is associated with a path from source node u_1 to a parent node of v. If one of suffixes of string x_i is a proper prefix of pattern P, then the suffix of x_i is called a *potential occurrence at node* v. The length of a potential occurrence at node v is called a *candidate at node* v.

In the case of $n_v \geq m$ for all v, an occurrence of the pattern can be divided by at most one link. If there exists a potential occurrence at node v then the start position of potential occurrence must be in text T_u where u is one of the parent nodes of v. All potential occurrences which exist in a parent node u of v can be represented by a single value, i.e., the prefix value q_u stored at the end position of text T_u of node u. Since all candidates at v whose start positions are in a parent node u are $q_u, F(q_u), F(F(q_u)), \ldots$, we can find them by using q_u and failure function F of pattern P.

The algorithm in this subsection first traverses all the nodes, and for each text it encounters, runs the KMP algorithm on the text, and saves the prefix value q_v at the end of the text in v. By this we have found all occurrences of the pattern inside texts and also we have all the prefix values of the ends of texts T_v for all node v. To find occurrences of the pattern across links, the algorithm traverses the graph once again and applies witness and duel.

In the pattern preprocessing stage, we compute the KMP failure function $F(i)$ of each location i in pattern $P[1..m]$. To perform duels in the text matching stage we compute the witness distance array $W(i)$ for each shift i in pattern P. Given the reverse pattern P^R, we can find the maximum prefix of P^R starting at each shift i of P^R in linear time by [16]. If the maximum prefix length for location i in reverse pattern P^R is j, we set the witness distance array $W(i)$ as j. If $i + W(i) \geq m$, then there is no witness and we set the witness distance $W(i)$ as m.

Lemma 1. *Suppose that the witness distance array W of pattern $P[1..m]$, and the two candidates c_1, c_2 at node v such that $k = c_2 - c_1 > 0$ are given. If $0 \leq W(k) < m - c_2$ then at least one of c_1 and c_2 has no occurrence in node v.*

Proof. (See Fig. 2.) Let $m - c_1$ and $m - c_2$ be the locations of text T_v. Consider a duel between $m - c_1$ and $m - c_2$. If $W(k) = m$, then there is no witness against k and thus we cannot duel. If $m - c_2 \leq W(k) < m$, we have $P[c_2 + 1..m] = P[c_1 + 1..m - k]$. It means that overlapping segments of two copies of P match completely in the region of text T_v. Since we have no witness in T_v, we cannot duel. But if $0 \le W(k) < m - c_2$, then we have the position $m - c_2 - W(k)$ in text T_v as a witness against k. Therefore, we can perform the duel between $m - c_1$ and $m - c_2$. Since we can eliminate one of $m - c_1$ and $m - c_2$ as end positions of possible occurrences of pattern P, one of candidates c_1 and c_2 has no occurrence in node v.

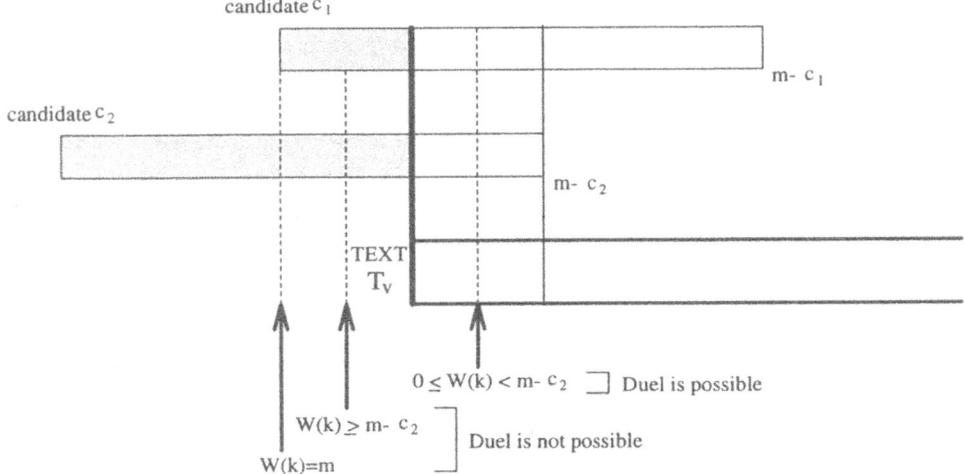

Fig. 2. Duel with the witness distance $W(k)$, where $k = c_2 - c_1$.

Given two candidates c_1, c_2 at node v such that $k = c_2 - c_1 > 0$, if $0 \le W(k) < m - c_2$ then we say that *duel is possible* for c_1 and c_2; otherwise (i.e., $W(k) \ge m - c_2$) we say that *duel is not possible* for c_1 and c_2. When duels are not possible for any pairs of candidates $c_1, c_2, ..., c_j$ at node v, we say that the candidates $c_1, c_2, ..., c_j$ are *consistent* at node v.

Lemma 2. *Let c_{max} be the maximum candidate among consistent candidates c_i's at node v. If c_{max} is eliminated by a duel as possibly having an occurrence at node v, then all candidates consistent with c_{max} cannot have occurrences in T_v.*

Proof. Since c_{max} is the maximum among consistent candidates c_i's, $P[c_{max} + 1..m] = P[c_i + 1..m - (c_{max} - c_i)]$ holds for every c_i. If c_{max} is eliminated by a duel as a possible occurrence at node v, there exist x, y such that $P[x] \ne T_v[y]$, $c_{max} < x \le m$ and $1 \le y \le m - c_{max}$. Since c_{max} is consistent with c_i, $P[x] = P[x - (c_{max} - c_i)] \ne T_v[y]$ and $c_i < x - (c_{max} - c_i) \le m - (c_{max} - c_i)$. Hence, every consistent candidate c_i cannot have an occurrence at node v.

Given consistent candidates c_i's at node v, we can check their matches in node v by scanning text T_v once. Let c_{min} be the minimum candidate among consistent candidates c_i's at node v. Let τ be the maximum value such that $P[c_{min} + 1..c_{min} + \tau] = T_v[1..\tau]$. Since c_i's are consistent, we have $P[c_i + 1..m] = P[c_{min} + 1..m - (c_i - c_{min})]$ for every c_i. If $c_i + \tau \geq m$ for each c_i then a pattern occurrence associated with c_i exists; otherwise it does not exist. Therefore, we can determine which of c_i's have occurrences in T_v with one scan of T_v.

The algorithm for the text matching stage is as follows.

[**Stage 1**] We traverse all the nodes by a depth-first-search traversal of the input graph G from the source node. For each node v encountered, do the KMP matching for text T_v in order to find all occurrences of the pattern inside texts. At the end of text T_v, save the last prefix value q_v for each node v.

[**Stage 2**] In this stage we traverse all the nodes once again in order to find all occurrences of the pattern across links. Set v as the source node and do the following steps.

1. We use a bit array B of length m to represent the set of candidates at node v. Initialize bit array B to $B(i) = 0$ for all i.
2. This step finds all the candidates at node v and sets $B(i)$ as 1 for each candidate i. Select one parent node u of v and set $j = q_u$. Do the following.
   2.1 If $j = 0$ or $B(j) = 1$ then do step 2.2. Otherwise, set $B(j) = 1$ and $j = F(j)$. Repeat step 2.1.
   2.2 Select another parent node as u. Set $j = q_u$ and repeat step 2.1. If no parent node is left, go to step 3.
3. This step performs duels among the candidates in bit array B and finds consistent candidates at node v.
   Set α as the smallest j such that $B(j) = 1$ and set β as the next smallest j such that $B(j) = 1$, and do the following.
   3.1 If α and β are consistent, set $\alpha = \beta$ and do step 3.2. Otherwise, perform a duel between $m - \alpha$ and $m - \beta$. If α is eliminated by the duel, set $B(i) = 0$ for all $i \leq \alpha$ by Lemma 2 and set $\alpha = \beta$. If β is eliminated, set $B(\beta) = 0$. Go to step 3.2.
   3.2 Set β as the next smallest j such that $B(j) = 1$ and repeat 3.1 until there is no other candidate in B.
4. This step finds pattern occurrences among consistent candidates from step 3 by naive text matching.
   Set q as the smallest j such that $B(j) = 1$. That is, q represents the minimum candidate among the consistent candidates that have possible occurrences in text T_v. Perform naive text matching from the start of T_v and from the $(q + 1)$st location of the pattern until a mismatch is found or the end of the pattern is reached. Set τ as the number of matching characters. Then there exists an occurrence associated with j if $q < j \leq m$, $j + \tau \geq m$ and $B(j) = 1$.
5. Find the next node in graph G and set v as the node. Repeat step 1 through step 5.

Theorem 3. *If $n_v = |T_v| \geq m$ for every node $v \in V$ in a general graph $G = (V, E)$, then finding all the occurrences of a pattern in graph G takes $O(N + |E|)$ time.*

Proof. In stage 1, each node v can be processed in $O(n_v)$ time. Steps 1 and 4 of stage 2 take $O(m)$ time. If we denote the number of parent nodes of v as d, step 2 takes $O(m + d)$ time. Step 3 also takes $O(m)$ time. Therefore, the time needed for node v is $O(n_v + m + d)$ and for all nodes it takes $O(N + |V|m + |E|)$ time. By the assumption that $n_v \geq m$ for all v, $|V|m = O(N)$ and therefore the total time complexity is $O(N + |E|)$.

4.2   The General Case

We consider the general case, i.e., it may be possible that n_v is less than m for some node v. If $n_v \geq m$, then all potential occurrences whose start positions are in T_v can be represented by prefix value q_v stored at the end of node v. But if $n_v < m$ and text T_v of node v is a substring of the pattern, then multiple prefixes (which cannot be represented by a single prefix value) of the pattern, of which T_v is a suffix, can have occurrences after the current node v. Therefore, we need to store all the prefix values at the end of a node v when T_v is a substring of the pattern. In this case we store the prefix values at the end of node v as a bit array Q_v of length m. When we start the matching at the start of a node w whose parent is v, we use either prefix value q_v or bit array Q_v to construct a bit array B for all candidates at node w and perform the duels.

DAG Structures. We first consider the case when the input graph G is a directed acyclic graph (DAG). When $n_v < m$ for some nodes v, there can be occurrences of the pattern that cross multiple links, and thus we need not only the witness distances of the whole pattern P but also the witness distances of arbitrary substrings of P. Hence in the pattern preprocessing stage we also construct the suffix tree of reverse pattern $P^R\#$ and use the suffix tree to compute the witness distances of arbitrary substrings of P. The prefix of P ending at position j corresponds to a leaf node of the suffix tree of $P^R\#$. This leaf node will be called *leaf j*.

In the text matching stage, we perform duels with the help of the suffix tree for reverse pattern $P^R\#$. Given the suffix tree, there exist algorithms [14, 19] for finding the lowest common ancestor (LCA) of two nodes in the tree in constant time with a linear time preprocessing. Recall that for a node x in the suffix tree $L(x)$ denotes the string obtained by concatenating the labels on the path from the root to x. We can compute the length of $L(x)$ for all nodes x by traversing the suffix tree in $O(m)$ time. For each node x, we compute $|L(x)|$ using $|L(y)|$ at the parent node y of x and the length of string on the edge from y to x.

Lemma 4. *Suppose that the suffix tree of reverse pattern $P^R\#$ and two candidates c_1 and c_2 at node v such that $c_1 < c_2$ and $c_2 + n_v < m$ are given. Let x be* *the LCA of the two leaves $n_v + c_1$ and $n_v + c_2$ in the suffix tree. If $|L(x)| < n_v$, then at least one of c_1 and c_2 cannot have an occurrence in T_v.*

Proof. (See Fig. 3.) From the definition of the suffix tree, node x has at least two children nodes and the first characters of the two labels must differ. So $|L(x)|$ can be used as the witness distance against $c_2 - c_1$ of the substring of pattern P (i.e., $P[1..c_2 + n_v]$). If $|L(x)| \geq n_v$, then there is no witness in T_v and we cannot duel. If $|L(x)| < n_v$, then a witness is in T_v and we can use the witness distance $|L(x)|$ to perform a duel. Hence in the case of $|L(x)| < n_v$, at least one of c_1 or c_2 cannot have an occurrence in T_v.

If $|L(x)| < n_v$, then duel is possible for the two candidates by Lemma 4; we use $|L(x)|$ as the witness distance, which can be computed in constant time by an LCA algorithm and the preprocessing of the suffix tree. If $|L(x)| \geq n_v$, then candidates c_1 and c_2 are consistent at node v and thus both or neither of the two candidates have occurrences that cross T_v. We can check the existence of such occurrences with one scan of T_v.

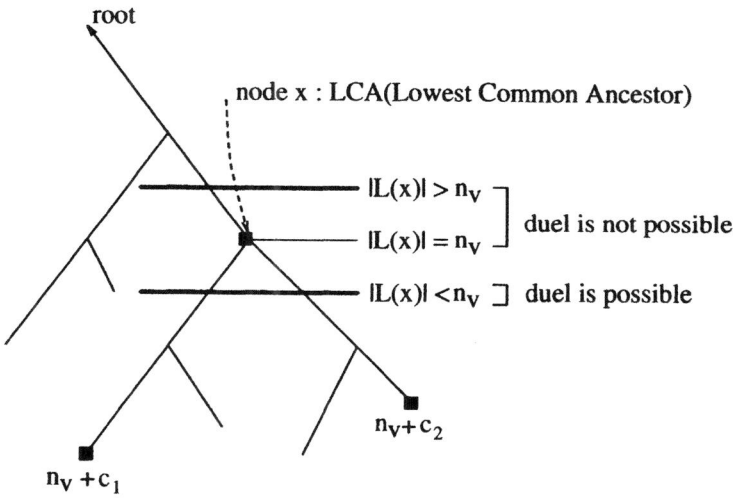

Fig. 3. Duel with suffix tree of reverse pattern $P^R\#$.

The following is the algorithm for DAG structures.

When the modeled graph $G = (V, E)$ is a DAG, we first use topological sorting [9] on graph G and then use the ordering to perform text matching. The stages below are performed on every node according to the topological order after setting v as the source node initially.

[**Stage 1**] In this stage we find all the candidates at node v. Compared with the algorithm of Section 4.1, we use not only prefix value q_u but also bit array Q_u of each parent node u of v to construct bit array B. Initialize bit array B as 0 and do the following for each parent node u of v.

1. If the prefix value at the end of u is set as q_u, set the appropriate bits in B as in the algorithm of Section 4.1.

2. Otherwise, the prefix values at the end of u are set as bit array Q_u. Perform a bitwise OR of Q_u and B and store the result in B.

[Stage 2] In this stage we perform duels with the witness distance array W and the suffix tree of $P^R\#$ and find consistent candidates among the candidates in B.

Set α as the smallest j such that $B(j) = 1$ and set β as the next smallest j such that $B(j) = 1$, and do the following.

1. If $\beta + n_v \geq m$, use the witness array W; otherwise $(\beta + n_v < m)$, use the suffix tree of $P^R\#$ to find the witness distance against $\beta - \alpha$.

2. By Lemmas 1 and 4, determine whether a duel is possible or not. If α and β are consistent, set $\alpha = \beta$ and do step 3. Otherwise, perform a duel between α and β. If α is eliminated by the duel, set $B(i) = 0$ for all $i \leq \alpha$ by Lemma 2 and set $\alpha = \beta$. If β is eliminated, set $B(\beta) = 0$. Go to step 3.

3. Set β as the next smallest j such that $B(j) = 1$ and repeat steps 1 and 2 until there is no other candidate in B.

[Stage 3] In this stage we find pattern occurrences across links among consistent candidates at node v by naive text matching and also find pattern occurrences inside text T_v using the KMP matching.

Set q as the smallest j with $B(j) = 1$ and perform naive text matching from the start of T_v and from the $(q+1)$st location of the pattern until a mismatch is found or the end of pattern is reached.

1. Set τ as the number of matching characters. Then there exists an occurrence associated with j if $q < j \leq m$, $j + \tau \geq m$ and $B(j) = 1$.

2. If $\tau < n_v$, no candidates at v contain T_v as a substring. We perform the KMP algorithm from location $\tau + 1$ of T_v and store the prefix value q_v at the end of v.

   Otherwise $(\tau = n_v)$, for all j with $B(j) = 1$ and $j+\tau < m$, set $Q_v(j+\tau) = 1$ and store bit array Q_v.

Theorem 5. *If graph $G = (V, E)$ is a DAG, the above algorithm takes $O(N + |E|m)$ time to find all the occurrences of a pattern in the hypertext.*

Proof. In stage 1, generating bit array B takes $O(dm)$ time for each node, where d is the number of parents nodes of v. In stage 2, since it takes $O(1)$ time to find the LCA of two nodes, we need $O(m)$ time to perform duels for a node. Whether we perform the naive matching or the KMP matching it is sufficient to scan the text T_v once, so stage 3 takes $O(n_v)$ time. Therefore, the time needed for node v is $O(n_v + dm)$, and the total time complexity is $O(N + |E|m)$.

General Graph Structures. We now consider a more general case when the graph $G = (V, E)$ modeling a hypertext has cycles. The algorithm to be described assumes that one pattern occurrence can include any node at most one time and finds all occurrences of the pattern, possibly crossing multiple links.

The edges in a directed graph G are partitioned into four categories by a DFS of G [2]: *tree edges* which are edges leading to new vertices during the search; *forward edges* which go from ancestors to proper descendants but are not tree edges; *back edges* which go from descendants to ancestors (possibly from a vertex to itself); *cross edges*, which go between vertices that are neither ancestors nor descendants of one another. A DAG is composed only of tree, forward, and cross edges. There are back edges in a graph if and only if the graph has cycles. The description of the matching algorithm for general graphs is as follows.

1. Start at the source node and traverse graph G in depth first order.
2. Construct graph G' which is G minus the back edges of G. Since G' is a DAG, perform the DAG algorithm on G'.
3. In step 2 we have stored one of prefix value q_v or bit array Q_v at the end of each node v. By the assumption that a pattern occurrence cannot include any node twice, we can find all occurrences by running once again the DAG algorithm on G including the back edges.

Consequently, this algorithm performs the DAG algorithm twice (the first time without back edges and the second time with back edges), and its time complexity is $O(N + |E|m)$ for finding all occurrences of the pattern in hypertext.

5 Concluding Remarks

Compared with the normal string matching problem, the hypertext has a non-linear structure and it should be considered that pattern occurrences can exist across links in a hypertext. We have presented an algorithm that takes linear time when $n_v \geq m$ for all nodes v. We also present a more involved algorithm that takes $O(N + |E|m)$ time when there exist nodes v with $n_v < m$.

Manber and Wu [17] solved the approximate string matching problem in hypertext. They proposed an algorithm for the case the modeled graph $G = (V, E)$ is a DAG. The time complexity is $O(N + |E|m + R \log \log m)$, where R is the total number of ordered pairs of positions at which the sequences match.

References

1. A.V. Aho, Algorithms for finding patterns in strings, in van Leeuwen, ed., *Handbook of Theoretical Computer Science* , North Holland, 1992.
2. A.V. Aho, J.E. Hopcroft and J.D.Ullman, The Design and Analysis of Computer Algorithms, *Addison-Wesley* , 187–189.

3. A. Amir, G. Benson and M. Farach, An alphabet independent approach to two-dimensional pattern matching, *SIAM J. Comput.* 23, 2 (1994), 313–323.

4. A. Apostolico, C. Iliopoulos, G.M. Landau, B. Schieber and U. Vishkin, Parallel construction of a suffix tree with applications, *Algorithmica* 3, (1988), 347–365.

5. R.S. Boyer and J.S. Moore, A fast string searching algorithm, *Comm. ACM* 20 (1977), 762–772.

6. D. Breslauer and Z. Galil, An optimal $O(\log\log n)$ time parallel string matching algorithm, *SIAM J. Comput.* 19 (1990), 1051–1058.

7. R. Cole, M. Crochemore, Z. Galil, L. Gąsieniec, R. Hariharan, S. Muthukrishnan, K. Park and W. Rytter, Optimally fast parallel algorithms for preprocessing and pattern matching in one and two dimensions, *Proc. 34th IEEE Symp. Found. Computer Science* (1993), 248–258.

8. J. Conklin, Hypertext : An introduction and survey, *Computer 20 Sep.* (1987), 17–41.

9. T.H. Cormen, C.E. Leiserson and R.L. Rivest, Introduction to Algorithms, *MIT Press*, 485–487.

10. M. Dubiner, Z. Galil and E. Magen, Faster tree pattern matching, *Proc. 31st IEEE Symp. Found. Computer Science* (1990), 145–150.

11. Z. Galil, On improving the worst case running time of the Boyer–Moore string matching algorithm, *Comm. ACM* 22 (1979), 505–508.

12. Z. Galil and K. Park, Alphabet-independent two-dimensional witness computation, to appear in *SIAM J. Comput.*

13. L. Gąsieniec and K. Park, Work-time optimal parallel prefix matching, *2nd European Symp. on Algorithms*, 1994.

14. D. Harel and R.E. Tarjan, Fast algorithms for finding nearest common ancestors, *SIAM J. Comput.* 13, (1984), 338–355.

15. D.E. Knuth, J.H. Morris, and V.B. Pratt, Fast pattern matching in strings, *SIAM J. Comput.* 6 (1977), 323–350.

16. M.G. Main and R.J. Lorentz, An $O(n\log n)$ algorithm for finding all repetitions in a string, *J. Algorithms* 5 (1984), 422–432.

17. U. Manber and S. Wu, Approximate string matching with arbitrary costs for text and hypertext, *Proc. International Workshop on Structural and Syntactic Pattern Recognition* (1992), 22–33.

18. E.M. McCreight, A space-economical suffix tree construction algorithms, *J. ACM* 23, (1976), 262–272.

19. B. Schieber and U. Vishkin, On finding lowest common ancestors: simplification and parallelization, *SIAM J. Comput.* 17, (1988), 1253–1262.

20. U. Vishkin, Optimal parallel pattern matching in strings, *Inform. and Control* 67 (1985), 91–113.

21. P. Weiner, Linear pattern matching algorithms, *Proc. 14th IEEE Symp. Switching and Automata Theory* (1973), 1–11.

Approximation algorithms for multiple sequence alignment under a fixed evolutionary tree[*]

R. Ravi[1] and John D. Kececioglu[2]

[1] DIMACS, Department of Computer Science, Princeton University, NJ 08544.
[2] Department of Computer Science, The University of Georgia, Athens, GA 30606.

Abstract. We consider the problem of aligning sequences related by a given evolutionary tree: given a fixed tree with its leaves labeled with sequences, find ancestral sequences to label the internal nodes so as to minimize the total cost of all the edges in the tree. The cost of an edge is the edit distance between the sequences labeling its endpoints. In this paper, we consider the case when the given tree is a regular d-ary tree for some fixed d and provide a $\frac{d+1}{d-1}$-approximation algorithm for this problem that runs in time $O(d(2kn)^d + n^2 k^{2d})$ where k is the number of leaves in the tree and n is the maximum length of any of the sequences labeling the leaves.

We also consider a new bottleneck objective in labeling the internal nodes. In this version, we wish to find the labeling of the internal nodes that minimizes the maximum cost of any edge in the tree. For this problem we provide a simple $2\delta + 1$-approximation algorithm where δ is the depth of the given undirected tree defined as the maximum over all internal nodes of the number of edges from the internal node to a closest leaf. For phylogenetic trees on n nodes that have no internal nodes of degree two, $\delta \leq \lg n$.

1 Introduction

We study the problem of aligning several biological sequences guided by a phylogenetic tree representing the evolutionary relationship between the species from which the sequences are derived. The input to the problem is the phylogenetic tree with its leaves labeled with sequences of extant species. The objective in the multiple sequence alignment is to derive ancestral sequences to label internal nodes of this tree with so as to minimize some objective function defined by the resulting costs on the tree edges. The cost on a tree edge is typically the edit distance between the sequences labeling its endpoints. Motivated by the maximum parsimony criterion in evaluating phylogenetic trees, one objective that has been proposed and studied in the literature before [1, 3, 4, 5] is the sum of the costs of the tree edges.

---

[*] Most of this work was performed while the authors were at the Department of Computer Science, University of California at Davis.

Sankoff [2] introduced the problem and proposed an algorithm to compute the optimal tree alignment using dynamic programming. Sankoff, Cedergren and Laplame [3] gave local improvement algorithms that start with a heuristic alignment labeling internal nodes with leaf sequences and improves these alignments by considering the stars around the internal nodes. They report that their method typically converges in very few iterations. Jiang, Lawler and Wang [1] showed that this problem is NP-hard even in the case when the fixed phylogenetic tree is binary, and gave a 2-approximation algorithm for the problem using a scheme that "lifts" the leaf sequences to label the internal nodes in the tree. The running time of their algorithm is $O(k^3 + k^2 n^2)$ when there are k leaf sequences each of length at most n. For the case when the input phylogenetic tree has bounded degree d, they also provided the first PTAS for the tree alignment problem that runs in time $O(k^{d^{t-1}+2} n^{\frac{d^{t-1}-1}{d-1}})$ to produce an alignment within a factor $(1 + \frac{3}{t})$ of optimal.

Regular trees of bounded degree

In this paper, we propose a natural heuristic for the tree alignment problem, and analyze it in the special case when the input phylogenetic tree is a regular d-ary tree for some fixed d, i.e., after arbitarily rooting the tree every internal node has exactly d children.

Theorem 1. *There is a polynomial-time approximation algorithm for multiple sequence alignment under a fixed evolutionary tree, that finds a solution of cost at most $\frac{d+1}{d-1}$ times the minimum in $O(d(2kn)^d + n^2 k^{2d})$ time, given a regular d-ary tree of k leaves, labeled by sequences of length at most n.*

Though our proof of the above theorem relies heavily on the symmetry of the input tree and does not extend to arbitrary bounded-degree trees, our result suggests that our algorithm will perform reasonably well in practice. Furthermore, the running time of our algorithm compares favorably with the prohibitive running time of the PTAS for aligning via bounded-degree trees in [1], while producing approximation guarantees better than two for trees with degree greater than four.

Bottleneck alignments

Finding alignments that minimize the total cost of the tree has some possible disadvantages: the alignment can be biased towards over-represented sequences, and to keep most of the edges in the resulting tree short a few can be made comparatively long. We propose the investigation of tree alignments under an alternative objective, the cost of the maximum edge in the tree. We call this *bottleneck alignment under a fixed evolutionary tree*. The solution to this problem may produce alignments with more uniform edge lengths.

Define the depth δ of an undirected tree as the maximum over all internal nodes, of the number of edges from the internal node to a closest leaf. For trees on n nodes that have no internal nodes of degree two, $\delta \leq \lg n$.

Theorem 2. *There is a simple polynomial-time $2\delta + 1$-approximation algorithm for bottleneck alignment under a fixed evolutionary tree, where δ is the depth of the tree. The algorithm runs in time proportional to the number of nodes in the tree, independent of the lengths of the sequences labeling the leaves.*

We also present an exact dynamic programming algorithm for the bottleneck objective function for special cost functions; we defer a full description until Section 3.

In the next section, we present our algorithm for tree alignment for d-ary trees and its analysis. In the following section, we present the algorithm for bottleneck tree alignment. We close with some open questions.

2 Alignment of sequences via a bounded-degree tree

First of all, observe that the alignment problem for a regular d-ary tree can be thought of as one on a complete d-ary tree by padding the incomplete portions of the tree with complete d-ary subtrees of the appropriate size. Each of the complete trees added in place of a leaf labeled with a sequence S has all its leaves labeled with S. It is straightforward to check that there is a minimum cost alignment over the the complete d-ary tree where every node in such a padded subtree is labeled with S. Thus, in our analysis, we consider the problem on complete d-ary trees. We use K to denote the total number of leaves in the padded complete d-ary tree, and k to denote the total number of leaves in the unpadded regular d-ary input tree. After padding, K may be exponential in k.

Let T denote the complete d-ary tree with root r. Denote the children of a node v in the tree by $v_1, \ldots, v_d$. For any internal node v, define $l(v)$ to be the set of leaves of T in the subtree of T rooted at v, and $L(v)$ to be the (multi)set of sequences labeling the leaves in $l(v)$. For any two sequences S and S', we use $d(S, S')$ to denote the edit distance between them. Define the *level* of any node in T to be the total height of T minus the depth of this node from the root. Alternatively, we can define the level recursively as follows: the level of any leaf is zero; the level of any internal node is one plus the level of any of its children.

The basic subproblem that we shall repeatedly use in our alignment algorithm is the *Steiner sequence problem* over a set of sequences, defined as follows: given p sequences $S_1, \ldots, S_p$, find a Steiner sequence S^* that minimizes the sum of the distances $\sum_{i=1}^{p} d(S^*, S_i)$. Each of the sequences S_i is called a *component* of the Steiner sequence S^*. We shall also refer to the sum $\sum_{i=1}^{p} d(S^*, S_i)$ as the *Steiner cost* of the sequence S^* and denote it by $c(S^*)$.

The approximation algorithm

Our algorithm first computes for every internal node, a candidate list of sequences that may be used to label this node and finds a minimum-cost tree in which every internal node is labeled with a sequence from its candidate list. The candidates at each node are selected from a set of valid labels at this node. For any internal node v in the tree T, recall that $v_1, \ldots, v_d$ denote the d children of v in T. Define a *valid* label for v to be any sequence that is a Steiner sequence for $S_1, S_2, \ldots, S_d$ where $S_i \in L(v_i)$ $\forall i \in \{1, \ldots, d\}$. For the choice of a good ancestral labeling, we not only choose the labels at each internal node from the set of valid labels at this node but also ensure consistency between the valid labels in the following way. Given a labeling of all the children of an internal node v with valid labels, the *consistent* set of valid labels at v is defined as the set of Steiner sequences for $S_1, S_2, \ldots, S_d$ where $S_i \in L(v_i)$ and the Steiner sequence labeling v_i has S_i as one of its components, $\forall i \in \{1, \ldots, d\}$. In other words, if the valid sequence labeling v_i is S_i^* which is a Steiner sequence for $S_{i,1}, \ldots, S_{i,d}$, then a valid label at v is consistent if it is a Steiner sequence for $S_{1,j_1}, \ldots, S_{d,j_d}$ for some j_i's in $\{1, \ldots, d\}$. A consistent valid labeling of the given d-ary tree is one where the label at every internal node is consistent and valid.

Our approximation algorithm for finding good ancestral sequences is simply to find the best consistent valid labeling of the internal nodes, namely, one that results in a tree of minimum cost. First we show that such a tree can be computed efficiently using dynamic programming. Then we show that this tree is a good approximation of the optimal tree.

Running time

Though the total number of valid consistent labelings of a tree may be large we can find the best among these using dynamic programming as follows: For any internal node v and any valid label l for v, let $C(v, l)$ denote the minimum cost of a consistent valid labeling of the subtree rooted at v in which v is labeled with l. To compute $C(v, l)$ for an internal node v from the C-values for its children, we consider each child v_i of v in turn and determine the best label l_i of v_i such that l is a consistent valid label of v and the distance $d(l, l_i)$ plus the cost $C(v_i, l_i)$ is minimized. This computation is independent for each of the children v_i of v.

We can bound the time used by the approximation algorithm on the unpadded input tree as follows. There are certainly at most k^d valid labels for nodes, as each label is a subset of size d from the k leaves. Computing the Steiner sequence for a label takes time $O(d(2n)^d)$ by the standard dynamic programming algorithm, given $O(ds^{d+1})$ time preprocessing for an alphabet of size s. Thus the time to determine all Steiner sequences for the labels is $O(k^d d(2n)^d) = O(d(2kn)^d)$. Notice that no two nodes have the same valid label. Thus the total time spent computing edit distances between labels, when determining the best valid consistent labeling of the tree, is certainly at most $O((k^d)^2 n^2) = O(n^2 k^{2d})$. This dominates the time to compute the optimum labeling bottom-up over the tree. Thus the time to find the best valid consistent labeling is $O(d(2kn)^d + n^2 k^{2d})$, which proves the running time claimed in Theorem 1.

Performance guarantee

In this section, we show that the best valid consistent labeling of the tree T is within a factor of $\frac{d+1}{d-1}$ of the minimum cost labeling. We do this by computing an upper bound on the total cost of all the consistent valid labelings of T and dividing out by the number of such labelings and using the fact that the minimum-cost labeling has at most this average value.

Our main tool in relating the cost of a valid labeling to the cost of an optimal tree is triangle inequality on the distances between sequences. Any edge in a consistent, valid labeled tree has its endpoints labeled by two Steiner sequences that share a component. We upper bound the cost of the edge between these two sequences by the sum of the distances from these sequences to the common component leaf in T. The symmetry in a complete d-ary tree allows us to account for all the edges in all valid consistent labelings of T by a certain weighted sum of all the Steiner sequence costs $\sum_{i=1}^{d} d(S^*, S_i)$ over all the Steiner sequences S^* labeling all the internal nodes of T. By a balancing argument, we can compute the multiplier for each of these costs in this weighted sum and argue that this is the same for all the labels for nodes at any given level. We then use this estimate of the total cost of all consistent valid labelings to complete the proof by averaging over the number of such labelings. We elaborate on this outline below.

First observe the following.

Lemma 3. *The total number of valid labelings of any node v at level i in T is exactly $d^{d(i-1)}$.*

Let ρ denote the level number of the root r of T. Note that ρ is exactly the logarithm to the base d of the total number of leaves in T. For any fixed valid label for the root r, let $\mathcal{N}$ denote the total number of valid consistent labelings of T that contain this label at the root. Let $\mathcal{T}$ denote the total number of consistent valid labelings of T. Since there are exactly $d^{d(\rho-1)}$ valid labels for the root r by Lemma 3, we have by symmetry

$$\mathcal{T} = d^{d(\rho-1)} \cdot \mathcal{N}. \tag{1}$$

Consider any internal node v of T at level $i < \rho$. Applying lemma 3 again, the total number of valid labels for v is exactly $d^{d(i-1)}$. Note that each of these labels appears an equal number of times over all consistent valid labelings of T. Thus the *usage* of any of these labels over all the consistent valid labelings of T, i.e., the total number of consistent valid labelings of T in which this node v is labeled with one of these labels is exactly

$$\frac{\mathcal{T}}{d^{d(i-1)}} = d^{d(\rho-i)}\mathcal{N} \tag{2}$$

using Equation (1).

We introduce some more definitions. Let OPT denote an optimal (minimum cost) labeling of T and let $c(OPT)$ denote the total cost of T under this labeling. For any internal node v of T, let $star(v)$ denote the sum of the distances of the edges $(v, v_1), \ldots, (v, v_d)$ in OPT. Further, for any level i, let $star(i) = \sum_{v \text{ in level } i} star(v)$.

Lemma 4. *For any level i in T*

$$\sum_{v \text{ in level } i} \sum_{\text{valid labels } S_v^*} c(S_v^*) \leq d^{d(i-1)}(star(i) + \frac{star(i-1)}{d} + \ldots + \frac{star(1)}{d^{i-1}}).$$

Proof. Consider a Steiner sequence S^* that is a valid labeling for an internal node v at level i in T. Let the components of this sequence be the leaves $S_1, \ldots, S_d$ in T. The key observation is that the cost $c(S^*) = \sum_{i=1}^{d} d(S^*, S_i)$ is at most the sum of the costs of all the edges in the paths between v and the leaves labeled with $S_1, \ldots, S_d$ in OPT. We can thus use these edges in OPT to upper bound the cost $c(S^*)$.

The number of valid labels for v is exactly $d^{d(i-1)}$ by Lemma 3. We can now sum over all the valid labels for v, the accounting paths for these labels in OPT. In this summation, every edge of the form (v, v_i) is used once for every valid label of v, for a total of $d^{d(i-1)}$ times. However, all the edges one level below v, such as an edge going from a child v_i of v to a grandchild of v, are used only in a fraction d of these accounting paths, i.e., $\frac{d^{d(i-1)}}{d}$ times. Extending this, any edge going from a node at j levels below v to a node $j+1$ levels below v is used exactly $\frac{d^{d(i-1)}}{d^j}$ times in the accounting. The cost of all the edges going from level i to level $i-1$ is $star(i)$ by definition. Thus, in the accounting $star(i-j)$ is used exactly $\frac{d^{d(i-1)}}{d^j}$ times. Summing over k gives the Lemma.

We are now ready to prove the performance guarantee of our approximation algorithm.

Lemma 5. *The best consistent valid tree has cost at most $\frac{d+1}{d-1}$ times the minimum.*

Proof. We use the strategy outlined in the beginning of this section. We first derive an upper bound on the cost of all consistent valid labelings of T. To do this, we first use a simple upper bound on the cost of any edge in any consistent valid tree. We can upper bound the cost of an edge in any consistent valid tree by the sum of the two distances from the Steiner sequences labeling its endpoints to the leaf component shared by them. Summing such pairs of distances over all edges over all consistent valid trees, we determine how many times each distance from a Steiner sequence S^* labeling a node v at level i to a component leaf, labeled with, say S_j, (i.e., $d(S^*, S_j)$) gets used in this upper bounding. The "usage" for any Steiner sequence computed in Equation (2) comes in handy here. We claim that every such cost $d(S^*, S_j)$ is used exactly $\frac{d+1}{d} \cdot d^{d(\rho-i)}\mathcal{N}$ times. The $\frac{d+1}{d}$ overhead is due to the fact that the distance $d(S^*, S_j)$ is used $d^{d(\rho-i)}\mathcal{N}$ times to account for edges going from node v to the child v_j containing the leaf S_j in all the consistent valid trees, while it is used an extra $\frac{d^{d(\rho-i)}\mathcal{N}}{d}$ times in accounting for edges going from v to its parent in T in all valid consistent trees.

We can now write an upper bound for the sum of the costs of all consistent valid trees.

$$\sum_{\text{consistent valid labelings } \mathcal{L} \text{ of } T} c(T \text{ labeled with } \mathcal{L})$$

$$\leq \sum_{\text{levels } i} \frac{d+1}{d} d^{d(\rho-i)} \mathcal{N} \sum_{v \text{ in level } i} \sum_{\text{valid labelings } S^* \text{ of } v} c(S^*)$$

$$\leq \sum_{\text{levels } i} \frac{d+1}{d} d^{d(\rho-i)} \mathcal{N} d^{d(i-1)} (star(i) + \frac{star(i-1)}{d} + \ldots + \frac{star(1)}{d^{i-1}})$$

$$\leq \sum_{\text{levels } i} \frac{d+1}{d} \mathcal{T} (star(i) + \frac{star(i-1)}{d} + \ldots + \frac{star(1)}{d^{i-1}})$$

$$\leq \sum_{\text{levels } i} \frac{d+1}{d} \mathcal{T} \frac{1}{1 - \frac{1}{d}} star(i)$$

$$\leq \frac{d+1}{d} \mathcal{T} \frac{1}{1 - \frac{1}{d}} c(OPT)$$

$$\leq \frac{d+1}{d-1} \mathcal{T} c(OPT)$$

Since the total number of consistent valid trees is $\mathcal{T}$, by averaging, the minimum-cost consistent valid tree has cost at most $\frac{d+1}{d-1} c(OPT)$ proving Theorem 1.

3 Bottleneck alignments

In this section, we present a simple approximation algorithm for the bottleneck alignment problem on a tree: given an arbitrary undirected tree T with leaves labeled with sequences, find a labeling of internal nodes with sequences that minimizes the maximum distance on any edge in the resulting labeled tree.

We use a simple lower bound to derive our approximation algorithm.

Lemma 6. *Suppose S and S' are sequences labeling two leaves that are k edges apart in T, then $\frac{d(S,S')}{k}$ is a lower bound on the bottleneck cost of any labeling of T.*

Proof. Consider the sum of the distances of the edges along the path from the leaf labeled S to the leaf labeled S' in any labeling of the internal nodes of T. The sum of these distances must be at least $d(S, S')$ by definition. Hence by averaging, one of the edges in this path must have cost at least $\frac{d(S,S')}{k}$. This is a lower bound on the bottleneck cost of this labeling. $\blacksquare$

The above lemma gives the following lower bound on the bottleneck cost of any labeling of T immediately.

$$\max_{\text{leaves labeled } S, S'} \frac{d(S, S')}{\text{number of edges between leaf}(S) \text{ and leaf}(S') \text{ in } T}$$

The approximation algorithm motivated by this lower bound simply labels every internal node with the sequence of a closest leaf in the tree. Such a labeling may be thought of a particular way to "lift" the leaf labels to the internal nodes where the criterion for lifting a leaf label to an internal node is that the leaf be closest. Note that "closest" refers to number of edges between the internal node and the leaf in the input tree. A different lifting criterion is used in [1].

Our lifting algorithm can be implemented in time proportional to the size of the tree (independent of the size of the sequences labeling the leaves), since the lifting criterion uses only distances using edges in the tree. This can be done by working inwards from the leaves of the tree and assigning labels as we proceed. A simple conceptual way to see how each node can be assigned its closest leaf in linear time is as follows: add a new node l to the tree with edges from l to all the leaves of the tree. A breadth-first tree computation in this augmented graph with l as source will give us the information about a closest leaf for every internal node.

To prove the performance guarantee, we show that the cost of any edge is at most $2\delta + 1$ times the lower bound presented above. Recall that the depth δ of an undirected tree is the maximum over all internal nodes, of the distance from the internal node to a closest leaf. Consider any edge (u, v) in the lifted tree where the endpoints are labeled with S_u and S_v. the cost of this edge is $d(S_u, S_v)$. Let l_u and l_v denote leaves in the tree whose labels are S_u and S_v respectively. By the criterion for the lifting method and the definition of δ, the number of edges between l_u and l_v is at most $2\delta + 1$. Applying the above lemma to this pair of leaves, we have a lower bound of $\frac{d(S_u, S_v)}{2\delta + 1}$ on the bottleneck cost of the tree. Since this argument was made for an arbitrary edge (u, v), this implies the $2\delta + 1$ performance guarantee for our algorithm, completing the proof of Theorem 2.

Note that this argument specialized to a star gives a performance ratio of two.

Dynamic programming

While exact algorithms for alignment via a known tree to minimize the total cost of the edges in the tree have been designed using dynamic programming [2], it is not obvious how such an algorithm can be designed for our bottleneck objective.

We first sketch an algorithm for a star tree when indels are not permitted in computing the edit distance. More precisely, we are given a star tree with k sequences each of length exactly n and we wish to compute a sequence S of length n such that the maximum number of mismatches of S with any of the given k sequences (maximum cost of any edge in the star) is minimized. Note that if the objective were the sum of the costs of the edges in the star, taking the plurality character at each column would result in an optimal sequence S. However, for the bottleneck objective, this can be shown to be nonoptimal.

For each prefix-length l from 1 to n, we maintain a bit-vector of length $O(n^k)$ where each position in the vector corresponds to a point in k-dimensional space having integer coordinates in the range 0 to l. The meaning of setting the bit for point $(p_1, \ldots, p_k)$ is that there is a sequence of length l whose distance (number of mismatches) to the ith sequence is exactly p_i. The vector for prefix-length $l+1$ can be obtained from the vector for prefix-length l by considering each character in the alphabet as a possibility for the $(l+1)$st position of the candidate sequence S. The final answer is obtained by looking at the vector for prefix-length n and finding the point that is set and has minimum value of maximum coordinate. The time for this method is $O(skn^{k+1})$ for an alphabet of size s.

It is not hard to extend this dynamic programming method to the unit-cost metric, allowing indels, by considering all "frontiers" of the k sequences and all their extensions and maintaining a bit vector for each of the frontiers. This gives a time complexity of $O(sk(2n)^{2k})$, arising from $O(n^k)$ frontiers with a bit-vector of length $O((2n)^k)$ maintained for each frontier, and taking time $O(sk(2n)^k)$ for each of the 2^k extensions. The approach could possibly be extended further to arbitrary trees as in [2] by assigning sequences to internal nodes of the tree as we progress through the frontiers of multiple alignment of the leaf sequences and maintaining a bit vector with number of dimensions equal to the number of edges in the tree. This is under investigation.

4 Open problems

We proposed a new approximation algorithm for aligning sequences via an evolutionary tree and provided an analysis of its worst-case performance guarantee on regular d-ary trees. The advantage of our algorithm is its improved performance ratio compared to [1] for values of d higher than three while using lower running time. Since our method uses labels that take into account interactions between the sequences labeling the leaves in the different subtrees, we expect our algorithm to perform reasonably well in practice. It remains an interesting open problem to analyze the worst-case performance ratio of our algorithm when applied to arbitrary trees.

We also proposed a bottleneck objective for alignment and presented a very efficient approximation algorithm for arbitrary trees as well as an exponential dynamic programming exact solution. The time and space requirements of the exact solution methods must be pruned down considerably before exact implementations are possible to investigate the solutions output by this objective function. This remains an important avenue for future research.

Acknowledgments

Research of the second author was supported in part by a DOE Human Genome Postdoctoral Fellowship.

References

1. T. Jiang, E. L. Lawler, and L. Wang, "Aligning sequences via an evolutionary tree: complexity and approximation," *Proc. 26th ACM Symposium on the Theory of Computing*, 760-769 (1994).
2. D. Sankoff, "Minimal mutation trees of sequences," *SIAM J. Appl. Math.*, 28(1), 35-42, (1975).

3. D. Sankoff, R. Cedergren and G. Laplame, "Frequency of insertion-deletion, transversion, and transition in the evolution of 5S ribosomal RNA," *J. Mol. Evol.* 7, 133-149, (1976).
4. D. Sankoff and R. Cedergren, "Simultaneous comparisons of three or more sequences related by a tree, in D. Sankoff and J. Kruskal (eds.) *Time warps, string edits and macromolecules: the theory and practice of sequence comparison*, 253-264, Addison Wesley, Reading MA, (1983).
5. M. S. Waterman and M. D. Perlwitz, "Line geometries for sequence comparisons," *Bull. Math. Biol.* 46, 567-577, (1984).

A New Flexible Algorithm for the
Longest Common Subsequence Problem*

Claus Rick

University of Bonn, Computer Science Department IV,
Römerstr. 164, 53117 Bonn, Germany, Email: rick@cs.uni-bonn.de

Abstract. A new algorithm that is efficient for both short and long
longest common subsequences is presented. It also improves on previous algorithms for longest common subsequences of intermediate length.
Thus, it is more flexible and can be used for a wider range of applications than others. The algorithm is based on the well-known paradigm
of computing dominant matches and was obtained through a kind of
dualization. Some experimental results are given, too.

1 Introduction

Let $A = a_1 a_2 \ldots a_m$ and $B = b_1 b_2 \ldots b_n$, $m \leq n$, be two sequences over an alphabet Σ of size s. A sequence $S = s_1 s_2 \ldots s_l$, $l \leq m$, is called a *common subsequence*
of A and B iff S can be obtained from both A and B by deleting a number of
(not necessarily consecutive) symbols. Finding a common subsequence of greatest
possible length is called the *longest common subsequence* (LCS) *problem*. Let $|A|$
denote the length of a string A and let A_i, $0 \leq i \leq |A|$, denote the length i pre-
fix of A. Define $L_{i,j} = \max\{|S| \; : \; S \text{ is a common subsequence of } A_i \text{ and } B_j\}$,
$0 \leq i \leq m, 0 \leq j \leq n$, to be the length of a LCS between prefixes A_i and B_j.

Then the LCS problem can be solved by a dynamic programming approach
[14, 16]. It works on the $(m+1) \times (n+1)$ L-matrix spanned up by the two strings
by computing all the values $L_{i,j}$ using the recursion

$$L_{i,j} = \begin{cases} 0 & \text{if } i = 0 \text{ or } j = 0 \\ L_{i-1,j-1} + 1 & \text{if } a_i = b_j \\ \max\{L_{i-1,j}, L_{i,j-1}\} & \text{if } a_i \neq b_j \end{cases}$$

in time and space $\Theta(mn)$. The asymptotically fastest general solution needs time
$O(n^2/\log n)$ [9] and uses the "Four Russians" trick. A lot of algorithms have been
developed that, although not improving the general $O(mn)$ time bound of the
dynamic programming approach, exhibit a much better performance by special-
izing on certain classes of pairs of sequences. These algorithms are sensitive to
other problem parameters, e.g. the length p of a LCS (the output parameter),
the number of (dominant) matches or the alphabet size s (see [12] for a survey).

---

* This work is dedicated to my father and to the memory of my mother.

Generally speaking, there are algorithms that suit for a class of pairs of sequences where a LCS can be expected to be very long. Their time complexity is $O(n(n-p))$ or $O(n(m-p))$. On the other hand, there are algorithms that are fast for short longest common subsequences. Their running time is roughly $O(pm)$. As pointed out in [11] we have to select one of the algorithms a priori depending on the kind of sequences we wish to compare. However, this might prove difficult because of insufficient knowledge of the nature of the sequences to be compared. Or the class of sequences is so heterogeneous that both short and long longest common subsequences are likely to occur, e.g. when searching a data base for entries similar to a query. Moreover, when dealing with sequences of approximately equal length consisting of symbols drawn from a small alphabet in a more or less uniform manner the length of a LCS can be expected to lie in the range between $\frac{1}{3}m$ and $\frac{2}{3}m$, depending on the alphabet size [4, 12, 14]. This situation is typical of the analysis of macromolecular sequences in biology. None of the algorithms mentioned above is well-suited for the cases described. Although some algorithms exist whose time analysis indicates somewhat more flexibility [1, 5, 6, 7], the constant factors employed by these methods seem rather high. No specific attention was given to the above problems so far.

In this paper, a new algorithm for the LCS problem is presented which exhibits a good performance for both short and long longest common subsequences and can deal with longest common subsequences of intermediate length in a more efficient manner than previous algorithms. The worst case running time of the algorithm is $O(ns + \min\{pm, p(n-p)\})$. Experimental results proved its practicability.

2   The Paradigm of Computing Dominant Matches

There are two main paradigms used by all the algorithms for the LCS problem. One is to compute a cheapest-cost path in an editgraph [10, 15, 17]. The other amounts to what Eppstein et al. [5] call sparse dynamic programming. It will be reviewed since it is the basis for the new algorithm, too. An ordered pair (i, j) of positions in A and B is called a *match* iff $a_i = b_j$, $1 \le i \le |A|, 1 \le j \le |B|$. Obviously the LCS problem can now be restated as the problem of finding a longest sequence of matches that is strictly increasing in both components. Thus, it suffices to compute the entries $L_{i,j}$ in the L-matrix where (i, j) is a match. For each match the value $L_{i,j}$ is called its *rank*. A match of rank k is said to be a k-match and k is the highest position in a common subsequence to which this match may correspond. We can collect matches of the same rank k in *classes* $C_k := \{(i,j) \mid a_i = b_j \wedge L_{i,j} = k\}$, $k \ge 0$, where $C_0 := \{(0,0)\}$. The goal is, starting with C_0, to determine the ranks of more and more matches with the aid of matches whose rank is already known. A match (i, j) is called a *successor (predecessor)* of match (i', j') if $i' < i \wedge j' < j$ ($i' > i \wedge j' > j$). The following observations are easy to see. For each match in class C_k there has to be a (*generating*) predecessor in the preceding class C_{k-1}. A match must have greater rank than each of his predecessors and therefore matches belonging to the same class cannot be successors or predecessors of each other. Figurally speaking, matches of the same class shift from right to left in the L-matrix for increasing rows and contours given by connecting matches of the same class do never cross or touch. A lot of matches are superfluous in the sense that they are not essential for the generation of matches in the next higher class. In fact it is well-known that we can concentrate on the so-called *dominant* matches, i.e. the minimal matches in each class according to the standard "$\leq$" vector ordering; there is always a LCS consisting solely of dominant matches. In the L-matrix these are the matches in each class where no other match of the same class lies to the left in the same row or above in the same column. Now denote the smallest column of the k-matches up to the i-th row by $T_{i,k}$, i. e.

$$T_{i,k} := \min\{\{t \mid (s,t) \in C_k, \ 0 \leq s \leq i\} \cup \{\infty\}\}, \quad 0 \leq i \leq m, \ k \geq 0 \ .$$

The main observation is that a match (i,j) in row i is in C_k if and only if $T_{i-1,k-1} < j \leq T_{i-1,k}$. Notice that a match with $j \leq T_{i-1,k-1}$ would have no predecessor in C_{k-1} and that $j > T_{i-1,k}$ would imply a rank of at least $k+1$ for (i,j). We can therefore state a recursion for $T_{i,k}$:

$$T_{i,k} = \begin{cases} \min\{j \mid a_i = b_j \wedge T_{i-1,k-1} < j \leq T_{i-1,k}\} \\ T_{i-1,k} \ \text{if there is no such } j \end{cases}$$

Note that $T_{i,k} < T_{i-1,k}$ implies that $(i, T_{i,k})$ is a dominant match in C_k. We even have that (i,j) is minimal in C_k if and only if $T_{i-1,k-1} < j < T_{i-1,k}$ and $j = T_{i,k}$. Initializing $T_{0,0}$ by 0 and $T_{0,k}$ by ∞, $k \geq 1$, we can compute all dominant matches row by row using the recursion above. Instead of defining $T_{i,k}$ as the smallest column of the k-matches up to row i we could also define $T_{j,k}$ as the smallest row of the k-matches up to column j and then, after making analogous observations, compute the dominant matches column by column.

Let us now look at an algorithm presented by Apostolico and Guerra [1] implementing this paradigm. They observed that it is very useful to precompute information about the first occurrence of a symbol in sequence $B = b_1 b_2 \ldots b_n$ to the right of some given position. Let $\Sigma = \{\sigma_1, \sigma_2, \ldots, \sigma_s\}$ be the relevant alphabet and define

$$CLOSEST : [\sigma_1, \ldots, \sigma_s] \times [0, \ldots, n+1] \longrightarrow [1, \ldots, n+1]$$
$$CLOSEST \ [\sigma_i, j] := \min\{\{j' > j \mid b_{j'} = \sigma_i\} \cup \{n+1\}\} \ .$$

If s is fixed and small, this function can be made available in form of a $s \times (n+1)$ matrix computable in time and space $O(sn)$. A compact representation can be computed in time and space $O(n)$ at the cost of $O(\log s)$ time for each lookup. Assume that we would like to compute a dominant k-match (i,j) in row i. Looking up $j := CLOSEST[a_i, T_{i-1,k-1}]$ we have three cases:

1. $j < T_{i-1,k} \implies (i,j)$ is a dominant k-match;
2. $j = T_{i-1,k} \implies (i,j) \in C_k$, but is no dominant match;
3. $j > T_{i-1,k} \implies (i,j)$ is a (dominant) match in the first class $C_{k'}$, $k' > k$, with $(T_{i-1,k'} > j) \ T_{i-1,k'} = j \ .$

Employing a row-by-row order of computation the following algorithm uses one vector $Thresh$ of size $O(m)$ to store the current leftmost column of dominant matches found so far for each class respectively. The correctness follows immediately from the preceding discussion and the invariant $Thresh[k] = T_{i-1,k}$, $k \geq 0$, at the beginning of the i-th iteration of the for-loop. **Algorithm 1** [1] takes as input the sequences $A = a_1 a_2 \ldots a_m$ and $B = b_1 b_2 \ldots b_n$, $m \leq n$, and outputs a LCS $S = s_1 s_2 \ldots s_p$.

```
(0)   Preprocessing: compute CLOSEST for sequence B
(1)   Thresh[0] := 0;
      for k := 1 to m do Thresh[k] := n + 1 od;
(2)   for i := 1 to m do;                      (* proceed row by row *)
         j := CLOSEST[a_i, 0];                 (* first match in row i *)
         k := 1;
(3)      while j ≠ n + 1 do                    (* still a match left in row i ? *)
            if j < Thresh[k] then
               temp := Thresh[k];
               Thresh[k] := j;
(4)            record new dominant match (i, j);
               j := CLOSEST[a_i, temp];        (* next relevant match *)
            fi;
            if j = Thresh[k] then j := CLOSEST[a_i, j] fi;
            k := k + 1;
         od;
      od;
(5)   recover LCS S
```

This algorithm's time complexity is $O(ns + pm)$. Time $O(ns)$ is needed for the preprocessing and in each iteration of the for-loop we check no more than $O(p)$ elements in $Thresh$ which equals the number of iterations of the while-loop. For each dominant match we dynamically create a new node and let it point to a generating predecessor (step 4) in order to be able to reconstruct a LCS in step 5 in $O(m)$ time (for details see [8]). Therefore the space complexity is $O(ns + d)$ where d denotes the number of dominant matches.

3 A New Flexible Algorithm

As we have argued above, a new dominant match is found if and only if a certain $Thresh$-value can be reduced in the current row. On the other hand the time complexity of all algorithms using the paradigm of computing dominant matches is given by an upper bound on the number of checks of the $Thresh$-values that are necessary to detect these reductions. Therefore it seems to be a good strategy to try to reduce the number of these checks as far as possible by avoiding unnecessary checks. A closer look at Algorithm 1 reveals the main drawback: once a class C_k has been identified not to be empty, the value $Thresh[k]$ has to be checked in each subsequent row. This means that a class is never recognized as being determined completely. For short longest common subsequences this does not matter too much since only a few *Thresh*-values have to be checked in each row. But as the longest common subsequences are getting longer the algorithm becomes inefficient. How could we make this general approach work for long longest common subsequences, too?

The first step is to observe that a match $s_k = (i, j) \in C_k$ – which is the k-th match in a LCS of length p between sequences A and B of length m and n respectively – can be sandwiched as follows:

$$k \leq i \leq k + m - p \text{ and } k \leq j \leq k + n - p . \tag{1}$$

Otherwise (i, j) could not be a k-match or we could not generate a LCS of length p by appending matches after (i, j). As a direct implication of this observation we would not need to check $Thresh[k]$ in rows $i > k + m - p$ because dominant k-matches that might occur in these rows could never be part of a LCS. But unfortunately we can't use this as a termination criterion for the computation of class C_k since we do not know p a priori; this is the value we would like to determine. Thus, our goal will be to formulate another criterion that does not depend on the knowledge of p although it allows us the identification of classes that need not be taken into consideration any further. Note that the above observation also makes a statement of existence: if there is a LCS of length p then in *each* class C_k there has to be at least one dominant k-match that satisfies (1) and which may be extended to a LCS. We call such a match a k-*witness*.

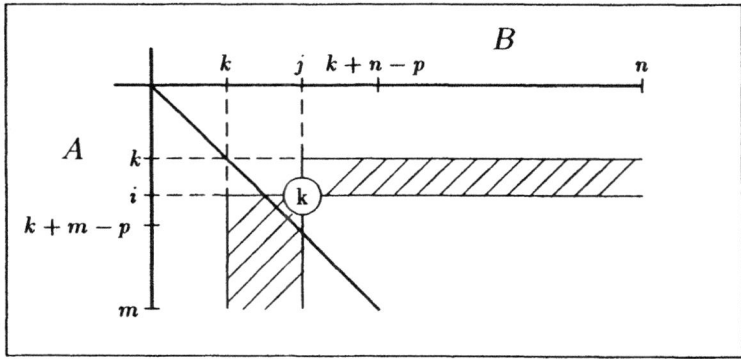

Fig. 1. Assuming a long LCS there are only two small stripes in the L-matrix where we have to search for dominant k-matches; (i, j) shall be a k-witness.

Suppose now the two input sequences are roughly of the same length ($m \approx n$) and there is a long LCS ($p \to m$). Then the components of a k-witness (i, j) must be close to k because of (1). In this case all dominant matches $(i', j') \in C_k$ must lie to the top right or bottom left of this match in the L-matrix, i.e. we either have $i' < i$ or $j' < j$. Of course it must also hold that $i' \geq k$ and $j' \geq k$. Therefore all dominant matches in C_k must lie in the two small stripes indicated in Figure 1. As a direct consequence each class C_k cannot have more than $(n-p)+(m-p)+1$ dominant matches and so we can give an upper bound on the total number d of dominant matches: $d \leq p(m+n-2p+1)$.

Further we can exploit this observation to devise a new efficient algorithm. Obviously all dominant k-matches have been determined if both stripes of Figure 1 have been searched. Using previous algorithms (which employ a row-by-row order of computation) only matches located in the horizontal stripe will be found quickly. The vertical stripe will be searched completely only after scanning the very last row; dominant matches are computed in order of increasing first components.

Therefore it makes sense to change the order of computation in such a way that dominant matches located in the vertical stripe are searched quickly, too. This can be achieved by computing the dominant matches in order of increasing smaller components, i.e. we first compute dominant matches in row 1 and column 1, followed by those in row 2 and column 2, etc. Of course in row i we only have to consider matches (i, j) with $j \geq i$ (to the right of and on the diagonal) since $j < i$ would imply that the match was found already when considering column j. Analogously, in column j we only have to consider matches (i, j) with $i > j$ (below the diagonal). Let (i_1, j_1), (i_2, j_2), ..., $(i_l, j_l) \in C_k$ be the dominant k-matches computed in this new order. Then we have $k \leq \min\{i_1, j_1\} \leq \min\{i_2, j_2\} \leq \ldots \leq \min\{i_l, j_l\}$. We can now give a sufficient criterion to detect that all dominant k-matches have been computed.

Theorem. *Let $D_{l,k}$ be the set of all dominant k-matches in rows $\leq l$ and columns $\leq l$, i.e. $(i, j) \in D_{l,k} \Rightarrow \min\{i, j\} \leq l$. Then take the minimum among the greater components of these matches as*

$$MinMax_{l,k} := \min\{\max\{i, j\} \mid (i, j) \in D_{l,k}\}$$

and define $l_{min}(k)$ to be the smallest l such that at least one match $(i, j) \in D_{l,k}$ has both components $\leq l$, i.e.

$$l_{min}(k) := \min\{l \mid l = MinMax_{l,k}\} \ .$$

Then we have:

1. *There is no dominant k-match (i', j') with $\min\{i', j'\} > l_{min}(k)$, i.e. the dominant k-matches have been determined completely after searching row and column $l_{min}(k)$;*
2. *$l_{min}(k) \leq k + n - p$, where p is the length of a LCS.*

Proof.

1. According to the definition we have at least one dominant k-match $(i, j) \in D_{l_{min}(k),k}$ with $i \leq l_{min}(k)$ and $j \leq l_{min}(k)$. This match is a predecessor for all matches (i', j') with $\min\{i', j'\} > l_{min}(k)$ which can be no k-matches for this reason. But according to the order of computation only such matches might be found subsequently.

2. Assume that $l_{min}(k) > k + n - p$ would hold. Then no dominant k-match (i, j) such that $i \leq k + m - p \leq k + n - p$ and $j \leq k + n - p$ would exist because at least one component of each dominant k-match would be $> k + n - p$. But this means that there is no k-witness and therefore no LCS of length p. A contradiction. □

Note that this criterion does not depend on the knowledge of p but only on the coordinates of the dominant matches found upto a certain point in the computation. On the other hand we can give an upper bound on the number of rows and columns to be searched for dominant k-matches, namely $(k + n - p) - k + 1$. Assuming that we can detect a dominant k-match in a certain row or column in time $O(1)$ this gives a time bound of $O(n - p + 1)$ to compute all dominant matches of one class and $O(p(n - p + 1))$ to determine all dominant matches in p classes.

The idea to implement the new order of computation consists in a kind of dualization of the method proposed by Apostolico and Guerra which allows us to proceed row by row as well as column by column. We now precompute two tables $CLOSEST\text{-}A$ and $CLOSEST\text{-}B$, at a cost of time and space $O(ns)$, containing information about the next occurrences of symbols in A and B respectively as defined in Section 2. We divide the set of dominant matches into two disjoint subsets of those to the right of and on the diagonal of the L-matrix and those below the diagonal. Intuitively one can imagine to slide down the diagonal, starting at point $(1, 1)$ and ending at point (m, m). Having reached point (l, l), $1 \leq l \leq m$, we can compute dominant matches (l, j), $j \geq l$, in row l with the aid of $CLOSEST\text{-}B$ and dominant matches found in previous rows, like in Algorithm 1. Analogously we can compute dominant matches (i, l), $i > l$, in column l with the aid of $CLOSEST\text{-}A$ and dominant matches found in previous columns. Information about dominant matches found in previous rows is recorded in a vector $row\text{-}Thresh$ containing for each class C_k the leftmost position of a dominant k-match found so far. Symmetrically $col\text{-}Thresh$ contains for each class the topmost position of a corresponding dominant match found in previous columns.

Having reached point (l, l) we can detect whether $l = l_{min}(k)$ for some class C_k, i.e. whether all dominant k-matches have been determined, by testing whether $l = row\text{-}Thresh[k]$ or $l = col\text{-}Thresh[k]$. Further notice that at each point (l, l) we have to perform this test only for one distinguished class C_{low} since no two classes may be completed at the same time; otherwise the contours of the classes would have to intersect or touch. In computing dominant matches in a certain row and column we therefore only have to look for dominant k-matches where $k \geq low$, thus saving checks of the values $row\text{-}Thresh[k']$ and $col\text{-}Thresh[k']$, $k' < low$, because all dominant k'-matches have been determined already.

Having in mind this discussion the following **Algorithm 2** [13] can easily be seen to implement exactly the ideas presented above. From this, the correctness follows immediately, too.

Algorithm 2

Input: Sequences $A = a_1 a_2 \ldots a_m$ and $B = b_1 b_2 \ldots b_n, m \leq n$.

Output: A LCS $S = s_1 s_2 \ldots s_p$.

Method:

(0)   Preprocessing:
    Compute $CLOSEST\text{-}A$ and $CLOSEST\text{-}B$

(1)   Initialization:
    $row\text{-}Thresh[0] := 0; col\text{-}Thresh[0] := 0;$
    for $k := 1$ **to** m
        do $col\text{-}Thresh[k] := m + 1; row\text{-}Thresh[k] := n + 1;$ **od;**
    $low := 1;$

(2)   **for** $l := 1$ **to** m **do**   (* slide down diagonal *)

(3)   **if** $col\text{-}Thresh[low] = l$ **then**   (* all dom. matches found in C_{low} ? *)
            $j := CLOSEST\text{-}B[a_l, row\text{-}Thresh[low]];$
            $low := low + 1;$
        else
            $j := CLOSEST\text{-}B[a_l, l - 1];$
        fi;
        $k := low;$

(4)   **while** $j \neq n + 1$ **do**   (* find dom. matches in row l *)
            if $j < row\text{-}Thresh[k]$ **then**
                $temp := row\text{-}Thresh[k];$
                $row\text{-}Thresh[k] := j;$
                record new dominant match $(l, j);$
                $j := CLOSEST\text{-}B[a_l, temp];$
            fi;
            if $j = row\text{-}Thresh[k]$ **then** $j := CLOSEST\text{-}B[a_l, j]$ **fi;**
            $k := k + 1;$
        od;

(5)   **if** $row\text{-}Thresh[low] = l$ **then**   (* all dom. matches found in C_{low} ? *)
            $i := CLOSEST\text{-}A[b_l, col\text{-}Thresh[low]];$
            $low := low + 1;$
        else
            $i := CLOSEST\text{-}A[b_l, l];$
        fi;
        $k := low;$

(6)   **while** $i \neq m + 1$ **do**   (* find dom. matches in column l *)
            if $i < col\text{-}Thresh[k]$ **then**
                $temp := col\text{-}Thresh[k];$
                $col\text{-}Thresh[k] := i;$
                record new dominant match $(i, l);$
                $i := CLOSEST\text{-}A[b_l, temp];$
            fi;
            if $i = col\text{-}Thresh[k]$ **then** $i := CLOSEST\text{-}A[b_l, i]$ **fi;**
            $k := k + 1;$
        od;
    od;

(7)   recover LCS S

With respect to the time complexity, recall the discussion after the proof of the theorem which shows the upper bound of $O(p(n-p+1))$ for the computation of all dominant matches. A second upper bound is given by $O(pm)$ since at each of the m positions on the diagonal at most $O(p)$ row-Thresh and col-Thresh values will be checked. Taking into account the preprocessing we have an overall time complexity of $O(ns + \min\{pm, p(n-p)\})$. It is easy to verify the space complexity $O(ns + d)$, where d is the number of dominant matches.

4 Experimental Results

In order to validate the efficiency of the new algorithm, some experiments were performed that compared it to previous methods. As a measure of efficiency that is independent of the actual implementation, the number of checks of elements in vector(s) Thresh was chosen (which is equivalent to the number of iterations of the main processing block in all algorithms using the paradigm of computing dominant matches). Omitting time $O(ns)$ for the preprocessing, the following six algorithms have been evaluated. First Algorithm 1 by Apostolico and Guerra [1] with running time $O(pm)$. Second the algorithm presented by Nakatsu et al. [11] which can be implemented in time $O(p(m-p))$ using the preprocessing introduced in Section 2. The original time bound was $O(n(m-p))$. Third the algorithm developed by Chin and Poon [2] with time complexity $O(\min\{pm, ds\})$. Fourth the new Algorithm 2 presented in this paper.

Also included in the comparison is another new algorithm which, in a certain sense, is equivalent to the Chin/Poon algorithm but uses a different order of computation and a different data structure. Again, the strategy is to avoid unnecessary checks of Thresh-values when searching dominant matches in a certain row. The main idea is as follows: from the calculation of dominant matches in row i_1 we can conclude that positions in B in intervals $[T_{i_1-1,k}+1, T_{i_1,k+1}-1]$ cannot match a_{i_1}. Now look at the smallest row $i_2 > i_1$ such that $a_{i_2} = a_{i_1}$. Then we can conclude from $T_{i_2-1,k} = T_{i_1-1,k}$ that $T_{i_2,k+1} = T_{i_2-1,k+1}$, i.e. we do not have to check $Thresh[k+1]$ in row i_2 because we know this value cannot change and so we can't find a new dominant $(k+1)$-match in row i_2. So $Thresh[k+1]$ may possibly change in row i_2 only if $Thresh[k]$ has been reduced in at least one of the rows $i_1, i_1 + 1, \ldots, i_2 - 1$. Thus, we have to remember in which row the individual Thresh-values changed for the last time. This is achieved by employing a doubly linked list. Each time an individual Thresh-value is reduced it is deleted at its old position and is inserted at the beginning of the list. Thus by scanning the list from the beginning we can always obtain those Thresh-values that changed most recently in preceding rows. This provides us exactly with the information needed to identify the Thresh-values that might change in the current row under consideration. For details on this algorithm see [13]. Since the underlying concepts of the two new algorithms are orthogonal to each other, they may also be used in combination, which was the last algorithm in the experiments.

The results are shown in Table 1 as the average of 100 pairs of input sequences. The new Algorithm 2 clearly beats the first two algorithms. Only the $O(ds)$ algorithms compare well for short longest common subsequences for quite a long time. The combined algorithm joins merits from both algorithms. Notice the very close approximation of dominant matches by the new Algorithm 2 for long longest common subsequences. This means that there are almost no superfluous checks of *Thresh*-values in this case.

Table 1. Frequency of checks of *Thresh*-values for six different algorithms. $|A| = 4000$, $|B| = 4000$, $|\Sigma| = 16$.

$|LCS|$	dominant matches	Apostolico/ Guerra $O(pm)$ **Algorithm 1**	Nakatsu et al. $O(p(m-p))$	Chin/Poon $O(ds)$	New $O(p(n-p))$ **Algorithm 2**	New $O(ds)$	Combined New $O(p(n-p))$
50	1,435	120,679	147,361	9,638	60,112	9,638	16,904
97	4,989	262,925	284,614	32,499	141,838	32,499	37,979
195	17,602	543,821	549,565	109,935	307,136	109,935	107,627
378	53,424	1,046,973	995,970	313,505	579,587	313,505	278,950
684	133,360	1,816,673	1,598,142	718,009	906,427	718,009	573,656
1,034	257,687	2,692,127	2,166,799	1,277,350	1,255,101	1,277,350	931,968
1,352	401,620	3,475,230	2,567,138	1,868,036	1,569,218	1,868,036	1,288,738
1,572	513,599	4,024,663	2,792,765	2,293,083	1,777,573	2,293,083	1,539,632
1,663	506,354	4,094,862	2,715,082	2,300,281	1,532,089	2,300,281	1,367,447
2,161	423,878	4,748,025	2,414,038	2,236,754	859,494	2,236,754	814,394
2,830	282,017	5,867,671	1,861,615	1,819,923	410,560	1,819,923	402,579
3,337	165,328	6,772,005	1,203,500	1,224,247	200,006	1,224,247	199,103
3,648	90,828	7,344,405	689,941	735,664	99,845	735,664	100,167
3,818	48,939	7,663,327	373,510	424,754	51,206	424,754	51,774
3,907	26,783	7,830,310	197,717	250,206	27,353	250,206	28,148

Comparing the actual running times of straightforward implementations of the algorithms resulted in Table 2. The algorithm by Wu et al. [17], using the paradigm of computing a shortest-path, seems to be the fastest for long longest common subsequences. So it was included in the comparison of running times, too. As can be seen, Algorithm 2 now even beats the Chin/Poon algorithm. The algorithm of Wu et al. proved to be very specialized. It is extremely fast for long longest common subsequences but its running time decreases dramatically for shorter ones. This is also due in part to a second specialization of this algorithm on situations when the input sequences differ in length considerably. However, in this case the sequences can't be similar, just because of their difference in length; remember that the application we have in mind was to find sequences in a database that are similar to a query.

Unfortunately, the combined algorithm is not as fast for longest common subsequences of intermediate length as it could be expected from Table 1. Here Algorithm 2 clearly performs best of all.

Table 2. Running times for seven algorithms in msec; $|A| = 4000, |B| = 4000, |\Sigma| = 16$.

$|LCS|$	Apostolico/ Guerra $O(pm)$ Algorithm 1	Nakatsu et al. $O(p(m-p))$	Chin/Poon $O(ds)$	New $O(p(n-p))$ Algorithm 2	New $O(ds)$	Combined New $O(p(n-p))$	Wu et al. $O(n(m-p))$
50	644	1,350	1,044	745	371	685	123,388
97	1,074	2,350	1,229	1,011	550	849	120,120
195	1,945	4,293	1,875	1,554	1,187	1,417	114,183
378	3,539	7,604	3,637	2,505	2,910	2,892	103,586
684	6,067	12,159	7,267	3,770	6,527	5,680	87,605
1,034	9,084	16,598	12,447	5,285	11,441	9,456	70,496
1,352	11,920	19,839	18,043	6,738	16,903	13,511	56,835
1,572	14,058	21,702	22,111	7,771	21,282	16,471	48,094
1,663	14,163	21,095	22,069	6,908	21,487	15,342	45,383
2,161	15,841	18,731	20,607	4,679	20,578	10,933	28,329
2,830	18,668	14,476	16,071	2,791	16,007	6,641	11,482
3,337	20,947	9,451	10,836	1,737	10,681	3,915	3,726
3,648	22,386	5,533	6,827	1,164	6,389	2,358	1,074
3,818	23,187	3,117	4,344	865	3,646	1,520	309
3,907	23,606	1,774	2,969	708	2,163	1,091	100

5 Conclusion

A new flexible algorithm for the LCS problem was presented which avoids a possibly difficult or even impossible a priori decision for some specialized algorithm. The new algorithm competes well with previous algorithms for both short and long longest common subsequences and can deal with longest common subsequences of intermediate length in a more efficient manner than any previous algorithm. The new algorithm is based on the well-known paradigm of computing dominant matches and was obtained through a kind of dualization. Its time and space complexity is $O(ns + \min\{pm, p(n-p)\})$ and $O(ns + d)$ respectively. It remains to give a linear space implementation of this algorithm and to generalize it to more than two sequences where the savings can be conjectured to be even larger.

Acknowledgment. I would like to thank Prof. Dr. N. Blum for drawing my attention to the LCS problem.

References

1. A. Apostolico, C. Guerra: *The Longest Common Subsequence Problem Revisited*, Algorithmica, Vol. **2**, 1987, 315–336.
2. F. Y. L. Chin, C. K. Poon: *A Fast Algorithm for Computing Longest Common Subsequences of Small Alphabet Size*, Journal of Information Processing, Vol. **13**, No. 4, 1990, 463–469.
3. F. Y. L. Chin, C. K. Poon: *Performance Analysis of Some Simple Heuristics for Computing Longest Common Subsequences*, Algorithmica, Vol. **12**, 1994, 293–311.

4. V. Dančík: *Expected Length of Longest Common Subsequences*, PhD thesis, University of Warwick, 1994.

5. D. Eppstein, Z. Galil, R. Giancarlo, G. F. Italiano: *Sparse Dynamic Programming I: Linear Cost Functions*, Journal of the ACM, Vol. **39**, No. 3, 1992, 519–545.

6. D. S. Hirschberg: *Algorithms for the Longest Common Subsequence Problem*, Journal ACM, Vol. **24**, Oct. 1977, 664–675.

7. W. J. Hsu, M. W. Du: *New Algorithms for the LCS Problem*, Journal of Computer and System Sciences **29**, 1984, 133–152.

8. J. W. Hunt, T. G. Szymanski: *A Fast Algorithm for Computing Longest Common Subsequences*, Comm. ACM, Vol. **20**, May 1977, 350–353.

9. W. J. Masek, M. S. Paterson: *A Faster Algorithm Computing String Edit Distances*, Journal of Computer and System Sciences **20**, 1980, 18–31.

10. E. W. Myers: *An $O(ND)$ Difference Algorithm and Its Variations*, Algorithmica, Vol. **1**, 1986, 251–266.

11. N. Nakatsu, Y. Kambayashi, S. Yajima: *A Longest Common Subsequence Algorithm Suitable for Similar Text Strings*, Acta Informatica **18**, 1982, 171–179.

12. M. Paterson, V. Dančík: *Longest Common Subsequences*, Proceedings of the 19th Intern. Symp. on Mathematical Foundations of Computer Science, Vol. **841** of LNCS, 1994, 127–142.

13. C. Rick: *New Algorithms for the Longest Common Subsequence Problem*, Research Report No. **85123-CS**, University of Bonn, 1994.

14. D. Sankoff, J. B. Kruskal (Ed.): *Time Warps, String Edits and Macromolecules: The Theory and Practice of Sequence Comparison*, Addison-Wesley: Reading, MA, 1983.

15. E. Ukkonen: *Algorithms for Approximate String Matching*, Information and Control **64**, 1985, 100–118.

16. R. A. Wagner, M. J. Fischer: *The String to String Correction Problem*, Journal ACM, Vol. **21**, 1974, 168–173.

17. S. Wu, U. Manber, G. Myers, W. Miller: *An $O(NP)$ Sequence Comparison Algorithm*, Information Processing Letters **35**, 1990, 317–323.

Smaller Representations for Finite-State Transducers and Finite-State Automata

Emmanuel Roche

Mitsubishi Electric Research Laboratories
201, Broadway, Cambridge, MA 02139,
roche@merl.com

Abstract. Finite-state transducers and finite-state automata are efficient and natural representations for a large variety of problems. We describe a new algorithm for turning a finite-state transducer into the composition of two deterministic finite-state transducers such that the combined size of the derived transducers can be exponentially smaller than other known deterministic constructions. As a consequence, this can also be used to build deterministic representations of finite-state automata smaller than the minimal finite-state automata computed by the classic determinization and minimization algorithms. We also report experimental results on large scale dictionaries and rule-based systems.

1 INTRODUCTION

Finite-state transducers and finite-state automata are used in a great variety of programs such as lexical analyzers. Some of these representations can contain more than a million states in applications such as Natural Language Processing. This points out the need for efficient compaction methods.

The problem is solved in the case of deterministic finite-state automata where the minimization algorithm is well known. A minimization procedure is also available in the case of subsequential transducers [11]. For the general case of rational functions, Reutenauer and Schützenberger [12] give a way to construct a bimachine which is minimal modulo a certain equivalence relation and in the case of subsequential functions, their construction is equivalent to the minimal subsequential transducer. Building a bimachine (introduced by [14], see also [4, 2]) can also be seen as building a decomposition of a rational function f into $\alpha \circ \beta$ where α (resp. β) is a right-sequential function (resp. a left-sequential function). Such a decomposition is possible for any rational function [5]. We give here an algorithm that, like in [12], builds a decomposition of any rational function into a right-sequential and a left-sequential transducer; this decomposition can be exponentially smaller than the one proposed in [12] and in the case of subsequential functions it can be exponentially smaller than the minimal subsequential transducer. In other words, there exists a family of finite-state transducers T_n such that, if $T_n = \alpha_n \circ \beta_n$ is the result of the factorization procedure presented here and if τ_n is the minimal subsequential transducer equivalent to T_n (when it exists), then $\|\alpha_n\| + \|\beta_n\| = O(log\|\tau_n\|)$.

As a curious consequence, this construction can lead to deterministic representations of finite-state automata smaller than minimal deterministic automata. A finite-state automaton can indeed be viewed as a constant rational function whose domain is the language recognized by this finite-state automaton.

In this paper, we informally describe the factorization algorithm and illustrate it by a step-by-step application on an example. We then show the correctness of the algorithm. We will briefly say how this can be used for building finite-state representations of finite-state automata that can be smaller than the minimal deterministic finite-state automata.

We will then present some experimental results on some large scale dictionaries and rule based systems. The result show that the method is efficient and it can, in some cases, represent a big improvement upon previously known representation.

2 FACTORIZATION ALGORITHM

The algorithm can be applied to any rational function, that is to any function represented by a transducer $T = (\Sigma, Q, i, F, E)$ where Σ is the alphabet, Q is the finite set of states, $i \in Q$ is the initial state, $F \subset Q$ is the set of final states and $E \subset Q \times \Sigma \times \Sigma^* \times Q$ is the set of edges. For each edge (q, a, b, q'), also called a transition, q is called the starting state of the transition, a is called its input label, b is called its output label and q' is called its arrival state. We also define the notion of transition function by $d(q, a) = \{q' \in Q | \exists (q, a, b, q') \in E\}$ and the notion of emission function by $\delta(q, a, q') = \{b \in \Sigma^* | \exists (q, a, b, q') \in E\}$. Both d and δ are extended to words in the classical way.

Let us first consider the particular case of a sequential function such as the one represented by the finite-state transducer T of Figure 1. The minimal sequential transducer τ representing the same function, that is $|\tau| = |T|$,[1] is given Figure 2. We will here show that it is possible to obtain a decomposition of T into a right-sequential transducer τ_{right} and a left-sequential transducer τ_{left} which can be smaller than the minimal sequential representation. In the case of T of Figure 1, the decomposition $T = \tau_{right} \circ \tau_{left}$ is given in Figure 3 and in Figure 4.

The algorithm is informally described on Figure 5. We shall now illustrate it by a step-by-step application on the example T of Figure 1 and show how it generates the decomposition of Figure 3. The core of this method and the reason why it improves upon previous results is introduced in STEP 6 which calls the function $GRAPH\_COLORING$. This function attempts to find a minimal number of color given a graph such that no two adjacent vertices share the same color.

STEP 1 and 2: The first step consists of identifying where is the left-to-right undeterminicity of the original transducer. More specifically, we look for the pairs of states (q_1, q_2) such that they both can be accessed by the same input

---

[1] If T is a finite-state transducer, $|T|$ denotes the function associated to it.

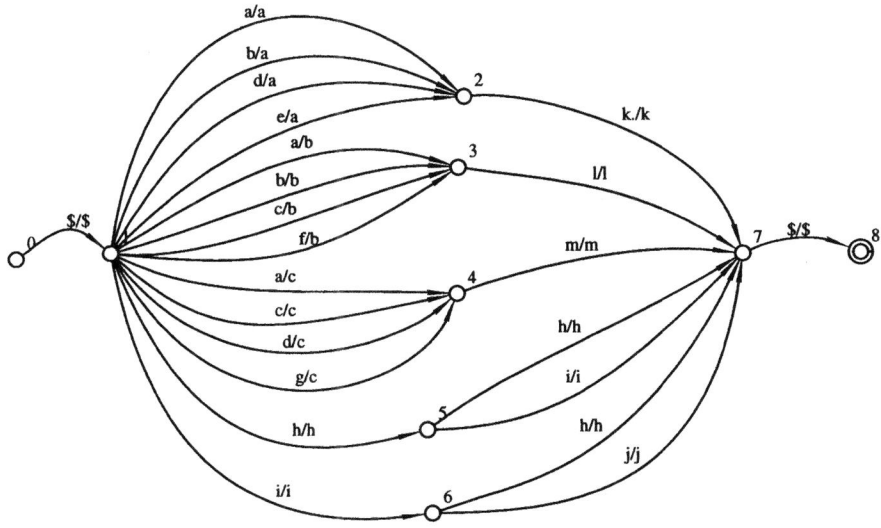

Fig. 1. Transducer T.

string. For instance, for T of Figure 1, since $\$ \cdot a$ can lead to the states 2, 3 and 4, then $(2,3), (2,4)$ and $(3,4)$ are such pairs. [2]

To identify these pairs we only have to consider the input part of the transducer, that is only the input labels (left labels) on each transition. We thus first build the automaton A obtained by removing from T all the output labels (step 1). An obvious way to identify the set of pairs is to determinize the automaton A according to the classical power set construction algorithm (see [1] for instance). In fact, the fact that two states q_1 and q_2 can be reached by the same word is equivalent to the fact that there is some state set $\bar{q}$ of the powerset construction such that $q_1 \in \bar{q}$ and $q_2 \in \bar{q}$. For instance, the determinization of A is given Figure 6, it shows that the set of pairs we are looking for is the symmetric closure of $\{(2,3), (3,4), (2,4)\}$. This method however has an exponential time and space complexity and the deterministic version of A isn't used later. It is thus possible to compute the square of A (function SQUARE(A) of step 2) and get the same set of pairs. In fact, if $A = (Q, i, F, d)$ then A^2 is defined by $A^2 = (Q_2 = Q \times Q, (i,i), F \times F, d_2)$ where $d_2((q,q'),a) = d(q,a) \times d(q,b)$ and there is an equivalence between the fact that a word w reaches two different states in A and the fact that it reaches a state (q,q'), with $q \neq q'$, in A^2. Therefore, the set of pairs we look for is $Q_2 - \{(q,q) | q \in Q\}$. SQUARE($A$) is partially

---

[2] With the notations of [4], we look for the pairs (q_1, q_2) s.t. $i^{-1}q_1 \cap i^{-1}q_2 \neq \emptyset$.

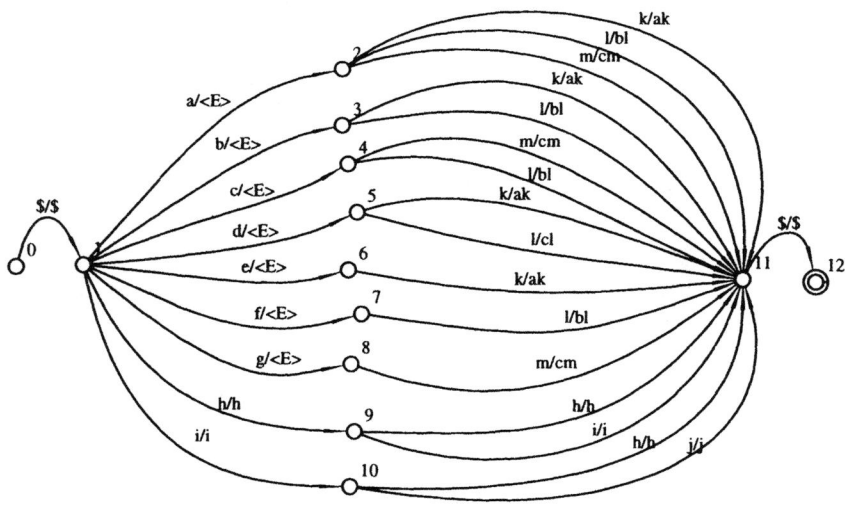

Fig. 2. Subsequential transducer τ, $< E >$ stand for the empty string ϵ.

represented in Figure 7, it leads to the same set of pairs, that is the reflexive closure of $\{(2,3),(2,4),(3,4)\}$.

STEP 3 : At this point, we just format the set of pairs as a graph whose vertices are the states of A and whose edges are the pairs (q, q') s.t. $q \neq q'$ and $(q, q') \in Q_2$ with Q_2 the states of SQUARE(Q). For our example, this leads to the graph G_1 of Figure 8. Two states are connected within this graph iff there exists a string that leads to these two states.

STEP 4 : This step computes the graph G_2 by applying the previous procedure (STEP 1, 2 and 3) to the reverse transducer $\tilde{T}$ obtained by reversing the edges. Two states are connected within this graph iff there exists a string w and a path from both of these states to a terminal state labeled by w. In other words, $(q_1, q_2) \in E_{G_2}$ iff $\exists w \in \Sigma^*$ s.t. $d(q_1, w) \cap F \neq \emptyset$ and $d(q_2, w) \cap F \neq \emptyset$.

STEP 5 : This step combines the two graphs G_1 and G_2 into a supergraph G of G_1. Two states q_1 and q_2 are connected within this graph iff they are connected within G_1 or there exists a state q' s.t. q_1 and q' are connected within G_1 and q' and q_2 are connected within G_2.

STEP 6 : The purpose of this step is to define an equivalence relation R_G (or R for short) on Q compatible with the graph G we just built. In other words, we look for an equivalence relation R_G such that $qR_Gq' \Rightarrow (q, q')$ is not in E_G where E_G is the set of edges of G. Moreover, we will see that in order to get a representation as small as possible, R_G should have as few equivalent classes as possible. This is exactly the definition of the extensively studied graph coloring

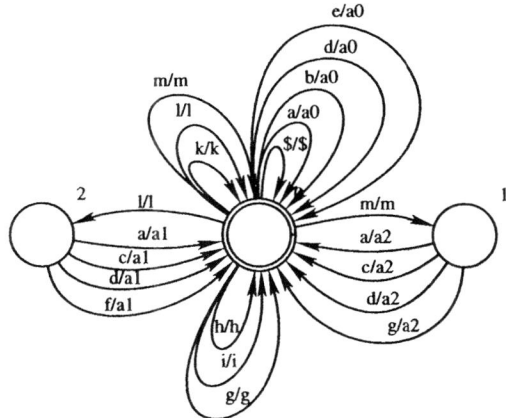

Fig. 3. τ_{right}

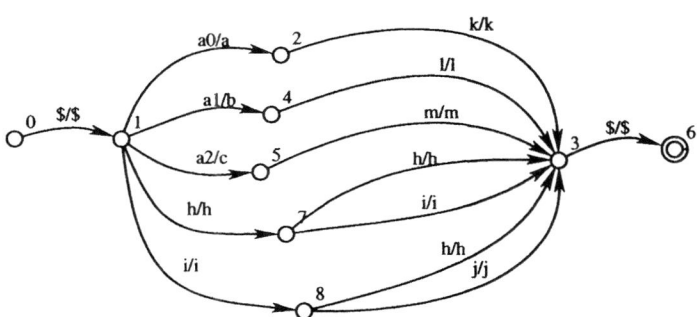

Fig. 4. τ_{left}

problem. Recall that given a graph $G = (Q, E_G)$, the graph coloring problem consists of finding a partition of Q into a minimum number of *color classes* $C_1, C_2, .., C_k$ where no two vertices q and q' can be in the same color class if there is an edge in E_G between them. Recall also that the graph coloring problem is NP-hard [8], thus no general optimization method for it is known. However, it is possible to use heuristics [3, 10, 7] to find a solution that, while not optimal, is not very far from the optimality (see [7] for experimental results). Here, we applied the simplest heuristic which consists in coloring the vertex in order. The

$(\tau_{right}, \tau_{left}) =$ FACTORIZE(T)

STEP 1.Call $A=$ EXTRAC_DOMAIN(T) to build the automaton A obtained from T by considering only the input symbols.

STEP 2. Call $A_2=$SQUARE(A) to build $A^2 = (Q_2 \subset Q \times Q, (i,i), F \times F, E_2)$.

STEP 3. Call $G_1=$BUILD_GRAPH_1(Q,Q_2) to generate the graph $G_1 = (Q, E_{G_1})$ such that $(q_1,q_2) \in E_{G_1}$ iff $(q_1,q_2) \in Q_2$.

STEP 4. Call $G_2=$BUILD_GRAPH_2(T) to generate the graph $G_2 = (Q, E_{G_2})$ such that $(q_1,q_2) \in E_{G_2}$ iff $\exists w \in \Sigma^*$ s.t. $d(q_1,w) \cap F \neq \emptyset$ and $d(q_2,w) \cap F \neq \emptyset$

STEP 5. Call $G=$BUILD_GRAPH1(G_1,G_2) to generate the graph $G = (Q, E_G)$ such that $(q_1,q_2) \in E_{G_1}$ iff $(q_1,q_2) \in E_{G_1}$ or $\exists q' \in Q$ s.t. $(q_1,q') \in E_{G_1}$ and $(q',q_2) \in E_{g_2}$.

STEP 6. Call $R_G=$GRAPH_COLORING(G) to build an equivalence relation R_G on Q as small as possible compatible with G, that is such that $q_1 R_G q_2 \Rightarrow (q_1,q_2)$ is not in E_G.

STEP 7. Call $A_1=$DET_MERGE(REV(A),R_G) to determinize the reverse of A into A_1 while identifying states that are in the same equivalence class in R_G.

STEP 8. Call $T_1=$ADD_STATE_CONTEXT(A_1) to transform A_1 into the transducer T_1 s.t. dom($|T_1|) = |A_1|$ and s.t. each output edge can be written $(q, a, (a, q), q')$.

STEP 9. Call $T_2=$ADJUST_LEFT(T,T_1) which builds the transducer T_2 such that $T = T_1 \circ T_2$.

STEP 10. Call $R_2=$MONOID(T_2) to compute the equivalence relation R_2 on $\Sigma \times Q$ such that $(a,q_1)R_2(b,q_2)$ iff $(q, (a,q_1), b, q') \in E_2 \Leftrightarrow (q, (b,q_2), b, q') \in E_2$.

STEP 11. Call $\tau_{right}=$REPLACE_OUTPUT(T_1,R_2) to replace each output symbol (a,q) by its equivalence class $R_2(a,q)$ and Call $\tau_{left}=$REPLACE_INPUT(T_2,R_2) to replace each input symbol of T_2 by its equivalence class in R_2.

Fig. 5. The factorization algorithm.

vertex q_1 is assigned the color C_1 and the vertex q_i is assigned the lowest indexed color C_j that contains no vertex adjacent to q_i. If no such color exists, a new one containing only q_i is created. In our example, it created the following partition :$\{\{0,1,2,5,6,7,8\}, \{3\}, \{4\}\}$. We thus end up with the equivalence relation R_G s.t. $|Q/R_G| = 3$ and such that $R_G(0) = \{0,1,2,5,6,7,8\}$, $R_G(3) = \{3\}$ and $R_G(4) = \{4\}$.

STEP 7: The purpose of this step is to build a right-to-left deterministic automaton that encodes just enough information about the right context to separate the states of A (and therefore T) than can be reached by the same input

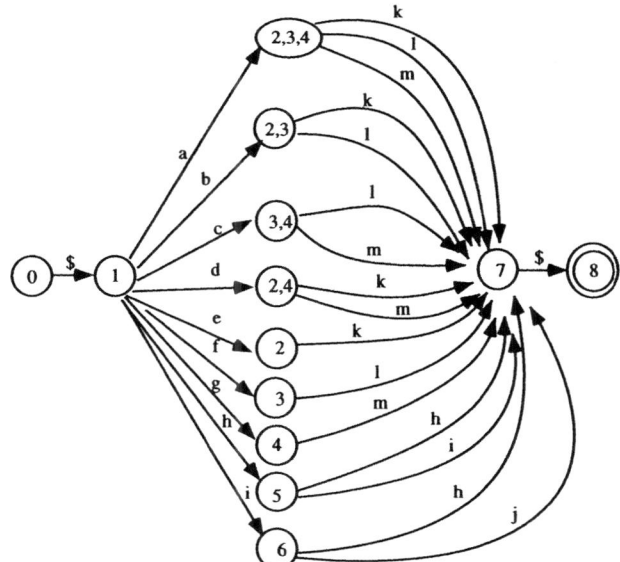

Fig. 6. Determinization of A

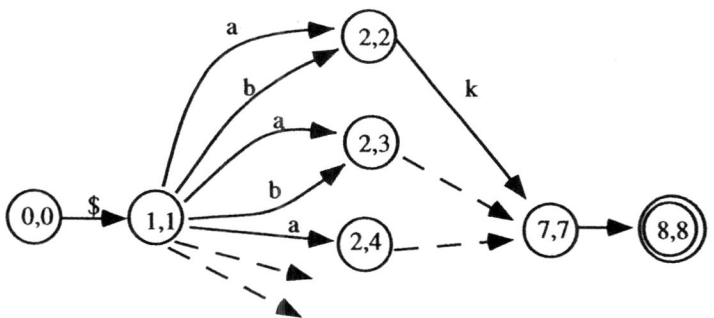

Fig. 7. $A_2 = \text{SQUARE}(A)$

word. In other words, we want to build A_1 such that, for each w such that w reaches two different states q and q' of A, for each $u, v \in \Sigma^*$ s.t. $w \cdot u, w \cdot v \in |A|$ then u and v lead to two different states in A_1. We will later show how this allows to do only deterministic transitions. To compute A_1 we first have to take the reverse automaton of A, denoted $\tilde{A}$, obtained by inverting the transitions $(q' \in d_A(q, a)$ iff $q \in d_{\tilde{A}}(q', a))$ and by having the starting states $\tilde{\imath}$ defined by $F = \{\tilde{\imath}\}$ and the final states set defined by $\tilde{F} = \{i\}$ [3]. The automaton A_1 is then defined by $A_1 = (2^{Q/R_G}, R_G(\tilde{\imath}), R_G(\tilde{F}), d_1)$ where the transition function d_1 is

---

[3] We can indeed assume that, without loss of generality, F only has one element.

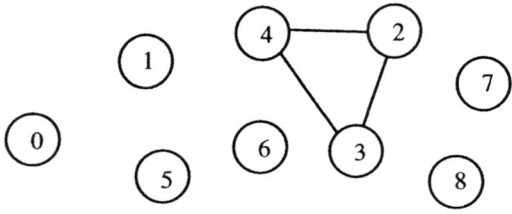

Fig. 8. $G_1 = \text{BUILD\_GRAPH\_1}(Q, Q_2)$ and $G = \text{BUILD\_GRAPH}(G_1, G_2)$

defined by

$$d_1(\overline{q}, a) = R_G(\bigcup_{R(q_i) \in \overline{q}} \bigcup_{q_j \in R(q_i)} d_{\tilde{A}}(q, a))$$

In our example, this operation leads to the automaton A_1 of Figure 9.

STEP 8: To compute the image of an input word w in the final decomposition, we will first read the input from right to left and go through A_1. While doing that, we should remember the path followed in A_1. A way to do that [2] is to turn A_1 into a transducer that, for each transition, reemits the input and adds a reference to the state at which the transition started. For instance, if one reads the letter a while being at the state number 2, one emits the pair $(a, 2)$. Therefore, if one takes the string $\$ \cdot a \cdot m \cdot d \cdot g \cdot \$$ and applies it on A_1, one gets the output $(\$, 0) \cdot (a, 0) \cdot (m, 0) \cdot (g, 1) \cdot (\$, 0)$. The transducer T_1 that realizes this transformation is obtained from A_1 by adding on each transition (q, a, q') the output (q, a) (function **ADD_STATE_CONTEXT**) . For our example, T_1 is represented Figure 10.

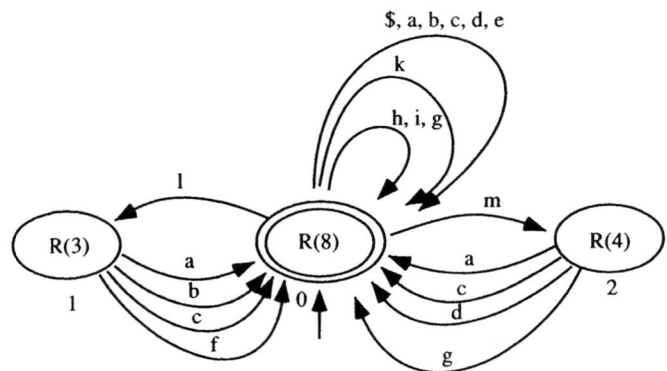

Fig. 9. $A_1 = \text{DET\_MERGE}(\text{REV}(A), R_G)$

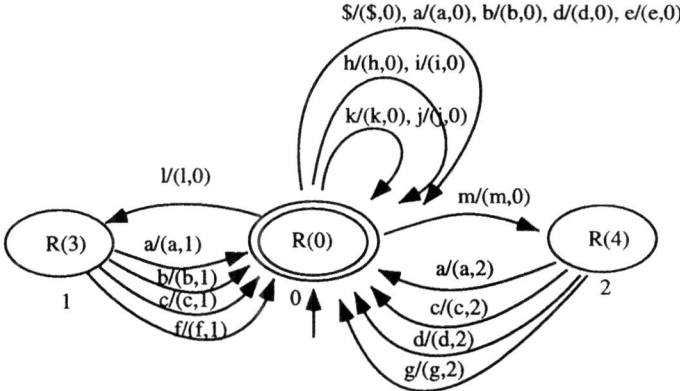

Fig. 10. $T_1 = \text{ADD\_STATE\_CONTEXT}(A_1)$

STEP 9 : This step consists of adjusting the original transducer T such that one can first apply the transducer T_1. More precisely, we want to compute the transducer T_2 such that $|T| = |T_1| \circ |T_2|$. T_2 is obtained by looking at each transition (q, a, b, q') of T, each transition $(q_1, a, (a, q_1), q_2)$ of T_1, and if there is one word $u \cdot a \cdot v \in \text{dom}(f)$ s.t. $q \in d(i, u)$ and $q_1 \in d(i_1, v)$ then replace a by (a, q_1) in T to produce the transition $(q, (a, q_1), b, q')$. For our example, such a transducer T_2 is represented Figure 11. The exact definition of T_2 is $T_2 = (Q, i, F, E_2)$ where E_2 is defined as follows: we first define the auxiliary function g_2 that to each state of T associates the set of states of T_1 defined by

$$g_2(q) = \bigcup_{w \in \Sigma^* | d(q,w) \cap F \neq \emptyset} d_1(i_1, w)$$

E_2 is then defined by

$$E_2 = \bigcup_{(q,a,b,q') \in E} \bigcup_{q'' \in g_2(q'), q''' \in g_2(q)} \bigcup_{(q'', a, (a, q''), q''') \in E_1} (q, (a, q''), b, q')$$

Obviously, $\|T_2\| = \|T\|$ and we will prove that $|T| = |T_1| \circ |T_2|$.

STEP 10 and 11: Note that on Figure 11 the input symbols $(a, 0)$ and $(b, 0)$ behave in exactly the same way. That is, they have the same output symbols from the same starting states and to the same arrival states. Therefore, if instead of outputting two different symbols in T_1 one outputs only one symbol, say $(a, 0)$, it reduces the number of transitions: the transition $(1, (b, 0), a, 2)$ disappears. More formally, we can define the equivalence relation R_2 on the input symbols of T_2 by $(a, q_1) R_2 (b, q_2)$ iff $(q, (a, q_1), c, q') \in E_1 \Leftrightarrow (q, (b, q_2), c, q') \in E_1$. The function MONOID returns this equivalence relation R_2. The function $\tau_{left} = \text{REPLACE\_INPUT}(T_2, R_2)$ then takes each input symbol and replaces it by its equivalence class, that is

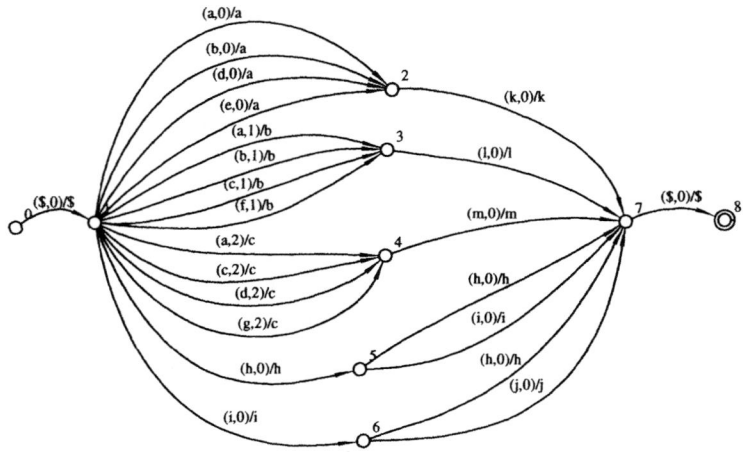

Fig. 11. $T_2 = \text{ADJUST\_LEFT}(T, T_1)$

$$E_{T_{left}} = \{(q, R_2((a, q_1)), b, q')|(q, (a, q_1), b, q') \in E_2\}$$

In a similar way, the function $\tau_{right}=\text{REPLACE\_OUTPUT}(T_1, R_2)$ takes each output symbol and replaces it by its equivalence class.

3 CORRECTNESS

Let us first show that the algorithm is complete, this is expressed by the following proposition:

Proposition: For any finite-state transducer T representing a function which is deterministic in the sense of automata, the algorithm terminates.

Proof: The only point at which the algorithm might not terminate would be if G contained an edge of the shape (q, q). Indeed, in that case it would be impossible to build an equivalence relation compatible with G. Suppose that there exists $q \in Q$ such than $(q, q) \in E_G$, then there exists q_2 such that $(q, q_2) \in E_{G_1}$ ($q_2 \neq q_1$) and $(q_2, q) \in E_{G_2}$. $(q, q_2) \in E_{G_1}$ means that there exists $u \in \Sigma^*$ s.t. $q \in d(q, u)$ and $q_2 \in d(q, u)$. On the other hand, $(q_2, q) \in E_{G_2}$ means that there exists $v \in \Sigma^*$ s.t. $d(q_2, v) \cap F \neq \emptyset$ and $d(q, v) \cap F \neq \emptyset$. Therefore, $w = u \cdot v$ is the label of two different paths in T, this contradicts the fact that T is a function. $\square$

The soundness is expressed by the following proposition:

Proposition: If T is a finite-state transducer representing a function, if τ_{right} and τ_{left} are the transducers built in the algorithm (steps 1 to 11) then τ_{right} and τ_{left} are sequential and $|T| = |\tau_{right}| \circ |\tau_{left}|$.

Proof: Let us first prove that the composition of τ_{right} and τ_{left} is equal to $|T|$. Since STEP 10 and STEP 11 keep the functional properties of the decomposition, it is sufficient to prove that $|T| = |T_1| \circ |T_2|$ as computed from STEP 1 to STEP 9. Without loss of generality, one also assumes that $F = \{f\}$.

Let $w = a_0 a_1 .. a_n$ be a string of $\mathrm{Dom}(|T|)$. There is then a unique path defined by $(q_i, a_i, b_i, q_{i+1}) \in E$ that describes the mapping $|T|(w) = b_0 b_1 .. b_n$. In addition, one has $q_0 = i$ and $q_n = f$ where f is the unique element of F. By construction of T_1, one has $\mathrm{Dom}(T) \subset \mathrm{Dom}(|T_1|)$. Therefore, there is a path $(\overline{q_{i+1}}, a_i, (a_i, \overline{q_{i+1}}), \overline{q_i})$ where

$$q_{i+1} = d_1(i_1, a_n a_{n-1} .. a_{i+1})$$

and

$$q_i = d_1(i_1, a_n a_{n-1} .. a_i)$$

Therefore, $\overline{q_{i+1}} \in g_2(q_{i+1})$ and $\overline{q_i} \in g_2(q_i)$. Hence $(q_i, (a_i, \overline{q_{i+1}}), b_i, q_{i+1}) \in E_2$, and therefore $b_0 b_1 .. b_n = |T|(w) \in |T_2|((a_0, \overline{q_1})..(a_n, \overline{q_{n+1}}))$, that is, $|T|(w) \in |T_2|(|T_1|(w))$. The fact that $|T_1| \circ |T_2|$ is a function will be deduced from the determinicity of T_1 and T_2.

Let us now prove that τ_{right} and τ_{left} are sequential. Since T_1 is sequential by construction and since the determinicity is preserved through STEP 10 and STEP 11 one only needs to show that T_2 is sequential.

Let $(q, a, b_1, q_1) \in E$ and $(q, a, b_2, q_2) \in E$ with $q_1 \neq q_2$. Let us show that $g_2(q_1) \cap g_2(q_2) = \emptyset$. Suppose indeed that $\overline{q} \in g_2(q_1) \cap g_2(q_2)$ then there exists $w_1 \in \Sigma^*$ s.t. $d(q_1, w_1) \cap F \neq \emptyset$ and $d_1(i_1, w_1) = \overline{q}$. One also has $w_2 \in \Sigma^*$ s.t. $d(q_2, w_2) \cap F \neq \emptyset$ and $d_1(i_1, w_2) = \overline{q}$. By definition of d_1, since $d(q_1, w_1) \cap F \neq \emptyset$, $q_1 \in d_{\tilde{A}}(i, w_1)$ then $R(q_1) \in \overline{q}$. Since $R(q_1) \in \overline{q}$ and $d_1(i_1, w_2) = \overline{q}$ then $\exists q' \in R(q_1)$ s.t. $d(q', w_2) \cap F \neq \emptyset$ and since $d(q_2, w_2) \cap F \neq \emptyset$ one has $(q', q_2) \in E_{G_2}$ (i.e. q' and q_2 are linked within the graph G_2). Moreover, since $(q, a, b_1, q_1) \in E$ and $(q, a, b_2, q_2) \in E$ then $(q1, q_2) \in E_{G_1}$ and therefore $(q_1, q') \in E_G$ which contradicts the fact that R is compatible with G. $\square$

Remark 1: Note that, in order to prove that $|T| = |T_1| \circ |T_2|$, the only requirement for T_1 is to verify $\mathrm{Dom}(|T|) \subset \mathrm{Dom}(|T_1|)$. It is therefore possible to use an even smaller $|T_1|$ (by building an equivalence relation with larger equivalence classes for instance or by building it by hand). The resulting smaller factorization might not be deterministic; but it can be less undeterministic than the original transducer T in the following sense: for any input string w the number of paths that one has to follow in T_1 and T_2 is smaller than or equal to the number of paths that one has to follow in T. There is therefore a "continuous" way to look at the deterministic-size tradeoff (that is, also, the space-size tradeoff).

Remark 2: Using G as built at STEP 4 guarantees the determinicity of T_2 but in practice we found it sufficient to use only G_1 (which is a subgraph of G and therefore leads to fewer equivalent classes), the resulting decomposition remained deterministic. The determinicity is checked after STEP 9. Note also that it is possible to try randomly a variety of equivalence classes compatible either with G or G_1 and keep the construction that provides the smallest representation.

4 EXPERIMENTS

In order to evaluate the factorization method we compared the size of the resulting factorization first with the minimal sequential transducer [11] when it exists and with the bimachine factorization as given in [2]. We give the size (number of states and number of transitions) of the sequential transducer on the first line. The size of the bimachine factorization is given on the second line and the size of the factorization we obtain with the method presented here is given on the third line.

We first took a dictionary, called *DELAPF*, that to each inflected word of French associates its phonetic transcription [9]. In that case, we obtain a factorization smaller that the sequential representation.

We then encoded two sets of syntactic constraints taken from two different syntactic classes from [6] as described in [13]. The finite functions that we represent take a simple sentence described by this part of the grammar, and associates to it the same simple sentence with parenthesis around the noun phrases. These functions are easy to represent by non-deterministic transducers but they are not sequential. Trying to convert such functions into a sequential transducer results in an infinite number of states. The complexity of these functions is due to the fact that the sentence structures are lexically very sensitive to the main verb. In that case, we observe that the factorization we present here is an order of magnitude smaller than the classical factorization. In that case, the factorization presented here is the only reasonable deterministic representation.

Note that the compression capabilities of this method relies on the availability of a compact non-deterministic representation. In these experiments, the input was in such a form: in the case of the phonetic dictionary it came from the shape of the phonemic rules and in the case of syntax, the natural description is highly non-deterministic.

Table 1. Results of experiments

	Delapf		SynCons-1		SynCons-2	
	nb of sta.	nb of trans.	nb of sta.	nb of trans.	nb of sta.	nb of trans.
Sequ. Trans.	48,224	134,726	∞	∞	∞	∞
Bima. Fact.	95,659	245,852	244,721	511,311	>1,000,000	>1,000,000.
Pres. Fact.	46,881	133,847	8,783	27,511	42,224	165,991

5 Generalization to NFA

Since any finite-state automaton A can also be defined as the domain of the rational function that maps each word in $L(A)$ to a constant word, this factorization procedure can also be applied on a finite-state automaton. In fact, note that the automaton obtained by removing the output labels from the transducer T of Figure 1 is a non-deterministic finite-state automaton (NFA) whose minimal deterministic automaton has the same number of states as the sequential transducer τ of Figure 2. Therefore, the same method leads to a deterministic decomposition smaller than the minimal deterministic automaton. Of course, looking up a word in this decomposition takes exactly (if the transducers are represented in an appropriate manner, that is, with a random access to the transitions) twice the time it takes to look it up in the minimal deterministic automaton.

6 CONCLUSION

We described an algorithm that given any finite-state function, computes a factorization of this function into the composition of a right-sequential transducer with a left-sequential transducer. In the particular case of sequential functions, this factorization can be exponentially smaller than the minimal sequential transducer. Moreover, since each factor is deterministic, the space and time complexity of computing the output of an input word w is linear in term of the length of w and is independent of the size and the complexity of the function.

The method can be applied to finite-state automata in order to build deterministic representation (deterministic factorization, in fact) of finite-state automata smaller than the minimal deterministic automaton.

Experiments performed on large scale dictionaries and large scale rule-based systems demonstrates the usefulness of the approach for various aspects of language processing.

References

1. Alfred Aho, John Hopcroft, and Jeffrey Ulllman. *The design and analysis of computer algorithms.* Addison Wesley, 1974.
2. Jean Berstel. *Transductions and Context-Free Languages.* Teubner, Stuttgart, 1979.
3. D. Brelaz. New methods to color vertices of a graph. *Comm. ACM*, 22:251–256, 1970.
4. Samuel Eilenberg. *Automata, languages, and machines, Volume A.* Academic Press, New York, 1974.
5. C. C. Elgot and G. Mezei. On relations dfined by generalized finite automata. *IBM J. of Res. and Dev.*, 9:47–65, 1965.
6. Maurice Gross. *Méthodes en syntaxe,régime des constructions complétives.* Hermann, 1975.

7. David S. Johnson, Cecilia R. Aragon, Lyle A. McGeoch, and Catherine Schevon. Optimization by simulated annealing: an experimental evaluation; part ii, graph coloring and number partitionning. *Operations Research*, 39(3):378–406, 1991.

8. R. M. Karp. *Reducibility among combinatorial problems*, in R. E. Miller and J. W. Thatcher (eds.) 'Complexity of Computer Computations'. Plenum Press, New York, 1972.

9. Eric Laporte. *Méthodes algorithmiques et lexicales de phonetisation de textes.* PhD thesis, Université Paris 7, 1988.

10. F. T. Leighton. A graph coloring algorithm for large scheduling problems. *J. Res. Natl. Bur. Standards*, 84:489–506, 1979.

11. Mehryar Mohri. Minimisation of sequential transducers. In *Proceedings of the Conference on Computational Pattern Matching 1994*, 1994.

12. Christophe Reutenauer and Marcel-Paul Schützenberger. Minnimization of rational word functions. *SIAM J. Comput.*, 20:669–685, 1991.

13. Emmanuel Roche. *Analyse Syntaxique Transformationelle du Français par Transducteurs et Lexique-Grammaire.* PhD thesis, Université Paris 7, January 1993.

14. Marcel Paul Schützenberger. A remark on finite transducers. *Information and Control*, 4:185–187, 1961.

Multiple Sequence Comparison:
A Peptide Matching Approach

Marie-France Sagot [1,3]   Alain Viari [1]   Henri Soldano [1,2]

[1] Atelier de BioInformatique
CPASO - URA CNRS 448
Section de Physique et Chimie de l'Institut Curie
11, rue P. et M. Curie 75005 - Paris - FRANCE

[2] LIPN - Université de Paris Nord
URA CNRS 1507
Avenue J.B. Clément 93430 - Villetaneuse

[3] Institut Gaspard Monge
Université de Marne la Vallée
2, rue de la Butte Verte 93160 - Noisy le Grand

e-mail: sagot@radium.jussieu.fr
fax: (33-1) 40-51-06-36

Abstract

We present in this paper a peptide matching approach to the multiple comparison of a set of protein sequences. This approach consists in looking for all the words that are common to q of these sequences, where q is a parameter.

The comparison between words is done by using as reference an object called a model. In the case of proteins, a model is a product of subsets of the alphabet Σ of the amino acids. These subsets belong to a cover of Σ, that is, their union covers all of Σ. A word is said to be an instance of a model if it belongs to the model.

A further flexibility is introduced in the comparison by allowing for up to e errors in the comparison between a word and a model. A word is said to be this time an occurrence of a model if the Levenshtein distance between it and an instance of the model is inferior or equal to e. Two words are said to be similar if there is at least one model of which both are occurrences. In the special case where $e = 0$, the occurrences of a model are simply its instances. If a model M has occurrences in at least q of the sequences of the set, M is said to occur in the set.

The algorithm presented here is an efficient and exact way of looking for all the models, of a fixed length k or of the greatest possible length k_{max}, that occur in a set of sequences. It is linear in the total length n of the sequences and proportional to k^{2e+1} where $k << n$ is a small value in all practical situations.

Models are closely related to what is called a consensus in the biocomputing area, and covers are a good way of representing complex relationships between the amino acids.

keywords: multiple comparison, cover, model, edit distance

1 Introduction

The comparison of more than two sequences at the same time is a very important problem in molecular biology. It is also a hard one that presents many difficulties, the first of which is deciding how to define the concept of multiple comparison. A natural way to proceed is to start from a well-established measure of pairwise comparison - either a distance or a similarity score - and to extend it to the case of a multiple comparison. For instance, the score of a multiple-sequence alignment can be defined as the sum of the scores of all projected pairwise alignments ([3]), or as the sum of the projected pairwise alignment score associated with each edge of an evolutionary tree ([18] [19]). In each case, a function of similarity is defined, and an optimal alignment is one which minimizes this function. Dynamic programming for instance is an algorithm that, in the special case of two sequences, provides an optimal solution in reasonable time and space. For more than two sequences, exhaustive algorithms become rapidly impracticable and various heuristics have been proposed ([1] [6] [7] [22] [25]).

Another approach to the problem of multiple comparison is that of peptide matching or block searching, where a block is an array of aligned individual sequence segments that are usually, but not necessarily, all of the same length ([8] [9] [15]). In this case, the definition of multiple rests on the comparison of the segments of a block. This comparison can be seen once again as an extension of a pairwise comparison and the same functions of similarity given above can be used locally to define the score of a block ([20]). Alternatively, to avoid having to rely on pairwise comparisons between segments in order to define a multiple comparison, one can work with the concept of *model*. In a general way, a model is an object against which all the elements of a block are compared. For instance, a model can be a word over the alphabet of amino acids (i.e, it is a peptide) and a block is then composed of segments related to this word (e.g. identical or similar in a certain way to it). In this paper, we work with a more general definition of a model that uses subsets of the alphabet. A model is then no longer a word but a product of these sets. Such models are closely related to what is called consensus in the biocomputing area.

Given the definition of a model, a brute-force approach to comparing a set of sequences consists in generating all possible models and looking in the sequences for the words that are similar to each one of them. This is of course extremely inefficient for long models. The purpose of this paper is to present algorithms that achieve the same goal without exploring the whole search space. Our interest will be in the local comparison of protein sequences either as an end in itself (establishment of a list of common words), or as a first step in a multiple alignment program.

It is not realistic, when working with protein sequences, to look for blocks composed of segments identical to a given model. Some flexibility is necessary and the segments of a block need only be similar in a certain way to the corresponding model.

A classical approach to introducing flexibility in sequence comparisons consists in grouping the amino acids into classes. More formally, an equivalence relation E is defined over the alphabet Σ of amino acids and each class of E constitutes then a new symbol of a reduced alphabet ([10]). This approach can be extended to the case of a reflexive, symmetric but not necessarily transitive relation R over Σ. R thus expresses a similarity relation between the amino acids and can be obtained by setting a threshold on a numerical matrix such as Dayhoff's PAM250 ([5]). This idea of a relation between the amino acids was independently used by Cobbs ([4]) and in a slightly different way by Brutlag ([2]) for pairwise comparisons, and by Soldano ([21]) for multiple comparisons. In this latter case, the authors use the concept of maximal cliques of R as an extension of the classes of E.

However the previous definitions are based on a single relation on Σ that may be insufficient to capture complex relationships between the amino acids. For instance, no single relation can faithfully preserve the web of such relationships present in Taylor's Venn diagram ([23]). The problem would become even more acute should one wish to compare proteins at various levels of the molecules structures simultaneously (primary, secondary and super-secondary).

The most natural way of modeling these complex relationships is by using a number of subsets of the alphabet with the condition that they cover all of Σ.

We start by formally defining models based on these subsets. Then we progressively introduce two relations between the words and the models. In the first and simpler case, a word is related to a model if the symbol at each position in the word belongs to the subset located at the same position in the model; in other words, if the word belongs to the model. This definition however does not allow for errors. Therefore a second and more general relation is proposed that authorizes errors by introducing a special form of the Levenshtein distance between words and models. The algorithm resulting from the first definition is called Poivre. Its time complexity is $O(n.g^k.k)$ where n is the sum of the lengths of the sequences, k is either a fixed length or that of the longest possible model and g is a characteristic of the subsets of the alphabet and is in general quite small. Poivre's space complexity is $O(n.k)$. The algorithm resulting from the second definition is called LePoivre. Its time complexity is $O(n.g^k.k^{2e+1}.$ $\mid \Sigma \mid^{e+1})$ where e is the number of errors allowed (mismatches and gaps) and the other variables are as before. Its space complexity is $O(n.g^k.(e+1)^2)$.

Section 2 formally establishes the problem. Section 3 presents Poivre. Section 4 describes LePoivre and two variants. Finally, results concerning the performance of both algorithms on biological sequences are given in annex.

2 Statement of the Problem

Definition 2.1 *Given an alphabet Σ, a cover of Σ is a set of subsets $S_1, S_2, ..., S_p$ of Σ such that:*

$$\bigcup_{i=1}^{p} S_i = \Sigma.$$

In some practical cases, one may wish to add the further condition that none of the subsets S_i be included in another.

Of course, classes and maximal cliques are just special cases of subsets of a cover.

Models have first been introduced in a recent paper where they were defined as words over the alphabet ([17]). That definition was appropriate for comparing nucleic acid sequences (since there exists no special relation between the symbols) but not for protein sequences. In this case, we have to give a new definition of a model based on sets.

Let $S = \{S_1, S_2, ..., S_p\}$ be a cover of Σ.

Definition 2.2 *A model M of length k is an element of S^k. We note $\mid M \mid$ the length of M.*

Notation 2.1 *Given $u = u_1...u_k \in \Sigma^k$ and a model $M = M_1...M_k \in S^k$, we say that u is an instance of M if, for all $i \in \{1,...,k\}$, $u_i \in M_i$.*
In other words, u is an instance of M if $u \in M$.

Notation 2.2 *Given $u \in \Sigma^*$ and a model $M = M_1...M_k \in S^k$, we call SL-distance between u and M, noted $d_{SL}(u, M)$, the minimum number of substitutions, deletions and insertions that is necessary to convert u into an instance of M (SL stands for Set-Levenshtein).*

Example 2.1 *See figure 1.*

Definition 2.3 *Given an integer e and a model M, M is said to be SL-present in s at location i if there exists at least one word u starting at position i in s such that $d_{SL}(u, M) \leq e$. We call u an SL-image of M.*
Let $d_s + d_d + d_i = d_{SL}(u, M)$ where d_s, d_d and d_i are the number of substitutions, deletions and insertions respectively between u and a nearest instance of M. Observe that this decomposition may not be unique. We say that (i, d_s, d_d, d_i) is an SL-occurrence of M in s and we have $\mid u \mid = \mid M \mid - d_d + d_i$.
Finally, we note $SL(M)$ the set of all SL-occurrences of M in s, that is:

$$SL(M) = \{(i, d_s, d_d, d_i) \mid d = d_s + d_d + d_i \text{ is a SL-distance and } d \leq e\}.$$

A model may be SL-present in s at location i more than once, with different SL-images. It could even be present for the same image u with different SL-occurrences.

Observe also that the relation that is established between a model and its images induces a relation between the images themselves, that is between the words of s, since the Levenshtein distance between any two of them is at most $2e$.

Statement of the problem *Given an alphabet Σ, a set $\{s1, s2, ..., sN\}$ of sequences over Σ, that is of elements of Σ^*, a constant e and another constant q between 2 and N, the problems that the algorithms in this paper are able to solve are:*

1. *find all models of a fixed length k that are SL-present in at least q of the sequences of the set;*

2. *find the greatest length k_{max} for which there exists at least one model of length k_{max} that is SL-present in at least q of the sequences of the set, and solve problem 1 for $k = k_{max}$.*

Example 2.2 *See figure 1.*

We start by presenting the algorithm for the special case where $e = 0$, that is, where no errors are allowed and the occurrences of a model are in fact its instances. Then we present the algorithm for the general case where up to $e > 0$ errors are authorized. We call the first algorithm Poivre and the latter one LePoivre.

Let us observe finally that looking for models SL-present in at least q of the sequences of the set $\{s1, s2, ..., sN\}$ can be done by searching for the models SL-present at least q times in the string $s = s1s2...sN \in \Sigma^*$, with a constraint imposed on the positions of the occurrences of each model so that there are at least q of them that correspond to different sequences of the set.

3 Poivre

3.1 Main Idea of the Algorithm

Poivre (and LePoivre) were initially inspired by an earlier, classical algorithm for finding all exact repetitions in a string. That algorithm, which we call KMR, was elaborated in 1972 by Karp, Miller and Rosenberg ([11]). Algorithms that were different extensions of KMR fulfilling other purposes were presented in previous papers ([16] [17] [21]).

KMR's principle for finding all exact repetitions of a length $2k$ in a string s is based on the idea that these repetitions can be decomposed into two juxtaposed repetitions of length k.

Let $\Sigma = \{A, C, D, E, F, G, H, I, K, L, M, N, P, Q, R, S, T, V, W, Y\}$, $e = 1$, q $= 100\%$ and C be the following cover of Σ:

$$S_1 = \{I, L, V, M\}$$
$$S_2 = \{A, G, S, C, T\}$$
$$S_3 = \{K, R, H\}$$
$$S_4 = \{K, R\}$$
$$S_5 = \{Y, W, F, H\}$$
$$S_6 = \{Y, W, F\}$$
$$S_7 = \{D, E\}$$
$$S_8 = \{Q, N\}$$
$$S_9 = \{P\}$$
$$S_{10} = \{C\}$$
$$S_{11} = \{W\}$$

Given the following four sequences:

$s1 = $ VMGPIKDM *VHITHGPIG*CSFYTWGGRRFKSKPENGTGLNFNEY VFSTDMQESDIVFGGVN,

$s2 = $ CAYAGSKGVVWGPIKDM*IHISHGPVG*CGQYSRAGRRNYYIGTT GVNAFVTMNFTSDFQEK,

$s3 = $ EFCGGHTHAISRYGLEDMLPAN *VRMIHGPGC*PVCVLPAGRIDMA IRLAMRPDIILCVYGD,

$s4 = $ MKLAHWMYAGPAHIGTLRVASSFKN *VHAIMHAPLG*DDYFNVMR SMLERERNFTPATASIV,

the longest models found in all four sequences are:

$$S_1 S_3 S_1 S_1 S_5 S_2 S_9 S_1 S_2$$

at positions 9 (with 1 mismatch), 18 (with 1 mismatch), 23 (twice, with 1 mismatch and with 1 deletion) and 26 (with 1 insertion) respectively , and:

$$S_1 S_3 S_1 S_1 S_3 S_2 S_9 S_1 S_2$$

at positions 9 (with 1 mismatch), 18 (with 1 mismatch), 23 (twice, with 1 mismatch and with 1 deletion) and 26 (with 1 insertion) respectively.

Figure 1: Example of models and occurrences

In a likewise manner, Poivre's principle for finding all the models of length k that are SL-present in a string s is based on an iterative construction of these models from those of a smaller length.

This construction can proceed by doubling the length of the models at each step in the same way as the length of the exact repetitions were doubled in KMR. This comes from the fact that the set of occurrences of a model $M = M_1 M_2$ with $| M | = 2 | M_1 | = 2 | M_2 | = k$ is the set of occurrences of M_1 whose images are juxtaposed with at least one image of M_2. In this case, we need to obtain and stock all the sets of occurrences of the models of length k SL-present in at least q of the sequences before we can construct the models of length $2k$. This approach corresponds to a breadth-first exploration of the search tree of all models, where not all levels are visited. At each step a lot of pruning may be realized, so we do not really need to traverse the whole tree. This comes from the fact that if a model M' of length k does not occur in more than q of the sequences, no model M of length $2k$ having M' as a prefix or suffix can itself occur in more than q sequences. The branch of the tree leading to M' can therefore be cut.

However, if errors are allowed, this approach can consume a lot of memory in the earlier stages of the exploration when almost all models occur at almost all positions.

A second, more space-parsimonious approach to constructing such models is to traverse the first level of the search tree of models and then realize a depth-first exploration of the rest of the tree, again with possible pruning along the way. Indeed, if instead of doubling the length of the models at each step, we extend each model separately to the right by just one unit at a time, then all we need to obtain the set of occurrences of a model M are the set of occurrences of the model M' of length $| M | - 1$ that is its prefix, together with the set of occurrences of the model of length 1 that is its suffix. In terms of memory then, all we need to stock at any time are the sets of occurrences of all models of length 1 together with the sets of occurrences of all the models that are prefixes of M.

Since the time complexity of both methods of construction varies only in that a $\log k$ factor for the first approach is changed into a k factor for the second, it is this second method we adopt. In most practical applications, $k << n$ and is generally small (typically around 10).

3.2 Algorithm

We start by presenting the Poivre algorithm. Before we show the lemma on which it rests, we introduce a simplified notation for the SL-occurrences of a model when $e = 0$.

Notation 3.1 *Given a string s and a model M, if (i, d_s, d_d, d_i) is an SL-occurrence of M in s such that $d_s = d_d = d_i = 0$, then we note it simply by: $i \in SL(M)$.*

Let s be a string of length n. Let M, M_1, M_2 be three models such that $M = M_1 M_2$ and $i \in \{1,...,n\text{-}|M|+1\}$ a position in s. Then the following lemma reproduces one of KMR's two lemmas:

Lemma 3.1

$$i \in SL(M = M_1 M_2)$$

$$\Longleftrightarrow$$

$$i \in SL(M_1), (i+ \mid M_1 \mid) \in SL(M_2).$$

This lemma gives a constructive property to build $SL(M)$ from $SL(M_1)$ and $SL(M_2)$.

Notation 3.2 *Given $SL(M)$ a set of SL-occurrences, we note:*

$$(SL(M))_{-b} = \{i \mid (i+b) \in SL(M)\}.$$

This is simply the set obtained from $SL(M)$ by subtracting b to each i in $SL(M)$.

Proposition 3.1 $SL(M) = SL(M_1) \cap SL(M_2)_{-|M_1|}.$

The above lemma and proposition allow us to build the models present at least q times in s in an iterative way (that is, by increasing lengths). A general idea of how this is actually done to solve the first of the problems presented in the previous section is given in figure 2. It is implemented as a recursion. It is the version that traverses the search tree of models in a depth-first way, so the value of b in notation 3.2 is equal to 1 and the models M_2 of proposition 3.1 are the sets of the cover of Σ. They are the models of length 1.

3.3 Time and Space Complexity

We first give the time complexity for a breadth-first exploration of the search tree of all models.

Let s be a string of length n over Σ and let u be a word of length k of s starting at position i. There are at most $N_{mod} = g^k$ models of length k of which i can be an SL-occurrence, for $k \geq e$, where g is the maximal number of sets of the cover of Σ to which a symbol may belong. Formally:

$$g = \text{Max} \{g_\alpha \mid \alpha \in \Sigma, g_\alpha = \text{number of sets to which } \alpha \text{ belongs}\}.$$

Since there are $n - k + 1$ words of length k in s, the total number of elements in the sets $SL(M)$ for all models M of length k is thus at most $n.N_{mod}$. As each of these elements cannot be extended in relation to more than g^k models, at most $g^k.n.N_{mod}$ operations are necessary to construct the models of length

```
/* Input */
        s = s1s2...sN : a string of length n = concatenation of N sequences
        S₁, S₂, ..., Sₚ : a cover C of Σ
        q : as indicated in the statement of the problem in section 2
        k : a fixed length
/* Output */
        All models M of length k that have instances in at least q of the
        sequences s1, s2, ..., sN
/* Main data structures */
        M = product of sets of C
        SL(M) = set of occurrences of a model M : implemented as a stack
        Models[i] = sets of C to which sᵢ belongs : implemented as a stack
        ExtensionPossible = set of models that verify quorum of q and may
        therefore be extended : implemented as a stack

Main {
        for i ∈ [1..n]
                for j ∈ [1..p] {
                        if (i ∈ Sⱼ)
                                SL(Sⱼ) = SL(Sⱼ) ∪ {i};
                        Models[i] = Models[i] ∪ Sⱼ;
                }
        for j ∈ [1..p]
                if (SL(j) verifies quorum of q)
                        Poivre(Sⱼ, 1, SL(Sⱼ));
}

Poivre(M, l, SL(M)) {
        if (l = k)                              /* end of recursion */
                store M and store SL(M);
        else {                                  /* body of recursion */
                for (i ∈ SL(M))
                        for (M₁ ∈ Models[i + l]) {
                                SL(MM₁) = SL(MM₁) ∪ {i};
                                ExtensionPossible = ExtensionPossible ∪ {M₁};
                        }
                for (M₁ ∈ ExtensionPossible)
                        if (SL(MM₁) verifies quorum of q)
                                Poivre(MM₁, l + 1, SL(MM₁));
        }
}
```

Figure 2: Poivre algorithm for solving problem 1.

$2k$ from those of length k. So Poivre's time complexity for finding all models of length k is $O(n.g^k. \log k)$ if the models length is doubled at each step.

In the depth-first approach, the models length is increased of just one unit at each step. Poivre's time complexity is then $O(n.g^k.k)$. This comes from the fact that each element M of length k cannot be extended in relation to more than g models, so at most $g.n.N_{mod}$ operations are necessary to construct the models of length $k+1$ from those of length k.

Poivre's space complexity is $O(n.g^k)$ for the breadth-first approach, it is $O(n.k)$ for the depth-first one. When no errors are allowed, the difference between the two approaches is slight.

Additional experimental results are given in annex.

4 LePoivre

4.1 Main Differences with Poivre

We present now the more general case where up to $e > 0$ errors (mismatches and gaps) are authorized between a model and its SL-occurrences. This is the LePoivre algorithm. The main differences with Poivre are:

1. given a model M, a position i of an occurrence of M in s may be associated with various triplets (d_s, d_d, d_i). Each one corresponds to an image u (some may be identical) and indicates a different combination of substitutions, deletions and insertions that permits to convert u into an instance of M in a minimal number of operations;

2. given a model M of length k and its SL-occurrences, the length of the corresponding SL-images may vary between $\mid M \mid$ - e and $\mid M \mid$ + e. So when we pick an occurrence (i, d_s, d_d, d_i) of a model M of length k and try to extend the model to which it belongs, we cannot simply look at position $i + k$ to see what is the next symbol in s as we did with Poivre. We have to take into account the fact that the image u corresponding to that occurrence (i, d_s, d_d, d_i) is such that $\mid u \mid = k$ - shift, where shift $= d_d - d_i$;

3. finally, when we juxtapose two SL-images u and u' from models M and M' respectively, all we know is that $d(uu', MM') \leq d(u, M) + d(u', M')$. We do not always have a strict equality. In dynamic programming terms, this means that the concatenation of two optimal paths does not necessarily produce an optimal path. What we can guarantee is that at least one concatenation does (Lemma 4.1 given below assures us of that). So uu' is indeed an SL-image of MM'. The SL-distance of uu' to a nearest instance of MM' is the minimal one among all those that are obtained.

These considerations lead to a different, more complex algorithm, than Poivre.

4.2 Algorithm

As before, let s be a string of length n. Let also M, M_1, M_2 be three models such that $M = M_1 M_2$ and $i \in \{1,...,n\text{-}|M|\text{+}1\}$ a position in s. Lemma 3.1 of the previous section has now to be written in two parts that are no longer symmetrical:

Lemma 4.1

$$(i, d_s, d_d, d_i) \in SL(M)$$

$$\Longrightarrow$$

$$\exists\, d_{s_1}, d_{d_1}, d_{i_1}, d_{s_2}, d_{d_2}, d_{i_2} \; such\, that$$

$$(i, d_{s_1}, d_{d_1}, d_{i_1}) \in SL(M_1), (i+\mid M_1 \mid -d_{d_1} + d_{i_1}, d_{s_2}, d_{d_2}, d_{i_2}) \in SL(M_2)$$

$$and\, d_s = d_{s_1} + d_{s_2}, \; d_d = d_{d_1} + d_{d_2}, \; d_i = d_{i_1} + d_{i_2}.$$

Lemma 4.2

$$(i, d_{s_1}, d_{d_1}, d_{i_1}) \in SL(M_1),\; (i+\mid M_1 \mid -d_{d_1} + d_{i_1}, d_{s_2}, d_{d_2}, d_{i_2}) \in SL(M_2)$$

$$and\, d_{s_1} + d_{d_1} + d_{i_1} + d_{s_2} + d_{d_2} + d_{i_2} \leq e$$

$$\Longrightarrow$$

$$(i, d_{s_1} + d_{s_2}, d_{d_1} + d_{d_2}, d_{i_1} + d_{i_2}) \in SS(M_1 M_2 = M) \supseteq SL(M)$$

$$where\, SS(M) = \{(i, d_s, d_d, d_i) \mid d_s + d_d + d_i \leq e\}.$$

Again, the previous lemmas give a constructive property to build $SL(M)$ from $SL(M_1)$ and $SL(M_2)$. Let us first introduce a notation, a filter operation and a special operation of intersection on the sets of SL-occurrences of two models M_1 and M_2.

Notation 4.1 *Given $SL(M)$ a set of SL-occurrences, we note:*

$$(SL(M))_{-b} = \{(i, d_s, d_d, d_i) \mid (i + b, d_s, d_d, d_i) \in SL(M)\}.$$

This is simply the set obtained from $SL(M)$ by subtracting b to each i in $SL(M)$.

Definition 4.1 *Given a set $SS(M) = \{(i, d_s, d_d, d_i) \mid d_s + d_d + d_i \leq e\}$, we define the operation $Filter(SS(M))$ by:*

$$Filter(SS(M))$$

$$=$$

$$\{(i, d_s, d_d, d_i) \in SS(M) \mid d = d_s + d_d + d_i\, is\, a\, SL - distance\}.$$

In other words:

$$Filter(SS(M))$$
$$=$$
$$\{(i, d_s, d_d, d_i) \in SS(M) \mid d_s + d_d + d_i \leq d'_s + d'_d + d'_i, \ \forall (i, d'_s, d'_d, d'_i) \in SS(M)$$
$$with \ d'_d - d'_i = d_d - d_i\}.$$

Definition 4.2 *Given two sets of SL-occurrences $SL(M_1)$ and $SL(M_2)$, we define the operation $\sqcap_{S_L}$ by:*

$$SL(M_1) \ \sqcap_{S_L} \ SL(M_2) = \{(i, d_s, d_d, d_i) \mid (i, d_{s_1}, d_{d_1}, d_{i_1}) \in SL(M_1),$$
$$(i + d_{d_1} - d_{i_1}, d_{s_2}, d_{d_2}, d_{i_2}) \in SL(M_2),$$
$$d_s = d_{s_1} + d_{s_2}, \ d_d = d_{d_1} + d_{d_2}, \ d_i = d_{i_1} + d_{i_2}$$
$$and \ d_s + d_d + d_i \leq e\}.$$

Then, from the lemmas, it comes:

Proposition 4.1 $SL(M) = Filter(SL(M_1) \ \sqcap_{S_L} \ SL(M_2)_{-|M_1|})$.

A general idea of the algorithm is given in figure 3.

```
/* Input */
        e : the maximal number of errors allowed
        The remaining as for Poivre
/* Output */
        All models M of length k that have occurrences in at least q of the
        sequences
/* Main data structures */
        Models[i] = sets of C to which occurrence i belongs, together with
        the number of substitutions, deletions and insertions between the
        occurrence and a nearest element of the set : implemented as a stack
        The remaining as for Poivre
```

4.3 LePoivre's Time and Space Complexity

The time complexity of LePoivre is calculated for the case where the search tree is explored in a depth-first way.

Let s be a string of length n over Σ and let i be a position in s. There are at most $N_{mod} =$

$$\sum_{j=0}^{e} (\sum_{l=k-j}^{l=k+j} (\sum_{\substack{d_i + d_d + d_s = j \\ l + d_i - d_d = k}} \binom{l}{d_i} \binom{l - d_i}{d_s} g^{l - d_s - d_d - d_i} (p-1)^{d_s} A_l^{d_d} (p-1)^{d_d}))$$

Main {
    for $i \in [1..n]$
        for $j \in [1..p]$
            if $(i \in S_j)$ {
                $SL(S_j) = SL(S_j) \cup \{(i,0,0,0)\};$
                Models$[i]$ = Models$[i] \cup (S_j,0,0,0);$
            }
            else {
                $SL(S_j) = SL(S_j) \cup \{(i,1,0,0)\} \cup \{(i,0,1,0)\} \cup$
                      $\{(i,0,0,1)\};$
                Models$[i]$ = Models$[i] \cup (S_j,1,0,0) \cup (S_j,0,1,0)$
                      $\cup (\emptyset,0,0,1);$
            }
        for $j \in [1..p]$
            if $(SL(j)$ verifies quorum of $q)$
                Poivre$(S_j, 1, SL(S_j));$
}

Poivre$(M, l, SL(M))$ {
    if $(l = k)$ /* end of recursion */
        store M and store $SL(M);$
    else {   /* body of recursion */
        for $((i,d_s,d_d,d_i) \in SL(M))$
            for $((M_1,d'_s,d'_d,d'_i) \in$ Models$[i + l - d_d + d_i])$
                if $(d'_s + d'_d + d'_i + d_s + d_d + d_i \le e)$ {
                    $SS(MM_1) = SS(MM_1) \cup$
                              $\{(i,d_s + d'_s, d_d + d'_d, d_i + d'_i)\};$
                    ExtensionPossible = ExtensionPossible $\cup$
                              $\{M_1\};$
                }
        for $(M_1 \in$ ExtensionPossible$)$
            if $(SL(MM_1) = $ Filter$(SS(MM_1))$ verifies quorum of $q)$
                Poivre$(MM_1, 1 + \mid M_1 \mid, SL(MM_1));$
    }
}

Figure 3: LePoivre algorithm for solving problem 1.

models of length k of which (i,d_s,d_d,d_i) can be an SL-occurrence, for $k \geq e$, where $A_l^{d_d} = \frac{l!}{(l-d_d)!}$ and p is the number of sets in the cover of Σ. Since there are $n - k + 1$ words of length k in s, the total number of elements in the sets $SL(M)$ for all models M of length k is thus at most $n.N_{mod}$. As each of these elements cannot be extended in relation to a model in more than p different ways, this results in no more than $p.n.N_{mod}$ operations to construct the models of length $k + 1$ from those of length k. Majoring N_{mod}, LePoivre's time complexity becomes $O(n.g^k.k^{2e+1}.p^{e+1})$. If the tree was explored in a breadth-first way, LePoivre's time complexity would be $O(n.g^k.\log k.k^{2e}.p^{e+1})$.

LePoivre's space complexity is $O(n.g^k.p^e.k^{2e})$ if the models length is doubled at each step, it is $O(n.k.(e + 1)^2)$ in the implementation we adopted.

Additional experimental results are given in annex.

4.4 Variants of LePoivre

4.4.1 Motivation

The LePoivre algorithm given above allows us to solve the problem stated at the end of section 2. From a practical point of view however, the definition of the problem as it was given there may appear too broad in the sense that what we find when we solve it may include cases that are not very significant.

For instance, all the errors of an occurrence against its model may appear at the end of the alignment of both. This may be biologically not very satisfying and can lead us to introduce constraints on the way the errors between a model and its occurrences are distributed. In particular, we can easily modify LePoivre so that this distribution is uniform. This produces a first variant of LePoivre.

On the other hand, a model may occur at least q times in the sequences yet never be instantiated in any of them. In the particular case of a database search, it may be biologically interesting to ask that a model be instantiated at least in the query sequence. This gives a second variant of the algorithm. Of course, both variants may be combined.

4.4.2 First Variant

This is the variant with a uniform distribution of errors. In this case, given a model M and any one of its images u, we ask that every subword of length $l \leq |M|$ of u have at most $f < e$ errors with the corresponding submodel of M, that is, at most a total of f substitutions, deletions or insertions with that submodel.

4.4.3 Second Variant

As observed, we may obtain models that are never SL-present in s with 0 errors, that is, that possess only SL-images with errors. In this variant, we ask that each model M have at least one occurrence i whose corresponding image $u \in$

M, i.e. is instance of M in s. This can considerably reduce the search space of models, and so speed up the algorithm execution time. The speed up will be the greater the more errors we allow between a model and its occurrences.

5 Perspectives

Covers and models allow for a clean definition of multiple comparison and the algorithms presented in this paper provide exact solutions for the problem of finding blocks of similar segments in a set of sequences with reasonable performing times.

One problem that remains to be solved to satisfaction is what to do with the results once we get them. The list of common segments is very informative in itself, but has to be presented in a nice way if it is to be of any practical use. If, however, one wishes to use this list to produce a multiple alignment of the sequences, so far only heuristical approaches have been proposed ([12] [13] [24]). This second problem is of course NP-complete, but improvements to it can still be realized.

Furthermore, the units of comparison on which the definition of multiple comparison given in this paper is based are the symbols of the alphabet, and we may wish for a greater degree of fuziness. In particular, it would be interesting to define a relation of similarity based on words instead of symbols. Work on this is currently under way.

Finally, it is possible to consider using the algorithms given here for scanning sequence databases from a multiple comparison point of view.

Acknowledgement

The authors would like to thank Maxime Crochemore for his very helpful suggestions during the composition of this paper. His comments resulted in numerous improvements in the presentation.

References

[1] G.J. Barton and M.J.E. Sternberg. A strategy for the rapid multiple alignment of protein sequences : confidence levels from tertiary structure comparisons. *J. Mol. Biol.*, 198:327–337, 1987.

[2] D.L. Brutlag, J-P. Dautricourt, S. Maulik, and J. Relph. Improved sensitivity of biological sequence database searches. *Comput. Applic. Biosc.*, 6:237–245, 1990.

[3] H. Carrillo and D.J. Lipman. The multiple sequence alignment problem in biology. *SIAM J. Appl. Math.*, 48:1073–1083, 1988.

[4] A.L. Cobbs. Fast identification of approximately matching substrings. In M. Crochemore and D. Gusfield, editors, *Combinatorial Pattern Matching*, pages 64–74. Springer Verlag, 1994.

[5] M.O. Dayhoff, R.M. Schwartz, and B.C. Orcutt. A model of evolutionary change in proteins. In M.O. Dayhoff, editor, *Atlas of Protein Sequence an Structure*, volume 5 suppl.3, pages 345–352. Natl. Biomed. Res. Found., 1978.

[6] D.F. Feng and R.F. Doolittle. Progressive sequence alignment as a prerequisite to correct phylogenetic trees. *J. Mol. Evol.*, 25:351–360, 1987.

[7] D. Gusfield. Efficient method for multiple sequence alignment with guaranteed error bounds. *Bull. of Math. Biol.*, 55:141–154, 1993.

[8] S. Henikoff. Comparative sequence analysis : finding genes. In D.W. Smith, editor, *Biocomputing. Informatics and Genome Projects*, pages 87–117. Academic Press, 1994.

[9] S. Henikoff and J.G. Henikoff. Automated assembly of protein blocks for database searching. *Nucl. Acids Res.*, 19:6565–6572, 1991.

[10] S. Karlin and G. Ghandour. Multiple alphabet amino acid sequence comparisons of the immunoglobulin κ-chain constant domain. *Proc. Natl. Acad. Sci. USA*, 82:8597–8601, 1985.

[11] R.M. Karp, R.E. Miller, and A.L. Rosenberg. Rapid identification of repeated patterns in strings, trees and arrays. pages 125–136. Proc. 4th Annu. ACM Symp. Theory of Computing, 1972.

[12] A. Landraud, J.F. Avril, and P. Chretienne. An algorithm for finding a common structure shared by a family of strings. *IEEE Trans. on Pattern Analysis and Machine Intelligence*, 11:890–895, 1989.

[13] H.M. Martinez. A flexible multiple sequence alignment program. *Nucleic Acids Res.*, 16:1683–1691, 1988.

[14] R.N. Pau. Nitrogenases without molybdenum. *Trends Biochem. Sci.*, 14:183–186, 1989.

[15] J. Posfai, A.S. Bhagwat, G. Posfai, and R.J. Roberts. Prediction motifs derived from cytosine methyltransferases. *Nucl. Acids Res.*, 17:2421–2435, 1989.

[16] M. F. Sagot, A. Viari, J. Pothier, and H. Soldano. Finding flexible patterns in a text - an application to 3D molecular matching. pages 117–145, Seattle, Washington, USA, 1994. First International IEEE Workshop on Shape and Pattern Matching in Computational Biology.

[17] M.F. Sagot, V. Escalier, A. Viari, and H. Soldano. Searching for repeated words in a text allowing for mismatches and gaps. Viñas del Mar, Chili, 1994. Second South American Workshop on String Processing.

[18] D. Sankoff. Minimum mutation trees of sequences. *SIAM J. Appl. Math.*, 28:35–42, 1975.

[19] D. Sankoff and R.J. Cedergreen. Simultaneous comparison of three or more sequences related by a tree. In D. Sankoff and J.B. Kruskall, editors, *Time Warps, String Edits, and Macromolecules. The Theory and Practice of sequence Comparison*, pages 253–263. Addison-Wesley, 1983.

[20] G.D. Schuler, S.F. Altschul, and D.J. Lipman. A workbench for multiple alignment construction and analysis. *Proteins : Struct., Func., and Genetics*, 9:180–190, 1991.

[21] H. Soldano, A. Viari, and M. Champesme. Searching for flexible repeated patterns using a non transitive similarity relation. *Pattern Recognition Letters*, 1994. in press.

[22] S. Subbiah and S.C. Harrison. A method for multiple sequence alignment with gaps. *J. Mol. Biol.*, 209:539–548, 1989.

[23] W.R. Taylor. The classification of amino acid conservation. *J. Theor. Biol.*, 119:205–218, 1986.

[24] A. Viari and J. Pothier. *SmartMulti: a tool for the multiple alignment of protein sequences using flexible blocks*. Atelier de BioInformatique, 11, rue P. et M. Curie - 75005 Paris, 1994. in preparation.

[25] A.K.C. Wong, S.C. Chan, and D.K.Y. Chiu. A multiple sequence comparison method. *Bull. Math. Biol.*, 55:465–486, 1993.

ANNEX

We present some experimental results on the performance of Poivre and LePoivre on biological data. In all experiments, we used the following cover of the alphabet of amino acids:

$S_1 = \{A, C, G, S, T\}$ - tiny
$S_2 = \{A, C, D, G, N, S, T, V\}$ - small
$S_3 = \{F, H, W, Y\}$ - aromatic
$S_4 = \{V, L, I, M\}$
$S_5 = \{K, R, H\}$ - basic
$S_6 = \{D, E\}$ - acidic
$S_7 = \{Q, N\}$ - amide
$S_8 = \{P\}$ - proline

This cover is strongly inspired on the clustering proposed by Taylor ([23]) which is based on the physico-chemical properties of the amino acids. Observe that this is a cover of Σ in the sense defined in section 2.2. In particular, this does not represent a partition of Σ nor maximal cliques of any relation.

Concerning the sequences, we chose a set of biologically related proteins from the nitrogenase family, namely the alpha and beta chains of component 1 ([14]). Nitrogenase is an oligomeric complex responsible for the fixation of nitrogen. This family contains at least two known consensus patterns (Prosite PS00699 and PS0090). One of them constitutes the active site and was already used in the example of figure 1. Finally, the proteins lengths are quite homogeneous, around 500 residues (range: 441-533).

Experiment 1

The first experiment concerns problem 1, that is, find all models of a given length k that occur on a quorum q of the input sequences.

Figure 4 displays the CPU times obtained on a Silicon Graphics Indigo Work-Station (R4000). k is fixed to 5 and the number of input sequences is increased from 2 to 30. The quorum is kept at 100%, the models are therefore requested to occur in all sequences. Because of the homogeneity of the sequences lengths, the X coordinate is practically proportional to n, the sum of the lengths of the input sequences. The plots are given for Poivre (figure 4a) and LePoivre with 1 and 2 errors (figure 4a-b). As expected, the running time is linear in n and increases quickly with the number of errors allowed. However, the increase is less then expected from the worst-case scenario discussed in the text. For $k = 5$ and $p = 8$, the ratio of the CPU times for 0 and 1 error or for 1 and 2 errors is in theory bounded by $k^2.p = 200$. The observed ratios are actually 50 and 10 respectively.

Experiment 2

The second experiment concerns problem 2, that is, find the models of greatest length k_{max}. Figure 5a gives the CPU times obtained under the same conditions as in the previous experiment. At least for a sufficient number of input sequences (around 10), the behavior is still linear in n. The reason for this can be seen in figure 5b which displays the value of the greatest length found k_{max} versus the number of sequences. k_{max} quickly decreases and tends to a constant value ($k_{max} \approx 10$) for a sufficient number of sequences. The same behavior is observed for random or functionally unrelated sequences but much smaller values of k_{max} are found in these latter cases.

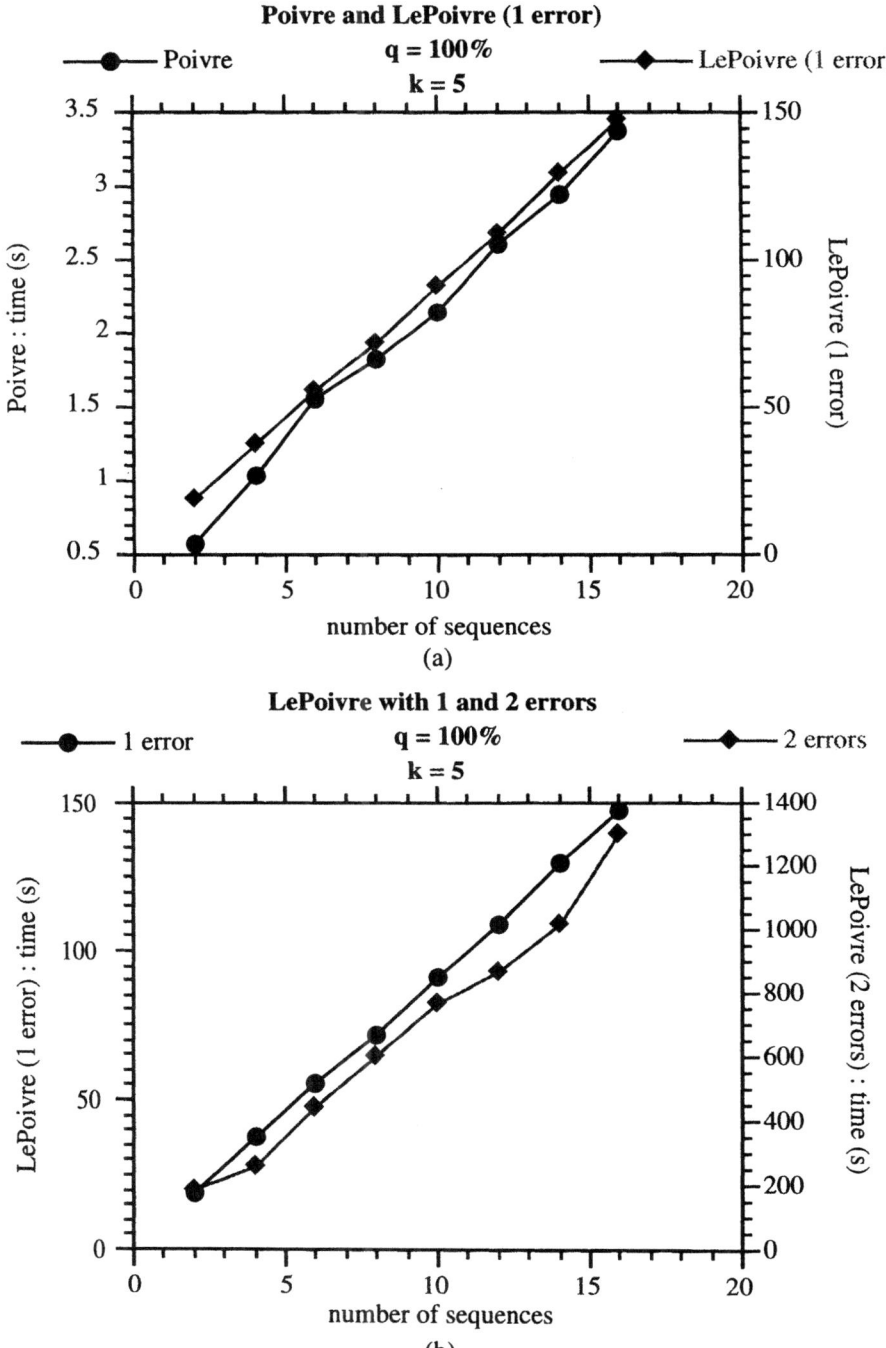

Figure 4: Experiment 1.

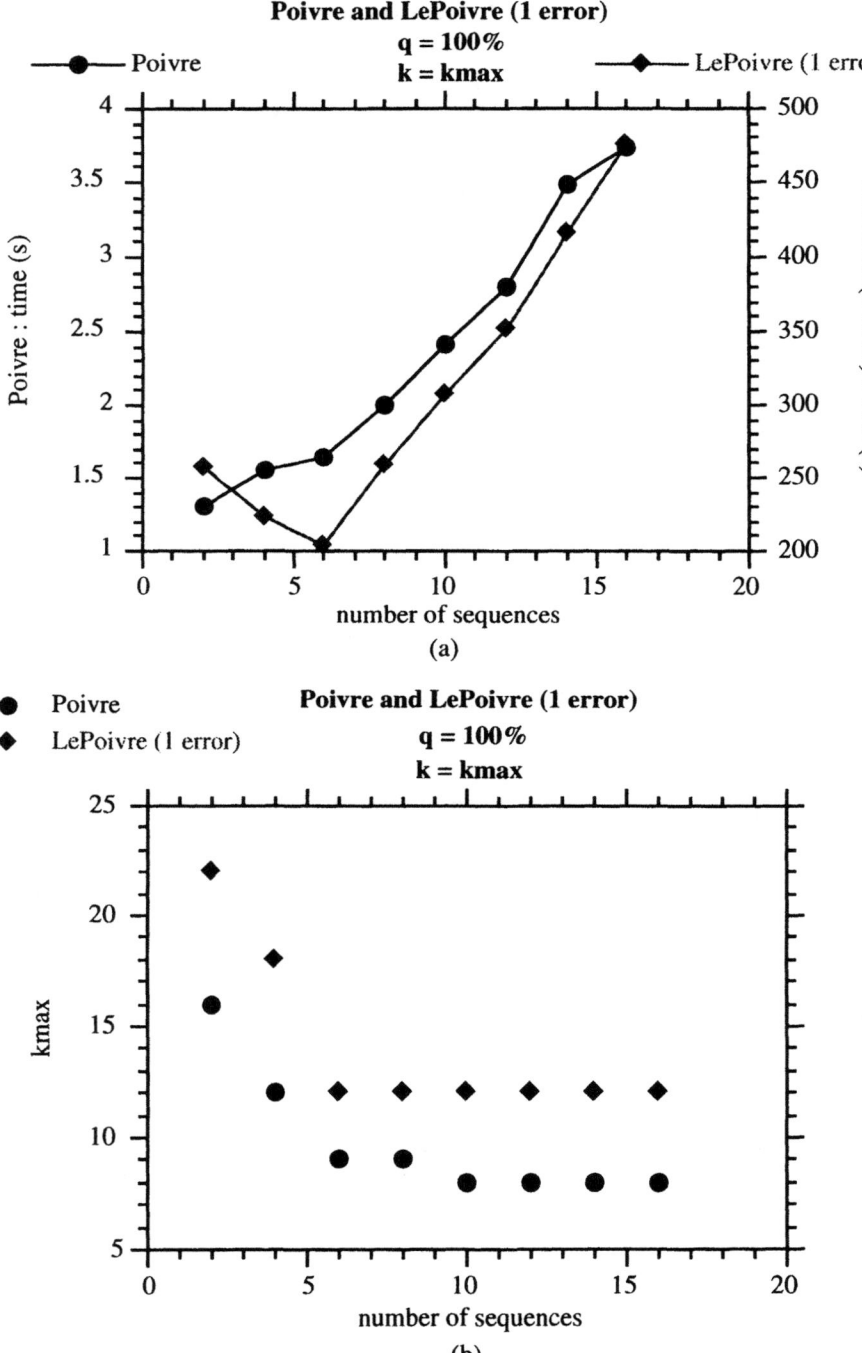

Figure 5: Experiment 2.

On a Technique for Parsing a String

Uzi Vishkin *

Abstract

Given a string S, we consider the problem of parsing it. Informally, an "ideal parsing" would: (i) (*Consistency in partition:*) Partition the string into short substrings of length 2 or 3 each, so that if identical substrings occur in different parts of S, then they are partitioned in the same way into short substrings. (ii) (*Consistency in labeling:*) Assign names to (or label) the short substrings so that identical short substrings get the same name, and thereby obtain a new string out of S. (iii) Apply items (i) and (ii) (recursively) to the new string. Motivated by this new problem, the talk will review the new technique presented in the ACM-STOC 1994 paper by Sahinalp and Vishkin. A fundamental new idea of that paper has been to use Cole-Vishkin's Deterministic Coin Tossing (DCT) technique to come close to ideal parsing. This has led to a novel and more efficient parallel algorithm for the key problem of suffix tree construction. Some more recent applications of that new technique will also be discussed.

*Institute for Advanced Computer Studies, and Department of Electrical Engineering, University of Maryland, College Park, MD 20742; and Department of Computer Science, Tel Aviv University, Tel Aviv, Israel; Partially supported by NSF grants CCR-9111348 and CCR-9416890.

Dictionary Look-Up with Small Errors

Andrew C. Yao[*] and Frances F. Yao[†]

Abstract

Let W be a set of n binary strings of length m each. We are interested in designing data structures for W that can answer d-queries quickly, that is, given a binary string α, decide whether there is any member of W within Hamming distance d of α. This problem, originally raised by Minsky and Papert [MP], remains a challenge in data structure design. In this paper, we make an initial effort towards a theoretical study of the small d case. Our main result is a data structure that achieves $O(m \log \log n)$ query time with $O(nm \log m)$ space for the $d = 1$ case.

1   Introduction

Let W be a set of n binary strings of length m each. We are interested in designing data structures for W that can answer d-queries quickly, that is,

> Given a binary string α, decide whether there is any member of W within Hamming distance d of α.

Many variants of this problem are possible; we focus on the decision problem here to simplify some of the details. Note that this data structure problem is related to but not the same as the approximate string matching problem discussed in recent literature (e.g. see Galil and Park [GP], Landau and Vishkin [LV], Ukkonen [U], Tarhio and Wood [TW], Ukkonen and Wood [UW]), where the task is to decide if an input string occurs as an approximate substring of another input string. The d-query problem we consider here was originally raised by Minsky and Papert [MP], and it has remained a challenge in data structure design. There has been renewed interest in this problem recently because of applications to DNA mapping and document retrieval. In Dolev et

---

[*]Department of Computer Science, Princeton University, Princeton, New Jersey 08544. This author was supported in part by the National Science Foundation under grant CCR-9301430.

[†]Xerox Palo Alto Research Center, 3333 Coyote Hill Road, Palo Alto, California 94304.

al ([DHLNP][DHP]) and in Greene, Parnas and Yao [GPY], some progress was made for the case when d is relatively large. Manber and Wu [MW] considered the $d = 1$ case from a more practical standpoint. In this paper, we make an initial effort towards a theoretical study of the small d case, focusing on the case $d = 1$.

One can formulate the above question in a general bitwise theoretical model, and ask what space/time bounds are achievable. For the standard dictionary problem, i.e., exact queries with $d = 0$, there is a solution with optimal space $O(nm)$ and optimal time $O(m)$. This is an immediate consequence of a result by Fredman, Komlós, and Szemerédi [FKS] when translated to the bitwise model. (We will refer to this data structure as the FKS-dictionary.) Thus, d-queries in general can be solved in optimal space $O(nm)$ by allowing $O(\binom{m}{d}m)$ query time to exhaustively search for $\binom{m}{d}$ exact queries. At the other extreme of the space/time spectrum, optimal query time $O(m)$ is achievable with space $O(nm\binom{m}{d})$. We will be interested in nontrivial techniques away from these extremes.

The cases of small d and large d seem quite distinct in nature, and may require different techniques for their solutions. In the former case, even $d = 1$ already poses a rather challenging problem. In this paper, we make an initial effort in establishing some nontrivial bounds for this case. (In Manber and Wu [MW], a 1-query is first transformed into $O(m)$ exact queries, and then solved by hashing techniques.) Our main result is a data structure that achieves $O(m \log \log n)$ query time with $O(nm \log m)$ space for the $d = 1$ case.

We consider a bitwise complexity model as suggested in Minsky and Papert [MP] and Elias [E] (or, equivalently, the cell-probe model in Yao [Y] with word size 1). A data structure for W consists of a binary string D_W called the *dictionary*, and a decision tree $T_{W,\alpha}$ for every d-query α. Each internal node of $T_{W,\alpha}$ branches according to the outcome of a question of the form "$D_W[j] =?$", while each leaf contains a yes or no answer to the d-query α. Such a data structure is said to use *space* s and *query time* t for W, if s is the size of the dictionary D_W and t is the maximum height of $T_{W,\alpha}$ for any α. The maximum of s and t over all sets W (with n binary strings of length m) defines the space and time complexity of the data structure, denoted by $s(n,m)$ and $t(n,m)$ respectively. Our main result is as follows.

Theorem 1 There is a data structure for the d-query problem that achieves query time $O(m \log \log n)$ and space $O(nm \log m)$ for the $d = 1$ case.

The next two sections of this paper give a proof for the above theorem. The last section lists some open problems.

2 Data Structure

Before describing the data structure, we consider some useful concepts. Assume W is a set of n binary strings of length m, where m is even. In the following definition and facts, we restrict u and v to binary strings of length $m/2$, i.e., $u, v \in \{0,1\}^{m/2}$.

Definition For any u, let $A_{W,u}$ be the set of all v such that $uv \in W$. For any v, let $B_{W,v}$ be the set of all u such that $uv \in W$. Furthermore, let W_L be the set of all u with $|A_{W,u}| > 0$, and let W_R be the set of all v with $|B_{W,v}| > 0$.

Thus, W_L and W_R are the left and right projections of the strings in W respectively.

Fact 1 If the answer to the 1-query $\alpha = uv$ is "yes" for W, then either $u \in W_L$ or $v \in W_R$.

Proof. If there is a match between α and some string of W within Hamming distance 1, then the match must be exact in either the first half or the second half of α; that is, either $u \in W_L$ or $v \in W_R$. □

Let $0 < \epsilon < 1/2$ be any fixed number. A string $\alpha \in \{0,1\}^m$ is said to be an ϵ-*bad query* for W (or simply ϵ-*bad*), if $\alpha = uv$ with both $|A_{W,u}| > n^{1-\epsilon}$ and $|B_{W,v}| > n^{1-\epsilon}$.

The next fact follows directly from the definition.

Fact 2 If $\alpha = uv$ is not an ϵ-bad query for W, then either $|A_{W,u}| \leq n^{1-\epsilon}$ or $|B_{W,v}| \leq n^{1-\epsilon}$.

One nice property about the bad queries is that there cannot be too many of them.

Fact 3 The number of ϵ-bad queries for W is less than $|W|^{2\epsilon}$.

Proof Since the sets $u \cdot A_{W,u}$ are disjoint for distinct u, the number of u satisfying $|A_{W,u}| > n^{1-\epsilon}$ is less than $n/n^{1-\epsilon} = n^\epsilon$. Similarly, there are less than n^ϵ strings v satisfying $|B_{W,v}| > n^{1-\epsilon}$. This immediately implies Fact 3. □

We now define our data structure by constructing a dictionary D_W recursively for a set W of n m-bit strings. Without loss of generality, assume that m is a power of two. If either n or m is equal to 1, use the obvious data structure. That is, for $n = 1$, let D_W contain the only string in W and answer a 1-query α by comparing D_W with α bit by bit; while for $m = 1$, the answer is always "yes". Thus,

$$s(1, m) = m, t(1, m) = m, s(n, 1) = 0, t(n, 1) = 0. \tag{1}$$

Let W be a set with $n, m > 1$. The dictionary D_W for W will consist of three parts:

1) an FKS-dictionary F_1 for the set of ϵ-bad queries for W, with an 'answer bit' 1 or 0 attached to each bad query α (indicating whether there is any $w \in W$ within distance 1 of α).

2) an FKS-dictionary F_2 for the set W_L, and attached to each $u \in W_L$ a data structure $D_{A_{W,u}}$ constructed recursively for the set $A_{W,u}$ of length $m/2$ strings.

3) an FKS-dictionary F_3 for the set W_R, and attached to each $v \in W_R$ a data structure $D_{B_{W,v}}$ constructed recursively for the set $B_{W,v}$ of length $m/2$ strings.

To answer a 1-query α, we write $\alpha = uv$ with $|u| = |v| = m/2$, and execute the following recursive algorithm **Lookup**(D_W, α). A global variable *answer*, which is initially set to be "no", will hold the answer to the query when the algorithm halts.

Algorithm Lookup(D_W, α)

If $n = 1$ or $m = 1$, do the obvious search, set the *answer* accordingly, and return; otherwise do the following:

Step 1. (*Is α a bad query?*)
Look in F_1 to see whether α is in the bad query list; if so, set *answer* to be the answer bit attached to α and return;

Step 2. (α *is not a bad query.*)
Look in F_2 to see whether $u \in W_L$; if so, use the attached data structure $D_{A_{W,u}}$ to execute **Lookup**$(A_{W,u}, v)$;

Step 3. Look in F_3 to see whether $v \in W_R$; if so, use the attached data structure $D_{B_{W,v}}$ to execute **Lookup**$(B_{W,v}, u)$;

Step 4. Return;

3 Analysis

We first establish the correctness of the above algorithm by induction on m. Let $\alpha = uv$ be the 1-query. Consider the case when there exists a string of W within distance 1 of α. If α is an ϵ-bad query for W, then Step 1 sets *answer* to "yes". Otherwise, Fact 1 and the inductive hypothesis for $m/2$ imply that *answer* will be set to "yes" in either Step 2 or Step 3. This proves that the algorithm behaves correctly in this case. The case when there is no string of W within distance 1 of α can be argued similarly.

We now analyze $t(n, m)$ and $s(n, m)$, the time and space requirements of the data structure. Suppose the FKS solution to the standard dictionary problem (for n strings of length m) use cm retrieval time and $c'mn$ space, where $c, c' \geq 1$ are positive constants.

The time for accessing F_1, F_2, and F_3 is $cm + cm/2 + cm/2 = 2cm$. By the definition of bad queries and Fact 2, we can write down the following recurrence relation for $t(n, m)$:

$$t(n, m) \leq 2cm + t(n, m/2) + t(\lfloor n^{1-\epsilon} \rfloor, m/2). \tag{2}$$

Similarly, Fact 3 leads to the following recurrence relation for $s(n, m)$:

$$\begin{aligned} s(n, m) \leq \; & c' \lfloor n^{2\epsilon} \rfloor m + c'(|W_L| + |W_R|)m/2 \\ & + \sum_{u \in W_L} s(|A_{W,u}|, m/2) + \sum_{v \in W_R} s(|B_{W,v}|, m/2). \end{aligned} \tag{3}$$

Note that in the last formula,

$$\sum_{u \in W_L} |A_{W,u}| = n, \tag{4}$$

$$\sum_{v \in W_R} |B_{W,v}| = n. \tag{5}$$

Let $\lambda = -4c/\log_2(1 - \epsilon/2)$. We use induction on the value of $n + m$ to prove the following query time bound:

$$t(n, m) \leq m + \lambda m \log_2 \log_2(2n). \tag{6}$$

Clearly, (6) is true for the base case of $n = 1$ or $m = 1$. For the inductive step, let $n > 1$, $m > 1$. We have from (2) and the induction hypothesis,

$$\begin{aligned} t(n, m) \; &\leq \; 2cm + t(n, m/2) + t(\lfloor n^{1-\epsilon} \rfloor, m/2) \\ &\leq \; 2cm + m/2 + \lambda(m/2) \log_2 \log_2(2n) + m/2 + \lambda(m/2) \log_2 \log_2(2n^{1-\epsilon}) \\ &\leq \; 2cm + m + \lambda(m/2) \log_2 \log_2(2n) + \lambda(m/2) \log_2 \log_2((2n)^{1-\epsilon/2}) \\ &= \; m + \lambda m \log_2 \log_2(2n) + m(2c + (\lambda/2) \log_2(1 - \epsilon/2)) \\ &= \; m + \lambda m \log_2 \log_2(2n). \end{aligned}$$

This completes the proof of (6).

To finish the analysis, we prove that

$$s(n,m) \leq 2c'nm(1 + \log_2 m). \tag{7}$$

Clearly, (7) is true for the base case of $n = 1$ or $m = 1$. For the inductive step, we have from (3), (4), (5) and the induction hypothesis,

$$
\begin{aligned}
s(n,m) \;\leq\; & c'n^{2\epsilon}m + (c'm/2)(|W_L| + |W_R|) \\
& + \sum_{u \in W_L} s(|A_{W,u}|, m/2) + \sum_{v \in W_R} s(|B_{W,v}|, m/2) \\
\leq\; & c'n^{2\epsilon}m + (c'm/2)2n \\
& + \sum_{u \in W_L} 2c'|A_{W,u}|(m/2)(1 + \log_2(m/2)) \\
& + \sum_{v \in W_R} 2c'|B_{W,v}|(m/2)(1 + \log_2(m/2)) \\
\leq\; & 2c'nm + 2c'mn(1 + \log_2(m/2)) \\
=\; & 2c'nm(1 + \log_2 m).
\end{aligned}
$$

This proves (7) and Theorem 1.

4 Open Problems

In this paper we have taken a step in studying rigorously the d-query problem for small d. There are many interesting questions that remain to be answered.

(a) We have given an $O(m \log \log n)$ query time and $O(nm \log m)$ space solution to the $d = 1$ case. Is there a better solution, such as an optimal solution with $O(m)$ query time and $O(nm)$ space?

(b) The bitwise model is used in this paper. To make a better comparison with the standard dictionary problem, it would be interesting to examine the problem in the cell-probe model. To be specific, suppose each word can store a string of length m. Can our solution be converted in some way to give $O(\log \log n)$ query time and $O(n \log m)$ space?

(c) One can consider the $d = 1$ problem where the output is the subset of strings in W within distance 1 to the query. It is straightforward to modify our scheme by making each word in the FKS dictionary in F_1 point to the desired words in W instead of just to a bit "1". This leads to an $O(m \log \log n +$ output size) query time and $O(nm \log m)$ space solution. Can one get an optimal query time and space solution in this case?

(d) It is a challenging problem to find efficient data structure for fixed $d > 1$. The natural approach to extend our present $d = 1$ solution does not work well even for $d = 2$. In that case, the obvious recursive design would consist of five substructures F_1, F_2, F_3, F_4, F_5, where F_1 is some bad query dictionary; F_2 is an FKS-dictionary for W_L with each $u \in W_L$ linking to a recursive data structure for 2-querying the set $A_{W,u}$; F_3 has a $d = 1$ dictionary for W_L identifying all the strings $u' \in W_L$ within distance 1 to u (as in (c)), and with each u' linking to its own $d = 1$ structure for $A_{W,u'}$; finally F_4 and F_5 are the duals to F_2, F_3 respectively for the set W_R. The search becomes costly when the search path goes through F_3, since it may have to examine many u' and their associated structures. A possible way to cope with the problem is to redesign the F_2 structure in some fashion to make the search easier for this situation. However, we have not found a good way to accomplish that.

Acknowledgments

Acknowledgments We thank the anonymous referees, who reviewed this paper for the CPM'95 program committee, for their helpful comments.

References

[DHLNP] D. Dolev, Y. Harari, N. Linial, N. Nisan, and M. Parnas, *Neighborhood preserving hashing and approximate queries*, Proceedings of Fifth ACM Symposium on Discrete Algorithms, 1994.

[DHP] D. Dolev, Y. Harari and M. Parnas, *Finding the neighborhood of a query in a dictionary*, Proceedings of Second Israel Symposium on Theory of Computing and Systems, 1993.

[E] P. Elias, *Efficient storage and retrieval by content and address of static files*, Journal ACM **21** (1974), 246-260.

[FKS] M. Fredman, M. Komlós, and E. Szemerédi, *Storing a sparse table with $O(1)$ worst case access time*, Journal ACM **31** (1984), 538-544.

[GP] Z. Galil and K. Park, *An improved algorithm for approximate string matchng*, SIAM J. on Computing **19** (1990), 989-999.

[GPY] D. Greene, M. Parnas and F. Yao, *Multi-index hashing for information retrieval*, Proceedings of 1994 IEEE Symposium on Foundations of Computer Science, November 1994, 722-731.

[LV] G. Landau and U. Vishkin, *Fast string matching with k differences*, J. Comp. Sys. Sci. **37** (1988), 63-78.

[MW]   U. Manber and S. Wu, *An algorithm for approximate membership checking with application to password security, Information Processing Lettters* **50** (1994), 191-197.

[MP]   M. Minsky and S. Papert, *Perceptrons*, MIT Press, 1969.

[TU]   J. Tarhio and E. Ukkonen, *Approximate Boyer-Moore string matching*, Report A-1990-3, Department of Computer Science, Univ. of Helsinki, March 1990.

[U]   E. Ukkonen, *Finding approximate patterns in strings, J. Algorithms* **6** (1985), 132-137.

[UW]   E. Ukkonen and D. Wood, *Approximate string matching with suffix automata, Algorithmica* **10** (1993), 353-364.

[Y]   A. Yao, *Should tables be sorted?, Journal ACM* **28** (1981), 615-628.

On the Editing Distance between Undirected Acyclic Graphs and Related Problems*

Kaizhong Zhang[1], Jason T. L. Wang[2] and Dennis Shasha[3]

[1] Department of Computer Science, The University of Western Ontario,
London, Ontario, Canada N6A 5B7 (kzhang@csd.uwo.ca)
[2] Department of CIS, New Jersey Institute of Technology,
Newark, NJ 07102, USA (jason@vienna.njit.edu)
[3] Courant Institute of Mathematical Sciences, New York University,
251 Mercer Street, New York, NY 10012, USA (shasha@cs.nyu.edu)

1 Introduction

Problem. We consider the problem of comparing $CUAL$ graphs (*C*onnected, *U*ndirected, *A*cyclic graphs with nodes being *L*abeled).[4] Suppose we define the *distance* between two CUAL graphs to be the weighted number (the user chooses the weighting) of edit operations (insert node, delete node and relabel node) to transform one graph to the other. By reduction from exact cover by 3-sets, one can show that finding the distance between two graphs is NP-hard. In view of the hardness of the problem, we propose a constrained distance metric, called the *degree-2 distance*, for graphs by requiring that any node to be inserted (deleted) have no more than 2 neighbors. As will become clear, this constraint is sensible in defining the edit operations on graphs. Further, the measure is a natural extension of the edit distance for strings [22] and Selkow's distance for trees [15].

Main Results. We develop algorithms to find the degree-2 distance between a class of limited graphs, including CUAL graphs, planar CUAL graphs, unordered trees and ordered trees. (A planar CUAL graph is one that can be embedded in the plane in such a way that the edges of the embedding intersect only at the nodes of the graph. An unordered tree is a rooted tree in which the order among siblings is unimportant.) Let G_1 and G_2 be two given graphs. Let N_i, $i = 1, 2$, represent the number of nodes in G_i. Let $deg(n)$ denote the number of neighbors of node n (in the rooted tree case, $deg(n)$ is defined as the number of n's children, excluding n's parent); $d_i = \max_{n \in G_i} deg(n)$. Table 1 summarizes the asymptotic time complexities of our algorithms for finding the degree-2 distance between

---

* This work was supported, in part, by the National Science Foundation under Grants IRI-9224601 and IRI-9224602, by the Office of Naval Research under Grant N00014-92-J-1719, by the Natural Sciences and Engineering Research Council of Canada under Grant OGP0046373, and by a grant from the AT&T Foundation.
[4] Such graphs are also known as labeled free trees. When the context is clear, we refer to CUAL graphs simply as graphs. Note that, in practice, edges of a graph may have labels. In that case, one can transform a labeled edge between two nodes u and v to a labeled node connecting u and v.

G_1 and G_2 for the general weighting and integral weighting edit operations, respectively.

Table 1. Time complexities of the proposed algorithms for the limited graphs; $D = \min\{d_1, d_2\}$.

Limited Graph	General Weighting	Integral Weighting
CUAL Graph	$O(N_1 N_2 D^2)$	$O(N_1 N_2 D \sqrt{D} \log D)$
Planar CUAL Graph	$O(N_1 N_2 \log D)$	$O(N_1 N_2 \log D)$
Unordered Tree	$O(N_1 N_2 D)$	$O(N_1 N_2 \sqrt{D} \log D)$
Ordered Tree	$O(N_1 N_2)$	$O(N_1 N_2)$

Significance of the Work. Undirected labeled graphs have long been used to represent two-dimensional (2-D) chemical compounds and molecules in chemical information systems [2, 3]. Figure 1(a) shows two examples of 2-D compounds; each node in the graphs represents an atom and each edge represents a bond. The compounds can be represented alternatively as two CUAL graphs (Figure 1(b)).

There are two common uses of the chemical information systems. The first, referred to as a *structure search*, is to recognize if a compound has been included in the data file previously, and if not, to register it in the file [12]. Here the root subroutine is based on graph isomorphism algorithms. The second, referred to as a *similarity search*, is to find compounds that are similar to a query structure [7, 23, 24]. Many similarity measures have been devised; they are usually calculated by considering the atom, bond, or ring-centered substructural fragments found in common in the query and in a compound.[5] While these measures are often useful, they don't capture many of the interesting topological differences between two compounds, which play a key role in identifying the difference in the compounds' functionalities.

Example 1. Consider the atom-centered fragment f and the three compounds G_1, G_2, G_3 in Figure 2. According to the aforementioned fragment weighting scheme, G_1 is closer to G_3 than to G_2 because f occurs twice in both of the G_1 and G_3, whereas it occurs only once in G_1 and G_2. On the other hand, according to the proposed degree-2 distance metric, G_1 is closer to G_2 than to G_3.[6] Visually, G_1 is closer to G_2, consistent with the degree-2 metric; thus, we argue that this is a plausible metric.

---

[5] For example, the much used fragment weighting scheme works by considering the number of occurrences of a particular fragment type within a compound [25]. The more frequently a fragment occurs, the greater weight it gains. Thus, a pair of molecules that had several occurrences of a given fragment in common would be considered to be more similar to each other than if they had only a single occurrence in common.

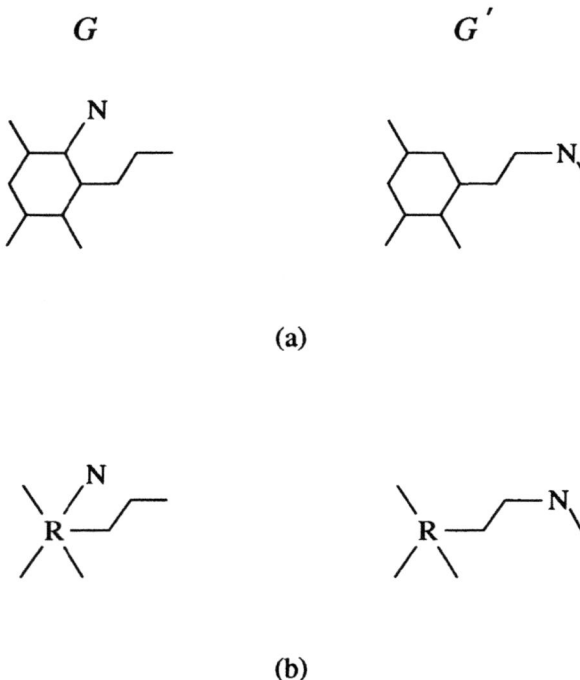

G G'

(a)

(b)

Fig. 1. (a) Two examples of chemical compounds [8]. N represents a nitrogen atom. Omitted node labels are carbon atoms (C). Hydrogen atoms (H) are not included in the graph representations since their presence or absence can be deduced from the other information. (b) The same compounds can be represented as CUAL graphs, with each ring being represented by a special node label R.

Thus our work provides a complementary measure capable of reflecting the structural differences between chemical compounds (except that our graphs must be acyclic, so rings must be reduced to single nodes). We believe the presented techniques can also contribute to comparison and search of 2-D and 3-D (macro)molecules in protein and DNA structures [14].

Comparison to Past Research. This paper generalizes the work on the edit distance between strings [6, 11, 13, 16, 20, 21, 26] and trees [19, 29, 30]. Various kinds of constrained and generalized edit distance on strings and trees have been developed [1, 9, 10, 17, 28]. Our degree-2 distance, when applied to unordered trees, is a restricted form of the constrained distance previously reported in [28]. When applied to ordered trees, the degree-2 distance is a generalized measure of the constrained distance originated from Selkow [17], though our algorithm

---

[6] Assume all edit operations have unit cost. The degree-2 distance between G_1 and G_2 is 1, obtained by relabeling the leftmost C in G_1 to N in G_2, whereas the distance between G_1 and G_3 is 7, obtained by inserting the seven Rs into G_3.

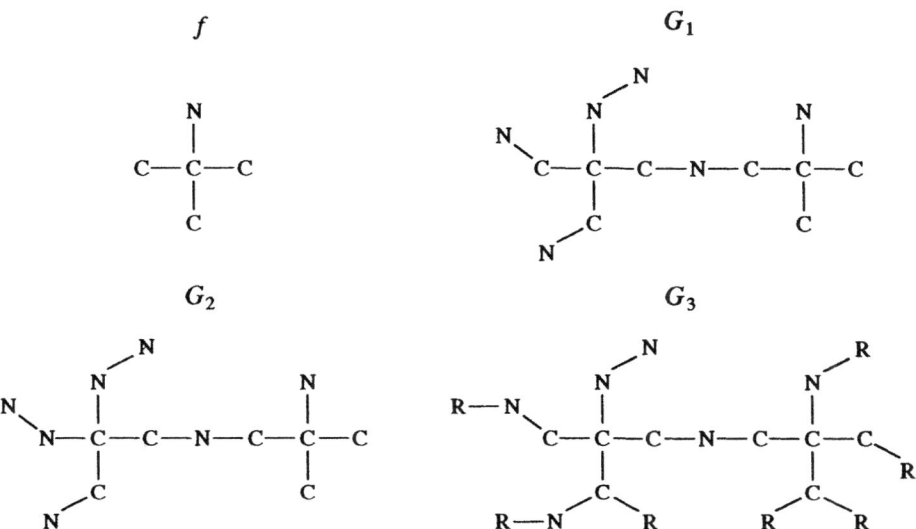

Fig. 2. An augmented atom (fragment) and three chemical compounds.

has the same asymptotic complexity as Selkow's algorithm. (Selkow's distance measure requires that all node deletions and insertions occur at leaves i.e., with degree 1.) In this extended abstract, we present algorithms for computing the degree-2 distance between two CUAL graphs and unordered trees. (The result for the latter is used as a subroutine to calculate the former.) The entire set of algorithms is in the full paper and is available from the authors. We have chosen to present these two results, because we believe they are the most useful of our results to date for approximate graph matching.

2 Preliminaries

2.1 Edit Operations on Graphs

There are three kinds of edit operations on graphs: relabel, delete and insert a node. Relabeling node n means changing the label on n. Deleting a node n means making the neighbors of n (except an arbitrarily specified neighbor n') become the neighbors of n' and then removing n. (This amounts to contraction of the edge between n and n' [4] and making the resulting node have label of n'.) Insert is the complement of delete. This means that inserting n as a neighbor of n' makes a subset of the current neighbors of n' become the neighbors of n. We represent an edit operation as a pair $(u, v) \neq (\Lambda, \Lambda)$, sometimes written $u \to v$. We call $u \to v$ a relabeling operation if $u \neq \Lambda$ and $v \neq \Lambda$; a delete operation if $v = \Lambda$; and an insert operation if $u = \Lambda$. Let G_2 be the graph that results from the application of an edit operation $u \to v$ to graph G_1; this is written $G_1 \Rightarrow G_2$ via $u \to v$. Let S be a sequence $s_1, s_2, \ldots, s_k$ of edit operations. S transforms graph G to graph G' if there is a sequence of graphs $G_0, G_1, \ldots, G_k$ such that $G = G_0, G' = G_k$ and $G_{i-1} \Rightarrow G_i$ via s_i for $1 \leq i \leq k$. Let γ be a cost function that assigns to each edit operation $u \to v$ a nonnegative real number $\gamma(u \to v)$. We require γ to be a metric. By extension, the cost of the sequence S, denoted $\gamma(S)$, is simply the sum of costs of the constituent edit operations. The *distance* from G to G', denoted $\Delta(G, G')$, is the minimum cost of all sequences of edit operations taking G to G'.

Theorem 1. *Finding $\Delta(G, G')$ is NP-hard.*

Proof. Similar to the NP-completeness proof given in [27]. □

2.2 Degree-2 Distance

In view of the hardness of the problem, we propose to impose the following constraint on the edit operations: a node n can be deleted (inserted) only when $deg(n) \leq 2$.[7] Intuitively one can delete either a leaf or a node n with two neighbors; in the latter case, after deleting n, we simply connect its two neighbors together. When inserting n between two nodes n' and n'', we remove the edge between n' and n'' and make n the neighbor of both n' and n''. These constrained edits will be referred to as the *degree-2* edit operations; they are natural in manipulating nodes and edges in the updated graphs. We define the degree-2 distance between graph G and graph G', denoted $\delta(G, G')$, to be the minimum cost of all sequences of the degree-2 edit operations transforming G to G'. Clearly δ is a metric.

2.3 Mappings

The degree-2 edit operations correspond to a *mapping*, which is a graphical specification of what edit operations apply to each node in the two graphs. For example, the mapping in Figure 3 shows a way to transform the CUAL graph G to the CUAL graph G' given in Figure 1. It corresponds to the sequence (delete (node with label N), insert (node with label N)).

To formalize the notion of mappings, we need some definitions. Let u, v, w be three nodes in a graph G; let $[u, v]$ denote the path between node u and node v. Define the *center* of the three nodes u, v, w, denoted $center(u, v, w)$, to be the intersection node of the three paths $[u, v]$, $[v, w]$ and $[w, u]$. Figure 4 illustrates the definition.

Let $g[i]$ represent the ith node of graph G according to some ordering (e.g., a depth-first search order). Formally, a mapping from G to G' is a triple (M, G, G') (or simply M when the context is clear), where M is any set of pairs of integers (i, j) satisfying the following conditions:

1. $1 \leq i \leq |G|, 1 \leq j \leq |G'|$.

---

[7] Thus, to delete a node n with $deg(n) > 2$, one has to first delete some of its neighbors to make its degree less than or equal to 2 before removing it.

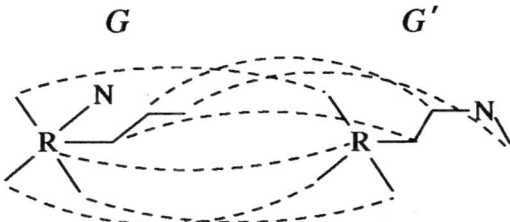

Fig. 3. A mapping from G to G'. Nodes in G not touched by a mapping line are to be deleted; nodes in G' not touched by a mapping line are to be inserted. The mapping shows a way to transform G to G'.

Fig. 4. Illustrations of the center, which is represented by the bullet $\bullet$.

2. For any two pairs (i_1, j_1) and (i_2, j_2) in M, $i_1 = i_2$ iff $j_1 = j_2$ (one-to-one).
3. For any three pairs (i_1, j_1), (i_2, j_2) and (i_3, j_3) in M, (i^*, j^*) is also in M where $g[i^*] = center(g[i_1], g[i_2], g[i_3])$ and $g'[j^*] = center(g'[j_1], g'[j_2], g'[j_3])$ (center relationship preservation).

The cost of M, denoted $\gamma(M)$, is the cost of deleting nodes of G not touched by a mapping line plus the cost of inserting nodes of G' not touched by a mapping line plus the cost of relabeling nodes in those pairs related by mapping lines with different labels.

Lemma 2. *Given S, a sequence $s_1, s_2, \ldots, s_k$ of degree-2 edit operations from G to G', there exists a mapping M from G to G' such that $\gamma(M) \leq \gamma(S)$. Conversely, for any mapping M, there exists a sequence of degree-2 edit operations S such that $\gamma(S) = \gamma(M)$.*

Hence, $\delta(G, G') = \min\{\gamma(M) | M$ is a mapping from G to $G'\}$. As an example, consider again the graphs G and G' in Figure 3. The mapping in the figure is a minimum cost mapping and $\delta(G, G') = 2$.

3   Algorithms

We first present the algorithm for finding the degree-2 distance between two rooted unordered trees and then extend the algorithm to handle graphs.

3.1   The Algorithm for Unordered Trees

For notational convenience, in this subsection we use T rather than G to represent a rooted unordered tree. Let $t[i]$ denote the ith node of T according to the depth-first search order. $T[i]$ represents the subtree rooted at $t[i]$ and $F[i]$ represents the forest obtained by deleting $t[i]$ from $T[i]$. Let T_1 and T_2 be two rooted unordered trees. We use $t_1[i_1], t_1[i_2], \ldots, t_1[i_{n_i}]$ to represent the children of $t_1[i]$ in $T_1[i]$ and use $t_2[j_1], t_2[j_2], \ldots, t_2[j_{n_j}]$ to represent the children of $t_2[j]$ in $T_2[j]$. When applied to the rooted unordered trees T_1 and T_2, the mapping M defined in § 2.3 is exactly the same as the edit distance mapping between the unordered trees with the following constraint: for any two pairs (i_1, j_1) and (i_2, j_2) in M, (i^*, j^*) is also in M where $t_1[i^*] = lca(t_1[i_1], t_1[i_2])$, $t_2[j^*] = lca(t_2[j_1], t_2[j_2])$ and $lca(.)$ represents the least common ancestor of the indicated nodes.[8] With this notion in mind, it's easy to develop a dynamic programming algorithm for rooted unordered trees. We now present several lemmas, which will be the basis of our algorithm.

Lemma 3. *For all* $1 \leq i \leq N_1$ *and* $1 \leq j \leq N_2$,
(i) $\delta(\emptyset, \emptyset) = 0$;
(ii) $\delta(T_1[i], \emptyset) = \delta(F_1[i], \emptyset) + \gamma(t_1[i] \to \Lambda)$;
(iii) $\delta(F_1[i], \emptyset) = \sum_{k=1}^{n_i} \delta(T_1[i_k], \emptyset)$;
(iv) $\delta(\emptyset, T_2[j]) = \delta(\emptyset, F_2[j]) + \gamma(\Lambda \to t_2[j])$;
(v) $\delta(\emptyset, F_2[j]) = \sum_{k=1}^{n_j} \delta(\emptyset, T_2[j_k])$.

Lemma 4. *For all* $1 \leq i \leq N_1$ *and* $1 \leq j \leq N_2$,

$$\delta(T_1[i], T_2[j]) = \min \begin{cases} \delta(T_1[i], \emptyset) + \min_{1 \leq s \leq n_i} \{\delta(T_1[i_s], T_2[j]) - \delta(T_1[i_s], \emptyset)\} \\ \delta(\emptyset, T_2[j]) + \min_{1 \leq t \leq n_j} \{\delta(T_1[i], T_2[j_t]) - \delta(\emptyset, T_2[j_t])\} \\ \delta(F_1[i], F_2[j]) + \gamma(t_1[i] \to t_2[j]) \end{cases}$$

Proof. Let M be a minimum-cost mapping from $T_1[i]$ to $T_2[j]$. There are four cases to be considered:

Case 1. $i \notin M$ and $j \in M$. Let (z, j) be in M. Thus $t_1[z]$ must be a node in $F_1[i]$. Let $t_1[i_s]$ be the child of $t_1[i]$ on the path from $t_1[z]$ to $t_1[i]$. Thus $\delta(T_1[i], T_2[j]) = \delta(T_1[i_s], T_2[j]) + \delta(T_1[i_1], \emptyset) + \ldots + \delta(T_1[i_{s-1}], \emptyset) + \delta(T_1[i_{s+1}], \emptyset) + \ldots + \delta(T_1[i_{n_i}], \emptyset) + \gamma(t_1[i] \to \Lambda)$. Since $\delta(T_1[i], \emptyset) = \gamma(t_1[i] \to \Lambda) + \sum_{k=1}^{n_i} \delta(T_1[i_k], \emptyset)$, we can rewrite the right hand side of the formula as $\delta(T_1[i], \emptyset) + \delta(T_1[i_s], T_2[j]) - \delta(T_1[i_s], \emptyset)$. The range of k is from 1 to n_i; therefore we take the minimum of the corresponding costs.

Case 2. $i \in M$ and $j \notin M$. This is analogous to Case 1.

Case 3. $i \in M$ and $j \in M$. By the mapping conditions, (i, j) must be in M. Thus $\delta(T_1[i], T_2[j]) = \delta(F_1[i], F_2[j]) + \gamma(t_1[i] \to t_2[j])$.

Case 4. $i \notin M$ and $j \notin M$. We would have $\delta(T_1[i], T_2[j]) = \delta(F_1[i], F_2[j]) + \gamma(t_1[i] \to \Lambda) + \gamma(\Lambda \to t_2[j])$. Since $\gamma(t_1[i] \to t_2[j]) \leq \gamma(t_1[i] \to \Lambda) + \gamma(\Lambda \to t_2[j])$ (the triangle inequality), we need not include this case in our formula. □

---

[8] An edit distance mapping M_e between two rooted unordered trees satisfies the node one-to-one relationship and preserves the ancestor relationship, i.e., supposing u is mapped to v and x is mapped to y in M_e, u is an ancestor of x iff v is an ancestor of y [18].

In calculating $\delta(F_1[i], F_2[j])$, notice that if two nodes $t_1[x_1]$ and $t_1[x_2]$ of $T_1[i_s]$ are in M, then by the mapping conditions there must exist an integer t such that the two nodes connected to $t_1[x_1]$ and $t_1[x_2]$, respectively, by the mapping lines of M are in $T_2[j_t]$. We try to find a best mapping between the children of $t_1[i]$ and the children of $t_2[j]$ by constructing a weighted bipartite graph BG as follows. Let $U = \{t_1[i_1], \ldots, t_1[i_{n_i}]\}$ and $V = \{t_2[j_1], \ldots, t_2[j_{n_j}]\}$. Assign the weight for each edge $(t_1[i_s], t_2[j_t])$, denoted $\omega((t_1[i_s], t_2[j_t]))$, $1 \leq s \leq n_i$ and $1 \leq t \leq n_j$, based on the formula

$$\omega((t_1[i_s], t_2[j_t])) = \delta(T_1[i_s], \emptyset) + \delta(\emptyset, T_2[j_t]) - \delta(T_1[i_s], T_2[j_t])$$

Without loss of generality, assume $n_i \leq n_j$. To better bound the complexity of our algorithm, for each node $u \in U$, we only pick the top n_i highest weighted edges touching on u and store these edges as well as their end nodes in BG. Thus BG has at most $n_i n_i$ edges and at most $n_i + n_i n_i$ nodes. Let Ma be the maximum weighted matching in BG.

Lemma 5.

$$\delta(F_1[i], F_2[j]) = \sum_{s=1}^{n_i} \delta(T_1[i_s], \emptyset) + \sum_{t=1}^{n_j} \delta(\emptyset, T_2[j_t]) - \sum_{(u,v) \in Ma} \omega((u, v))$$

Thus the problem of calculating $\delta(F_1[i], F_2[j])$ becomes that of finding the maximum weighted matching in BG. One can solve the problem by using Gabow and Tarjan's algorithm in [5]. Figure 5 summarizes the algorithm.

Algorithm A
Input: Unordered trees T_1 and T_2.
Output: $\delta(T_1[i], T_2[j])$ where $1 \leq i \leq N_1$ and $1 \leq j \leq N_2$;
$\quad\quad\quad \delta(T_1[N_1], T_2[N_2]) = \delta(T_1, T_2)$.
$\delta(\emptyset, \emptyset) := 0$;
for $i := 1$ to N_1 do
$\quad$ compute $\delta(F_1[i], \emptyset)$ and $\delta(T_1[i], \emptyset)$ as in Lemma 3 (ii) (iii);
for $j := 1$ to N_2 do
$\quad$ compute $\delta(\emptyset, F_2[j])$ and $\delta(\emptyset, T_2[j])$ as in Lemma 3 (iv) (v);
for $i := 1$ to N_1 do
$\quad$ for $j := 1$ to N_2 do
$\quad\quad$ compute $\delta(F_1[i], F_2[j])$ as in Lemma 5;
$\quad\quad$ compute $\delta(T_1[i], T_2[j])$ as in Lemma 4;

Fig. 5. Algorithm for computing $\delta(T_1, T_2)$ for two unordered trees T_1 and T_2.

Time Complexity. The complexity of computing $\delta(T_1[i], T_2[j])$ is, by Lemma 4, bounded by $O(n_i + n_j)$. In constructing BG for calculating $\delta(F_1[i], F_2[j])$, for each node $u \in U$, it takes $O(n_j)$ time to calculate the weights of the edges touching on u and pick the n_i edges with the highest weights. Thus, it takes a total of $O(n_i n_j)$ time to construct BG. Let V be the number of nodes in BG and let E be the number of edges in BG. The complexity of finding the maximum weighted matching in BG is $O(\min\{n_i, n_j\}(E + V \log V))$ when the edges have general weights and is $O(\sqrt{V}E \log(VW))$ when the edges have integral weights where W is the maximum weight [5]. Without loss of generality, assume $n_i \leq n_j$. Then V is at most $n_i + n_i n_i$ and E is at most $n_i n_i$. Thus for the general weighting case, the complexity of computing $\delta(T_1[i], T_2[j])$ for any pair of i and j is bounded by $O(n_i n_j + n_i(n_i n_i + V \log V))$. $V = \min\{n_i + n_j, n_i + n_i^2\}$, and therefore $\log V = \log n_i$. Hence the complexity is bounded by

$$O(n_i n_j + (\min\{n_i, n_j\})^3 + ((\min\{n_i, n_j\})^2 + \min\{n_i, n_j\} \max\{n_i, n_j\}) \log n_i)$$
$$= O(n_i n_j \log(\min\{n_i, n_j\}) + (\min\{n_i, n_j\})^3)$$
$$= O(n_i n_j \min\{n_i, n_j\})$$

For the integral weighting case, suppose the node label alphabet for unordered trees is finite. Then the maximum edit cost is finite. As a consequence, the maximum weight W in BG is bounded by cV for some constant c. Thus, the complexity is bounded by

$$O(n_i n_j + \sqrt{n_i + n_j}\, n_i n_i \log(n_i + n_i^2))$$
$$= O(n_i n_j + \sqrt{2n_j}\, n_i n_i \log n_i)$$
$$= O(n_i n_j + n_i n_j n_i / \sqrt{n_j} \log n_i)$$
$$= O(n_i n_j + n_i n_j \sqrt{n_i}\sqrt{n_i/n_j} \log n_i)$$
$$= O(n_i n_j + n_i n_j \sqrt{n_i} \log n_i)$$
$$= O(n_i n_j \sqrt{\min\{n_i, n_j\}} \log(\min\{n_i, n_j\}))$$

Therefore for the general weighting case, the complexity of Algorithm A is

$$\sum_{i=1}^{N_1} \sum_{j=1}^{N_2} O(n_i n_j \min\{n_i, n_j\})$$
$$\leq \sum_{i=1}^{N_1} \sum_{j=1}^{N_2} O(n_i n_j \min\{d_1, d_2\})$$
$$\leq O(\min\{d_1, d_2\} \sum_{i=1}^{N_1} n_i \sum_{j=1}^{N_2} n_j)$$
$$\leq O(N_1 N_2 \min\{d_1, d_2\})$$

Likewise, for the integral weighting case, the complexity of Algorithm A is $O(N_1 N_2 \sqrt{\min\{d_1, d_2\}} \log(\min\{d_1, d_2\}))$.

3.2 The Algorithm for CUAL Graphs

Let G_1 and G_2 be two CUAL graphs. By the definition, if (i, j) is in a minimum cost mapping from G_1 to G_2, then we can assign $g_1[i]$ as the root of G_1 and assign $g_2[j]$ as the root of G_2, resulting in two rooted unordered trees. By applying Algorithm A to the two trees, we can find $\delta(G_1, G_2)$. This naive algorithm runs in time $O(N_1^2 N_2^2 \min\{d_1, d_2\})$ when the edit operations have general cost, and in time $O(N_1^2 N_2^2 \sqrt{\min\{d_1, d_2\}} \log(\min\{d_1, d_2\}))$ when the edit operations have integral cost.

A more careful analysis leads to a faster algorithm. Let us choose an arbitrary node, say r, in G_1 and assign r as the root of G_1. Thus the graph G_1 can be considered as a rooted unordered tree. For any node u in G_1, we use $T_1^r[u]$ to represent the unordered tree rooted at u with respect to G_1's root r. Let M be a minimum cost mapping from G_1 to G_2. The distance is the minimum of the following two cases: in the rooted G_1, (i) there exists a node x such that x is touched by a line of M and all nodes touched by lines of M are in the tree $T_1^r[x]$; (ii) there exist two nodes x_1 and x_2 such that both x_1 and x_2 are touched by lines of M and all nodes touched by lines of M are either in the tree $T_1^r[x_1]$ or in the tree $T_1^r[x_2]$.[9]

Case 1. In this case, $\delta(G_1, G_2)$ can be obtained by trying each node of G_2, in turn, as the root and running a modified version of Algorithm A in each trial. We can show that this case requires at most $O(N_1 \, N_2 \, (\min\{d_1, d_2\})^2)$ time when the edit operations have general cost and $O(N_1 \, N_2 \, \min\{d_1, d_2\} \, \sqrt{\min\{d_1, d_2\}} \log(\min\{d_1, d_2\}))$ time when the edit operations have integral cost.

Case 2. Let y_1, y_2 be in G_2 such that $(x_1, y_1) \in M$ and $(x_2, y_2) \in M$. Then we can find an arbitrary edge (v_1, v_2) on the path connecting y_1 and y_2 and split G_2 at the edge into two rooted unordered trees $T_2^{v_2}[v_1]$ and $T_2^{v_1}[v_2]$. Each of $T_1^r[x_1]$, $T_1^r[x_2]$, $T_2^{v_2}[y_1]$, $T_2^{v_1}[y_2]$ is a rooted unordered tree (see Figure 6). The best mapping from $T_1^r[x_1]$ to $T_2^{v_2}[y_1]$ and the best mapping from $T_1^r[x_2]$ to $T_2^{v_1}[y_2]$ can be obtained during the computation of Case 1 above. We can show that this case requires at most $O(N_1 N_2)$ time.

Thus, the algorithm for calculating $\delta(G_1, G_2)$ for two CUAL graphs G_1 and G_2, referred to as Algorithm B, runs in time $O(N_1 N_2 \, (\min\{d_1, d_2\})^2)$ when the edit operations have general cost and in time $O(N_1 N_2 \min\{d_1, d_2\} \sqrt{\min\{d_1, d_2\}} \log(\min\{d_1, d_2\}))$ when the edit operations have integral cost. Note that the gap between the running times of Algorithm A and Algorithm B is $\min\{d_1, d_2\}$. If one of the CUAL graphs has a bounded degree, then the running time of both algorithms is $O(N_1 N_2)$.

---

[9] Note that there cannot be more than two nodes. If that were true (say there were three nodes x_1, x_2, x_3), the center of the three nodes would be their least common ancestor. By the mapping conditions, this ancestor would also be in the mapping, contradicting the fact that all the nodes touched by mapping lines are in the trees rooted at x_1, x_2, x_3.

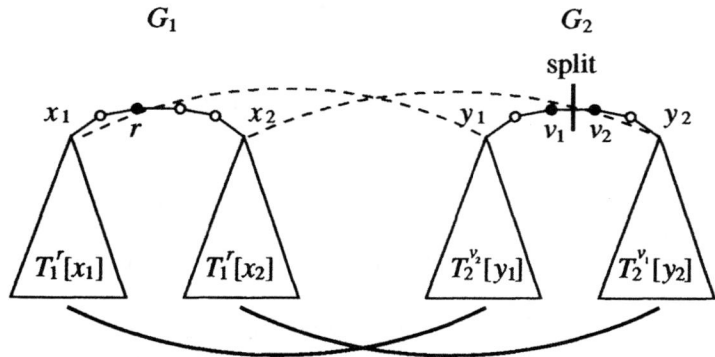

Fig. 6. Illustration of Case 2 in computing $\delta(G_1, G_2)$ for two CUAL graphs G_1 and G_2.

4 Conclusion

Using these simple, efficient algorithms, a user can submit a query structure and obtain those data structures approximately matching the query. To our knowledge, this work gives the first polynomial time algorithm ever presented to solve the edit distance problem between undirected acyclic graphs. We will have this algorithm implemented within a few months and will make it available to the community.

Acknowledgements

We thank the helpful comments of the referees. Thanks are also to Professor Dave Kristol and Dawn Luo and to Jim Kaminski and Karen Pysniak of the Schering-Plough Research Institute for inspirational discussions concerning the work.

References

1. K. Abrahamson. Generalized string matching. *SIAM J. Comput.*, 16:1039–1051, 1987.
2. J. E. Ash, P. A. Chubb, S. E. Ward, S. M. Welford, and P. Willett. *Communication, Storage and Retrieval of Chemical Information.* Ellis Horwood, Chichester, England, 1985.
3. J. E. Ash and E. Hyde, editors. *Chemical Information Systems.* Ellis Horwood, Chichester, England, 1975.
4. H. N. Gabow, Z. Galil, and T. H. Spencer. Efficient implementation of graph algorithms using contraction. In *Proceedings of the 25th Annual IEEE Symposium on Foundations of Computer Science*, pages 347–357, 1984.
5. H. N. Gabow and R. E. Tarjan. Faster scaling algorithms for network problems. *SIAM J. Comput.*, 18(5):1013–1036, 1989.

6. Z. Galil and K. Park. An improved algorithm for approximate string matching. *SIAM J. Comput.*, 19(6):989–999, Dec. 1990.

7. M. A. Johnson and G. M. Maggiora, editors. *Concepts and Applications of Molecular Similarity*. Wiley, New York, 1990.

8. J. Kaminski, B. Wallmark, C. Briving, and B.-M. Andersson. Antiulcer agents. 5. inhibition of gastric H^+/K^+-ATPase by substituted imidazo[1,2-a]pyridines and related analogues and its implication in modeling the high affinity potassium ion binding site of the gastric proton pump enzyme. *Journal of Medicinal Chemistry*, 34:533–541, 1991.

9. R. L. Kashyap and B. J. Oommen. An effective algorithm for string correction using a generalized edit distance – I. Description of the algorithm and its optimality. *Information Sci.*, 23(2):123–142, 1981.

10. P. Kilpelainen and H. Mannila. Query primitives for tree-structured data. In M. Crochemore and D. Gusfield, editors, *Combinatorial Pattern Matching, Lecture Notes in Computer Science, 807*, pages 213–225. Springer-Verlag, 1994.

11. G. M. Landau and U. Vishkin. Fast parallel and serial approximate string matching. *Journal of Algorithms*, 10(2):157–169, 1989.

12. Y. C. Martin, M. G. Bures, and P. Willett. Searching databases of three-dimensional structures. In K. B. Lipkowitz and D. D. Boyd, editors, *Reviews in Computational Chemistry*, pages 213–263. VCH Publishers, New York, 1990.

13. W. J. Masek and M. S. Paterson. A faster algorithm for computing string editing distances. *J. Comput. System Sci.*, 20:18–31, 1980.

14. E. M. Mitchell, P. J. Artymiuk, D. W. Rice, and P. Willett. Use of techniques derived from graph theory to compare secondary structure motifs in proteins. *Journal of Molecular Biology*, 212:151–166, 1989.

15. A. S. Noetzel and S. M. Selkow. An analysis of the general tree-editing problem. In D. Sankoff and J. B. Kruskal, editors, *Time Warps, String Edits, and Macromolecules: The Theory and Practice of Sequence Comparison*, pages 237–252. Addison-Wesley, Reading, MA, 1983.

16. P. A. Pevzner and M. S. Waterman. A fast filtration algorithm for the substring matching problem. In A. Apostolico, M. Crochemore, Z. Galil, and U. Manber, editors, *Combinatorial Pattern Matching, Lecture Notes in Computer Science, 684*, pages 197–214. Springer-Verlag, 1993.

17. S. M. Selkow. The tree-to-tree editing problem. *Information Processing Letters*, 6(6):184–186, Dec. 1977.

18. D. Shasha, J. T. L. Wang, K. Zhang, and F. Y. Shih. Exact and approximate algorithms for unordered tree matching. *IEEE Transactions on Systems, Man and Cybernetics*, 24(4):668–678, April 1994.

19. K.-C. Tai. The tree-to-tree correction problem. *J. ACM*, 26(3):422–433, 1979.

20. E. Ukkonen. Finding approximate patterns in strings. *Journal of Algorithms*, 6:132–137, 1985.

21. E. Ukkonen. Approximate string-matching over suffix trees. In A. Apostolico, M. Crochemore, Z. Galil, and U. Manber, editors, *Proc. of the 4th Annual Symposium on Combinatorial Pattern Matching*, pages 228–242. Lecture Notes in Computer Science, 684, Springer-Verlag, 1993.

22. R. A. Wagner and M. J. Fischer. The string-to-string correction problem. *J. ACM*, 21(1):168–173, Jan. 1974.

23. W. E. Warr. *Chemical Structures*. Springer-Verlag, Berlin, 1988.

24. P. Willett. *Similarity and Clustering Methods in Chemical Information Systems*. Research Studies Press, Letchworth, 1987.

25. P. Willett. Algorithms for the calculation of similarity in chemical structure databases. In M. A. Johnson and G. M. Maggiora, editors, *Concepts and Applications of Molecular Similarity*, pages 43–61. John Wiley & Sons, Inc., 1990.

26. S. Wu and U. Manber. Fast text searching allowing errors. *Communications of the ACM*, 35(10):83–91, Oct. 1992.

27. K. Zhang. *The Editing Distance between Trees: Algorithms and Applications*. PhD thesis, Courant Institute of Mathematical Sciences, New York University, 1989.

28. K. Zhang. A new editing based distance between unordered labeled trees. In A. Apostolico, M. Crochemore, Z. Galil, and U. Manber, editors, *Combinatorial Pattern Matching, Lecture Notes in Computer Science, 684*, pages 254–265. Springer-Verlag, 1993; journal version is to appear in *Algorithmica*.

29. K. Zhang and D. Shasha. Simple fast algorithms for the editing distance between trees and related problems. *SIAM J. Comput.*, 18(6):1245–1262, Dec. 1989.

30. K. Zhang, D. Shasha, and J. T. L. Wang. Approximate tree matching in the presence of variable length don't cares. *Journal of Algorithms*, 16(1):33–66, Jan. 1994.

Author Index

Lecture Notes in Computer Science

For information about Vols. 1–865
please contact your bookseller or Springer-Verlag

Vol. 901: R. Kumar, T. Kropf (Eds.), Theorem Provers in Circuit Design. Proceedings, 1994. VIII, 303 pages. 1995.

Vol. 902: M. Dezani-Ciancaglini, G. Plotkin (eds.), Typed Lambda Calculi and Applications. Proceedings, 1995. VIII, 443 pages. 1995

Vol. 903: E. W. Mayr, G. Schmidt, G. Tinhofer (Eds.), Graph-Theoretic Concepts in Computer Science. Proceedings, 1994. IX, 414 pages. 1995.

Vol. 904: P. Vitányi (Ed.), Computational Learning Theory. EuroCOLT'95. Proceedings, 1995. XVII, 415 pages. 1995. (Subseries LNAI).

Vol. 905: N. Ayache (Ed.), Computer Vision, Virtual Reality and Robotics in Medicine. Proceedings, 1995. XIV, 567 pages. 1995.

Vol. 906: E. Astesiano, G. Reggio, A. Tarlecki (Eds.), Recent Trends in Data Type Specification. Proceedings, 1995. VIII, 523 pages. 1995.

Vol. 907: T. Ito, A. Yonezawa (Eds.), Theory and Practice of Parallel Programming. Proceedings, 1995. VIII, 485 pages. 1995.

Vol. 908: J. R. Rao Extensions of the UNITY Methodology: Compositionality, Fairness and Probability in Parallelism. XI, 178 pages. 1995.

Vol. 909: H. Comon, J.-P. Jouannaud (Eds.), Term Rewriting. Proceedings, 1993. VIII, 221 pages. 1995.

Vol. 910: A. Podelski (Ed.), Constraint Programming: Basics and Trends. Proceedings, 1995. XI, 315 pages. 1995.

Vol. 911: R. Baeza-Yates, E. Goles, P. V. Poblete (Eds.), LATIN '95: Theoretical Informatics. Proceedings, 1995. IX, 525 pages. 1995.

Vol. 912: N. Lavrac, S. Wrobel (Eds.), Machine Learning: ECML – 95. Proceedings, 1995. XI, 370 pages. 1995. (Subseries LNAI).

Vol. 913: W. Schäfer (Ed.), Software Process Technology. Proceedings, 1995. IX, 261 pages. 1995.

Vol. 914: J. Hsiang (Ed.), Rewriting Techniques and Applications. Proceedings, 1995. XII, 473 pages. 1995.

Vol. 915: P. D. Mosses, M. Nielsen, M. I. Schwartzbach (Eds.), TAPSOFT '95: Theory and Practice of Software Development. Proceedings, 1995. XV, 810 pages. 1995.

Vol. 916: N. R. Adam, B. K. Bhargava, Y. Yesha (Eds.), Digital Libraries. Proceedings, 1994. XIII, 321 pages. 1995.

Vol. 917: J. Pieprzyk, R. Safavi-Naini (Eds.), Advances in Cryptology - ASIACRYPT '94. Proceedings, 1994. XII, 431 pages. 1995.

Vol. 918: P. Baumgartner, R. Hähnle, J. Posegga (Eds.), Theorem Proving with Analytic Tableaux and Related Methods. Proceedings, 1995. X, 352 pages. 1995. (Subseries LNAI).

Vol. 919: B. Hertzberger, G. Serazzi (Eds.), High-Performance Computing and Networking. Proceedings, 1995. XXIV, 957 pages. 1995.

Vol. 920: E. Balas, J. Clausen (Eds.), Integer Programming and Combinatorial Optimization. Proceedings, 1995. IX, 436 pages. 1995.

Vol. 921: L. C. Guillou, J.-J. Quisquater (Eds.), Advances in Cryptology – EUROCRYPT '95. Proceedings, 1995. XIV, 417 pages. 1995.

Vol. 922: H. Dörr, Efficient Graph Rewriting and Its Implementation. IX, 266 pages. 1995.

Vol. 923: M. Meyer (Ed.), Constraint Processing. IV, 289 pages. 1995.

Vol. 924: P. Ciancarini, O. Nierstrasz, A. Yonezawa (Eds.), Object-Based Models and Languages for Concurrent Systems. Proceedings, 1994. VII, 193 pages. 1995.

Vol. 925: J. Jeuring, E. Meijer (Eds.), Advanced Functional Programming. Proceedings, 1995. VII, 331 pages. 1995.

Vol. 926: P. Nesi (Ed.), Objective Software Quality. Proceedings, 1995. VIII, 249 pages. 1995.

Vol. 927: J. Dix, L. Moniz Pereira, T. C. Przymusinski (Eds.), Non-Monotonic Extensions of Logic Programming. Proceedings, 1994. IX, 229 pages. 1995. (Subseries LNAI).

Vol. 928: V.W. Marek, A. Nerode, M. Truszczynski (Eds.), Logic Programming and Nonmonotonic Reasoning. Proceedings, 1995. VIII, 417 pages. 1995. (Subseries LNAI).

Vol. 929: F. Morán, A. Moreno, J.J. Merelo, P. Chacón (Eds.), Advances in Artificial Life. Proceedings, 1995. XIII, 960 pages. 1995 (Subseries LNAI).

Vol. 930: J. Mira, F. Sandoval (Eds.), From Natural to Artificial Neural Computation. Proceedings, 1995. XVIII, 1150 pages. 1995.

Vol. 931: P.J. Braspenning, F. Thuijsman, A.J.M.M. Weijters (Eds.), Artificial Neural Networks. IX, 295 pages. 1995.

Vol. 932: J. Iivari, K. Lyytinen, M. Rossi (Eds.), Advanced Information Systems Engineering. Proceedings, 1995. XI, 388 pages. 1995.

Vol. 933: L. Pacholski, J. Tiuryn (Eds.), Computer Science Logic. Proceedings, 1994. IX, 543 pages. 1995.

Vol. 934: P. Barahona, M. Stefanelli, J. Wyatt (Eds.), Artificial Intelligence in Medicine. Proceedings, 1995. XI, 449 pages. 1995. (Subseries LNAI).

Vol. 935: G. De Michelis, M. Diaz (Eds.), Application and Theory of Petri Nets 1995. Proceedings, 1995. VIII, 511 pages. 1995.

Vol. 936: V.S. Alagar, M. Nivat (Eds.), Algebraic Methodology and Software Technology. Proceedings, 1995. XIV, 591 pages. 1995.

Vol. 937: Z. Galil, E. Ukkonen (Eds.), Combinatorial Pattern Matching. Proceedings, 1995. VIII, 409 pages. 1995.

Vol. 938: K.P. Birman, F. Mattern, A. Schiper (Eds.), Theory and Practice in Distributed Systems. Proceedings, 1994. X, 263 pages. 1995.

Vol. 939: P. Wolper (Ed.), Computer Aided Verification. Proceedings, 1995. X, 451 pages. 1995.

Vol. 940: C. Goble, J. Keane (Eds.), Directions in Databases. Proceedings, 1995. X, 277 pages. 1995.

Vol. 941: M. Cadoli, Tractable Reasoning in Artificial Intelligence. XVII, 247 pages. 1995. (Subseries LNAI).

Vol. 942: G. Böckle, Exploitation of Fine-Grain Parallelism. IX, 188 pages. 1995.

Vol. 943: W. Klas, M. Schrefl, Metaclasses and Their Application. IX, 201 pages. 1995.